Mitochondria 1983

Mitochondria 1983

Nucleo-Mitochondrial Interactions

Proceedings of a Conference held in
Schliersee, Germany, July 19-23, 1983

Editors
R. J. Schweyen · K. Wolf · F. Kaudewitz

Walter de Gruyter · Berlin · New York 1983

Editors

Rudolf J. Schweyen, Prof. Dr.
Klaus Wolf, Prof. Dr.
Fritz Kaudewitz, Prof. Dr.

Universität München
Genetisches Institut
Maria-Ward-Straße 1a
D-8000 München 19
Federal Republic of Germany

CIP-Kurztitelaufnahme der Deutschen Bibliothek

Nucleo-mitochondrial interactions: proceedings of a conference held in Schliersee, Germany, July 19 – 23, 1983 / Mitochondria 1983. Ed. R. J. Schweyen . . . –
Berlin; New York: de Gruyter, 1983.
ISBN 3-11-009871-7
NE: Schweyen, Rudolf J. [Hrsg.]; Mitochondria <1983, München>

Library of Congress Cataloging in Publication Data

Mitochondria 1983, nucleo-mitochondrial interactions.
Sponsored by the Deutsche Forschungsgemeinschaft.
Bibliography: p.
Includes index.
1. Mitochondria--Congresses. 2. Cell nuclei--Congresses.
I. Schweyen, R. J. (Rudolf J.), 1941– . II. Wolf, K. (Klaus), 1944– . III. Kaudewitz, F. (Fritz), 1921– . IV. Deutsche Forschungsgemeinschaft.
QH603.M5M53 1983 574.87'342 83-26341
ISBN 3-11-009871-7

Printing: Gerike GmbH, Berlin. – Binding: Dieter Mikolai, Berlin. Printed in Germany.

Welcome Address

I am glad to welcome you as participants of the Schliersee Meeting on Nucleo-Mitochondrial Interactions. For most of you this is the second, for some even the third time that we have met in the relaxing atmosphere of Schliersee to discuss our scientific work. In 1976, it was the International Conference on Yeast Genetics and Molecular Biology; in 1977 it was the Meeting on Mitochondria.

In 1977 we were right at the beginning of a scientifically very fruitful period of studies on mitochondrial genes both in genetic and physical terms, and on the organization and replication of mitochondrial genomes. This period made mitochondria a respectable system in molecular biology.

Many of the questions raised in 1977 are now answered, others are still the subject of great interest, e.g. transcription and RNA processing, and these will be discussed in the present meeting. Now, six years after the 1977 meeting, we feel a second successful period of research is beginning: the analysis of nuclear genes whose products are essential for mitochondrial biogenesis. This dual system of encoding mitochondrial functions in both nuclear DNA and mtDNA and the mechanisms of interactions of cytoplasmic and mitochondrial translation products has interested scientists from the beginning of mitochondrial research. After making considerable progress in the characterization of mitochondrial molecular mechanisms, we now have to increase our understanding of the nuclear influence on these processes. This has been facilitated by rapid advances in molecular techniques such as gene cloning and transformation.

As in 1977, the 1983 meeting has been organized in connection with an International Training Course for young scientists, which was financed by the European Communities and held in our Institute in Munich. Our meeting is sponsored by the Deutsche Forschungsgemeinschaft, to whom we owe our sincerest gratitude.

F. Kaudewitz

CONTENTS

REVIEWS

Mitochondrial Genes, Mutants, Maps - A Review
B. Dujon 1

Mitochondrial Gene Expression 1983
L.A. Grivell 25

Import of Proteins into Mitochondria - A Survey
M. Yaffe 47

REPLICATION AND TRANSCRIPTION OF mtDNA

The Replication of Yeast Mitochondrial DNA
G. Baldacci, G. Bernardi 57

Some Ionophores Act Preferentially on Mitochondria in Intact Cells of Saccaromyces cerevisiae and Induce Respiration-Deficient Mutants
L. Kovác, V. Klobucnîkova 65

Initiation and Transcription of Yeast Mitochondrial RNA
J.C. Edwards, Th. Christianson, D. Mueller, T.K. Biswas, D. Levens, D. Li, J. Wettstein, M. Rabinowitz 69

Initiation of Transcription of Yeast Mitochondrial DNA
H.F. Tabak, K.A. Osinga, E. de Vries, A.M. van der Bliek, G.T.J. van der Horst, M.J.A. Groot Koerkamp, G. van der Horst, E.C. Zwarthoff, M.E. MacDonald 79

Functions of Non-Genic DNA in Yeast Mitochondria
H.P. Zassenhaus, R.A. Butow 95

Regulation of Ribosomal RNA and Messenger RNA Synthesis in Human Mitochondria
G. Attardi, G. Gaines, J. Montoya 107

Regulation of Mitochondrial Transcription in Rat Liver: Determination of Mitochondrial Transcript Concentration in Adult and Developing Rat
P. Cantatore, Z. Flagella, G. Pesole, M.N. Gadaleta, C. Saccone 127

Heterologous in Vitro Transcription of Mitochondrial DNA Segments
G. Görtz, R. Grosskopf, H. Feldmann 141

RNA Processing and Splicing

Transcription of Yeast Mitochondrial tRNA Genes and Processing of tRNA Gene Transcripts
D.L. Miller, K. Underbrink-Lyon, D.R. Najarian, J. Krupp, N.C. Martin 151

Expression of Mitochondrial Genes in *Saccharomyces cerevisiae*: Release from Glucose Repression and Transcription of the tRNA Genes
G. Baldacci, S. Francisci, C. Palleschi, E. Zennaro, L. Frontali 165

Implications of Fungal Mitochondrial Intron RNA Structure for the Mechanism of RNA Splicing
R.W. Davies, R.B. Waring, T.A. Brown, C. Scazzocchio 179

COB Intron 1 and 4: Studies on Mutants and Revertants Uncover Functional Intron Domains and Test the Validity of Predicted RNA-Secondary Structures
G. Rödel, J. Holl, C. Schmelzer, C. Schmidt, R.J. Schweyen, B. Weiss-Brummer, F. Kaudewitz 191

Splicing of Cytochrome b mRNA by in Vitro Complementation of Mitochondrial Extracts from Splicing Deficient Yeast Mutants

R. Schroeder, P.P. Slonimski 203

Mitochondrial Contributions to Hybrid Enzyme Complexes for Mitochondrial RNA Processing and Electron Transport

W.C. MacKellar, D.K. Hanson, Ph.M. Sass, C.D. Zarley, H.R. Mahler, P.S. Perlman 211

The First Intron of the Gene OXI3 in Yeast Mitochondria Encodes a mRNA-Maturase

O. Groudinsky, G. Carignani, E. Schiavon, D. Frezza, E. Bergantino, P.P. Slonimski 227

Genetic and Biochemical Studies of the Nuclear Suppressor NAM2: Extraneous Activation of a Latent Pleiotropic Maturase

G. Dujardin, M. Labouesse, P. Netter, P.P. Slonimski 233

Nuclear-Mitochondrial Interactions Cause Senescence in the Filamentous Fungus _Podospora anserina_

K. Esser, H.D. Osiewacz, U. Stahl, P. Tudzynski 251

DNA-Splicing of Introns in the Gene: A General Way of Reverting Intron Mutations

A. Gargouri, J. Lazowska, P.P. Slonimski 259

Organization and Expression of Mitochondrial Genes

Mitochondrial Genes and Their Expression in Higher Plants

C.J. Leaver, E. Hack, A.J. Dawson, P.G. Isaac, V.P. Jones 269

Gene Expression and Organization in Trypanosome Mitochondria

R. Benne, M. Agostinelli, B.F. de Vries, J. Van den Burg, B. Klaver, P. Borst 285

Organization of the Aspergillus nidulans Mitochondrial Genome

C. Scazzocchio, T.A. Brown, R.B. Waring, J.A. Ray, R.W. Davies 303

Sequence of the Mitochondrial DNA, Arrangement of Genes and Processing of Their Transcripts in Schizocaccharomyces pombe

B.F. Lang, F. Ahne, S. Distler, H. Trinkl, F. Kaudewitz, K. Wolf 313

Nucleotide Sequence and Transcript Mapping of a mtDNA Segment Comprising CO 1, $tRNA_{arg}$, and Several Unidentified Reading Frames in Neurospora crassa

G. Burger, S. Werner 331

The Expression of the Mitochondrial Genes for Subunit 1 of Cytochrome c Oxidase and for an ATPase Proteolipid in Neurospora crassa: Nucleotide Sequences and Transcript Analysis

H. de Vries, J.C. de Jonge, A. Arnberg, Ad A.C.M. Peijnenburg, E. Agsteribbe 343

Expression of Split Genes of the Neurospora crassa Mitochondrial Genome

M. Helmer Citterich, G. Morelli, M.A. Nelson, G. Macino .. 357

Mutations of the Rat Mitochondrial Genome

K. Koike, M. Taira, Y. Kuchino, K. Yaginuma, T. Sekiguchi, M. Kobayashi 371

Organization of the 21S rRNA Gene in Saccharomyces cerevisiae: Mutants of the Peptidyl Transferase Centre and the Nature of the ω-Locus

B. Dujon, A. Jacquier 389

The Corrected Sequence of the First Intron of the Cob-Box Gene in the Yeast Strain 777-3A

J. Lazowska, A. Gargouri, P.P. Slonimski 405

NUCLEAR CONTROL OF MITOCHONDRIAL FUNCTIONS

Nuclear Mutants Defective in Processing of cob and oxi 3 Transcripts in Saccharomyces cerevisiae
T. Pillar, J. Kreike, B.F. Lang, F. Kaudewitz 411

Interference of S. cerevisiae Nuclear Mutations with Mitochondrial rRNA Biosynthesis and Oxi 3-Transcript Maturation
J. Naggert, E. Schweizer 425

Processing of the Oxi-3 Pre-Messenger RNA in Yeast
G. Faye, M. Simon 433

Mitochondrial Inheritance of a Variant Form of the oxi2 Transcript in Saccharomyces cerevisiae
Th.D. Fox, M.K. Reif 441

A Precursor to Subunit II of Cytochrome Oxidase in Saccharomyces cerevisiae
G. Mannhaupt, G. Michaelis, E. Pratje, E. Schweizer, D.C. Hawthorne 449

Mitochondrial Ribosomal RNA Mutations and Their Nuclear Suppressors in Yeast
M. Bolotin-Fukuhara, F. Sor, H. Fukuhara 455

Cytoplasmic Synthesis of Proteins of the Mammalian Mitochondrial Ribosome
G.L. Schieber, Th. W. O'Brien 469

Diuron Resistance in Saccharomyces cerevisiae
A.-M. Colson, A. Fauconnier, B. Meunier 481

Nuclear Mutants Affected in the Recombination of Mitochondrial DNA in Saccharomyces cerevisiae
F. Foury, J. Kolodynski 487

Genetic Control of the Biogenesis of Cyanide Insensitive Respiration in *Neurospora crassa*
H. Bertrand, C.A. Argan, N.A. Szakacs 495

The Role of Heme and Oxygen in the Regulation of Mitochondrial Cytochromes
Th. Mason, H.N. Little, C. VanSickle, L. Guarente 509

About the Regulation of Protoheme Synthesis in the Yeast *Saccharomyces cerevisiae*
R. Labbe-Bois, D. Urban-Grimal, Ch. Volland, J.M. Camadro, P. Dehoux 523

Monoclonal Anti-H^+-ATPase Antibodies in the Study of the Assembly of the Yeast Mitochondrial H^+-ATPase
S. Marzuki, R.G. Hadikusumo, W.M. Choo, L. Watkins, H.B. Lukins, A.W. Linnane 535

IMPORT OF PROTEINS INTO MITOCHONDRIA

Transport of Cytochrome *c* into Mitochondria: Involvement of Specific Receptors and of Cytochrome *c* Heme Lyase
B. Hennig, H. Köhler, W. Neupert 551

Mitochondria and the Import Problem. Genes for Imported Subunits of the Ubiquinol-Cytochrome *c* Reductase
L.A. Grivell, A.P.G.M. Van Loon, M. De Haan, A.C. Maarse, A. Dekker 563

A Model System for the Study of Transport of Proteins into Mitochondria
E. Marra, S. Passarella, M. Greco, C. Saccone, E. Quagliariello, S. Doonan 579

Biosynthesis and Topogenesis of Cytosolic and Mitochondrial Aspartate Aminotransferase
Ph. Christen, R. Jaussi, P. Sonderegger, H. Gehring, U. Graf-Hausner, R. Behra, J. Flückiger, R.C. Skoda 593

Import of ATP2-LacZ Gene Fusion Proteins into Mitochondria
B.L. Geller, M.L. Britten, C.M. Biggs, M.G. Douglas, S.D. Emr 607

Assembly of the Mitochondrial ATPase in Saccharomyces cerevisiae
A.S. Lewin, L.J. Wells, D.K. Norman 621

AUTHOR INDEX 637

SUBJECT INDEX 639

PARTICIPANTS

Agsteribbe, E., Laboratory of Physiological Chemistry, State University, Groningen, Bloemsingel 10, 9712 KZ Groningen, The Netherlands

Ahne, F., Institut für Genetik und Microbiologie der Universität München, Maria-Ward-Str. 1a, D- 8000 München 19, Germany

Arnold, E. Universität Regensburg, Universitätsstr. 31, D-8400 Regensburg, Germany

Attardi, G., Division of Biology, Californian Institute of Technology, Pasadena, California 91125, U.S.A.

Bajwa, W., Friedrich Miescher Institut, P.O.Box 2543, 4000 Basel, Schweiz

Baldacci, G., Institut J. Monod, Tour 43, 2, Place Jussieu, 75005 Paris, France

Bandlow, W., Institut für Genetik und Microbiologie der Universität München, Maria-Ward-Str. 1a, D-8000 München 19, Germany

Benne, R., Dept. of Biochemistry, University of Amsterdam, P.O.Box 60000, 1005 GA Amsterdam, The Netherlands

Bernardi, G., Institut Jacques Monod, 2 Place Jussieu, 75005 Paris, France

Bertrand, H., Biology Dept., University of Regina, Regina, Saskatchewan S4S OA2, Canada

Blanc, H. Centre de Génétique Moléculaire du CNRS, Gif-sur-Yvette, France

Bolotin-Fukuhara, M., Laboratoire des Biologie Générale, Bât. 400, Université Paris-Sud, 91405 Orsay, France

Borkhardt, B., University of Copenhagen, Institute of Genetics, Øster Farimagsgade 2A, 1353 Copenhagen, Denmark

Burger, G., Institut für Physiologische Chemie und Physikalische Biochemie, Goethestr. 33, D-8000 München 2

Butow, R., Health Science Center at Dallas, The University of Texas, 5323 Harry Hines Boulevard, Dallas, Texas 75235, U.S.A.

Cantatore, P., Institute of Biological Chemistry, University of Bari, Bari, Italy

Carignani, G., Istituto di Chimica Biologica, Via Marzolo, 3, Padova Italia,

Christen, Ph., Biochemisches Institut der Universität Zürich, Zürichbergstr. 4, 8028 Zürich, Schweiz

Colson, A.-M., Laboratoire de Génêtique, Université de Louvain, Louvain-la-Neuve, Belgium

Connerton, I., University of Warwick, Dept. of Chem. & Molec. Science, Coventry CV- 7 AL. United Kingdom

Davies, R., The University of Manchester, Institute of Science and Technology, P.O.Box 88, Manchester M60 1QD, UK

Distler, S., Institut für Genetik und Microbiologie der Universität München, Maria-Ward-Str. 1a, D-8000 München 19, Germany

Dobres, M., Institut für Genetik und Microbiologie der Universität München, Maria-Ward-Str. 1a, D-8000 München 19, Germany

Douglas, M., The University of Texas, Health Science Center at San Antonio, 7703 Floyd Curl Drive, San Antonio, Texas 78284, U.S.A.

Dujardin, G., Centre de Génêtique Moleculaire du CNRS, 91190 Gif-sur-Yvette, France

Dujon, B., Centre de Génêtique Moléculaire du CNRS, 91190 Gif-sur-Yvette, France

Edwards, J., The University of Chicago, Dept. of Medicine, 950 East 59th Street, Chicago, Illinois 60637, U.S.A.

Egel, R., Universitetets Genet. Institut, Øster Farimagsgade 2A, Copenhagen, Denmark

Egel-Mitani, N., Novo Research Institute, Nova Alle, 2880 Bagsvaerd Denmark

Faye, G., Institut Curie, Section de Biologie, Centre Universitaire, 91405 Orsay-Cedex, France

Feldmann, H., Institut für physiologische Chemie der Universität München, Goethestr. 33, D-8000 München, Germany

Foury, F., Laboratoire dÉnzymologie, Université de Louvain, Place du Sud 1 Bte 8, 1348 Louvain-La-Neuve, France

Fox, T., Cornell University, Section of Genetics and Development, Bradfield-Emerson-Halls, Ithaca, New York, 14853 U.S.A.

Forsbach, V., Institut für Genetik und Microbiologie der Universität München, Maria-Ward-Str. 1a, D-8000 München 19, Germany

Frontali, L., Università degli Studi di Roma, Facoltà di Scienze M.F.N. Instituto Di Fisiologia Generale, Roma, Italia

Fukuhara, H., Institut Curie, Section de Biologie-Bat. 110, Centre Universitaire, 91405 Orsay, France

Garbrecht, S., Institut für Genetik und Microbiologie der Universität München, Maria-Ward-Str. 1a, D-8000 München 19, Germany

Gargouri, A., Centre de Génétique Moléculaire du CNRS, Gif-sur-Yvette, France

Ghislain, M., Laboratory of Enzymology, Place Croix du Sud I, Bte 8 1348 Louvain la Neuve, France

Goursot, R., Laboratoire de Génétique Moléculaire, Institut Jaques Monod, Paris, France

Grivell, L., Universiteit van Amsterdam, Afdeling Moleculaire Biologie, Laboratorium voor Biochemie, Kruislaan 318, 1098 SM Amsterdam, The Netherlands

Haid, A., Institut für Genetik und Microbiologie der Universität München, Maria-Ward-Str- 1a, D-8000 München 19, Germany

Halfmann, H., Institut für Medizin AG., Membranforschung, KFA Jülich

Hase, T., Biozentrum der Universität, Klingelbergstr. 17, 4056 Basel, Schweiz

Hennig, B., Institut für Biochemie, Humboldtallee 7, 3400 Göttingen, Germany

Hensgens, L.A.M. Dept. of Biochemistry, University of Amsterdam, P.O.Box 6000, 1005 GA Amsterdam, The Netherlands

Kaudewitz, F., Institut für Genetik und Microbiologie der Universität München, Maria-Ward-Str. 1a, D-8000 München 19, Germany

Kohlwein, S., Technische Universität, Institut für Biochemie und Lebensmittelchemie, Schlögelgasse 9, 8010 Graz, Austria

Koike, K., Cancer Institute, Komi-Ikebukuro, Toshima-ku, 170 Tokyo, Japan

Kovać, L., Centre National de la Recherche Scientifique, Centre de Génétique Moléculaire, 91190 Gif-sur-Yvette, France

Kreike, J., Institut für Genetik und Microbiologie der Universität München, D-8000 München 19, Maria-Ward-Str. 1a, Germany

Kuiper, M.T.R. Laboratory of Physiological Chemistry, University of Groningen, Bloemsingel 10, 9712 KZ Groningen, The Netherlands

Labbé-Bois, R., Université Paris VII, Laboratoire de Biochimie des Porphyrines, 2, Place Jussieu, Tour 43, 75005 Paris, France

Lang, B., Institut für Genetik und Microbiologie der Universität München, Maria-Ward-Str. 1a, D-8000 München 19, Germany

Leaver, C., University of Edingburgh, Dept. of Botany, The Kings Buildings, Mayfield Road, Edinburgh EH9 3JH

Lewin, A., University of Indiana, Dept. of Chem. Chemistry Building, Bloomington, Indiana 47405, U.S.A.

Linnane, A., Monash University, Dept. of Biochemistry, Clayton, Victoria 3168, Australia

Macino, G., Università degli Studi di Roma, Facoltà di Roma, Facoltà di Scienze M.F.N. Instituto di Fisiologia Generale, Roma, Italia

MacKellar, W.C. Indiana Univ., Dept. of Chemistry, Chemistry Building, Bloomington, Indiana 47405, U.S.A.

Mannhaupt, G., Universität Bielefeld, Fakultät für Biologie III, D-4800 Bielefeld, Germany

Marra, E., Università di Bari, Instituto di Chimica Biologica, Facoltà di Scienze, 70126 Bari, Italia

Mason, T., University of Massachusetts, Amherst 01003, Dept. of Biochemistry, Amherst, Massachusetts, U.S.A.

Mecklenburg, K., The Ohio State University, Dept. of Genetics, 1735 Neil Avenue, Columbus, Ohio 43210, U.S.A.

Michaelis, G., Universität Bielefeld, Fakultät für Biologie, Postfach 8640, D-4800 Bielefeld 1, Germany

Miller, D., Health Science Center at Dallas, The University of Texas, 5323 Harry Hines Boulevard, Dallas, Texas 75235, U.S.A.

Nelson, M. Dept. of Genetics and Molecular Biology, University of Rome, "La Sapienza", Rome, Italy

Neupert, W., Institut für Biochemie, Humboldtallee 7, Göttingen, Germany

O'Brien, W., University of Florida, Dept. of Biochemistry and Molecular Biology, Gainesville, Florida, U.S.A.

Ohta, S., Biozentrum der Universität Basel, Klingelbergstr. 70, 4056 Basel, Schweiz

Osinga, K., Dept. of Biochemistry, University of Amsterdam, P.O.Box 600000, 1005 GA Amsterdam, The Netherlands

Osiewacz, H.-D., Lehrstuhl für Allgemeine Botanik, Ruhr Universität, D-4630 Bochum Germany

Pajot, P., Centre de Génêtique Molêculaire du CNRS, 91190 Gif-sur-Yvette, France

Perlman, P., Ohio State University, Dept. of Genetics, 1735 Neil Avenue Columbus, Ohio 43210, U.S.A.

Pillar, T., Institut für Genetik und Microbiologie der Universität München, Maria-Ward-Str. 1a, D-8000 München 19, Germany

Pocklington, M., University of Leicester, Dept. of Genetics, Adrain Building, Leicester 7R11, England

Poyton, R., University of Colorado, Boulder, Dept. of Molecular, Cellular and Developmental Biology, Campus Box 347, Boulder, Colorado 80309, U.S.A.

Pratje, E., Universität Bielefeld, Fakultät für Biologie, D-4800 Bielefeld, Germany

Rödel, G., Institut für Genetik und Microbiologie der Universität München, Maria-Ward-Str. 1a, D-8000 München 19, Germany

Sadler, J., Institut für Genetik und Microbiologie der Universität München, Maria-Ward-Str. 1a, D-8000 München 19, Germany

Scazzocchio, C., University of Essex, Dept. of Biology, Wivenhoe Park, Colchester CO4 3SQ, England

Sebald, W., Gesellschaft für Biotechnologische Forschung mbH, Mascheroder Weg 1, D-3300 Braunschweig

Simchen, G., The Hebrew University, Department of Genetics, Jerusalem, Israel

Slonimski, P., Centre Gênétique Moléculaire du C.N.R.S., 91190 Gif-sur-Yvette, France

Stahl, U., Lehrstuhl für Allgemeine Botanik, Ruhr Universität, D-4630 Bochum, Germany

Schinkel, A., University of Amsterdam, Section for Molecular Biology, Laboratory of Biochemistry, Kruislaan 318, 1098 Amsterdam, The Netherlands

Schleyer, M., Institut für Biochemie, Humboldtallee 7, Göttingen, Germany

Schmelzer, C., Institut für Genetik und Microbiologie der Universität München, Maria-Ward-Str. 1a, D-8000 München 19, Germany

Schmidt, C., Institut für Genetik und Microbiologie der Universität München, Maria-Ward-Str. 1a, D-8000 München 19, Germany

Schroeder, R., Institut für Allgemeine Biochemie, Währingerstr. 38, 1090 Wien, Austria

Schweizer,E., Institut für Microbiologie und Biochemie der Universität Erlangen-Nürnberg, Egerlandstr. 7, D-8520 Erlangen, Germany

Schweyen, R., Institut für Genetik und Microbiologie der Universität München, Maria-Ward-Str. 1a, D-8000 München 19, Germany

Tabak, H., Dept. of Biochem., Section for Molecular Biology, University of Amsterdam, Kruislaan 318, 1098 Amsterdam, The Netherlands

Trinkl, H., Institut für Genetik und Microbiologie der Universität München, Maria-Ward-Str. 1a, D-8000 München 19, Germany

Van der Horst,G., University of Amsterdam, Section for Molecular Biology, Laboratory of Biochemistry, Kruislaan 318, 1098 Amsterdam, The Netherlands

Van Loon, A., University of Amsterdam, Section for Molecular Biology, Laboratory of Biochemistry, Kruislaan 318, 1098 Amsterdam, The Netherlands

Van der Veen, R., University of Amsterdam, Section for Molecular Biology, Laboratory of Biochemistry, Kruislaan 318, 1098 Amsterdam, The Netherlands

Vries de, H., Laboratory of Physiological Chemistry, University of Groningen, Bloemsingel 10, 9712 KZ Groningen, The Netherlands

Werner, S., Institut für Physiologische Chemie und Physikalische Biochemie, Goethestr. 33, D-8000 München 2

Wolf, K., Institut für Genetik und Microbiologie der Universität München, Maria-Ward-Str. 1a, D-8000 München 19, Germany

Yaffe, M., Biozentrum der Universität Basel, Klingelbergstr. 70, 4056 Basel, Schweiz

Zassenhaus, P., The University of Texas, Health Science Center at Dallas, 5323 Harry Hines Blvd., Dallas, Texas 75235 U.S.A.

Zonneveld, B., Genetisch Laboratorium der RUL, Kaiserstraat 63, Postbus 9516, 2300 RA Leiden, Holland

Zwizinski, C., Institut für Biochemie, Humboldtallee 7, Göttingen, Germany

1

2

3

4

5

6

7

1. Gérard Faye; 2. Tom Fox; 3. Leslie Grivell;
4. Krainsberger Hof, Schliersee; 5. Chris Leaver;
6. Claude Scazzocchio; 7. Giorgio Bernardi

(photography: Jan Kreike)

1

3

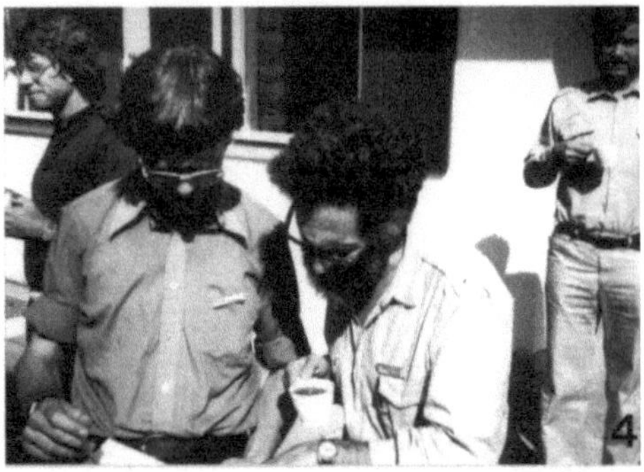
4

6

7

1. Giuseppe Attardi, Fritz Kaudewitz; 3. Piotr Slonimski, Lambert Hensgens; 4. Andreas Körte, B. Franz Lang, Claudio Scazzocchio, Ulf Stahl; 5. Bernard Dujon; 6. Shigeo Ohta, Walter Neupert, Ron Butow, Craig Zwizinski, Etienne Agsteribbe; 7. Kirk Mecklenburg, Rudolf Schweyen, Phil Perlman; 8. Renée Schroeder, Piotr Slonimski

(photography: Jan Kreike)

MITOCHONDRIAL GENES, MUTANTS AND MAPS: A REVIEW

Bernard Dujon

Centre de Génétique moléculaire du C.N.R.S., Laboratoire associé à l'Université Pierre et Marie Curie
91190 Gif sur Yvette, France

Introduction

The biogenesis of mitochondria is the result of an intricate interplay between the nuclear genome and the mitochondrial genome. Mitochondrial genomes of various organisms have been the objects of many detailed molecular and genetic studies in recent years (see 1-5 for recent Symposia and 6,7 for reviews) and a clear picture of the mitochondrial genetic functions now emerges. All mitochondrial genomes studied so far encode a small number of ubiquitous mitochondrial translation products involved in respiration and ATP synthesis. Those are the three larger subunits of the cytochrome oxidase complex, the cytochrome b and, depending on the organism, two or three subunits of the ATP synthetase complex. Their final destinations are membrane bound complexes in which they are assembled with nuclear encoded polypeptides. Mitochondrial genomes also encode other mitochondrial translation products in some organisms but not in others. In fungi, for example, a polypeptide associated with the mitochondrial ribosome is mitochondrially translated (8-10) while plants mitochondria synthesize a number of, so far, unidentified polypeptides which seem to have no equivalent in other systems (see Leaver et al. this volume). In addition, a number of open reading frames, the products of which are unknown (URFs), have been found from direct sequencing of mtDNA in various organisms. Mammalian mtDNAs, for example, contain eight such URFs (see Attardi et al. this volume). In yeast, most of the URFs are localized inside introns (11-15) and some of them have been found to encode mRNA-maturases (15-19). In Neurospora, intronic URFs have also been found (Burger et al., Helmer-C. et al., DeVries et al. this volume) but one of them encodes a ribosomal protein (20). Other URFs in fungi may resemble mammalian URFs (see Schazzocchio et al. and Helmer-C. et al., this volume). All mitochondrial genomes also encode the RNA molecules involved in the mitochondrial protein synthetic machinery, namely the two rRNAs (three in the case of plants (21)) and a complete set of tRNAs. Perhaps the clearer lesson of the mitochondrial genetic systems, as we view them at present, is the complete absence of exchange of RNA molecules between the mitochondria and the cytoplasm (or the nucleus). It follows that all

Mitochondria 1983

mitochondrial translation products must be encoded by mitochondrial genes while all nuclear gene products, eventually located in mitochondria, are synthesized in the cytoplasm and transported into the organelle.
Figure 1 cartoons the major nucleo-mitochondrial interactions involved in the expression of the mitochondrial genetic information and the biogenesis of mitochondria. Transcription of the mitochondrial genes and processing of the transcripts require both nuclearly encoded enzymes (e.g. the RNA polymerase) and, at least in fungi, mitochondrially encoded products. In yeast, for example, the splicing of introns from the primary transcripts of split genes needs the presence of intron encoded maturases (15-19). But, at the same time, intron splicing also requires functional nuclear gene products (22-24 and Pillar et al, this volume). In addition, a specific mitochondrial transcript is needed, also in yeast, for the maturation of tRNAs (see Miller et al. this volume). Translation of mature mRNAs (or some of their precursors) also requires both nuclear and mitochondrial genes products. Mitochondrial ribosomes are themselves a mosaic of mitochondrial genes products (the RNAs and , at least in fungi, one protein) and of nuclear genes products synthesized on cytoplasmic ribosomes (the bulk of the mitochondrial ribosomal proteins). Other factors necessary for protein synthesis in mitochondria are again of dual origin (e.g. the tRNAs are coded for by the mitochondrial DNA while the activating enzymes are from nuclear origin (25)). In most cases, nuclear encoded mitochondrial enzymes differ from their cytoplasmic analogues. Recently, however, there has been report of the same tRNA modification enzyme being active both in mitochondria and in the cytoplasm (26). Two other major functions for the biogenesis of mitochondria seem to involve nuclear encoded products only: the import of cytoplasmically made polypeptides into mitochondria (see 27 for review) and the replication of mtDNA (28). To complete this picture, nuclear genes also control the processing of some mitochondrially synthesized polypeptides before their final assembly into the membrane bound complexes (29).
In some systems, a number of mitochondrial mutations conferring a variety of phenotypes have been instrumental in characterizing mitochondrial genes, in mapping the mitochondrial genomes and in determining the rules for segregation and recombination of mitochondrial genes (30). Mitochondrial mutants always show mitotic segregations during the cell divisions that follow a cross and do not show segregation at meiosis. Mitotic segregation, however, occurs at very different rates in different organisms. For example, homoplasmic cells (those with only one type of mtDNA molecules) are sorted out very rapidly in Saccharomyces cerevisiae (see 6), slowly or moderately slowly in Paramecium tetraurelia (depending on the phenotype conferred by the mutation (31)) and extremely slowly in mammalian cells in cultures (32). What mechanism is responsible for such

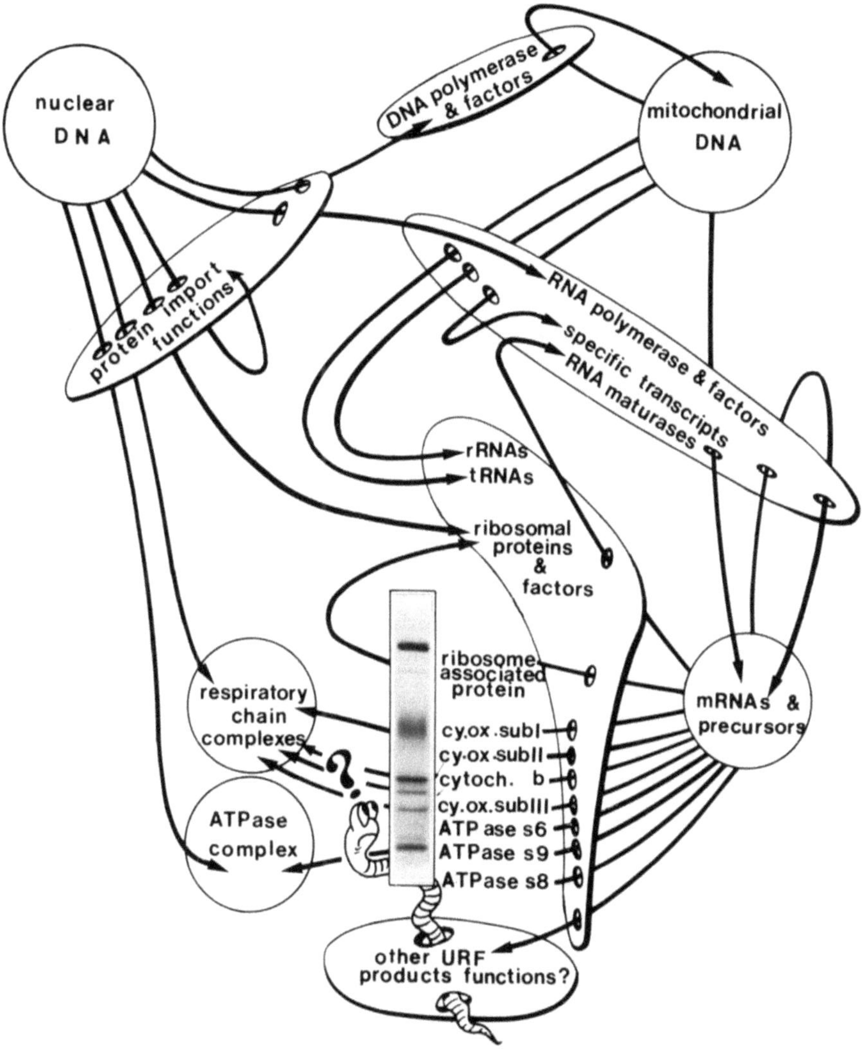

Figure 1: Intricate interplay of major genetic functions involved in mitochondrial biogenesis.
The cartoon symbolizes transfer and trimming of information (arrows) through and by the major functions (disks) involved in the biogenesis of mitochondria. The electrophoretic pattern of mitochondrial translation products is that of <u>S.cerevisiae</u> (kindly provided by A.Jacquier). Subunits 6 and 8 of ATPase are not visible. All polypeptides are ubiquitous mitochondrial translation products with the exception of the ribosome associated protein and the subunit 9 of ATPase. Additional mitochondrial translation products are found in other groups of organisms (e.g. plants).

tremendous differences is unclear but a high segregation rate is probably one of the conditions that determines the facility with which mitochondrial mutants can be isolated. Mitochondrial genetic recombination is not less variable between different organisms. Very efficient in yeast (see 6) and also probably in plants, genetic recombination has been found in a variety of fungi (33,34,35) as well as in somatic cells hybrids (36) but it could not be detected at all in Paramecium (37). In addition, in many multicellular organisms a strictly uniparental inheritance of mitochondrial genomes is the rule, hence the absence of genetic recombination.

It is obviously only in the few organisms in which there are enough mitochondrial mutants of various sorts (point mutations, deletions) that genetic maps could be correlated with the molecular maps of their mitochondrial genomes. For others, mapping mostly relies on gene sequence comparisons. Several mitochondrial genomes have now been entirely sequenced (e.g. human (38), mouse (39), and bovine (40)), others are nearing completion (e.g. Schizosaccharomyces pombe Lang et al. this volume and Aspergillus nidulans Scazzocchio et al. this volume) and fragments of mtDNA from Saccharomyces cerevisiae are gradually being aligned properly (see 93A and Lazowska et al. this volume). Depending upon the organism, the assignation of mitochondrial genes to mitochondrial gene products in constructing maps can be either direct or indirect. A best example is perhaps the gene for the subunit II of cytochrome oxidase. In bovine, this gene and its product were directly correlated by comparison of the DNA and protein sequences (40,41). In yeast, the equivalent gene was identified by detailed genetic analysis of mutants (42-45). But in other species, sequence homology with the genes for subunit II of cytochrome oxidase of the previous species has been sufficient to assign gene functions (this is the case of plant (46) or human (38)).

This short review is devoted to a brief introduction to the mitochondrial genomes of various organisms presented at this meeting, with emphasis on mutants, genes and maps, to help the non familiar readers to orient themselves through the various chapters of this book. A number of other organisms (e.g. Xenopus, birds, insects) have been disregarded although they also show considerable interest for studies on mitochondrial genes, and the reader is refered to Benne et al. (this volume) for the amazing beauties of kinetoplast DNA.

Mitochondrial genes, mutants and map of Saccharomyces cerevisiae.

Complete deficiency in the synthesis of any of the ubiquitous mitochondrial translation products results in the absence of ATP synthesis in mitochondria, a situation which, for most organisms is lethal to the cell, hence the difficulty to find mitochondrial mutants. Being an

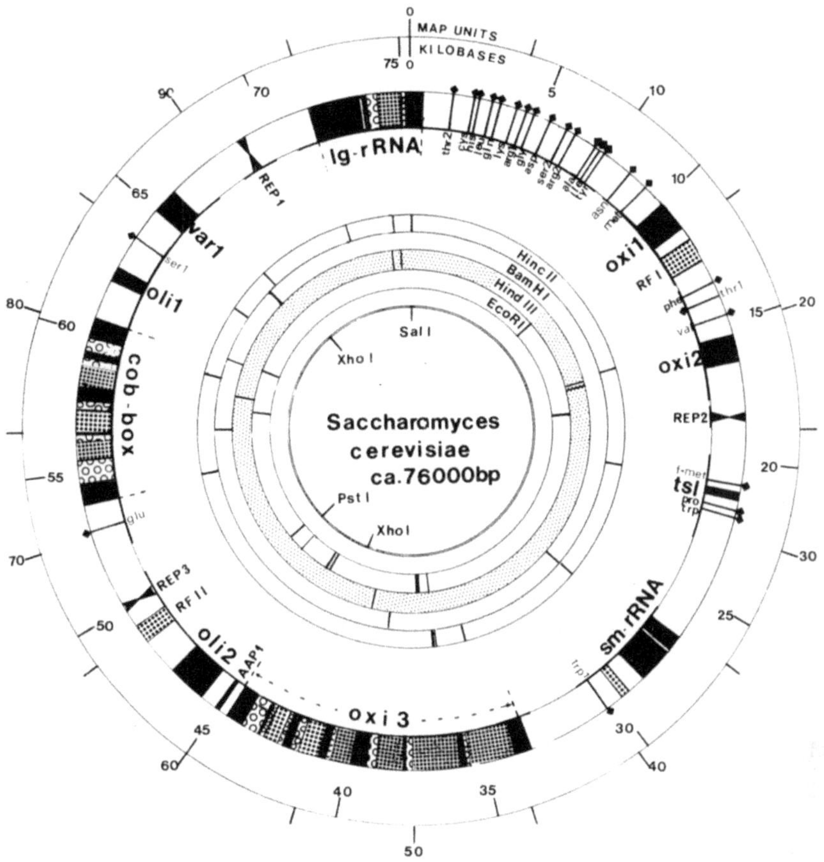

Figure 2: Physical and genetic map of the mitochondrial genome of the yeast Saccharomyces cerevisiae

The map has been adapted and updated from (6). Restriction map (inner circles) is redrawn from the original map (63) of the "long" strain (KL14-4A). For comparison with the "short" strain, see (64-65).

The genes have been placed on the map (outer circle) on scale with respect to restriction sites, using for their alignment, the complete nucleotide sequences. Because no complete DNA sequence is available from a unique strain of S.cerevisiae, the genetic map drawn is a patchwork of segments issued from a variety of strains. Sectors in which complete nucleotide sequence is available are indicated by heavy line.

References are (14, 57, 66, 67) for the rRNA genes, (68-84) for the tRNA genes (designated here by the corresponding amino acid), (11, 12, 42, 45, 71, 85) for the cytochrome oxidase genes (oxi 1 for subunit II, oxi 2 for subunit II and oxi 3 for subunit I), (86-91) for the ATPase genes (oli 1 for subunit 9, oli 2 for subunit 6 and aap 1 for subunit 8), (13, 15,92 and Lazowska et al. this volume) for the cytochrome b gene (cob-box), (93A) for the var 1 gene, (28,93) for the rep sequences and (94 and Miller et al. this volume) for the tRNA synthesis locus (TSL). Transcription is clockwise for all mitochondrial genes with the exception of tRNA thrl. exons open reading frames intronic sequences blocked in all frames.

For comparison, the mitochondrial gene organizations of other yeasts can be found in (Lang et al. this volume) for Schizosaccharomyces pombe, (95, 96) for Kluyveromyces lactis and Saccharomycopsis lypolityca and (97) for Torulopsis glabrata.

exception to this rule, the yeast S.cerevisiae deserves a special mention by the number and variety of the mitochondrial mutants which have been isolated and by the existence of the rho$^-$ mutation. Rho$^-$ mutants occur spontaneously at high frequency and are the result of extremely large deletions of the mtDNA, leaving only minute segments of genes (regularly amplified along the rho$^-$ mtDNA molecules) or even no mtDNA at all (the rho^0 type in the latter case, see 6 for review). The segments of mtDNA which remain are faithfully replicated even though they may not always contain a normal replication origins (28,47,48). They may be transcribed but the genes they contain are never expressed into proteins because essential genes for mitochondrial protein synthesis are not all present. Rho$^-$ mutants offer a virtually unlimited source of various deletions which have been essential in the construction of the genetic map in this species (see 6 for review). In addition, each rho$^-$ mutant provides a source of purified mtDNA fragments, extremely useful for both molecular and genetic studies on recombination, gene expression or gene amplification.

The other type of mitochondrial mutations in S.cerevisiae are point mutations or small alterations of mtDNA, affecting a specific gene or function (a compilation of such mutations can be found in the preceding publication of this series (49)). Many respiration deficient mutants resulting from point mutations (mit$^-$) have now been mapped in the mitochondrial genes encoding the cytochrome oxidase subunits I, II or III (genes oxi 3, oxi 1 or oxi 2 respectively), in the gene encoding cytochrome b (cob-box) and in the genes encoding subunits 9, 6 or 8 of ATPase (genes oli 1, oli 2 and aap 1 respectively - notice, however, that the assignation of subunit 6 to oli 2 is disputed (50) -). In the cases of the cob-box gene and the oxi 3 gene which are mosaic, mit$^-$ mutations have also been found within the introns. Intron mutants define special complementation groups (51,52) which reflect different functions for the different parts of a given intron (53,54 and Groudinsky et al.; MacKellar et al. this meeting). Other point mutations which map at a few specific loci within the previous genes confer resistance to various drugs or inhibitors instead of leading to respiratory deficiency. Those include resistance to antimycin A, mucidin, diuron or funiculosin for the cob-box gene or resistance to oligomycin (venturicidin) for the oli 1 and oli 2 genes (reviewed in 49).

Respiration deficient mutants may also result from point mutations in the genes of the protein synthetic machinery. Such mutants (syn$^-$) lead to complete or partial deficiency in mitochondrial protein synthesis, hence affecting the expression of all mitochondrial genes at once. Most syn$^-$ mutants occur in tRNA genes (reviewed in 6, 49) but a number of temperature dependent mutants have also been characterized in the rRNAs genes (55,56). Again, other point mutations mapping at specific loci in

the rRNA genes can confer resistance to antibiotics that affect the mitochondrial ribosomes (e.g. chloramphenicol, erythromycin or paromomycin (14,57,58 and Dujon and Jacquier, this volume)).
A last type of mitochondrial point mutations is made of those which restore respiratory capacity to other mitochondrial mutants but do not, themselves, confer a mutant phenotype (called mim). Two mitochondrial loci containing mim mutations have been described. Mim 1 is located within the large rRNA gene and probably restores the (pseudo) wild-type phenotype of some mit⁻ mutants by decreasing the fidelity of translation (59). An informational suppressor (ochre suppressor) has also been found, mapping in or near the small rRNA gene (60). Mim 2, on the contrary, is a genetic locus within intron 4 of the oxi 3 gene which is only able to suppress mutants of intron 4 of the cob-box gene, probably by activating an otherwise dormant maturase (61).
Using all these mutations in deletion mapping and recombination mapping experiments, the genetic map of the mitochondrial genome of S.cerevisiae could be established. This map (figure 2) has been correlated with the restriction maps of mtDNA from various strains and the DNA sequences of all known mitochondrial genes are now available. Whether the map is saturated is not, however, certain. There is no complete DNA sequence from a unique strain, as yet, and the present map is a patchwork of fragments from different strains. In addition, a few new mutants and a few unassigned reading frames which have been found in intergenic segments may well lead to the discovery of new functions.
Natural polymorphism also offers a means to investigate the mitochondrial genome of S.cerevisiae. Polymorphic variations are frequent between various strains. They mainly concern the presence or absence of introns (see 62 for review) and of G+C rich clusters, but protein genes can also exhibit intriguing size variations (e.g. var1 Zassenhaus et al. this volume).

Mitochondrial genes, mutants and maps in other yeasts.
Some mitochondrial mutants have been characterized in the fission yeast Schizosaccharomyces pombe. Although this organism is a petite negative yeast that cannot survive complete loss of mtDNA and, therefore, does not give rise to mutants equivalent to the rho⁻ of S.cerevisiae, a number of drug resistant mutants and respiratory deficient mutants (which are considered the mit⁻ equivalent) could be found in this organism. Most of them result from the action a specific mutator (ana 8), itself mitochondrially inherited, which increases 100 to 1000 times the frequency of spontaneous drug resistant mitochondrial mutants and accumulates up to several percent of respiratory deficient mutants (98). The latter mutants are small deletions or point mutations leading to deficiencies in

cytochrome oxidase, in cytochrome b or in the ATPase that can only survive in the mutator strain.

The 19kb mitochondrial DNA of S.pombe is now almost completely sequenced (Lang et al. this volume) and many transcripts have been precisely mapped. In addition to the rRNA and tRNA genes and to the genes coding for the known ubiquitous mitochondrial translation products (and mainly determined from their sequence homology with S.cerevisiae genes) two URFs have been found. One of them shows homology to the S.cerevisiae aap 1 gene and possibly also to the human URF A6L.

A few other yeast species have also provided mitochondrial mutants. This is the case, for example, of Kluyveromyces lactis in which a few mitochondrially inherited oligomycin resistant mutants have been described (99). This species, although closely related to S.cerevisiae, has a mitochondrial genome only half the size that of S.cerevisiae (100). Torulopsis glabrata, another yeast in which mitochondrial mutations have been isolated (101) and some mitochondrial genes have been mapped, has a mitochondrial genome of only one quarter that of S.cerevisiae (97). The significance of such extensive size variations among yeasts is not clear.

Mitochondrial genes, mutants and maps in filamentous fungi.

Neurospora crassa, Aspergillus nidulans and Podospora anserina offer major interest for mitochondrial genetics. In filamentous fungi, like in yeasts, mitochondrial genomes exhibit extensize size variations although their overall properties look similar (A. nidulans is half the size of N.crassa and a third of P.anserina). The mitochondrial genome of N.crassa has been most extensively studied both genetically and molecularly and several genes are now completely sequenced (20,106,109-113 and Burger et al., Helmer-C. et al. and DeVries et al. this volume). It shows the same ubiquitous genes as the yeast genome (fig.3). The genes for the two rRNAs and many tRNAs are clustered (20,108) and, interestingly, tRNAs genes were found flanked by short G+C-rich palindromic sequences, containing a common 18 bp long core sequence (119). Such palindromes are also found flanking other genes or even inside introns (20) and it is estimated that the G+C-rich palindromes make up to 5-10% of the total N.crassa mtDNA. These palindromes could serve as signals for RNA processing (119) and they may be also involved in intramolecular recombination. Another unique feature of N.crassa is the existence, in its mitochondrial genome, of a sequence (MAL) which potentialy codes for the subunit 9 of ATPase (113,120) although this subunit is a cytoplasmic translation product encoded by a nuclear gene (121). No evidence that the mitochondrial gene can be expressed exists so far and the reason for its maintenance is unclear.

The 33 kb mitochondrial genome of A.nidulans has now been almost completely sequenced (122,122A and Schazzochio et al. this volume).

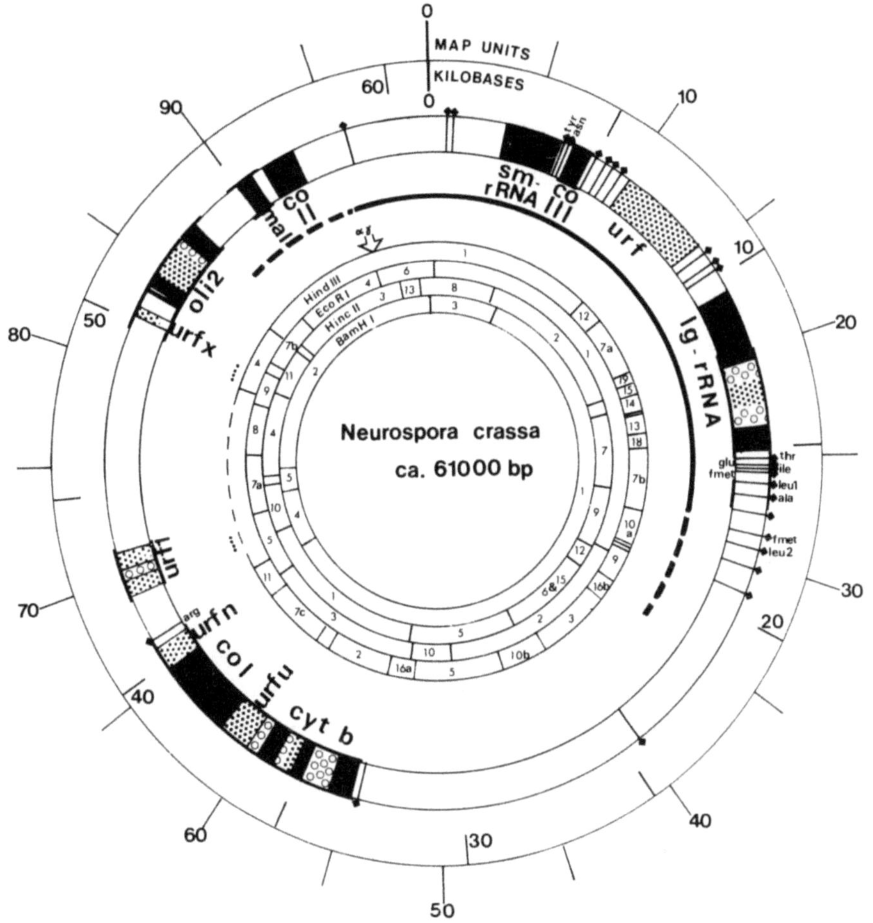

Figure 3: Physical and genetic map of the mitochondrial genome of the filamentous fungi Neurospora crassa.
The restriction map (inner circles) has been redrawn from original maps of type II laboratory strains of N.crassa (102-107) and genes have been placed on the map (outer circle) on scale with respect to the restriction sites. Sectors in which complete nucleotide sequences are available are indicated by heavy lines.
References are (20, 102,103,104,108) for the rRNA and tRNA genes (designated by the corresponding amino acid), (106,109,110) for the cytochrome oxidase genes (COI, COII and COIII respectively for subunits I, II and III) and their flanking open reading frames, (111,112) for the cytochrome b gene (cyt b), (113) for the mitochondrial ATPase proteolipid like gene (MAL). See also Burger et al., Citterich et al. and DeVries et al. this volume. Transcription is clockwise for all known genes.
... variable zones (see 107) and ⇩αγ site of the alpha and gamma tandem reiterations (105,114) The segment of mtDNA which is retained in some poky mutants and some stopper mutants (114-116) is indicated by the bold line. The precise end points of this segment varies, according to the mutant, in the region indicated by the dashed bold line.
For comparison, the mitochondrial gene organizations of other filamentous fungi can be found in (Schazzochio et al. this volume) for Aspergillus nidulans and (117,118) for Podospora anserina.

Perhaps the major finding in this genome is the fact that it contains, in addition to the genes for ubiquitous mitochondrial translation products, eleven URFs, four of which being homologous to the mammalian URFs, hence making a link between mammals and yeasts mitochondrial genomes. A few mitochondrially inherited mutations have been characterized in this species (mostly antibiotic resistance and cold sensitivity, see Schazzochio et al. this volume) and genetic mapping has been attempted. Finally, a few extranuclearly inherited mutants have also been reported in Podospora anserina (123).

A class of mutants apparently specific to the filamentous fungi deserves a specific attention in this chapter. They are large alterations of mtDNA often found associated with an arrest of vegetative growth of the mycelium. Several such mutants are known in N.crassa (called poky or stopper (115,116)). In the E35 stopper mutant for example, no intact mtDNA is detected (115). Instead, two different populations of abnormal molecules are present, a first one in which a ca. 4 kb fragment has been deleted and a second one, in which only a third of the genome remains and is amplified (see fig.3), with one edge of the deletion being common to the deletion point of the first class. Molecules of the second class tend to outreplicate those of the first class. The segment which is amplified is common to a variety of similar mutants although the end points of the deletion events vary (24,116). In Aspergillus amstelodami, like in N.crassa, extranuclearly inherited mutants (called ragged) have been found that are associated with excision and head to tail amplification of mtDNA sequences originating mainly from two regions of the genome. One of them possibly contains the origin of replication (125).

In Podospora anserina, a similar phenomenon occurs spontaneously in wild type laboratory strains. This phenomenon, called senescence because it results in the irreversible arrest of the vegetative growth of the mycelium, has been discovered many years ago (126). But recently, it has been correlated with large alterations of the mtDNA, comparable by their extent, to the rho^- mutation of S.cerevisiae. The mtDNA isolated from senescent mycelia consists of small circular molecules arranged in multimeric series of sizes (127,128). Those molecules result from the amplification of some specific segments of young mtDNA and behave as if they were independently replicating molecules (117,129-132). Different types of senescent events have been described that differ by the fragment of mtDNA which is amplified. The alpha senDNA is a ca. 2.6kb fragment while beta senDNAs represent a family of molecules containing a common central part but differing by their deletion end points. A gamma type has also been found. Most interestingly, it has recently been shown by DNA sequencing (133) that one of these molecules, excised from mtDNA in the alpha type of senescent event, shares distinctive features with class II mitochondrial introns of S.cerevisiae (see 134,135). Furthermore, this molecule shows excision points that corresponds exactly, to the nucleotide level, to the limits of the predicted intron (Esser et al. this volume and 136). The mechanism by which such a precise excision event could take place is not yet clear but it is tempting to assume that excised introns RNA could be copied to a circular DNA molecule which, provided it contains a sequence for replication, could then outnumber resident mtDNA (in yeast, class II introns are excised as circular RNAs (137)). Whether these molecules are the cause of senescence or its consequence is not yet clear.

Mitochondrial genes, mutants and maps in mammals.
The complete DNA sequence from the human mitochondrial genome (fig 4) has been determined two years ago (38), that of mouse and bovine are now also available (39,40) and that of rat is close to completion (138-141). Among all interesting features that were discovered, only a few can be mentioned in this short review. Genes on mammalian mitochondrial genomes are tightly packed and do not show a single intron. They lie on both DNA strands and transcription is complete and symmetrical from a limited number of promoters (see 142). Size constraints are apparently so drastic as to restrict rRNA genes to their minimal form, to alter the structures of some tRNAs and even to eliminate termination codons at the end of genes. The genetic code differ from both the nuclear one in mammals and the mitochondrial one in yeast and fungi (143,144). Eight URFs have been recognized from the DNA sequence, the functions of which remain unknown for most of them (see Attardi et al. this volume).
Mouse, rat and human cells in cultures have allowed the isolation of only a few mitochondrial mutants (32, 146-149). One of the major difficulty with these systems seems to be the high number of mitochondria per cell and the extremely slow mitotic segregation which limits the isolation of homoplasmic cell lines and the possibilities of genetic tests. Using a technique called cybrid, especially developped for this purpose (see 32), it was, however, possible to show that some chloramphenicol, erythromycin , rutamycin or efrapeptin resistant mutants are cytoplasmically inherited (32, 146-149). The mitochondrial nature of the C^R mutations has recently been directly demonstrated by DNA sequencing of part of the large mitochondrial rRNA gene from some mutants (150-152). Two sites, similar to those defined in S.cerevisiae (see 14 and Dujon and Jacquier, this volume), were found and a third and new site in the peptidyl transferase center has just been described in rat (Koike et al. this volume).
If mutants are scarce, natural polymorphism, on the contrary, provides a direct tool to study mitochondrial genetics in Mammals. As a rule, the paternal contribution to the mitochondrial genotype of the offspring is negligeable, if at all existent (153-158). Whether this simply results

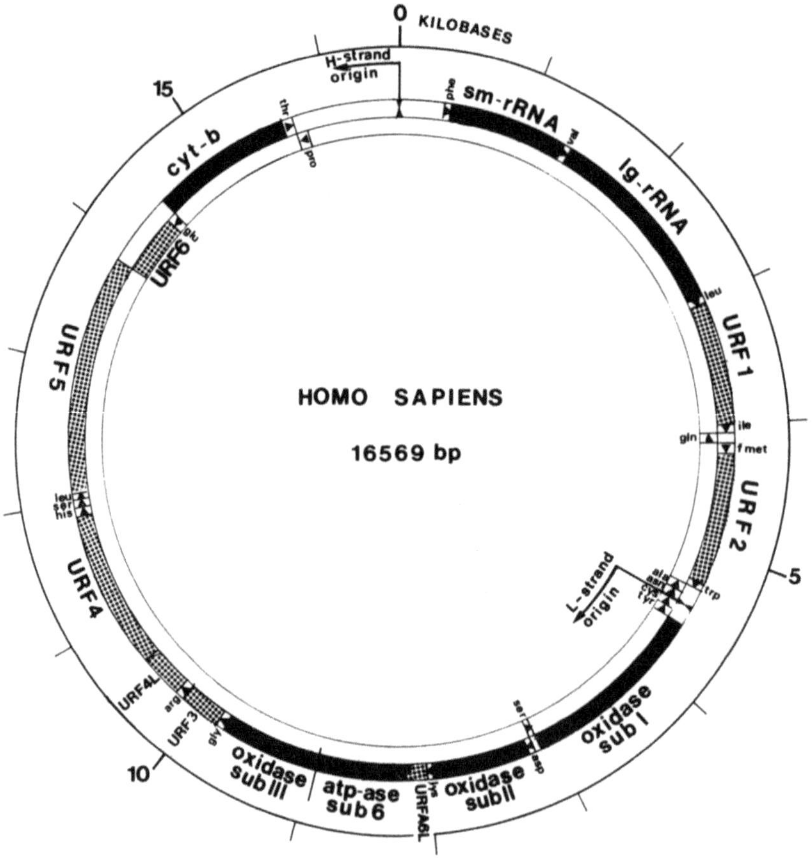

Figure 4: The genetic map of the human mitochondrial genome
The circular map has been redrawn to scale from the complete nucleotide sequence (38) The map has been numbered from the 5' end of the nascent H strand (the 7S DNA found in the D-loop).
exons, unassigned reading frames (URFs), tRNA genes (designated by the corresponding amino acid). Transcription is clockwise for those genes shown on the outer circle (the L strand), counterclockwise for the others.
For comparison, the complete mtDNA sequences of other mammals can be found in (39) for the mouse and (40) for bovine. For sequences of rat mitochondrial genes see (138-141).

from dilution of the copies of mtDNA from the spermatozoon during egg development or from a specific mechanism aimed at inactivating these copies is still a matter for discussion. But, it follows that Mammals contain mtDNA of the maternal type only, a situation which has allowed pedigree studies and evolution analyses, using restriction sites polymorphism as genetic markers (159-164). An example of heterogeneity within a single animal has, however, been reported (165). Similarly, polymorphism between animals of the same maternal lineage has been described (166). It is possible that both phenomena reflect mitochondrial DNA variations which are rapid with respect to the animal generation time (see 160).

Mitochondrial genes and mutants in ciliates.

Paramecium tetraurelia has been the object of extensive mitochondrial genetics studies a few years ago. Mitochondrial erythromycin resistant mutations and chloramphenicol resistant mutations have been isolated (31, 167). Homoplasmic mutant cell lines can be obtained relatively easily by direct selection in the drug (31, 168) but other mutations, including temperature sensitivity or slow growth, could also be isolated without selective pressure, using low doses of nitrosoguanidine as the inducing agent, a procedure that may be of general interest to isolate mitochondrial mutants in other organisms (37). Another class of mitochondrial mutations with great potential interest is represented by those mutations which suppress the deleterious effect of a specific nuclear mutation and restore nucleo-mitochondrial harmony (169). Microinjection of entire mitochondria in the cell offers a very powerful tool unique to this organism (170). However, the complete absence of genetic recombination between mitochondrial genes (37) in Paramecium imposes serious limits to the use of this organism for mitochondrial genetics. Mitochondrial mutants have also been reported in a few other ciliates such as Tetrahymena (171). But it is perhaps their peculiar mode of mtDNA replication, associated with the fact that the ciliates Tetrahymena and Paramecium represent the only examples of linear mtDNA molecules with cross linked termini, that should retain more attention to these organisms (172-176).

Mitochondrial genes and mutants in plants.

It is probably plants which, at present, offer the greatest potentialities for future investigations of the mitochondrial genomes. Mitochondrial DNA of higher plants has not yet been mapped in details since regular preparations of DNA from mitochondria yield highly heterogeneous populations of molecules (see 177,178 for reviews). The major part is made of linear DNA molecules of varying sizes with a small proportion of circular molecules ranging in size from very small (ca 1 kb) to relatively large (ca 100 kb) (see 178A,178B and Leaver et al.,this volume). Restriction mapping is also not facilitated because numerous fragments with various stoichiometry are often found (179). But it is clear, both from restriction enzyme analysis and from renaturation kinetics that the mitochondrial genome of plants is much larger than that of other groups of organisms (200 to 600 kb in general or even 2400 kb for some Cucurbitacae, 180). In addition, important variations in the organization of the mitochondrial genomes are observed between species (or families) or even inside members of the same species (181). Part of the heterogeneity probably results from intense intramolecular recombination. This should perhaps be correlated with the occurence of repeats of up to a few kilobases long, representing ca. 5-10% of total plant mtDNA (180). In striking contrast to the genomic gigantism of higher plants seems to be the algae Chlamydomonas reinhardtii in which a linear DNA molecule of 15 kb only (possibly the result of breakage at a specific point) is the only candidate found so far to represent the mitochondrial genome (182).

If size is variable, the G+C content of plant mt DNA is remarkably constant (47%) for all species examined (see 178). This is in contrast to what is found in other groups of organisms (eg. the human mt DNA is 44% G+C while the mouse mtDNA is 37% G+C, although their gene organisations are exactly the same).

A major question which is considered at present is obviously to determine whether the large size of plant mitochondrial genomes reflects the existence of novel functions, not found in other organisms. Only a few genes have been characterized so far on the mtDNA of some plant species. Those include the genes encoding the 28S and 16S ribosomal RNAs (183-186), the 5S rRNA, a situation unique to plant mtDNA (21) and some tRNAs. In addition, the gene for the subunits I and II of cytochrome oxidase and the gene for the cytochrome b (46 and see Leaver et al. this volume) have been detected by heterologous hybridization with yeast and/or mammalian equivalents. Determination of their DNA sequences has revealed a high degree of conservation (40 to 70% of sequence homology between the three groups for cytochrome b and for subunits of cytochrome oxidase). But it is the recent finding (187 and Leaver et al. this volume) that the alpha subunit of ATPase is a mitochondrial translation product in plants (hence likely the product of a mitochondrial gene) that provides the first evidence for the existence of additional genetic functions in plant mtDNA since this subunit is nuclearly encoded in all other systems. Also highly suggestive of additional functions is the fact that translation products synthesized by isolated plant mitochondria amount to some 20-30 major polypeptides, most of which being inner membrane bound (see Leaver et al. this volume). This represents ca. three times more than for other

organisms. Among those mitochondrial translation products, subunits I and II of cytochrome oxidase as well as subunit 9 of ATPase have already been identified. But even this number of polypeptides is far from accounting for the estimated size of the plant mitochondrial genomes and both the function and the nature of the extra sequences remain unknown. Perhaps the intriguing finding (188) of a 12 kb fragment of chloroplast DNA (a part of the inverted repeat of chloroplast DNA including rRNA and tRNA genes) stably integrated into mitochondrial DNA of maize and other species, reflects an essential mechanism (whose biological significance is not yet clear) responsible for the large size of plant mitochondrial genome.
Plants mitochondria offer another important interest as it seems likely that large size genomes do not allow the genetic code to evolve rapidly as do small size genomes. In plants, it is suspected that the CGG codon specifies tryptophan instead of arginin (46), a situation which might have been prevailing at some point of the evolution of the genetic code. Plants, in addition, deviate from other mitochondrial genetic codes by the fact that the UGA codon is very likely a stop codon and do not code for tryptophan.
The major drawback of plants is, certainly, the lack of mitochondrial mutants and the isolation of such mutants may always remain difficult since plants do not survive complete loss of respiration. Some non-mendelian mutants affecting mitochondrial structure and functions have been described in Chlamydomonas (189) but their growth is limited to minute colonies and they die in the dark, rendering difficult molecular analyses. There exists, however, a set of maternally inherited traits (called cytoplasmic male sterility or CMS) that result in the failure of mature plants to produce functional pollen (see 190 for review) and has been directly correlated to specific alterations of mtDNA (191-193). CMS is found in a wide range of plants, both monocots and dicots. Different male sterile lines can be restored to fertility by different and specific restorer genes which are nuclearly inherited (see 190). CMS lines can be classified on the basis of their ability to be restored by each nuclear restorer. In some maize CMS lines, two linear plasmid-like DNA molecules (ca. 5300 and 6200 bp long respectively) with inverted terminal repeats are found together with a few additional DNA species (194,195). One of them has been recently sequenced (196). In Sorghum CMS lines, the two plasmids are 5700 and 5300 bp long respectively (192) and were shown to have a protein bound covalently at their 5' ends (194). Other plasmid-like elements are found in other species (200). In maize, these plasmids are integrated in the mtDNA in normal fertile lines (197-199). The exact relationships between these DNA alterations and the abnormal pollen formation are not clear but the phenomenon is perhaps reminiscent of the excision of mtDNA fragments in filamentous fungi during mycelial growth

arrest. Variant polypeptides of larger or smaller apparent molecular weight are synthesized by mitochondria of male sterile lines (see Leaver et al. this volume), suggesting that male sterility may be related to abnormal mitochondrial gene expression. However, similar plasmid-like elements have also been found in some male fertile lines as well and, in addition, similar DNA alterations are also found in plant tissue culture or during regeneration of plants from culture and are then stably inherited during sexual cycle (see leaver et al. this volume). How such mtDNA alterations arise and become stably inherited in plants is not clear.

Concluding remarks

This short survey of mitochondrial genomes through the variety of present day living organisms reveals both the fundamental unity in the functions encoded by the autonomous mitochondrial genome and the conspicuous diversity in the solutions adopted for the organization and expression of this small set of genes. Genomes vary from tightly packed to largely expanded. The same genes are found split by many introns in some organisms but not in others, apparently without clear rules for the presence and location of such introns. Transcription or replication processes are not less diverse. Even universal molecular objects such as the rRNAs and the tRNAs exhibit extreme variations in their appearance from fully developped in fungi to heavily despoiled in kinetoplast or even in mammals. Diversity and not universality is also the case for the genetic code with codon assignments varying between mitochondria of different organisms in a manner that will, perhaps, allow us to trace back major steps in the evolution of the genetic code. The recent discoveries that fragments of mtDNA may be found in the nucleus and that fragments of chloroplasts DNA may be found in mitochondria raise more questions than it provides answers to the reason for an autonomous genetic system with an autonomous protein synthesis machinery in organelles. Finally, how is the harmony maintained in the intricate interplay between the nuclear and organelle genomes remains unclear. The nucleus sends genetically informative signals at almost every stage of mitochondrial gene expression, but evidence for reciprocity is still lacking. Perhaps in many cases, it is enough if mitochondria only return physiological signals to the cell, but in plants the two organelles provide partly equivalent physiological functions to the cell and it seems likely that specific regulations are needed.

Acknowledgments

I am indebted to A.Jacquier for his fruitful help in preparing the figures of this manuscript. This work has been supported by grants from the CNRS (ATP Biologie moléculaire du gène) and from INSERM (PRC 134004).

References

1 - "The Genetics Functions of Mitochondrial DNA (1976). C. Saccone and A.M. Kroon Eds North Holland Pub. Co, Amsterdam.

2 - "Genetics and Biogenesis of Chloroplasts and Mitochondria" (1976) T. Bucher, W. Neupert, W, Sebald and S. Werner Eds. North Holland Pub. Co. Amsterdam.

3 - "Mitochondria 1977 : Genetics and Biogenesis of Mitochondria" (1977) W. Bandlow, R.J. Schweyen, K. Wolf and F. Kaudewitz Eds, W. de Gruyter, Berlin

3A - "Extrachromosomal DNA" (1979) D.J.Cummings, P.Borst, I.B.Dawid, B.H.Weissman and C.F.Fox Eds. Acad.Press N.Y.

4 - "The Organization and Expression of the Mitochondria genome" (1980) C. Saccone and A.M. Kroon, Eds. North-Holland, Amsterdam.

5 - "Mitochondrial Genes" P.P. Slonimski, P. Borst, G. Attardi, Eds. Cold Spring Harbor, 1982, Monograph 12.

6 - B. Dujon (1981) in "The Molecular Biology of the Yeast Saccharomyces cerevisiae. J. Strathern, E. Jones, J.R. Broach, Eds. Cold Spring Harbor.(1981) pp. 505-635.

7 - L. A. Grivell (1983) Scientific American, 248, 78-89.

8 - Terpstra P.E. Zanders, R.A. Butow (1979) J. Biochem. Chem. 254, 12653 .

9 - G.S.P. Groot, T. Mason, N. Van-Harten-Loosbroek (1979) Mol. Gen. Genet. 174, 339

10 - R.J. La Polla and A.M. Lambowitz (1982) J. Cell Biol.95, 267-277.

11 - S.G. Bonitz, G. Coruzzi, B.E. Thalenfeld, A. Tzagoloff, G. Macino. J. Biol.Chem. (1980), 255, 11927-11941.

12 - L.A.M. Hensgens, L. Bonen, M. de Haan, G.V.D. Horst and L. A. Grivell. (1983), Cell 32, 379-389.

13 - F.G. Nobrega, A. Tzagoloff. (1980) J. Biol.Chem. 255, 9828-9837.

14 - B. Dujon (1980) Cell 20, 185-197.

14A - A.Jacquier and B.Dujon (1983) Molec.Gen.Genet. (in press)

15 - J. Lazowska, C. Jacq, P.P. Slonimski. (1980) Cell, 22, 333-348.

16 - G.Church and W.Gilbert (1980) in "Mobilization and Reassembly of Genetic Information" D.R.Joseph et al. eds Acad.press N.Y. pp 379-394

17 - H.R.Mahler, D.K.Hanson, M.R.Lamb, P.S. Perlman, P.Q.Anziano, K.R.Glaus and M.L. Haldi (1982) in "Mitochondrial genes "Slonimski et al. Eds Cold Spring Harbor Laboratory pp 185-199

18 - H.De La Salle, C.Jacq and P.P.Slonimski Cell (1982), 28, 721-732

19 - B.Weiss-Brummer, G.Rôdel, R.J.Schweyen and F.Kaudewitz Cell (1982), 29, 527-536

20 - J.M.Burke and U.L.RajBhandary Cell (1982) 31, 509-520

21 - C.J. Leaver, M.A. Harmey (1976) Biochem. J. 157, 275-277.

22 - C.L.Dieckmann, L.K.Pape and A.Tzagoloff Proc.Natl.Acad.Sc.US (1982) 79,1805-1809

23 - G.Faye and M.Simon Comptes Rendus Séances Acad.Sc. Paris (1982) 294, 245-248

24 - H. Bertrand, P. Bridge, R.A. Collins, G. Garriga, A.M. Lambowitz, Cell (1982) 517-526.

25 - J.-M. Schneller, C. Schneider, A. J.C. Stahl. (1978) Biochem. Biophys. Res. 85, 1392-1399.

26 - A. K. Hopper, A.H. Furukawa, H. Dinh Pham, N.C. Martin (1982) Cell, 28, 543-550.

27 - P.Boehni, S.Gaser, C.Leaver and G;Schatz (1980) in "The organization and expression of the mitochondrial genome" Kroon and Saccone Eds. pp 423-433

28 - H. Blanc, B. Dujon (1980) Proc. Natl Acad. Sci. (USA) 77, 3942-3946.

29 - E. Pratje, G. Mannhaupt, G., Michaelis, K. Beyreuther (1983), The EMBO J. 1049-1054.

30 - B.Dujon and P.P.Slonimski (1976) in "Genetics and biogenesis of Chloroplasts and Mitochondria " Bucher et al. eds North-Holland Amsterdam pp 393-403

31 - A. Adoutte, J. Beisson (1972) Nature 235, 393-396.

32 - D.C. Wallace, N. A. Oliver, H. Blanc, C.W. Adams (1982) in "Mitochondrial Genes" Cold Spring Harbor Laboratory (P.P. Slonimski et al. Eds) pp.105-116

33 - R.T.Rowlands and G.Turner Molec.Gen.Genet. (1974) 133, 151-161

34 - G.Steitz-Mayr, K.Wolf and F.Kaudewitz Molec.Gen.Genet. (1978) 164, 309-320

35 - A.Brunner, A.T.De Cobos and D.Griffiths Molec.Gen.Genet. (1977), 152, 183-191

36 - I.Horak, H.Coon and I.B.Dawid (1974) 71, 1828-1832

37 - A. Adoutte, J. Knowles, A. Sainsard-Chanet (1979) Genetics, 93, 797-831.

38 - S. Anderson, A.T. Bankier, B.G. Barrell, M.H.L. de Bruijn, A.R. Coulson, J. Drouin, I.C. Epernon, D.P. Nierlich, B.A. Roe, F.Sanger, P.H. Schreier, A.J.H. Smith, R. Staden, I.G. Young (1981) Nature, 290, 457-465.

39 - M.J. Bibb, R.A. Van Etten, C.T. Wright, M. W. Malberg, D.A. Clayton. (1981) Cell 26, 167-180.

40 - S. Anderson, M.H.L. de Bruijn, A.R. Coulson, I.C. Eperon, F. Sanger and I.G. Young. (1982) J. Mol. Biol. 156, 683-717.

41 - G.J. Steffens, G. Buse (1979) Hoppe-Seyler's Z. Physiol. Chem. 360, 613-

42 - T.D. Fox. (1979)Proc. Natl. Acad. Sci. (USA), 76, 6534-6538.

43 - B. Weiss-Brummer, R. Guba, A. Haid, R.J. Schweyen (1979) Current Genet. 1, 75

44 - F. Cabral, M. Solioz, Y. Rudin, G. Schatz, L. Clavilier, L., P.P. Slonimski (1978), J. Biol. Chem. 253, 297-

45 - G. Corruzzi, A. Tzagoloff, (1979) J. Biol. Chem.254, 9324-9330.

46 - T.D. Fox, C.J. Leaver (1981) Cell 26 315-323.

47 - M. de Zamaroczy, G. Faugeron-Fonty and G. Bernardi (1983) Gene, 21 193-202.

48 - P. Borst, C. Heyting, J.P. M. Sanders (1976) in "Genetics and Biogenesis of Chloroplasts and Mitochondria (TH. Bücher *et al.*,

Elsevier North Holland Biomed. Press. Amsterdam. pp.525-533

49 - B. Dujon, A.M. Colson, P.P. Slonimski (1977) In "Mitochondria 1977" (W. Bandlow et al. Eds) De Gruyter, Berlin pp. 579-669.

50 - M. Somlo, J. Cosson, L. Clavilier, M. Krupa, I. Laporte (1982) Eur. J. Biochem. 122, 369-374.

51 - A. Lamouroux, P. Pajot, A. Kochko, A. Halbreich, P.P. Slonimski (1980), in "The Organization and Expression of the mitochondrial genome" (Kroon and Saccone, Eds) Elsevier North Holland, pp. 153-156.

52 - A. Haid, R.J. Schweyen, H. Bechmann, F. Kaudewitz, M. Solioz, G. Schatz (1979) Eur. J. Biochem. 94, 451-464.

53 - P.Q. Anziano, D.K. Hanson, H.R. Mahler, P.S. Perlman (1982) Cell, 30, 925-932.

54 - C. Jacq, P. Pajot, J. Lazowska., G. Dujardin, M. Claisse, O. Groudinsky, H. de la Salle, C. Grandchamp, M. Labouesse, A.Gargouri, B. Guiard, A. Spyridakis, M. Dreyfus, P.P. Slonimski (1982), In "Mitochondrial Gnes" (PP. Slonimski et al. Eds) Monograph 12,pp. 155-183.

55 - R.J. Devenish, K.J. English, R.M. Hall, A,W. Linnane, H.B. Lukins (1978) Mol. Gen. Genet. 161, 251-259.

56 - M. Bolotin-Fukuhara (1979) Mol. Gen. Genet. 177, 39-46.

57 - M. Li, A. Tzagoloff, K. Underbrink-Lyon, N.C. Martin (1982)J. Biol. Chem. 257, 5921-5928.

58 - F. Sor, H. Fukuhara (1982) Nucleic Acids Res. 10, 6571-6577.

59 - G. Dujardin, P. Pajot, O. Groudinsky, P.P. Slonimski (1980) Mol. Gen. Genet. 179, 469-482.

60 - T. D. Fox, S. Staempfli (1982) Proc. Natl Acad. Sci. (USA) 70, 1583-1587.

61 - G. Dujardin, C. Jacq, P.P. Slonimski (1982) Nature 298, 628-632.

62 - P. Borst in "Proceedings of Mosbach Colloquia" (1980) (Th. Bücher et al., Eds) Springer Verlag, Berlin.

63 - P. Borst, J.L. Bos, L.A. Grivell, G.S.P. Groot, C. Heyting, A.F.M. Moorman, J.P.M. Sanders, J.L. Talen, C.F. Van Kreijl, G.J.B. Van Omnen in "Mitochondria 1977" Eds. W. Bandlow et al. p. 213. De Gruyter, Berlin.

64 - R. Morimoto, M. Rabinowitz, Mol. Gen. Genet. (1979) 170, 11-23

65 - R. Morimoto, M. Rabinowitz, Mol. Gen. Genet. (1979) 170, 25-48

66 - F. Sor and H. Fukuhara. Nucleic Acids Res. (1983), 11, 339-348.

67 - F. Sor and H. Fukuhara Comptes Rendus Acad Sci. (Paris) (1980), 291, 933-936.

68 - R.E. Berlani, C. Pentella, G.Macino, A. Tzagoloff. J. Bact. (1980), 141, 1006-1097.

69 - R.E. Berlani, S.G. Bonitz, G. Coruzzi, M. Nobrega, A. Tzagoloff. Nucleic Acids Res. (1980), 8, 5017-5030.

70 - S. Bonitz, A. Tzagoloff. J. Biol. Chem. (1980) 255, 9075-9081.

71 - G. Coruzzi, S. G. Bonitz, B.E. Thalenfeld, A. Tzagoloff. J. Biol. Chem. (1981) 256, 12780-12787.

72 - J.L. Bos, K.A. Osinga, G. Van der Horst, P. Borst. Nucleic Acids Res. (1979), 10, 3255-3266.

73 - Bonitz, G. Homison, B.E. Thalenfeld., A. Tzagoloff, F.G. Nobrega. J. Biol. Chem. (1982) 257, 6268-6274.

74 - M. Li and A. Tzagoloff. (1979), Cell, 18, 47-53.

75 - D. Newman, H.D. Pham, K. Underbrink-Lyon, N.C. Martin. Nucleic Acids Res. (1980), 8, 5007-5016.

76 - F.G. Nobrega, A. Tzagoloff. FEBS Letters (1980), 113, 52-54.

77 - D.L. Miller, N.C. Martin, H.D. Phan, J.E. Donelson. J. Biol. Biochem. (1979), 254, 11735-11740.

78 - R.P. Martin, A.P. Sibler, R. Bordonné, J. Canaday, G. Dirheimer in "The organisation and expression of the mitochodrial genome" (1980) Saccone and Kroon Eds. North-Holland Amsterdam pp311-314

79 - N.C. Martin, D.L. Miller, J. Donelson, C. Sigurdson, J. Hartley, P.S. Mayniham, H. D. Pham in "Extrachromosomal DNA" Cummings et al. Eds. Acad.Press N.Y. (1979) pp.357-375

80 - D.L. Miller, N.C. Martin. Current Genetics (1981), 4, 135-143.

81 - N.C. Martin, D.L. Miller, J. Hartley, P. Moynihan, J.E. Donelson, Cell (1980), 19, 339-343.

82 - N.C. Martin, H.D. Pham, K. Underbrink-Lyon, D.L. Miller, J.E. Donelson Nature (1980) 285, 579-581.

83 - D.L. Miller, C. Sigurdson, N.C. Martin, J.E. Donelson, Nucleic Acid Res. (1980) 8, 1435-1442.

84 - D.L. Miller, D.R. Najarian, J.R. Folse, N.C. Martin J. Biol. Chem. (1981) 256, 9774-9777.

85 - B.E. Thalenfeld, A. Tzagoloff. J. Biol. Biochem. (1980), 255, 6173-6180.

86 - C.E. Novitski, I.G. Macreadie, R. J. Maxwell, H. B. Lukins, A.W. Linnane, P. Nagley in "Manipulation and expression of genes in Eukaryotes" (P. Nagley, A.W. Linnane, W.J. Peacock, J. A. Pateman, Acad. Press (1983), pp. 297-

87 - I.G. Macreadie, C.E. Novitski, R.J. Maxwell, U. John, B.G. Ooi, G.L.McMullen, H. Lukins, A.W. Linnane, P. Nagley, (1983) Nucleic Acids Res. 11, (in press)

88 - G. Macino, A. Tzagoloff. J. Biol. Biochem. (1979), 254, 4617-4623.

89 - G. Macino, A. Tzagoloff. Cell (1980), 20, 507-517.

90 - L.A.M. Hensgens, L.A. Grivell, P. Borst, J.L. Bos. Proc. Natl Acad. Sci. (USA) (1979), 76, 1663-1667.

91 - A. Tzagoloff, M. Nobrega, A. Akai, G. Macino. Current Genetics (1980), 2, 149-157.

92 - J. Lazowska, C. Jacq, P.P. Slonimski. Cell (1981), 27, 12-14.

93 - M. de Zamaroczy, R. Marotta, G. Faugeron-Fonty, R. Goursot, M. Mangin, G. Baldacci, G. Bernardi. Nature (1981), 292, 75-78.

93A - M.E.S. Hudspeth, W.M. Ainley, D.S. Shamard, R.A. Butow, L.I. Grossman (1982) Cell 30, 617-626.

94 - N.C. Martin, K. Underbrink-Lyon Proc. Natl Acad. Sci. (USA) (1981) 78, 4743-4747.

95 - G. S.P. Groot, N. Van Harten-Loosbrooek (1980) Current Genetics, 1, 133-135

96 - Wesolowski, M., A. Algeri, H. Fukuhara (1981) Current Genetics, 3,

157-162.

97 - G.D. Clark-Walker, K.S. Sriprakash (1983) The EMBO J. 2 1465-1472.

98 - G. Seitz-Mayr, K. Wolf (1982) Proc. Natl Acad. Sci. (USA) 79, 2618-2622.

99 - A. Brunner, L. Alba Tuena de Cobos (1980) Mol. Gen. Genet. 178, 351-355.

100 - J.P.M. Sanders, P.J. Weijers, G.S.P. Groot, P. Borst (1974) Bioch. Bioph. Acta, 374, 136-144.

101 - K.S. Sriprakash, C. Batum (1981) Current Genetics 4, 73-80.

102 - U. Hahn, C.M. Lazarus, H. Lünsdorf, H. Küntzel, Cell (1979) 17, 191-200.

103 - J.E. Heckman, U.L. RajBhandary, Cell (1979) 17, 583-595.

104 - H. de Vries, J.C. de Jonge, H. Bakker, H. Meurs, A. Kroon. Nucleic Acid Res. (1979) 6, 1791-1803.

105 - R.A. Collins, A.M. Lambowitz. Current Genetics (1981) 4, 131-133.

106 - K.S. Browning, U.L. RajBhandary J. Biol. Chem. (1982) 257, 5253-5256.

107 - R.A. Collins, A.M. Lambowitz. Plasmid (1983) 9, 53-70.

108 - J.E. Heckman, S. Yin, B. Alzner-Deweerd, U.L. RajBhandary. J. Biol. Chem. (1979) 254, 12694-12700.

109 - G. Burger, C. Scriver, W. Machleidt, S. Werner. EMBO J. (1982) 1, 1385-1391.

110 - J. de Jonge, H. de Vries, Current Genetics (1983) 7 21-28.

111 - J. M. Burke, C. Breitenberger, J. E.Heckman, B. Dujon and U.L. RajBhandary (1983) J. Biol. Chem. (in press).

112 - M.H. Citterich, G. Morelli, G. Macino. EMBO J. (1983) 2, 1235-1242.

113 - P. Van den Boogaart, J. Samallo, E. Agsteribbe. Nature (1982) 298, 187-189.

114 - C.A. Manella. R.R. Goewert, A.M. Lambowitz Cell (1979), 18,1197-1207.

115 - H. de Vries, J.C. de Jonge, P. Van't Sant, E. Agsteribbe, A. Arnberg Current Genet. (1981) 3, 205-211.

116 - H. Bertrand, R.A. Collins, L.L. Stohl, R.R. Goewert, A.M. Lambowitz (1980) Proc. Natl Acad. Sci. (USA) 77, 6032-6036.

117 - R.M. Wright, M.A. Horrum, D.J. Cummings Cell (1982) 29, 505-515.

118 - U. Kück, K. Esser Current Genet. (1982) 5, 143-147.

119 - S. Yin, J. Heckman, V.L. RajBhandary Cell (1981) 26, 325-332.

120 - W. Sebald, J. Hoppe, E. Wachter in "Function and Molecular Aspects of Biomembrane Transport" (1979) Quagliariello Eds Elsevier North Holland, Amsterdam, pp.63-74.

121 - G. Jackl and Sebald, W. Eur. J. Biochem. (1975) 54, 97-106.

122 - M.A. Köchel, H. Küntzel. Nucleic Acids Res. (1982), 10, 4783-4795

122A - R.Netzker, H.G.Köchel, N.Basak, H.Kuntzel Nuc.Acids Res. (1982) 10, 4783-4794

123 - L.Belcour and O.Begel Molec.Gen.Genet (1977) 153, 11-21

125 - C.M.Lazarus and H.Kuntzel Curr.Genet. (1981) 4,99-107

126 - G.Rizet Comptes Rendus Séances Acad.Sc.Paris (1953) 237, 828-840

127 - D. J. Cummings, L. Belcour, C. Grandchamp Mol. Gen. Genet. (1979) 171, 239-250.

128 - U. Stahl, P.A. Lemke, P. Tudzynski, U. Kück, K. Esser Mol. Gen. Genet. (1978) 162, 341-343.

129 - C. Jamet-Vierny, O. Begel, L. Belcour Cell (1980) 21, 189-194.

130 - C. Vierny, A.M. Keller. O. Begel, L. Belcour Nature (1982) 297, 157-159.

131 - I.B. Lazdins, D.J. Cummings. Current Genetics (1982) 6 173-178.

132 - U.Kück, U.Stahl and K.Esser Curr.Genet. (1981), 3,151-156

133 - H.D.Osiewacz and K.Esser Curr.Genet. (1983) 7,219-223

134 - F.Michel, A.Jacquier and B.Dujon Biochimie (1982) 64,867-881

135 - F.Michel and B.Dujon The EMBO J. (1983) 2, 33-38

136 - D.J.Cummings and R.Wright Nuc.Acids Res. (1983) 11, 2111-2119

137 - L.A.M.Hensgens, A.C.Arnberg, E.Roosendaal, G.Van der Horst, R.Der Veen. G-J. B. Van Ommen and L.A.Grivell J.Mol.Biol. (1983) 164, 35-58

138 - P.Cantatore, C.De Benedetto, R.Gallerani, A.M.Kroon, M.Holtrop, C.Lanave, G.Pepe, C.Quagliariello and E.Sbisa Nuc.Acids Res. (1982) 10, 3279-3289

139 - C.Saccone, P.Cantatore, G.Gadaleta, R.Gallerani, C.Lanave, G.Pepe and A.M.Kroon Nuc.Acids Res. (1981) 9,4139-4148

140 - G.Görtz and H.Feldmann Curr.Genet. (1982) 5, 221-225

141 - R.Grosskopf and H.Feldmann Curr.Genet. (1981) 4, 151-158

142 - G.Attardi, P.Cantatore, A.Chomyn, S.Crews, R.Gelfard, C.Merkel, J.Montoya and D.Ojala (1982) in "Mitochondrial genes" Slonimski et al. eds Cold Spring Harbor Laboratory, N.Y. pp51-71

143 - B.G.Barrel, A.T.Bankier and J.Drouin Nature (1979) 282, 189-194

144 - B.G.Barrel, S.Anderson, A.T.Bankier, M.H.L. DeBruijn, E.Chen, A.R.Coulson, J.Drouin, I.C.Eperon, D.P.Nierlich, B.A.Roe, F.Sanger, P.H.Schreier, A.J.H.Smith, R.Staden and I.C.Young Proc.Natl.Acad.Sc.US (1980) 77, 3164-3166

146 - C.J.Doersen and E.J.Stanbridge (1982) in "Mitochondrial genes" Slonimski et al. Eds. Cold Spring Harbor Laboratory N.Y. pp129-132

147 - C.L.Bunn, D.C.Wallace and J.M.Eisenstadt Proc.Natl.Acad.Sc.US (1974) 71, 1681-1685

148 - T.Lichtor and G.Getz Proc.Natl.Acad.Sc.US (1978) 75, 324-328

149 - A.Wiseman and G.Attardi Molec.Gen.Genet. (1982) 186, 364-371

150 - H.Blanc, T.Wright, M.J.Bibb, D.C.Wallace and D.A.Clayton Proc.Natl.Acad.Sc.US (1981) 78,3789-3793

151 - S.E.Kearsey and I.W.Craig Nature (1981) 290, 607-608

152 - H.Blanc, C.W.Adams and D.C.Wallace Nucl.Acids Res. (1981) 9, 5785-5795

153 - C.A.Hutchinson III, J.E.Newbold, S.S.Potter, M.H.Edgell Nature (1974) 251, 536-538

154 - A.M.Kroon, W.M.DeVos and H.Bakker Bioch.Biophys. Acta (1978) 519, 269-273

155 - J.I.Hayashi, H.Yonekawa, O.Gotsch, J.Watanabe and Y.Tagashima Biochem.Biophys.Res.Comm. (1978) 83,1032-1038

156 - J.F.Francisco, G.G.Brown and M.V.Simpson Plasmid (1979) 2, 426-436

157 - J.C.Avise, R.A.Lansman and R.O.Shade Genetics (1979) 92, 279-295

158 - R.E.Giles, H.Blanc, H.M.Cann and D.C.Wallace Proc.Natl.Acad.Sc.US (1980) 77,6715-6719

159 - M.Denaro, H.Blanc, M.J.Johnson, K.H.Chen, E.Wilmsen, L.L.Cavalli-Sforza and D.C.Wallace Proc.Natl.Acad.Sc.US (1981), 78, 5768-5772

160 - W.W.Hauswirth, P.J.Laipis (1982) in "Mitochondrial genes" Slonimski et al. Eds. cold Spring Harbor Laboratory N.Y. pp 137-141

161 - W.M.Brown, M.George Jr. and A.C.Wilson Proc.Natl.Acad.Sc.US (1979) 76, 1967-1971

162 - S.D.Ferris, A.C.Wilson and W.M.Brown Proc.Natl.Acad.Sc.US (1981) 78, 2432-2436

163 - S.D.Ferris, W.M.Brown, W.S.Davidson and A.C.Wilson Proc.Natl.Acad.Sc.US (1981) 78, 6319-6323

164 - W.M.Brown, E.M.Prager, A.Wang and A.C.Wilson J.Mol.Evol. (1982) 18, 225-239

165 - J.L.Coote, G.Szabados and T.S.Work FEBS letters (1979) 99, 255-260

166 - W.M.Hauswirth and P.J.Laipis Proc.Natl.Acad.Sc.US (1982) 79, 4686-4690

167 - G.H.Beale, J.K.C.Knowles and A.Tait Nature (1972) 235,396-397

168 - R.Perasso and A.Adoutte J.Cell.Sc. (1974) 14, 475-497

169 - A.Sainsard-Chanet and J.Knowles Genetics (1979) 93, 833-859

170 - J.K.C.Knowles Expt.Cell.Res. (1972) 88, 79-87

171 - J.J.Curgy, R.Perasso, E.Boissonneau, F.I.Iftode, N.Stelly and J.André Curr.Genet. (1981) 4,121-130

172 - A.Arnberg, E.Van Bruggen, R.Clegg, W.Upholt and P.Borst Bioch.Biophys.Acta (1974) 361, 266-276

173 - J.Goddard and D.Cummings J.Mol.Biol. (1977) 109, 327-344

174 - A.E.Pritchard and D.J.Cummings Proc.Natl.Acad.Sc.US (1981) 78, 7341-7345

175 - A.E.Pritchard, J.L.Laping, J.J.Seilhamer and D.J.Cummings J.Mol.Biol. (1983) 164, 1-15

176 - R.W.Goldbach, J.E.Bollen de Boer, E.F.J.Van Bruggen and P.Borst Bioch.Biophys.Acta (1979) 562, 400-417

177 - C.S.Levings III, D.R.Pring in Stadler Symposium vol. 10 (1978) University of Missouri. pp 77-93

178 - C.J.Leaver, M.W.Gray (1982) Ann.Rev.Plant Physiol. 33, 373-402

178A - R.M.K.Dale (1981) Proc.Nat.Acad.Sc.USA 78, 4453-4457

178B - P.Lebacq and F.Vedel Plant Sc.Letters (1981) 23,1-9

179 - F.Quetier, F.Vedel, (1977) Nature 268, 365-368

180 - B.L.Ward, R.S.Anderson, A.J.Bendich (1981) Cell 25 793-803

181 - R.R.Sederoff, C.S.LevingsIII, D.H.Timothy and W.W.L.Hu Proc.Natl.Acad.Sc. (1981) 78, 5953-5957

182 - D.Grant, K.S.Chiang (1980) Plasmid 4, 82-96

183 - D.B. Stern, T.A. Dyer, D.M. Lonsdale Nucleic Acids Res. (1982) 10, 3333-3340

184 - K.P. Iams J.H. Sinclair (1982) Proc.Nat.Acad.Sc. USA. 79, 5926-5929

185 - L.Bonen and M.Gray Nucl.Acids Res. (1980) 8, 319-325

186 - R.S.Cunningham, L.Bonnen, W.F.Doolittle, M.W.Gray (1976) Febs lett. 69 116-122

187 - M.Boutry, M.Briquet and A.Goffeau J.Biol.Chem. (1983) 258,8524-8526

188 - D.B.Stern, D.M.Lonsdale (1982) Nature 299 698-702

189 - A.Wiseman, N.W.Gillham, J.E.Boynton (1977) Molec.Gen.Genet. 150,109-117

190 - C.J.Leaver (1980) Trends in Bioch.Sc. 5, 248-252

191 - C.S.Levings III and D.R.Pring Science (1976) 193, 158-160

192 - D.R.Pring, M.F.Conde, K.F.Schetrz and C.S.Levings III Molec.Gen.Genet. (1982) 186,180-184

193 - R.J.Kemble and J.R.Bedbrook Nature (1980) 284, 565-566

194 - R.J.Kemble and R.D.Thompson Nucl.Acids Res. (1982) 10, 8181-8190

195 - A.K. Weissinger, D.H.Timothy, C.S.Levings III, W.W.L. HU and M.M.Goodman Proc.Natl.Acad.Sc.U.S. (1982) 79,1-5

196 - C.S.Levings III and R.R.Sederoff Proc.Natl.Acad.Sc.US (1983), 80,4055-4059

197 - C.S. Levings III, B.D.Kim, D.R.Pring, M.F.Conde, R.J.Mans, J.R.Langhman and S.J.Gabay-Langman Science (1980) 209,1021-1023

198 - R.D.Thompson, R.J.Kemble, R.B.Flavell Nucl.Acids Res. (1980) 8, 1999-2008

199 - D.M.Lonsdale, R.D.Thomson and T.P.Hodge Nucl.Acids Res. (1981) 9, 3367-3369

200 - J.D.Palmer, C.R.Shields, D.B.Cohen, T.J.Orton Nature (1983) 301, 725-728

MITOCHONDRIAL GENE EXPRESSION 1983*

L.A.Grivell

Section for Molecular Biology, Laboratory of Biochemistry, University of Amsterdam, Kruislaan 318, 1098 SM Amsterdam, The Netherlands

Introduction

Most mtDNAs are simple; their contribution to mitochondrial biogenesis small. Nevertheless, few other genomes in nature display such a variety of gene organization and gene expression. Few other genomes have taught us so much about the working and evolution of genetic systems. In the six years that have elapsed since the field was reviewed in the last book in this series [1], three mammalian mtDNAs have been completely sequenced, others are nearing completion. Amongst the fungi, sequences of large areas of the mitochondrial genomes in *Aspergillus*, *Neurospora* and the yeasts *Saccharomyces cerevisiae* and *Schizosaccharomyces pombe* are known and no doubt, other protists will shortly follow. The purpose of this brief review is to survey what is known about these fascinating DNAs and the expression of their genes anno 1983.

Mitochondrial DNA: Main features

mtDNAs are diverse in size and shape (Table I). Animals, insects and man can make do with a mitochondrial genome of 14-18 kb, other organisms require considerably more complex genomes, ranging up to 200-2700 kb in higher plants. Differences in complexity can be dramatic even for closely related

---

Abbreviations: kb, kilo base pair(s); mRNA, messenger RNA; rRNA, ribosomal RNA; URF, unassigned reading frame.

* Dedicated to my Father.

Mitochondria 1983

TABLE I

SIZE AND STRUCTURE OF VARIOUS MITOCHONDRIAL DNAs

Modified from ref. 71.

Organism	Structure	Size (kb)
Animals, insects, man	Circular	14-18*
Fungi:		
<u>Torulopsis</u>, <u>Schizo-saccharomyces</u>	Circular	19
<u>Aspergillus</u>	Circular	29-36*
<u>Kluyveromyces</u>	Circular	37
<u>Neurospora</u>	Circular	60-73*
<u>Saccharomyces</u>	Circular	74-82*
<u>Podospora</u>	Circular	94
<u>Brettanomyces</u>	Circular	108
Protozoa:		
Malaria (<u>Plasmodium</u>)	Circular	27
<u>Acanthamoeba</u>	Circular	41
<u>Paramecium</u>	Linear	41
<u>Tetrahymena</u>	Linear	45-54*
<u>Trypanosoma</u>	Network** Maxi-circle Mini-circle	6000 21 1
<u>Crithidia</u>	Network** Maxi-circle Mini-circle	30,000 35 3
Algae:		
<u>Chlamydomonas</u>	Circular (?)	16
Higher plants:		
Cabbage	Circular	200
Maize	Circular	600
Cucurbits	Circular	370-2700*

* The molecular weights of the DNAs in different representatives of this group differ; only a single type of molecule is found in any one organism.

** This mtDNA is usually called 'kinetoplast' DNA.

members of a single group, like the yeasts [2] or certain plants [3], raising the question of whether the extra DNA is in fact informational. Circular genomes predominate, but linear mtDNAs are found in certain protozoa [4,5]. In trypanosomes, mtDNA is enmeshed in networks with mini-circular DNAs of unknown function [6]. In higher plants, the complexity of electron microscopic and restriction mapping data has led to problems in constructing physical maps of the mitochondrial genome [7]. For maize, however, these have been largely circumvented by cosmid cloning of the DNA. Genome complexity is now put at 500-600 kb (Table I), organized either as a single circle of about twice this size or as a collection of three smaller molecules generated by recombination at repeated sequences [8, Leaver et al., this volume].

Mitochondrial gene organization in different organisms is no less diverse (see also Dujon, this volume). Genes may be tightly packed and even overlapping, as in mammalian mtDNAs [9], or spread out and intron-infested as in yeast and the filamentous fungi [10]. There is no preferred gene order or obvious functional clustering of genes although order tends to be conserved over short evolutionary distances.

Mitochondrial genes

All mtDNAs characterized to date code for components of the mitochondrial translation system and for a limited number of proteins belonging to respiratory enzymes of the mitochondrial inner membrane. Genes identified in the yeast (*Saccharomyces*) and mammalian genomes are listed in Table II. Gene complements are similar, but not identical, with the most obvious differences involving the genes for ATPase subunit 9 and a ribosome-associated protein, both of which are present in yeast but absent from the mammalian mtDNAs. Other differences concern the unassigned reading frames (URFs). In yeast most of these are located in the introns of split genes and there is a growing body of evidence that their translation

TABLE II

MITOCHONDRIAL GENES IN YEAST AND MAMMALS

Mitochondrial component	Mitochondrial gene product in	
	Yeast	Mammals
Cytochrome *c* oxidase:		
Subunit I	+	+
Subunit II	+	+
Subunit III	+	+
Ubiquinol-cytochrome *c* reductase:		
Apocytochrome *b*	+	+
ATPase complex:		
Subunit 6	+	+
Subunit 8	+	+
Subunit 9	+	-
Ribosome:		
Large and small rRNAs	+	+
Ribosome-associated protein	+	-
RNA maturases*:		
bI2, bI3, bI4 (apocyt. *b* gene)	+	-
aI1, aI2, aI4 (coxI gene)	+	-
tRNAs	about 25	22
URFs	≥6**	7***

* Intron-encoded proteins for which there is genetic evidence indicating an involvement in RNA splicing.

** Mostly intronic, displaying homology with RNA maturases. Direct evidence for such a function is so far lacking.

*** No homology to the URFs in yeast.

products play a role (as yet undetermined) in RNA splicing. Genes in animal mtDNAs lack introns; such proteins are unnecessary, their genes are absent. On the other hand, the nucleotide sequence of these DNAs has revealed eight reading frames, of which one (A6L) has recently been shown to synthesize a protein identified as ATPase subunit 8, the product of the AAP-1 gene in yeast [11,12] and the URFx genes in Aspergillus [13,14] and Neurospora (Helmer-C. et al., this volume). The remainder, which code for proteins of quite different characters, are still without a known function. None of these have been seen in the region of yeast mtDNA sequenced so far, but four of them (URFs 1, 3, 4 and 5) have recently turned up in conserved form in Aspergillus mtDNA (Scazzocchio et al., this volume). URFs 1 and 5 are also found in Neurospora (Helmer-C. et al.) and URF 1 has been reported in maize (Scazzocchio et al.), indicating that they must fulfil important roles in mitochondrial function or biogenesis. The search for their analogues in yeast should not be abandoned just yet.

Most mitochondrially coded proteins have been relatively well-conserved in evolution, the homology for cytochrome b and the subunits of cytochrome c oxidase in man and yeast lying at around 50%. Most are also hydrophobic, but this is not a pre-requisite. Some of the human URFs [9], the yeast intronic URFs [10] and the var1 gene [15] code for hydrophilic proteins with a high proportion of charged residues.

Do the complexer mtDNAs contain genes absent from their simpler brethren? For yeast, the answer is probably not, since the 5-fold increase in size relative to the human genome can be almost entirely accounted for by differences in gene organization (see below). For higher plants, the situation is still anyone's guess. In maize, one gene not found in other mtDNAs is that for the α-subunit of F_1-ATPase (Leaver et al., this volume) and to judge from the complex pattern of mitochondrial translation products compared with yeast or man, there are more genes to be found. Even allowing for this, however, much of the genome remains unaccounted for. Together with the wide variation in mtDNA complexity between even closely-related plants, this has prompted the suggestion [3] that much of the genome may either fulfil a sequence-independent function or even no function at all.

Split genes

In direct contrast to the compact organization of their mammalian counterparts, several mitochondrial genes in other organisms contain introns. The unusual features of those in fungal mtDNAs have received extensive review [10] so that only a few new aspects deserve mention now:

1. Intron sequences are related. They can be grouped into two families (I and II) whose members display extensive similarities in potential RNA secondary structure [16] and amino acid sequence of the proteins they encode [17]. Each is thus likely to have arisen by duplication and dispersion of an ancestral gene. Group II introns have so far been found only in yeast. Their sequences accumulate after excision as stable, covalently-closed circles.

2. In yeast mtDNA, the introns show a highly asymmetric distribution in being restricted to only three genes (cytochrome b, cytochrome oxidase subunit I and the large ribosomal RNA). Other organisms appear more adventurous, however, with introns in the genes for cytochrome oxidase subunit II (maize [18]), ATPase 6 and even an URF (Neurospora; Helmer-C. et al., this volume). From an evolutionary standpoint, such a seemingly arbitrary distribution could mean that ancestral mitochondrial genes were riddled with many more introns than is now the case, with random loss in different evolutionary lines accounting for the present-day situation. More likely, however, is that many introns represent relatively recent sequence acquisitions resulting from the insertion of transposon-like elements at preferred sites in the genome. The findings of dissimilar introns at comparable sites and of similar introns at different sites in the split genes of different fungal lines [14,19, Helmer-C. et al., this volume] also point to a comparatively recent origin.

3. Many introns are optional in the sense that their occurrence is strain-dependent [20,21]. Some, however, like bI4, the fourth intron of the cytochrome b gene in yeast, may

only be dispensable in certain genetic contexts, because of the involvement of the proteins they encode in the splicing of other introns [22]. Whether introns confer any selective advantage on strains possessing them is still obscure. Laboratory yeast strains differing in their intron complements do not display radically different growth characteristics [23], so that advantages may be subtle or only pertain in situations met outside the laboratory.

Genetic code

Mitochondria use their own variations on the genetic code. The most striking deviation from the once 'universal' version is the use of UGA to specify tryptophan and this is found in all mitochondrial genetic systems studied so far. Besides this, however, many organisms use additional, non-standard codon assignments as shown in Fig. 1. It is sometimes assumed [24,25] that these unusual assignments are relics of the primitive origin of the mitochondrial genetic system. In fact, quite the opposite seems to be true: mammals and insects decode AGA(G) codons as stop [9] and serine [26], respectively, and this difference must reflect a change that has occurred since their divergence 680 million years ago. Similarly, AUA codes for methionine in yeast [15], insects [26] and mammals [24], but for isoleucine in all other organisms (including trypanosomes, see Benne et al., this volume). The fact that within the fungi the AUA-methionine assignment is restricted to yeast argues strongly for a relatively late fixation in evolution. This probably also holds for the aberrant CUN-threonine assignment in this organism, which may have simply been caused by mutation in a $tRNA_1^{Thr}$ gene [27].

Why have such changes occurred? One answer comes from sequence comparisons of mammalian mtDNAs, which show that they fix mutations 5-10 times faster than their nuclear counterparts and that in contrast to other genetic systems, ribosomal RNAs (rRNAs) and tRNAs are not spared from change [28,29]. Quite

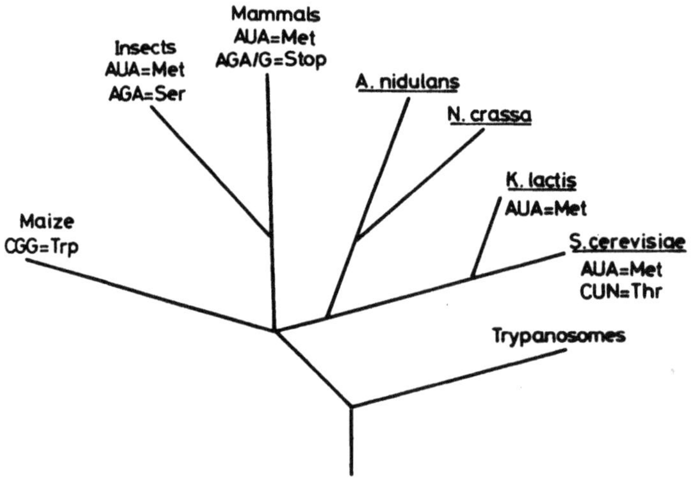

Fig. 1. Evolutionary variations on the genetic code. UGA specifies tryptophan in all cases and only additional unusual codon assignments are shown. For K. lactis, the AUA-methionine assignment is based on unpublished sequence data (De Haan, M., Grivell, L.A.); assignment of the CUN family in this organism is uncertain, since this family is not used in genes sequenced so far. Remaining assignments are taken from refs 9, 13, 15, 18, 26-28 and Benne et al. (this volume).

possibly, the simplicity of the mitochondrial system - the fact that it synthesizes a limited set of proteins in relatively constant amounts - means that it can tolerate changes that could be deleterious or lethal in a more finely-tuned system. In addition, most mitochondrial translation products are hydrophobic, trans-membrane proteins. Amino acid sequence comparisons of these [29] show that serine, alanine and threonine tend to be readily interchangeable both with each other and with leucine, isoleucine, methionine and valine, probably because relative hydrophobicity is a major criterion for their selection. Changes in the code affecting these amino acids could, therefore, be quite tolerable, especially if they involve infrequently used codons.

Transcription

In human mitochondria (and probably other mammals too) both DNA strands are completely transcribed from sites in a region near the origin of H-strand replication [30]. Mature, gene-sized RNAs are then generated from the nascent transcripts by cleavage at the boundaries of the tRNA sequences that flank virtually every gene [31]. Nevertheless, levels of the two rRNAs are 30-60 fold higher than other transcripts and pulse-labelling studies [32] show that much of this difference results from a higher rate of synthesis. How is this achieved? An early solution was provided by the suggestion [33] that H-strand transcripts simply have a high probability of premature termination at a transcription-attenuator site at or near the 3'-end of the large rRNA gene. More recent observations (Attardi et al., this volume) imply, however, that the controls involved are likely to be much more intricate. They show that the rRNA region of human mtDNA is the site of two overlapping transcription processes, each of which is initiated at a distinct promoter. As shown in Fig. 2, initiation of RNA synthesis at a site just upstream of the $tRNA^{Phe}$ gene occurs frequently and gives rise to transcripts terminating at the 3'-end of the 16S rRNA. Initiation at the second site, which is located downstream of the first, near the 5'-end of the small rRNA gene, is much less frequent and results in the

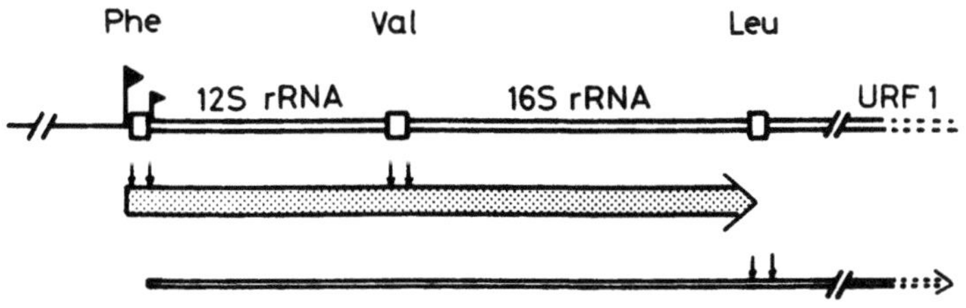

Fig. 2. The 'bypass' model for regulation of human mitochondrial rRNA and mRNA synthesis. Black flags represent promoters; vertical arrows are processing sites. See text and Attardi et al. (this volume) for explanation.

synthesis of a polycistronic RNA corresponding to almost the entire H-strand. An implicit consequence of this situation must be that the fate of each transcript is determined from the start by either the involvement of two different RNA polymerases or the extra sequences present at the 5'-end. As Attardi *et al*. point out, the latter alternative raises some intriguing questions. How, for example, is the signal for transcription termination transmitted some 2675 nucleotides downstream when the bulk of the transcripts have already undergone processing to mature rRNAs? One solution might be that signal transmission involves sequential formation of secondary structures and binding of ribosomal proteins, but clearly, the matter is far from settled.

In yeast, the only lower eukaryote characterized in sufficient detail so far, scope for differential transcription control is far more extensive (Fig. 3). Some 19 different RNAs are cappable *in vitro* with the enzyme guanylyl transferase and these presumably represent the products of *de novo* transcription-initiation at distinct promoters (Edwards *et al*., this volume). 13 of these have been precisely mapped in sequenced regions of the genome, either upstream of known genes or in putative origins of DNA replication (*ori* or *rep* sequences). Important points so far are:

1. Initiation of RNA synthesis occurs in all cases within a short conserved sequence which is either

```
ATATAAGTA →
-9       +1
```

- the nonanucleotide originally noted by Osinga *et al*. [34] - or a close variant of it (Edwards *et al*., Tabak *et al*., this volume).

2. Several genes conspicuously lack this motif and available evidence suggests that these are not transcribed independently. They form part of larger transcriptional units initiated (far) upstream and their mature RNAs are subsequently generated by processing (see below). The genes transcribed in this way form a rather motley collection (Fig. 4), so that the

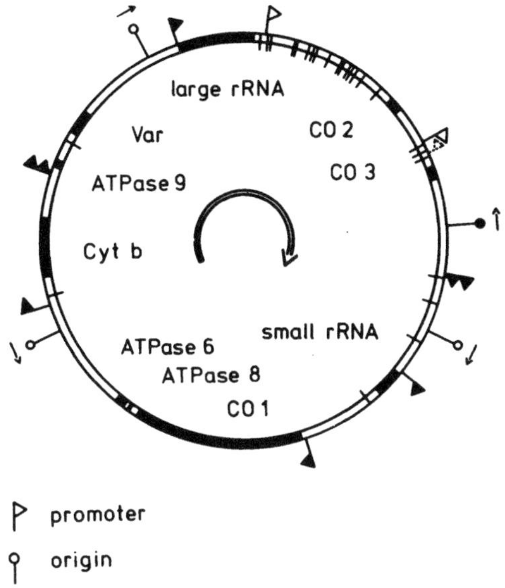

Fig. 3. Map of yeast mtDNA showing sites for initiation of transcription and origins of DNA replication. See Tabak et al. (this volume) and text for discussion.

possible advantages of this organization are not immediately apparent.

What role does the nonanucleotide motif play in transcription-initiation? Its position and strong evolutionary conservation [34] support the original idea that it is an important element of a mitochondrial promoter. It is equally clear, however, that it alone is insufficient for transcription and that other (downstream) sequences must also be involved: some genes contain tandem copies spaced 10-80 nucleotides apart, yet only the upstream box directs high level transcription (Tabak et al., Edwards et al.); the *Kluyveromyces lactis* rRNA genes contain perfectly conserved copies of the sequence, yet are not transcribed by the RNA polymerase from *S. cerevisiae* (Tabak et al.). A simple, if somewhat unsatisfying solution to this paradox is that like certain elements of eukaryotic

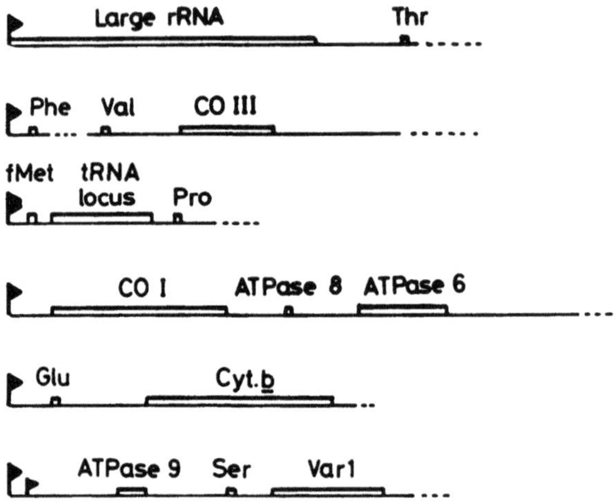

Fig. 4. Possible multi-gene transcription units in yeast mtDNA. Stippled lines indicate uncertainty in the exact limits of primary transcripts, which are not drawn to scale. See refs 36, 38, 72 and 73 together with the articles by Edwards et al., Miller et al. and Tabak et al. (this volume) for further details.

promoters [35], the box has a role *in vivo* that escapes detection in the *in vitro* run-off systems so far used. What this role is and which other sequences are involved in promoter specificity are questions that more extensive mutagenesis studies will no doubt resolve.

RNA processing

With the exception of *S. pombe*, which with its close packing of genes and tRNA-punctuated processing of RNA (Lang *et al*., this volume) has opted for the elegant simplicity of the human mitochondrial genome, mitochondrial RNA processing in the fungi is likely to be a complex affair. In yeast, extensive processing is required for the maturation of tRNAs (Miller *et al*., Baldacci *et al*., this volume), for the cleavage of separate RNAs from primary multi-gene transcripts (Fig. 4) and for the maturation of the individual gene transcripts.

Little use seems to be made of tRNA punctuation. In the few cases where tRNA and protein-coding genes are co-transcribed [36; Edwards et al.], the 5' termini of the mature messenger RNAs (mRNAs) are located some distance downstream of the 3'-end of the tRNA [37,38; Zassenhaus and Butow, this volume]. Even if tRNA secondary structure signals initial cleavage, other steps must still be required for maturation and the signals for these - as for other 5'- and 3'-trimming steps - are far from well-defined. In the var1 gene, for example (Zassenhaus and Butow), the major transcript (presumed to be mRNA) is generated from a precursor-RNA containing sequences of the ATPase subunits 9 and $tRNA^{Ser}$ genes by cleavage at the heptanucleotide AAUAUAA. This non-descript sequence also occurs frequently at gene-internal positions, but only two of these are used to generate minor amounts of sub-genic RNAs with unknown function. How cleavage at the remaining sites is avoided is not clear, but factors like RNA secondary structure and ribosome-loading could well be involved.

As in Escherichia coli, processing of yeast mitochondrial tRNAs at their 5'-ends is dependent on the action of an enzyme which requires a specific RNA component. This RNA is the product of a mitochondrial gene (the tRNA synthesis locus [39]), the loss of which results directly in an inability to make ripe tRNAs. Similarity of the processing mechanism with that in E. coli is suggested by the fact that yeast pre-tRNAs containing 5' extensions can be accurately processed by E. coli RNase P (Miller et al., this volume).

RNA processing also plays an important role in the control of mitochondrial gene expression in Neurospora, but the signals involved seem less diverse than in yeast. Evidence has been presented for co-transcription of rRNA and tRNA genes [40] and it has been suggested that mature transcripts arise by processing at GC-rich palindromic sequences which flank each gene [41]. These palindromes, which often contain two closely-spaced PstI sites, flank several other genes and their position relative to mapped transcripts strengthens the view that they form recognition sites for RNA processing enzymes (Helmer-C. et al., this volume).

RNA splicing

One aspect of mitochondrial RNA processing that has attracted considerable attention is splicing and with good reason. Excision of many mitochondrial introns is dependent on self-encoded proteins (maturases) [42-45], as well as numerous nuclear-encoded products [46,47]; on critical sequences far from the splice junctions [22,43,45,48,49] and on the probable formation of complex internal secondary structures [16,50].

Mitochondrial introns are like those in nuclear-coded rRNA and tRNA genes in lacking a GT...AG consensus sequence at their junctions [10]. They can be broadly divided into two main groups. Group I introns are the more widespread. They share a common potential secondary structure and contain a number of conserved sequence elements, some of which have been defined by mutation as being essential for splicing [16,43,45, 48,50]. Their reading frames, when present, contain a region of sequence homology bounded by distinctive decapeptide blocks [17,51]. Elements of both secondary structure and conserved sequences of group I introns can be identified in the nuclear-coded genes for large rRNA in Tetrahymena and Physarum and in chloroplast rRNA and tRNA genes [52-54]. Such strong evolutionary conservation suggests that all members of the group share elements of a common splicing mechanism, irrespective of whether or not they encode a maturase.

Group II introns, so far present only in yeast, are characterized by a distinctive potential secondary structure [52], by extensive sequence homology at their 3'-ends [10,49] and by similarities in reading frames when present [55]. Their sequences accumulate after excision as covalently-closed circular RNA [17,56,57], but these may not be direct products of splicing: circle formation can be uncoupled from splicing in certain growth phases [58,59] and the circles themselves may be closed by an unusual bond (Bonen et al., manuscript in preparation). As in group I, intron-internal sequences are also required for splicing (Carignani et al. and Mecklenberg et al., this meeting), but these have not yet been characterized in as much detail.

Details of current splicing models are described more fully elsewhere in this book (Davies et al.). Main elements are the use of intron folding to form a 'core' structure which brings 5' and 3' exons close together, the requirement for an intron-internal guide sequence for precise alignment of the splice junctions and the involvement of unpaired, conserved sequence elements in binding of either proteins or other cofactors. Evidence for the core structure comes from the finding of compensatory base changes in the various mitochondrial introns examined [50,52,60] and more recently for the Tetrahymena ribosomal intron from direct mapping of single- and double-stranded nuclease cleavage sites [54]. Evidence for the internal guide is less compelling at present. In some cases, base pairings are rather unstable and stabilization by tertiary structure or protein binding has to be invoked [50]. The critical unpaired sequences are the box2 and box9 loci or their analogues, which correspond to parts of the conserved elements P, Q, R and S identified by Davies et al. [50]. These elements have very different locations within the various introns, but can be brought together by intron folding. Destabilization of folding by mutation and sequence changes in the elements themselves lead to disruption of splicing.

What do intron-encoded maturases do? There is no direct answer to this question, since not a single maturase has yet been isolated in active form. A catalytic role seems likely since levels in the cell are normally extremely low, but there is no evidence that their present function is that of a spligase, even though they could have had such a function in the past. One appealing suggestion [50] is that maturases are specifically required in mitochondria because RNA synthesis, processing and protein synthesis occur within the same compartment. Splicing and translation are likely to be mutually exclusive events because ribosomes reading through into intron sequences will tend to disrupt folding. Maturases may, therefore, promote splicing either by stabilizing secondary structure in such a way as to prevent further translation or, if endowed with specific endonuclease activity, they could stop translation by cleavage at the 5'-end of the intron. In this context it is interesting that nuclear and some *cis*-acting mutants blocked in mitochondrial splicing (but still having active maturases) often accumulate 'half' molecules, which have arisen by cleavage at this point [20,45,46].

Concluding remarks: What next?

Considering the speed of developments in the past six years, any attempt to predict future trends over a similar period could be regarded as incautious if not downright reckless. A few topics deserve mention, however, if for no other reason than that they seem likely to carry the field beyond the Orwellian year of 1984.

1. *mtDNA and evolution* - Comparisons of mitochondrial genetic systems have begun to reveal the enormous variety in solutions to the age-old problems of rRNA and tRNA design, decoding, ribosome structure and mRNA recognition. Full physical characterization of some of the extremes, like the mini-rRNAs of *Trypanosomes* (Benne *et al*., this volume), the aberrant tRNAs and minimal mRNAs of human mitochondria [9,61] should establish the limits of such variation and tell us why it is possible. Similar comparisons, carried out on mitochondrially-encoded proteins of the respiratory chain should lead to the identification of conserved domains and the minimal requirements for their function.

2. *Nuclear genes* - Maintenance of a mitochondrial genetic system requires considerable investment on the part of the nucleus, but study of the interactions between the two genomes is still in early stages. In yeast, characterization of *pet*$^-$

mutants [62] and nuclear suppressors of mitochondrial mutations [63] has exposed an unexpected degree of complexity in the controls exerted by the nucleus, with nuclear-encoded proteins being required for splicing of individual introns [46,47], translation of specific mRNAs (see also Fox *et al.*, this volume) and processing of specific proteins [64]. Transformation-complementation has already been applied to identify some of the nuclear genes involved in mitochondrial splicing [46,47]. Availability of these and of those required for other steps in gene expression will permit characterization of the proteins they encode and eventually reconstruction *in vitro* of the reactions catalysed.

3. *Migratory mtDNA* - Although the nuclear, mitochondrial and chloroplast genomes are generally regarded as being maintained discretely in their respective organelles, such isolation may be more apparent than real. Sequences homologous to chloroplast DNA have been found stably integrated in plant mtDNA [65]; mtDNA sequences, sometimes scrambled, have been identified in the nuclear genomes of yeast [66], sea urchin [67] and locust [68]; transposition of mtDNA into the nucleus has even been suggested to be part of the process of senescence in the fungus *Podospora* [69]. In all instances, apart from the last, which still awaits further confirmation, migration appears to have been accidental, the transferred DNA nonfunctional. However, the observations provide obvious support for present scenarios of the endosymbiotic origin of organellar DNA [70] and they give some idea of the extent to which evolution of the eukaryotic nuclear genome has been interwoven with that of cytoplasmic DNA. Further characterization of such migratory DNA may, therefore, throw light on the paths followed during the emergence of the present-day genomes and perhaps even tell us why mitochondrial DNA has been retained at all.

Acknowledgements

I am indebted to students and colleagues in the Amsterdam laboratory for stimulating discussions and their comments on the manuscript. Experimental work in our laboratory was supported by a grant to LAG from The Netherlands Foundation for Chemical Research (SON) with financial aid from The Netherlands Organization for the Advancement of Pure Research (ZWO).

References

1 Borst, P., Bos, J.L., Grivell, L.A., Groot, G.S.P., Heyting, C., Moorman, A.F.M., Sanders, J.P.M., Talen, J.L., Van Kreijl, C.F., Van Ommen, G.J.B.: in: Mitochondria 1977: Genetics and Biogenesis of Mitochondria (Bandlow, W., Schweyen, R.J., Wolf, K., Kaudewitz, F., Eds), De Gruyter, Berlin, pp. 213-254 (1977).

2 Clark-Walker, G.D., Sriprakash, K.S.: in: Mitochondrial Genes (Slonimski, P., Borst, P., Attardi, G., Eds), Cold Spring Harbor Laboratory, Cold Spring Harbor, N.Y., pp. 349-354 (1982).

3 Ward, B.L., Anderson, R.S., Bendich, A.J.: Cell 25, 793-803 (1981).

4 Goddard, J.M., Cummings, D.J.: J.Mol.Biol. 97, 593-609 (1975).

5 Goldbach, R.W., Arnberg, A.C., Van Bruggen, E.F.J., Defize, J., Borst, P.: Biochim.Biophys.Acta 477, 37-50 (1977).

6 Borst, P., Hoeijmakers, J.H.J.: Plasmid 2, 20-40 (1979).

7 Levings III, C.S.: Cell 32, 659-661 (1983).

8 Lonsdale, D.M., Hodge, T.P., Fauron, C.M.-R., Flavell, R.B.: in: Plant Molecular Biology, UCLA Symp. on Molecular and Cellular Biology, New Series Vol. 12 (Goldberg, R.B., Ed.), Liss, New York, in press (1983).

9 Anderson, S., Bankier, A.T., Barrell, B.G., De Bruijn, M.H.L., Coulson, A.R., Drouin, J., Eperon, I.C., Nierlich, D.P., Roe, B.A., Sanger, F., Schreier, P.H., Smith, A.J.H., Staden, R., Young, I.G.: Nature 290, 457-465 (1981).

10 Grivell, L.A., Bonen, L., Borst, P.: in: Horizons in Biochemistry and Biophysics (Palmieri, F., Quagliariello, E., Eds), Vol. 7, Wiley, Chichester, in press (1983).

11 Marriottini, P., Chomyn, A., Attardi, A., Trovato, D., Strong, D.D., Doolittle, R.F.: Cell 32, 1269-1277 (1983).

12 Macreadie, I.G., Novitski, C.E., Maxwell, R.J., Ulrik, J., Ooi, B.-G., McMullen, G.L., Lukins, H.B., Linnane, A.W., Nagley, P.: Nucl.Acids Res. 11, 4435-4451 (1983).

13 Grisi, E., Brown, T.A., Waring, R.B., Scazzocchio, C., Davies, R.W.: Nucl.Acids Res. 10, 3531-3539 (1982).

14 Netzker, R., Köchel, H.G., Basak, N., Küntzel, H.: Nucl.Acids Res. 10, 4783-4794 (1982).

15 Hudspeth, M.E.S., Ainley, W.M., Shumard, D.S., Butow, R.A., Grossman, L.I.: Cell 30, 617-626 (1982).

16 Michel, F., Jacquier, A., Dujon, B.: Biochemie 64, 867-881 (1982).

17 Hensgens, L.A.M., Bonen, L., De Haan, M., Van der Horst, G., Grivell, L.A.: Cell 32, 379-389 (1983).

18 Fox, T.D., Leaver, C.J.: Cell 26, 315-323 (1981).

19 Burke, J.M., RajBhandary, U.L.: Cell 31, 509-520 (1982).

20 Hensgens, L.A.M., Arnberg, A.C., Roosendaal, E., Van der Horst, G., Van der Veen, R., Van Ommen, G.J.B., Grivell, L.A.: J.Mol.Biol. 164, 35-58 (1983).

21 Collins, R.A., Lambowitz, A.M.: Plasmid 9, 53-70 (1983).

22 De la Salle, H., Jacq, C., Slonimski, P.P.: Cell 28, 721-732 (1982).

23 Labouesse, M., Slonimski, P.P.: EMBO J. 2, 269-276 (1983).

24 Barrell, B.G., Bankier, A.T., Drouin, J.: Nature 282, 189-194 (1979).

25 Ohno, S., Epplen, J.T.: Proc.Natl.Acad.Sci.U.S. 80, 3391-3395 (1983).

26 De Bruijn, M.H.L.: Nature 304, 234-241 (1983).

27 Li, M., Tzagoloff, S.: Cell 18, 47-53 (1979).

28 Cantatore, P., De Benedetto, C., Gadaleta, G., Gallerani, R., Kroon, A.M., Holtrop, M., Lanave, C., Pepe, G., Quagliariello, E., Saccone, C., Sbisa, E.: Nucl.Acids Res. 10, 3279-3289 (1982).

29 Anderson, S., De Bruijn, M.H.L., Coulson, A.R., Eperon, I.C., Sanger, F., Young, I.G.: J.Mol.Biol. 156, 683-717 (1982).

30 Murphy, W.I., Attardi, B., Tu, C., Attardi, G.: J.Mol.Biol. 99, 809-814 (1975).

31 Ojala, D., Montoya, J., Attardi, G.: Nature 290, 470-474 (1981).

32 Gelfand, R., Attardi, G.: Mol.Cell.Biol. 1, 497-511 (1981).

33 Attardi, G., Cantatore, P., Chomyn, A., Crews, S., Gelfand, R., Merkel, C., Montoya, J., Ojala, D.: in: Mitochondrial Genes (Slonimski, P., Borst, P., Attardi, G., Eds), Cold Spring Harbor Laboratory, Cold Spring Harbor, N.Y., pp. 51-71 (1982).

34 Osinga, K.A., De Haan, M., Christianson, T., Tabak, H.F.: Nucl.Acids Res. 10, 7993-8006 (1982).

35 Grosveld, G.C., Rosenthal, A., Flavell, R.A.: Nucl.Acids Res. 10, 4951-4971 (1982).

36 Christianson, T., Rabinowitz, M.: J.Biol.Chem. in press (1983).

37 Grivell, L.A., Hensgens, L.A.M., Osinga, L.A., Tabak, H.F., Boer, P.H., Crusius, J.B.A., Van der Laan, J.C., De Haan, M., Van der Horst, G., Evers, R.F., Arnberg, A.C.: in: Mitochondrial Genes (Slonimski, P., Borst, P., Attardi, G., Eds), Cold Spring Harbor Laboratory, Cold Spring Harbor, pp. 225-239 (1982).

38 Thalenfeld, B.E., Hill, J., Tzagoloff, A.: J.Biol.Chem. 258, 610-615 (1983).

39 Martin, N.C., Underbrink-Lyon, K.: Proc.Natl.Acad.Sci. U.S. 78, 4743-4747 (1981).

40 Green, M.R., Grimm, M.F., Goewert, R.R., Collins, R.A., Cole, M.D., Lambowitz, A.M., Heckman, J.E., Yin, S., RajBhandary, U.L.: J.Biol. Chem. 256, 2027-2034 (1981).

41 Yin, S., Burke, J., Chang, D.D., Browning, K.S., Heckman, J.E., Alzner-De Weerd, B., Potter, M.J., RajBhandary, U.L.: in: Mitochondrial Genes (Slonimski, P., Borst, P., Attardi, G., Eds), Cold Spring Harbor Laboratory, Cold Spring Harbor, N.Y., pp. 361-373 (1982).

42 Lazowska, J., Jacq, C., Slonimski, P.P.: Cell 22, 333-348 (1980).

43 Weiss-Brummer, B., Rödel, G., Schweyen, R.J., Kaudewitz, F.: Cell 29, 527-536 (1982).

44 Dujardin, G., Jacq, C., Slonimski, P.P.: Nature 298, 628-632 (1982).

45 Anziano, P.Q., Hanson, D.K., Mahler, H.R., Perlman, P.S.: Cell 30, 925-932 (1982).

46 Dieckmann, C.L., Pape, L.K., Tzagoloff, A.: Proc.Natl.Acad.Sci.U.S. 79, 1805-1809 (1982).

47 Faye, G., Simon, M.: Cell 32, 77-87 (1983).

48 Netter, P., Jacq, C., Carignani, G., Slonimski, P.P.: Cell 28, 733-738 (1982).

49 Schmelzer, C., Schmidt, C., Schweyen, R.J.: Nucl.Acids Res. 10, 6797-6808 (1982).

50 Davies, R.W., Waring, R.B., Ray, J.A., Brown, G.A., Scazzocchio, C.: Nature 300, 719-724 (1982).

51 Waring, R.B., Davies, R.W., Scazzocchio, C., Brown, T.A.: Proc.Natl. Acad.Sci.U.S. 79, 6332-6336 (1982).

52 Michel, F., Dujon, B.: EMBO J. 2, 33-38 (1983).

53 Waring, R.B., Scazzocchio, C., Brown, T.A., Davies, R.W.: J.Mol.Biol. in press (1983).

54 Cech, T.R., Tanner, N.K., Tinoco Jr., I., Weir, B.R., Zuker, M., Perlman, P.S.: Proc.Natl.Acad.Sci.U.S. 80, 3903-3907 (1983).

55 Bonitz, S.G., Coruzzi, G., Thalenfeld, B.E., Tzagoloff, A., Macino, G.: J.Biol.Chem. 255, 11927-11941 (1980).

56 Arnberg, A.C., Van Ommen, G.J.B., Grivell, L.A., Van Bruggen, E.F.J., Borst, P.: Cell 19, 313-319 (1980).

57 Halbreich, A., Pajot, P., Foucher, M., Grandchamp, C., Slonimski, P.: Cell 19, 321-329 (1980).

58 Baldacci, G., Zennaro, E.: Europ.J.Biochem. 127, 411-416 (1982).

59 Schroeder, R., Breitenbach, M., Schweyen, R.J.: Nucl.Acids Res. 11, 1735-1746 (1983).

60 Weiss-Brummer, B., Holl, J., Schweyen, R.J., Rödel, G., Kaudewitz, F.: Cell 33, 195-202 (1983).

61 De Bruijn, M.H.L., Klug, A.: EMBO J. 2, 1309-1321 (1983).

62 Michaelis, G., Mannhaupt, G., Pratje, E., Fischer, E., Naggert, J., Schweizer, E.: in: Mitochondrial Genes (Slonimski, P., Borst, P., Attardi, G., Eds), Cold Spring Harbor Laboratory, Cold Spring Harbor, N.Y., pp. 311-321, (1982).

63 Groudinsky, O., Dujardin, G., Slonimski, P.P.: Mol.Gen.Genet. 184, 493-503 (1981).

64 Pratje, E., Mannhaupt, G., Michaelis, G., Beyreuther, K.: EMBO J. 2, 1049-1054 (1983).

65 Stern, D.B., Lonsdale, D.M.: Nature 299, 698-702 (1982).

66 Farrelly, F., Butow, R.A.: Nature 301, 296-301 (1983).

67 Jacobs, H.T., Posakony, J.W., Grula, J.W., Roberts, J.W., Xin, J.-H., Britten, R.J., Davidson, E.H.: J.Mol.Biol. 165, 609-632 (1983).

68 Gellissen, G., Bradfield, J.Y., White, B.N., Wyatt, G.R.: Nature 301, 631-634 (1983).

69 Wright, R.M., Cummings, D.J.: Nature 302, 86-88 (1983).

70 Borst, P.: in: Cellular Controls in Differentiation (Lloyd, C.W., Rees, D.A., Eds), Academic Press, London, pp. 231-252 (1981).

71 Borst, P., Tabak, H.F., Grivell, L.A.: in: Eukaryotic Genes: Their Structure, Activity and Regulation (Maclean, N., Gregory, S.P., Flavell, R.A., Eds), Butterworth, London, in press (1983).

72 Locker, J., Rabinowitz, M.: Plasmid 6, 302-314 (1981).

73 Frontali, L., Palleschi, C., Francisci, S.: Nucl.Acids Res. 10, 7283-7293 (1982).

IMPORT OF PROTEINS INTO MITOCHONDRIA
A SURVEY

Michael Yaffe
Biocenter, University of Basel, CH-4046 Basel
Switzerland

Most mitochondrial proteins are encoded in the nucleus, synthesized in the cytoplasm, and imported into mitochondria (Schatz & Mason, 1974). How are these polypeptides targeted to specific sub-mitochondrial locations, transferred into a membrane or across one or two membranes, and assembled into functional protein complexes? Thus far, some key steps in the import process have been identified (Neupert & Schatz, 1981; Schatz & Butow, 1983). Current research largely focuses on the molecules involved in these steps. A long-range goal is the reconstitution of mitochondrial protein import from the purified components.

Pathways for the import of a number of mitochondrial proteins have been elucidated by two basic approaches: pulse-labeling studies with intact cells in the presence and absence of inhibitors, followed by subcellular fractionation and the incubation of isolated mitochondria or mitochondrial fractions with radiolabeled protein percursors synthesized in vitro. The following key steps have been identified: 1) synthesis of a precursor in the cytoplasm, 2) recognition of the mitochondrial surface by the precursor, 3) transfer of the precursor into or across one or both mitochondrial membranes, 4) conversion of the precursor to the mature form by proteolytic cleavage, conformational change, or other modification, and 5) assembly of the polypeptide into an active protein complex. Within this framework, however, there is considerable variation.

Most proteins destined for the mitochondrial matrix or inner membrane are synthesized as larger precursors (Maccec-

Mitochondria 1983

chini et al., 1979; Neupert & Schatz, 1981). These precursors bind to the mitochondrial outer membrane, are translocated across the inner membrane in an energy-dependent step, and are cleaved to their matur size by a metallo-protease located in the matrix. This pathway was first demonstrated for the largest subunits of the yeast F_1-ATPase and seems to apply to a large number of proteins imported into or across the inner membrane (Schatz & Butow, 1983).

The import of several proteins of the intermembrane space initially follows the same pathway as that for matrix and inner membrane proteins. An example is cytochrome b_2 of yeast which is made as a larger precursor, binds to the mitochondrial surface, is at least partially translocated across the inner membrane in an energy dependent step, and is cut by the matrix protease (Gasser et al., 1982a). The resulting "intermediate" form of the protein is subsequently cut by a second protease (probably located on the cytoplasmic side of the inner membrane) and released into the intermembrane space.

A second pathway for import of intermembrane space proteins is followed by cytochrome c. This protein is not made as a larger precursor, but rather as a heme-free apo-enzyme (Korb et al, 1977). Apocytochrome c binds to the mitochondrial outer membrane and is transferred to the intermembrane space, where ist is converted to the holoenzyme (Henning & Neupert, 1981).

Proteins of the outer membrane are imported by still another pathway. They are not made as larger precursors and are inserted directly into the membrane by a process independent of energy (Freitag et al., 1982; Mihara et al., 1982; Gasser & Schatz, 1983).

Nuclear-encoded mitochondrial proteins are synthesized in the extramitochondrial cytoplasm, predominantly on free polysomes (Zimmermann et al., 1979; Raymond & Shore, 1979; Suissa & Schatz, 1982). While polysomes bound to the mitochondrial surface are enriched in mRNA for mitochondrial proteins (Ades & Butow, 1980; Suissa & Schatz, 1982), no evidence exists that they are obligate intermediates in the import process. Indeed, when cells are pulse-labeled for short periods and subsequently fractionated, mitochondrial percursors are first found in the cytoplasm and only later chase into the mitochondria (Mori et al., 1981; Reid & Schatz, 1982b). Additionally, protein import into isolated mitochondria can occur in an entirely post-translational manner (Korb et al., 1978; Maccecchini et al., 1979; Gasser & Schatz, 1983).

Most proteins destined for the matrix, the inner membrane, or the intermembrane space are synthesized as precursors of larger molecular weight (Maccecchini et al., 1979; Neupert & Schatz, 1981). Some of these were shown to contain an amino terminal extension of the polypeptide chain, accounting for their larger size (Lewin et al., 1980; Mihara & Blobel, 1980; Viebrock et al., 1982; Kaput et al., 1982; Faye & Simon, 1983). The role of these extensions has yet to be established. They may serve to target proteins to their correct destination, be involved in the mechanism of transmembrane protein movement, aid in the stability or solubility of precursors in the cytoplasm, or facilitate a conformational change in the protein after arrival at its functional site. These extensions are clearly required for import, however, since pretreatment of in vitro synthesized precursors with isolated processing protease blocks binding of precursors to the outer membrane (Riezman et al., 1983) as well as uptake into isolated mitochondria (Gasser et al., 1982b). Several imported proteins found in different mitochondrial locations are not made as larger precursors. These proteins may undergo modifications other than proteolysis upon import into mitochondria, or they may undergo a non-covalent conformational change upon reaching their destination.

The first site of interaction of precursors with the mitochondria is the outer membrane. Precursors synthesized in vitro can bind to the outer membrane (Henning & Neupert, 1981; Riezman et al., 1983). A mild protease treatment of this membrane abolishes this interaction. Similiar treatment does not abolish the insertion of outer membrane proteins (Gasser & Schatz, 1983), suggesting that these species either bind to a different receptor or partition directly into the membrane bilayer. Isolated, mature mitochondrial proteins and non-mitochondrial proteins do not bind specifically to the outer membrane. There appear to be at least several diffenrent outer membrane receptors for mitochondrial proteins: unlabeled apocytochrome competes with radiolabeled apocytochrome c for binding to isolated mitochondria, but fails to compete for binding with precursors to several other mitochondrial proteins (Zimmermann et al., 1981) The specific polypeptides (or lipids) which function as import receptors have yet to be identified.

The mechanism by which proteins are translocated across the mitochondrial membrane remains unknown. A component of this mechanism is an electrochemical potential across the inner membrane which is required for import of matrix-, inner membrane-, an (many) intermembrane space proteins (Gasser et al., 1982b; Schleyer et al., 1982). Dissipation of this potential by an uncoupler such as carbonyl cyanide m-chlorophenylhydrazone (CCCP) inhibits the import of proteins and leads to the accumulation of precursors of some mitochondrial proteins in intact yeast cells (Nelson & Schatz, 1979; Reid & Schatz, 1982a).

Mitochondrial proteins made as larger percursors are cleaved to their mature size following membrane translocation (Gasser et al., 1983; Reid et al., 1982). A matrix-localized protease, specific for this cleavage, has been characterized from yeast (Böhni et al., 1980; McAda & Douglas, 1982) and rat liver (Miura et al., 1982). This enzyme cuts precursors destined for the matrix and inner membrane to their mature size. It also catalyzes the first cleavage of those intermembrane space proteins which are made as larger precursors. The enzyme responsible for the second proteolytic cleavage of these precursors remains to be characterized.

The goal of much recent research has been the preparation and characterization of pure components of protein import:

the precursor polypeptides, the import receptors, and cytoplasmic factors factors necessary for import.

Studies of precursors have involved purification of the polypeptides and isolation of the genes for mitochondrial proteins. Two precursors of mitochondrial proteins have been purified. Apo-cytochrome c can be derived by chemical cleavage of the heme group from the holoenzyme (which can be purified in large amounts), an has been used to study import and binding to isolated mitochondria (Zimmermann et al., 1979; Henning & Neupert, 1981). The precursor of the F_1-ATPase ß-subunit accumulates in large amounts in rho- yeast cells poisoned with CCCP (Reid & Schatz, 1982a), and has been purified to homogeneity from such cells (S. Ohta, unpublished data). This precursor can be importet into isolated mitochondria and has been used to identify a cytoplasmic factor involved in protein import (see below).

Precursors of mitochondrial proteins must contain information for directing the proteins to their correct sub-mitochondrial locations. Gene cloning and recombinant DNA technology provide powerful tools for analyzing this information. An initial analysis can be made from amino acid sequences deduced from the nucleotide sequence of a cloned gene. Comparison of sequences of a number of proteins may reveal targeting signals. Further analysis involves alteration of a gene *in vitro*, followed by reintroduction into the cell and examination of the import and localization of the altered protein.

The genes for a number of mitochondrial proteins have been cloned. These include (among others) the genes for cytochrome c (Montgomery et al., 1978), yeast adenine nucleotide translocator (O'Malley, et al., 1982), subunit 9 of the ATP synthase of *Neurospora crassa* (Viebrock et al., 1982), yeast cytochrome c peroxidase (Goltz et al., 1982), three subunits of the cytochrome bc_1 complex of yeast (van Loon et al., 1982), a 70-kilodalton polypeptide of the mitochondrial outer membrane (Riezman et al., 1983b), and several other mitochondrial proteins (M. Suissa, G. Reid, & G. Schatz, unpublished results).

Import receptors have yet to be identified on the molecular level, but at least several different receptors must exist (as discussed above). An initial step in the purification of one receptor in the outer membrane has been its solubilization with detergent und subsequent reconstitution into vesicles (Riezman et al., 1983a). These vesicles exhibit binding activity similar to that of the isolated outer membrane.

Are extra-mitochondrial components involved in protein import? Recently, cytoplasmic factors have been reported to be necessary for the uptake of proteins into mitochondria (Argan et al., 1983; Miura et al., 1983; S. Ohta, personal communication). One such factor is involved in the import of ornithine carbamyltransferase into mammalian mitochondria and is found in reticulocyte lysates. A factor which stimulates the import of the purified ß-subunit precursor into isolated yeast mitochondria has been detected in both reticulocyte lysates and yeast cytosol. All of the reported factors are non-dialyzable and sensitive to protease treatment.

The major unsolved questions about protein import include: How do proteins traverse the mitochondrial membranes? What other cellular components, factors, or structures are involved in the import process? How is protein import regulated?

We have begun a new approach to address these problems. We have isolated temperature-sensitive yeast mutants in which the import of proteins is blocked at the high temperature. Genetic and biochemical analysis of these mutants should confirm the role of known components and identify new factors involved in the import process. The mutants can also be used to obtain the genes of proteins which are part of the import machinery and should aid our analysis of the molecular basis of mitochondrial protein import.

REFERENCES

Ades, I.z., & Butow, R. A. (1980) J. Biol. Chem. 255, 9918-9924.

Argan, C., Lusty, C.J., & Shore, G.C. (1983) J. Biol. Chem. 258, 6667-6670.

Böhni, P.C., Gasser, S., Leaver, C., & Schatz, G., (1980) In "The Organization and Expression of the Mitochondrial Genome" (A.M. Kroon & C. Saccone, eds.), pp. 423-433. North-Holland Biomedical Press.

Faye, G., & Simon, M. (1983) Cell 32, 77-87

Freitag, H., James, M., & Neupert, W. (1982) Eur. J. Biochem. 126, 197-202.

Gasser, S.M., Ohashi, A., Daum, G., Böhni, P.C., Gibson, J., Reid, G.A., Yonetani, T., & Schatz, G. (1982a) Proc. Nalt. Acad. Sci. USA 79, 267-271

Gasser, S.M., Daum, G., & Schatz, G. (1982b) J. Biol. Chem. 257, 13034-13041.

Gasser, S.M., & Schatz, G. (1983) J. Biol. Chem. 258, 3427-3430.

Goltz, S., Kaput, J., & Blobel, G. (1982) J. Biol. Chem. 257, 11186-11190.

Hennig, B., & Neupert, W. (1981) Eur. J. Biochem. 121, 203-212.

Kaput, J., Goltz, S., & Blobel, G. (1982) J. Biol. Chem. 257, 15054-15058.

Korb, H., & Neupert, W. (1978) Eur. J. Biochem. 91, 609-620.

Lewin, A., Gregor, I., Mason, T.L., Nelson, N., & Schatz, G. (1980) Proc. Natl. Acad. Sci. USA 77, 3998-4002.

McAda, P.C., & Douglas, M.G. (1982) J. Biol. Chem. 257, 3177-3182.

Maccecchini, M.-L., Rudin, Y., Blobel, G., & Schatz, G. (1979) Proc. Natl. Acad. Sci. USA 76, 343-347.

Mihara, K., & Blobel, G. (1980) Proc. Natl. Acad. Sci. USA 77, 343-347.

Mihara, K., Blobel, G., & Sato, R. (1982) Proc. Natl. Acad. Sci. USA 79, 7102-7106.

Miura, S., Mori, M., Amaya, Y, & Tatibana, M. (1982), Eur. J. Biochem. 122, 641-647.

Miura, S., Mori, M., & Tatibana, M. (1983) J. Biol. Chem. 258, 6671-6674.

Montgomery, D.L., Hall, B.D., Gillam, S., & Smith, M.(1978) Cell 14, 673-680.

Mori, M., Morita, T., Ikeda, F., Amaya, V., Tativana, M., & Cohen, P.P. (1981) Proc. Natl. Acad. Sci. USA 78, 6056-6060.

Nelson, N., & Schatz, G. (1979) Proc. Natl. Acad. Sci. USA 76, 4365-4369.

Neupert, W., & Schatz, G. (1981) Trends Biochem. Sci. 6, 1-4

O'Malley, K., Pratt, P., Robertson, J., Lilly, M., & Douglas, M.G. (1982) J. Biol. Chem. 257, 2097-2103.

Raymond, Y., & Shore, G.C. (1979) J. Biol. Chem 254, 9335-9338

Reid, G.A., & Schatz, G. (1982a) J. Biol. Chem. 257, 13056-13061.

Reid, G.A., & Schatz, G. (1982b) J. Biol. Chem. 257, 13062-13067.

Reid, G.A., Yonetani, T., & Schatz, G. (1982) J. Biol. Chem. 257, 13068-13074.

Riezman, H., Hay, R., Witte, C., Nelson, N., & Schatz, G. (1983a) EMBO J. 2, 1113-1118.

Riezman, H., Hase, T., van Loon, A.P.G.M., Grivell, L.A., Suda, K., & Schatz, G. (1983) manuscript submitted.

Schatz, G., & Mason, T.L. (1974) Ann. Rev. Biochem. 43, 51-87

Schatz, G., & Butow, R.A. (1983) Cell 32, 316-318.

Schleyer, M., Schmidt, B., & Neupert, W. (1982) Eur.J. Biochem. 125, 109-116

Suissa, M., & Schatz, G. (1982) J. Biol. Chem. 257, 13048-13055

van Loon, A.P.G.M., de Groot, R.J., van Eyk, E., van der Horst, G.T.J. & Grivell, L.A. (1982) Gene 20, 323-337

Viebrock, A., Perz, A., & Sebald, W. (1982) EMBO J. 1, 565-571.

Zimmermann, R., Paluch, U., & Neupert, W. (1979) FEBS Lett. 108, 141-146.

Zimmermann, R., Hennig, B., & Neupert, W. (1981) Eur. J. Biochem. 116, 455-460.

THE REPLICATION OF YEAST MITOCHONDRIAL DNA

Giuseppe Baldacci and Giorgio Bernardi
Laboratoire de Génétique Moléculaire
Institut Jacques Monod
2 Place Jussieu, 75005 Paris, France.

Introduction

Investigations from our laboratory have shown the presence of seven _ori_ sequences in the mitochondrial genome of wild-type _Saccharomyces cerevisiae_ (Fig.1). _Ori_ sequences have been mapped and oriented on the mitochondrial genome and their primary structure has been determined (see ref.1 for a brief review). So far, evidence for the involvement of _ori_ sequences in DNA replication, has been strong, yet indirect. Essentially, it rests on the following findings : a) the _ori_ sequences of the mitochondrial genome of yeast show structural similarities with those of animal cells (heavy

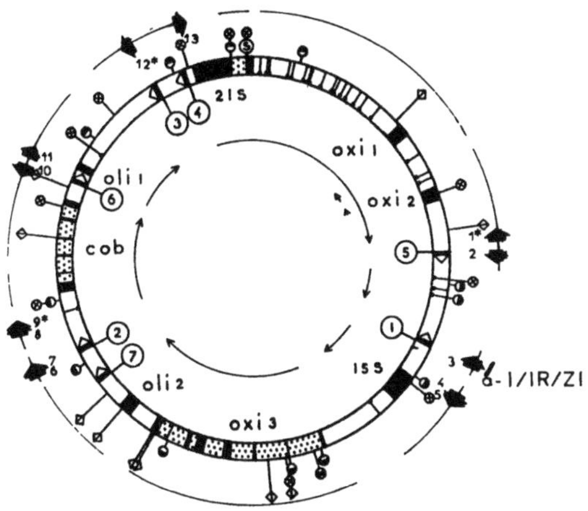

Figure 1. Physical map of the mitochondrial genome of wild-type _S. cerevisiae_ strain A. The inner circle indicates stable transcripts. Circled numbers refer to the _ori_ sequences. The outer circle indicates the sequenced regions of this genome. The location of the repeat unit of the mitochondrial genome of a-1/1R/Z1 is indicated. The location and the orientation of sequences _r_ are indicated by arrowheads pointing in the direction of its TAT end. For further details, see ref. 2.

Mitochondria 1983

strands) ; b) at least one ori sequence is present on the repeat units of the mitochondrial genome of the vast majority of spontaneous petites ; it should be stressed, however, that ori 7, was never found alone in a petite genome and ori 4 and ori 6, were only rarely found alone ; c) a correlation exists between the suppressivity of a petite and the intact or altered state of the ori sequence present in its repeat units ; ori^+ petites, containing an intact ori sequence, are highly suppressive (less so, however, in the case of ori 6 ; ori 4 and ori 7 were never tested because not available in haploid petites) ; ori^- petites, deleted in cluster A or cluster C of their ori sequences (see Fig.2) were, respectively, slightly and very much lower in suppressivity than the corresponding ori^+ oetites ; ori^r petites, having repeat units with inverted ori sequences and ori^o petites lacking altogether any of the seven canonical ori sequences were low in suppresivity ; d) ori sequences contain on one strand a 16-bp sequence, r, where transcription is initiated (2) ; such sequence is interrupted by the insertion of a GC-cluster, γ, and is not operational in ori 4, 6 and 7.

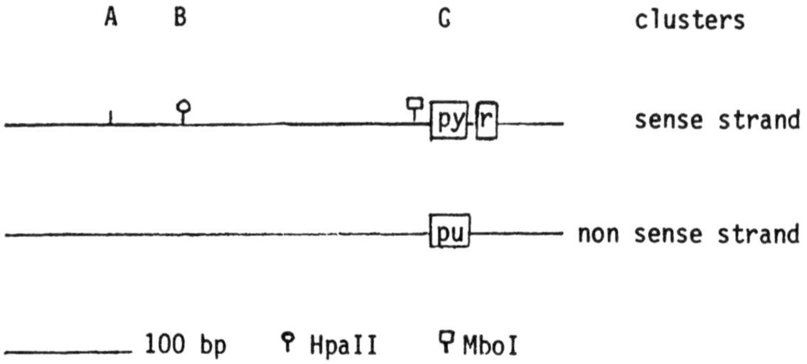

Figure 2. Schematic map of the repeat unit of the mitochondrial genome of petite a-1/1R/Z1. The position of clusters A, B and C of ori 1, and of some restriction sites are indicated. Boxes py, pu refer to the oligopyrimidine and oligopurine tracts of cluster C ; box r to the sequence of transcription initiation.

To sum up, the indirect evidence that ori sequences are involved in DNA replication rests on their similarity with bona fide replication origins, on their expected presence in the repeat units of the vast majority of petites, on the correlation between the intact or altered state of the ori sequence and the level of transmission of the petite mitochondrial genome to the progeny of crosses with wild-type cells, and on the expected association between the sites of initiation of DNA replication and RNA transcription.

Here we report the first direct evidence that, indeed, ori sequences of the mitochondrial genome of yeast are origins of DNA replication, since sites located on both strands in the neighborhood of cluster C are starting points for nascent DNA strands which are preceeded by short primer RNA sequences. The study reported here was done using the mitochondrial genome of petite a-1/1R/Z1. Figure 2 provides a schematic map of the repeat unit, containing ori 1, of this petite.

Materials and Methods

Petite a-1/1R/Z1, culture media and cell growth, preparations of mitochondrial nucleic acids, enzymatic treatment, techniques for DNA labeling and DNA-DNA and DNA-RNA hybridizations were previously described (2-4). Strand separation and DNA sequencing were performed according to Maxam and Gilbert (5).

Results

Mitochondrial DNA from petite a-1/1R/Z1 was degraded with Hpa II, an enzyme cutting every repeat unit at a site located in cluster B of the ori 1 sequence (Fig.2). The fragments so obtained were 5' labelled with P^{32} ; their strands were separated and used in hybridization experiments with total nucleic acids, RNA, or DNA, as present, respectively, in untreated, pancreatic DNas-treated, and pancreatic-treated mitochondrial phenol extracts from petite a-1/1R/Z1. Hybrids were then degraded with S1-nuclease, heat-denatured and submitted to high-voltage electrophoresis in parallel with partial chemical depurinates of the labelled strands.

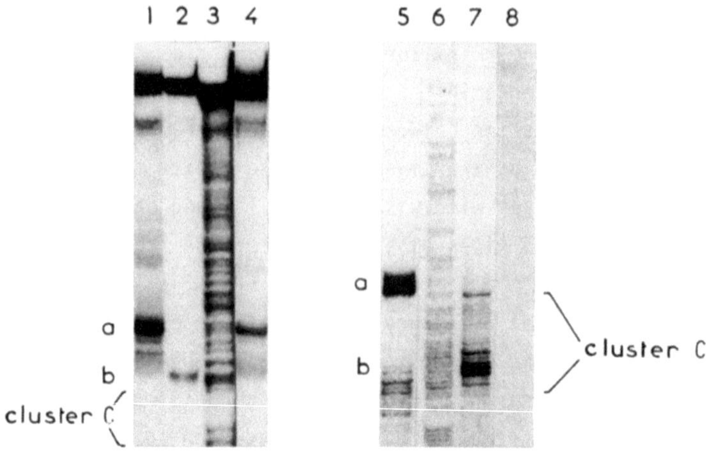

Figure 3. Autoradiograms of S1-nuclease degraded hybrids formed by end-labelled separated strands of Hpa II-degraded DNA from petite a-1/1R/Z1 with total nucleic acids, RNA and DNA from mitochondrial of the same petite. See text for further explanations.

The results (Fig.3) were different with hybrids obtained when the labelled strand was the sense strand used as a template in RNA transcription, namely the strand containing the oligopyrimidine tract of cluster C (2 ; Fig.2), or the other strand.

In the first case, hybrids with RNA yielded a major band (Fig.3 ; lane 4) corresponding to a fragment about 240-bp in size, which ended about 20 nucleotides 3' of cluster C (Fig.4). The same band was found when total nucleic acids were used in hybridization (Fig.3 ; lane 1) , showing that the protection of the labelled strand in this experiment was essentially due to RNA molecules. In contrast, when mitochondrial DNA was used in hybridization, the S1-protected labelled strand ended only very few nucleotides 3' from cluster C. (Fig.3 ; lane 2 and Fig.4).

In the second case, no detectable protected strands were found when DNase-treated mitochondrial nucleic acids were used in hybridization (Fig.3 ; lane 8). This is in agreement with our previous study that the strand carrying the oligopurine tract of cluster C is not detectably transcribed (2). When total nucleic acids were used in hybridization, the protected strand ended just before cluster C (Fig.3 ; lane 5; Fig.4).

```
     320                                                380
      .       .       .       .       .       .       .

                <-- DNA          <-- RNA

5' ATCACCCACCCCCTCCCCCTATAAAACTTAGTTTATTACTTATATATTTATAAATATAAATCTAACT
3' TAGTGGGTGGGGGAGGGGGATATTTTGAATCAAATAATGAATATATAAATATTTATATTTAGATTGA

RNA -->      DNA -->
```

Figure 4. Primary structure of cluster C and sequence r of the ori 1 sequence. The position corresponding to the 5' ends of hybrids resistant to S1-nuclease degradation are indicated by arrows showing the 5' to 3' direction of newly synthesized RNA and DNA molecules.

Finally, when RNase-treated extracts were used in hybridization, the strong band of lane 5 (Fig.3) just described was replaced by another band (Fig.3 ; lane 7) corresponding to a shorter fragment ending inside cluster C (Fig.4).

Several controls were performed. First, self-annealed probes were digested with S1-nuclease to detect any material protected by foldback. Second, the activity of S1-nuclease was tested by showing that the enzyme was able to remove the end label from hybrids reconstituted from the two separated probes. Third, hybrids were reconstituted by using strands derived from an MboI digest and complementary strands derived from an end-labelled Hpa II digest ; after S1-nuclease digestion, the protected hybrids had the expected lengths.

Discussion

The present results show the existence in the mitochondrial of petite a-1/1R/Z1 of a population of DNA molecules having the size of the repeat units and identical ends. The presence of DNA molecules having such a size in DNA preparations from petite mutants is well-known (2). Their identical ends indicate that they are initiated at specific sites. Our experiments map the 5' ends at the level of cluster C of ori. Analogous experiments aiming at detecting RNA molecules allowed us to find two starting points of transcription, one on each strand, about 20 nucleotides upstream of the sites where 5' ends of DNA molecules map.

The fact that the non-sense strand (i.e, the strand having the oligopurine tract in cluster C does not produce long transcripts, but only short ones which are degraded by RNase, indicates that its replication starts with an RNA primer, which is still present in the majority of monomer DNA molecules. RNase treatment allows the identification of the transition points (RNA → DNA) and the demonstration of intermediate-size nascent DNA molecules in which some ribonucleotides are replaced by deoxyribonucleotides (see the band ladders of Fig.3 ; lane 7).

As far as the sense strand is concerned, the situation is complicated by the presence of transcripts having the size of the repeat units and which start at sequence r. These transcripts are the only ones evident in untreated phenol extracts ; upon RNase treatment, they disappear to be replaced by slightly shorter DNA fragments. The sense strand appears, therefore, to produce both full-length transcripts and RNA primers for DNA replication, the start points being the same.

To sum up, both strands act as templates for a DNA replication which involves short RNA primers having their 5' ends in the neighborhood of cluster C. The critical importance of cluster C and of the neighboring sequences had already been stressed by the dramatic drop in suppressivity and transcription of petites which have lost these segments of their ori sequence (1,2). It is conceivable that the stem-and-loop structure formed by the cluster A-cluster B region acts instead as a recognition sequence. In any case, the loss of cluster A does not affect replication and transcription to any comparable extent (1,2). Interestingly, it is possible that the putative recognition sequence formed by the cluster A-cluster B region is close in space to the initiation of replication and transcription sequences formed by cluster C and sequence r, because of the interaction between two complementary inverted 15-bp sequences located at positions 167-181 and 364-378, respectively (see Fig.4 and ref.1). These sequences were also found (6) in the ori sequence of the mitochondrial DNA of *Candida utilis*, and (in one copy only) in yeast ars 2 (7).

In turn, the RNA initiation sequence r can be folded in a short loop-and-stem structure, also found in the other r sequence of the mitochondrial genome (2).

Results similar to those discussed in this paper have been obtained in the case of petite mutants containing ori 5 in their mitochondrial repeat units.

REFERENCES

1. Bernardi, G. : Trends in Biochem. Sciences 7, 1-5 (1982)
2. Baldacci, G., Bernardi, G. : EMBO Journal 1, 987-994 (1982)
3. Faugeron-Fonty, G., Culard, F., Baldacci, G., Goursot, R., Prunell, A., Bernardi, G. : J. Mol. Biol. 134, 493-537 (1979)
4. Gaillard, C., Bernardi, G. : Mol. Gen. Genet. 174, 333-337 (1979)
5. Maxam, A.M., Gilbert, W. : Proc. Natl. Acad. Sci. USA 74, 560-564 (1977)
6. Tikhomirova, L.P., Kryukov, V.M., Strizhov, N.I., Bayev, A.A. : Mol. Gen. Genet. 189, 479-484 (1983)
7. Tschumper, G., Carbon, J. : J. Mol. Biol. 156, 293-307 (1982)

Acknowledgement
One of us (G. Baldacci) thanks EMBO for a long term fellowship

SOME IONOPHORES ACT PREFERENTIALLY ON MITOCHONDRIA IN INTACT CELLS OF SACCHAROMYCES CEREVISIAE AND INDUCE RESPIRATION-DEFICIENT MUTANTS

Ladislav Kováč and Vlasta Klobučníková

Institute of Animal Physiology, Slovak Academy of Sciences
900 28 Ivanka pri Dunaji, Czechoslovakia

Introduction

Little is known of the role that gradients of chemical and electrical potentials established across cellular membranes may have in biological functions of nucleic acids. It appears useful to study this role by means of ionophores, especially if the ionophores can selectively affect the gradients across particular cell membranes. Divalent cation ionophores A 23187 and oleficin were found to act preferentially on mitochondria in yeast S. cerevisiae and the growth of the yeast in the presence of oleficin resulted in the disintegration of the mitochondrial genome (1). This paper shows that valinomycin and nigericin, which function as ionophores of monovalent cations in isolated yeast mitochondria (2, 3) and possibly also in mitochondria in situ but not in the cytoplasmic membrane of the intact yeast cells (4), can induce a mass formation of respiration-deficient mutants.

Methods

Diploid wild-type yeast S. cerevisiae DT XII was cultured aerobically at 30^{0} in a standard medium containing 2% glucose, 0.5% $(NH_4)_2SO_4$, 0.05% $MgSO_4.7H_2O$, 0.01% NaCl, 0.01% $CaCl_2$, 0.5% peptone (Imuna), 0.5% yeast extract (Difco) and 0.1% KH_2PO_4. Medium without K^+ contained 0.1% NaH_2PO_4 instead of KH_2PO_4 and medium with high K^+ was the standard medium containing in addition 0.5 M KCl. The antibiotics were added into the media as methanolic solutions. Cells were plated onto solid medium containing

Mitochondria 1983

2% glycerol, 0.1% glucose, 1% peptone and 1% yeast extract and the fraction of respiration-deficient mutants was determined from the number of small colonies as well as of colonies that did not stain with triphenyltetrazolium chloride. Viability was estimated from the total numbers of colonies. Respiration was measured manometrically.

Results and Discussion

As shown in Fig. 1, yeast cells growing in the presence of valinomycin or nigericin were converted into respiration-deficient mutants. The mutants were undoubtedly of the <u>rho</u>$^-$ type since nuclear mutants could not be formed at such a high frequency in the diploid yeast. Lower concentrations of the antibiotics were efficient in mutant formation in the high K^+ medium than in a medium containing no K^+.

When the yeast cells were incubated with nigericin or valinomycin in a buffer containing KCl under non-growth conditions with or without glucose the formation of the mutants also became apparent albeit less efficient (Table 1). The fraction of the mutants was lower in the absence of KCl.

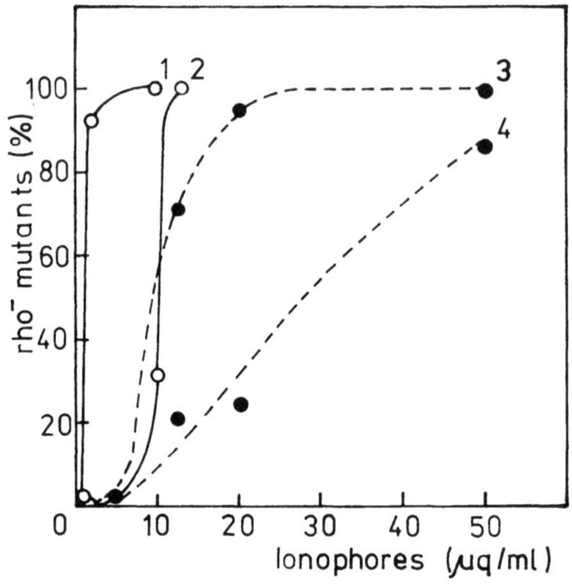

Figure 1.

Figure 1. Induction of respiration-deficient mutants by valinomycin and nigericin. S. cerevisiae DT XII was cultured for 42 h in the presence of the antibiotics at the concentrations indicated on the abscissa in a medium containing no K^+ ions or in a medium containing a high concentration of KCl (0.5 M). The ordinate indicates percentage of mutants present in the culture at the end of cultivation. 1 - nigericin, high K^+; 2 - valinomycin, high K^+; 3 - nigericin, no K^+; 4 - valinomycin, no K^+.

As shown in the Table, the cells incubated with the antibiotics, especially with nigericin, under non-growing conditions were losing viability.

The cultures that had been produced on glucose in the standard medium in the presence of either valinomycin or nigericin at concentrations of 10 to 20 µg/ml were able to respire. When allowance was made for the fraction of non-respiring rho^- mutants formed in the cultures, the respiratory capacity of the cells was calculated to be the same as of those grown in the absence of the antibiotics. Accordingly, the antibiotics present in the medium during growth on glucose did not prevent the synthesis of respiratory enzymes and their assembly in mitochondria.

Table 1. Viability and mutant formation of non-growing cells
Cells of S. cerevisiae DT XII, grown for 24 h in the standard medium, were washed and incubated for 24 h at $2x10^6$ cells/ml in a solution containing 33 mM Tris/maleate pH 6.4, 0.7 M KCl and valinomycin or nigericin (66 µg/ml) as shown below in the presence (+) or absence (-) of 0.3 M glucose.

Ionophore	Glucose	Viability (%)	Mutants (%)
Valinomycin	-	52	5
	+	108	1
Nigericin	-	0.2	13
	+	1.1	33

The results suggest that an undisturbed balance of monovalent cations in mitochondria is a prerequisite for normal replication of yeast mitochondrial DNA, and partly also for its maintenance in non-growing cells. This has also been shown to be true for the balance of divalent cations (1). Viability of non-growing cells may also be dependent on the normal ionic balances across the mitochondrial membrane.

References

1. Kováč, L., Poliachová, V. and Horváth, I.: Biochim. Biophys. Acta 721, 349-356 (1982).
2. Kováč, L., Groot, G. S. P. and Racker, E.: Biochim. Biophys. Acta 256, 55-65 (1972).
3. Kolarov, J., Šubík, J. and Kováč, L.: Biochim. Biophys. Acta 267, 465-478 (1972).
4. Kováč, L., Böhmerová, E. and Butko, P.: Biochim. Biophys. Acta 721, 341-348 (1982).

INITIATION AND TRANSCRIPTION OF YEAST MITOCHONDRIAL RNA

John C. Edwards, Thomas Christianson, David Mueller, Tapan K. Biswas, David Levens, Dean Li, Janice Wettstein, and Murray Rabinowitz

The Departments of Medicine, Biochemistry, Biology, and Microbiology, University of Chicago, Chicago, Illinois 60637

RNA synthesis in yeast mitochondria is an important step in the expression of the mitochondrial (mt) genome and one which is still only partially understood. Although only about one quarter of the coding capacity of yeast mitochondrial DNA codes for known products (1), nearly a full single strand equivalent is transcribed into RNA (2). RNA blot hybridization analysis of mitochondrial transcripts reveal that multiple size transcripts of each particular gene are present, many of which are much larger than the size necessary to contain the coding region of the gene (3). This pattern suggests that the genome is transcribed into large precursers which are processed to yield the smaller mature RNAs. The processing events must include trimming of 5' and 3' extensions as well as splicing out of introns (4,5).

We have begun to investigate the interactions leading to correct transcription of the mitochondrial genome using three methods. First, we have characterized the sites of transcriptional initiation by capping primary transcripts with guanylyl transferase and [α^{32}P]-GTP. Second, we have investigated the components required for correct initiation of yeast mitochondrial transcripts using a homologous *in vitro* transcription system. Third, we have begun to characterize genetically the requirements for correct transcription of the mitochondrial genome using a collection of temperature sensitive nuclear mutants that affect mitochondrial function .

Capping Analysis. The presence of low abundance long transcripts and extensive processing has obscured the pattern of transcriptional initiation

Mitochondria 1983

in yeast mitochondria. In such a heavily processed system, two classes of 5' ends are present: those originating from initiation and those which are products of processing reactions. However RNA 5' ends resulting from initiation are unique in that they retain the 5' triphosphate of their initiating nucleotide.

To label only RNA molecules retaining a 5' triphospate, we used the enzyme guanylyl transferase from vaccinia virions. Since the reaction catalyzed by this enzyme is specific for 5' di or tri phosphate terminated RNA molecules (6) and since the only known source of such molecules in yeast mitochondria is transcriptional initiation, this method uniquely labels RNA molecules which retain their 5' initiating nucleotide. We have previously reported the use of this method for the determination of the initiation sites of the large and small rRNAs (7,8). In addition to these two abundant transcripts, many much less abundant species are labeled with guanylyl transferase. Because of low yield and overwhelming contamination with degradation products of the more abundant species, it has been impossible to characterized these minor species by the same straightforward methods used for the rRNAs.

We used two methods to enrich for low abundance transcripts (9). First, RNA was isolated from a collection of genetically characterized petite deletion mutants which together cover the entire genome. Thus, transcripts from each initiation site should be represented in the entire collection and yet RNA isolated from each petite is free of contamination by transcripts from any initiation site not contained in the particular petite. Second, after labeling petite derived RNA with guanylyl transferase, the products were digested to completion with the ribonuclease T_1 which only cuts 3' to guanlyl residues in RNA. The products were then separated on a sequencing gel and visualized by autoradiography. With this method, all labeled transcripts from a particular initiation site collapse into a single band determined by the distance from the transcriptional initiation site to the site of the first G residue in the transcript. Since the abundance of GC

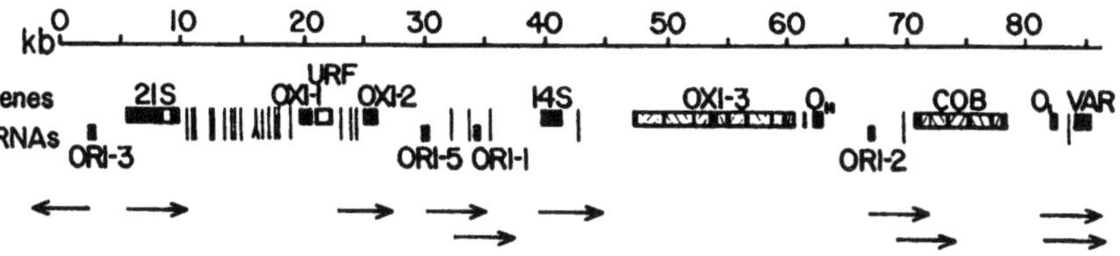

Figure 1. Initiation sites on the mitochondrial genome. The arrows indicate the position of transcriptional initiation sites on this physical map of the mitochondrial genome as described in the text.5

base pairs in mitochondrial DNA is so low, we expected the 5' terminal T_1 fragments to be unique in most cases and to be long enough to provide a sequence which we could unambiguously localize on the mitochondrial genome. As expected, using the above method of analysis, each petite exhibits a characteristic pattern of bands. The most abundant of the labeled bands were eluted from the gels and their sequences determined directly using partial digestion with base specific RNases. In this manner we have determined the sequence of 19 unique 5' terminal T_1 fragments, each representing transcriptional initiation at a different site on the mitochondrial genome. Thirteen of these initiation sites have been localized precisely by matching the sequence of the 5' terminal T_1 fragment to available DNA sequences.

The position of ten of these sites are illustrated in figure 1. We find transcriptional initiation sites at the 5' end of the 21S rRNA and 14S rRNA precurser, in ori loci 2, 3, and 5, upstream of $tRNA^{phe}$, upstream of $tRNA^{f-met}$, upstream of $tRNA^{glu}$, and at two sites separated by 78 nucleotides upstream of oli-I. Comparison of the DNA sequences at the sites of initiation for these 10 transcripts reveals a remarkable degree of homology. In each case, the initiating nucleotide is at the same postition within a highly conserved sequence which can be summarized as follows: 5' A/T·A/T·a·T·A·T·A·a·G·T·A*·a·t·a 3'. The asterisk marks the first transcribed base. The upper case letters represent invariant bases present at every initiation site identified to date. The lower case letters represent bases present at most but not all sites. The bases at postions -9 and -10 are always an A or a T. The presence of this sequence at each of these ten sites suggests that the sequence is an essential component of the yeast mitochondrial promoter.

In addition to the 10 identified sites which initiate at the same position within the consensus sequence listed above, 3 minor transcripts, each associated with a major cappable species, initiate at an unusual position. We find minor transcripts which have initiated two nucleotides downstream from the major initiation site for the 14S rRNA, the site lying in ori-2, and the site further upstream of the oli-1 gene. The initiation nucleotide for these three transcripts is underlined in the consensus sequence listed above. Although this position is not conserved among all ten major transcriptional initiation sites, at the three sites where we find transcription initiating at this position, the base is a T. The RNA sequencing data indicates that the initiating nucleotide for these three minor transcripts is UTP. To confirm the identity of the initiation nucleotide, the capped 5' terminal T_1 fragment corresponding to each secondary initiation site was digested to completion with nuclease P_1 which digests 5'-3' phosphodiester bonds producing 5' nucleotide monophosphates. The 5'-5' cap structure is not digested by this enzyme. The products of digestion were separated by thin layer chromatography and compared with unlabeled markers. The markers were visualed by U.V. absorption and the labeled caps were visualized by autoradiography. The cap analysis of both major and minor transcripts from ori-2 plus the major transcript of the 14S rRNA gene are shown in figure 2. The labeled caps from both of the major transcripts comigrate with GpppA as expected. However, the labeled cap from the secondary transcript does not comigrate with any of the cap markers. Thus,

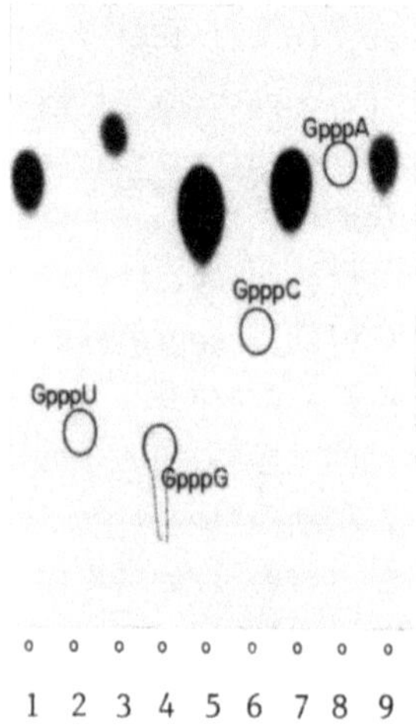

Figure 2. Cap analysis. P_1 digestion and thin layer chromatography were carried out as previously described (7). Lanes 2, 4, 6 and 8: unlabeled cap markers as indicated detected by U.V. absorbance. Lane 3 cap analysis of the minor transcript from ori-2. Lane 5: cap analysis of the major transcript from ori-2. Lane 7: cap analysis of 14S rRNA major primary transcript. Lanes 1 and 9: cap analysis of transcripts 5c and 14, respectively, as described in reference 9.

for the secondary transcripts, the 5' initiating nucleotide is predicted to be UTP by the DNA sequence, is digested by base specific RNases as if it is a uridylyl residue, by does not appear to be UTP when cap structures are characterized by thin layer chromatagraphy. The identity of these initiating nucleotides has not been characterized further. The data to date indicate that the 5' nucleotide may be a modified uridylyl residue which has altered chromatographic properties but which is still capable of interacting with base specific RNases.

Besides the 13 transcripts whose position has been precisely localized by matching the RNA sequence to available DNA sequences, we have determined the 5' sequence of several primary transcripts whose location on the mitochondrial genome has not yet been discovered, presumably because the region from which they arise has not yet been sequenced. Of particular interest is the absence of initiation sites between some genes in which the spacer DNA has been completely sequenced. For instance, we find no initiation site between oli-1, $tRNA^{ser}$ and var-1, suggesting that these genes are transcribed together from a common initiation site leading to a multigene transcript which must be subsequently processed to yield mature RNA. Similarly, we find no initiation site between $tRNA^{glu}$ and cob or between oxi-III and oli-2, again suggesting that these genes are part of multigene transcription units.

In vitro Transcription. A second approach we have taken to the investigation of transcription in yeast mitochondria is the use of a homologous transcription system comprised of purified mtRNA polymerase and mtDNA fragments inserted in bacterial plasmids. We have reported the ability of this system to correctly initiate transcription at the rRNA initiation sites (10). Recently, we have extended the analysis to tRNA and mRNA initiation sites.

One initiation site we investigated lies upstream of the $tRNA^{glu}$ gene[12]. Since no cappable RNA species map between this initiation site and the downstream cob gene, this site probably serves for transcription of both the tRNA and the cob genes. Plasmid pC8 contains a 700 bp Sau 3a fragment carrying the initiation site. This plasmid was was used as a template for transcription using the purified mtRNA polymerase with [$\gamma^{32}P$]-ATP in the absence of CTP. Under these conditions, correct initiation should yield a product uniquely labeled at the 5' end stretching from the initiation site to the first C in the transcript. Analysis of the DNA sequence[11] reveals that the first CTP residue is at position 34 in the predicted transcript. The results of transcription are shown in figure 3a. The major transcript is about the expected size of 33 nucleotides. To confirm the identity of the transcript, the band was cut from the gel and its sequence determined by partial digestion with base specific RNases (13,14) as shown in figure 3b. The deduced sequence matches the predicted sequence exactly. Thus, the purified mtRNA polymerase is capable of correct initiation at this promoter.

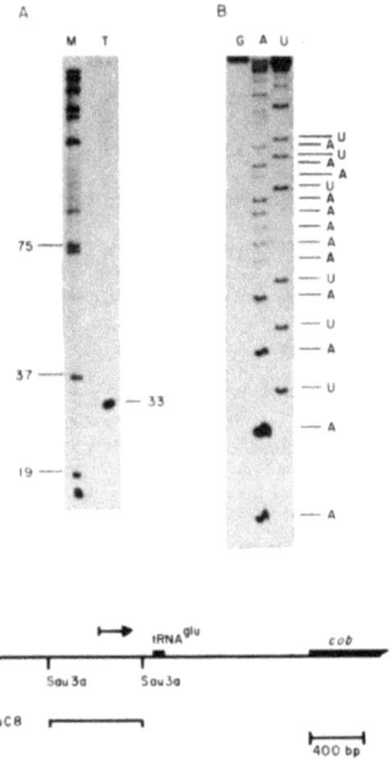

Figure 3. A: In vitro transcription from tRNA^{glu} promoter. M: molecular weight markers as indicated. T: products of transcription. B: Sequence analysis of transcripts. Lanes G, A, and U: products of partial digestion with enzymes T_1, U2, and B. cereus, respectively. Map: the arrow marks the site of transcriptional initiation.

In a similar manner, we have found that the purified polymerase is capable of correct initiation at several other sites which have been shown by capping analysis to serve as initiation sites in vivo. The sites investigated include two initiation sites upstream of oli-I and sites upstream of the tRNA$^{f\text{-}met}$ gene.

As well as investigating promoters shown to function in vivo, we analyzed two regions which appear not to contain an in vivo initiation site. Capping analysis fails to reveal an initiation site lying between the oxi-3 and oli-II genes. Plasmid pBLT778 contains EcoRI fragment 7 from strain MH41-7B[15]. This restriction fragment contains the 3' end of the oxi-3 gene, the 5' end of the oli-II gene and the entire intergenic region. pBLT778 was used as a template for transcription using either transcription in the absence of CTP or run off transcription as an assay for specific initiation . No evidence was found for initiation at a unique site on this fragment of DNA. In a similar manner, we investigated transcription from the region between the oli-I, $tRNA^{ser}$ and var-I genes. There is no evidence from capping analysis for initiation within this region in vivo. Plasmid 21D, a gift from D. Miller, contains the region from upstream of oli-I through the tRNA gene and to the middle of var-I. We have found no evidence for initiation at unique sites within this region using in vitro transcription as a functional promoter assay.

The template dependent transcription system allows an unique opportunity to test directly which sequences are required for correct initiation. We used the exonuclease Bal 31 to generate a collection of deletion mutants of a 14S rRNA promoter carrying plasmid. The details of the construction of the deleted plasmids will be reported elsewhere. The analysis rests on three of the generated mutants. Mutant 114, deleted to position -12 from the transcriptional initiation site with the insertion of an EcoRI linker, is capable of directing correct initiation, either when intact or when digested with EcoRI at the site of deletion. Thus no sequences upstream of position -12 are required for correct initiation. Indeed, even the physical presence of DNA is not required since the EcoRI digested plasmid can serve as a template. In contrast, plasmid 13 , deleted to position -9 from the transcriptional, initiation site with the insertion of a SmaI linker, is incapable of directing correct initiation, either when intact or when digested with SmaI at the site of deletion. These results, in good agreement with the interpretation of the consensus promoter seqence, suggests that the border of the sequences required for promoter function lies between position -9 and -12.

To test whether sequences upstream of -12 could affect transcription, a GC rich fragment of DNA was inserted at the EcoRI site of plasmid 114 to generate a plasmid denoted D31. This clone, although containing all the sequences present in 114, is incapable of supporting correct transcriptional initiation. Thus, while sequences upstream of -12 are not required for correct initiation, certain sequences in this position can prevent correct transcriptional initiation.

The results of capping analysis and in vitro transcription allow us to draw several conclusions about the nature of yeast mitochondrial transcription. First, the major sites of transcriptional initiation are at a few well defined sites spread widely throughout the genome. Second, the conservation of a short sequence at each of the initiation sites identifed to date plus the results of in vitro transcription analysis of deleted promoter-bearing plasmids suggests that this short sequence is an essential component of the yeast mitochondrial promoter. Third, the fact that the same purified polymerase preparation is capable of correct initation at sites serving for transcription of rRNA, tRNA, and mRNA plus the conservation of the same sequence at all sites of initiation indicate that a single mitochondrial RNA polymerase is used for transcription of all classes of mt RNA. Finally, the absence of initiation sites identified either by capping or by a functional promoter assay in vitro indicates that certain sets of genes are transcribed from a single site to yield multi-gene transcripts which must be processed to mature RNA species.

Mutant Analysis. A third approach we have been taking to the study of yeast mitochondrial transcription is through the generation and analysis of a collection of nuclear mutants which are temperature sensitive (ts) for mitochondrial function. Using a modification of the inositol selection procedure of Henry (16), we have generated a collection of about 1000 independent ts pet mutants. Of approximately 1000 independent ts pet mutants screened by northern analysis, 21 were found to have major alterations in the RNA content when grown at restrictive temperature. These 21 isolates fall into 14 complementation groups. The mitochondrial RNA from one representive of each complementation group grown on galactose at both permissive and retrictive temperature is illustrated in figure 4. Twelve of the complementation groups appear to lose all RNA in roughly equal a mounts,

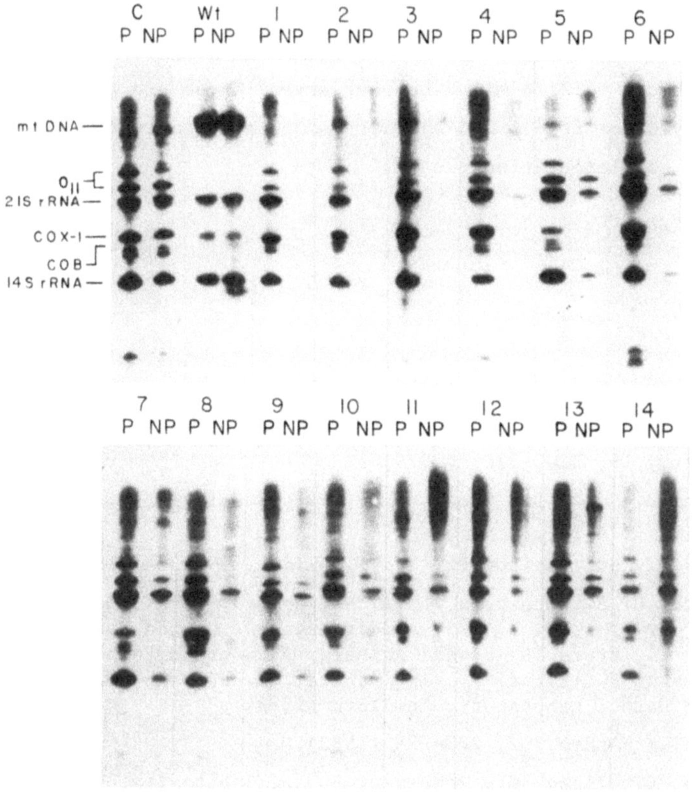

Figure 4. Northern hybridization analysis of RNA from representatives of each complementation group. Total cellular nucleic acids were isolated from each mutant after growth on 2% galactose at permissive or non-permissive temperature. Samples of equal O.D. at =260 were loaded on a 1.5% agarose, 6 M urea gels. After transfer, the filters were probed with labeled total mitochondrial DNA. C: ts pet mutant which show no effect on mtRNA levels. Wt: wild type. 1-14: representatives of each complementation group which has altered RNA pattern after growth at high temperature. P and NP: permissive and non-permissive growth conditions, respectively.

one group, labeled 14, represented by 4 independent isolates in our collection, appears to specifically lose the 14S rRNA transcript with other transcripts remaining about normal. One group labeled 5 represented by 2 independent isolates in our collection appears to specifically lose the cob transcript. None of the isolated mutants are quantitatively converted to petites during growth at high temperature as would be predicted for a polymerase mutant assuming the polymerase is involved in DNA replication.

The rate of petite formation is increased in a few of the isolates. We are currently characterizing the RNA polymerase activity isolated from these mutants in order to determine whehther any of these are actually mutants in the RNA polymerase itself and whether specificity factors may be altered in some of the complementation groups.

References

1. Grivell, L. A.: Sci. Am. 248, 78-89 (1983).
2. Jakovcic, S., Hendler, F., Halbreich, A. and Rabinowitz, M.: Biochemistry 15, 3200-3205 (1979).
3. van Ommen, G. J. B., Groot, G. S. P. and Grivell, L. A.: Cell 18, 511-523 (1979).
4. Locker, J. and Rabinowitz, M.: Plasmid 6, 302-314 (1981).
5. Grivell, L. A., Hensgens, L. A. M., Osinga, K. A., Tabak, H. F., Boer, P. H., Crusius, J. B. A., van der Laan, J. C., de Haan, M., van der Horst, G., Evers, R. F. and Arnberg, A. C.:in Mitochondrial Genes, Slonimski, P., Borst, P. and Attardi, G. Eds. pp. 225-239, Cold Spring Harbor Laboratory, New York (1982).
6. Moss, T.: Nature 302, 223-228 (1983).
7. Levens, D., Ticho, B., Ackerman, E. and Rabinowitz, M.: J. Biol. Chem. 256, 5226-5232 (1981).
8. Christianson, T., Edwards, J., Levens, D., Locker, J. and Rabinowitz, M.: J. Biol. Chem. 257, 6494-6500 (1982).
9. Christianson, T. and Rabinowitz, M.: J. Biol. Chem. in press (1983).
10. Edwards, J. C., Levens, D. and Rabinowitz, M.: Cell 31, 337-346 (1982).
11. Goursot, R., Mangis, M. and Bernardi, G.:EMBO J. 1, 705-711 (1982).
12. Christianson, T., Edwards, J. C., Mueller, D. and Rabinowitz, M.: Proc. Natl. Acad. Sci. USA in press (1983).
13. Donis-Keller, H., Maxam, A. M. and Gilbert, W.: Nucl. Acids Res. 4, 2527-2538 (1977).
14. Silberklang, M., Gilliam, A. M. and RajBhandary, J. L.: Meth. Enzymol. 59, 58-109 (1979).
15. Berg. P. E., Lewin, A., Christianson, T. and Rabinowitz, M.: Nucl. Acids Res. 6, 2133-2150 (1979).
16. Henry, S. A., Donahue, T. F. and Culbertson, M. R.: Molec. Gen. Genet. 143, 5-11 (1975).

INITIATION OF TRANSCRIPTION OF YEAST MITOCHONDRIAL DNA

Henk F.Tabak, Klaas A.Osinga, Eric De Vries, Alexander M.Van der Bliek, Gijsbertus T.J.Van der Horst, Marian J.A.Groot Koerkamp, Gerda Van der Horst, Ellen C.Zwarthoff and Marcy E.MacDonald

Section for Molecular Biology, Laboratory of Biochemistry, University of Amsterdam, Kruislaan 318, 1098 SM Amsterdam, The Netherlands

Introduction

We are interested in regulatory aspects of yeast mtDNA transcription and have initially selected the ribosomal RNA (rRNA) genes as examples for our studies. We expected that at a certain stage of gene expression coordination in synthesis of mature rRNA could be exerted to obtain equal contributions of both rRNAs for assembly in the mitochondrial ribosome. The genes for the large and small rRNA are far apart so that processing from a common precursor RNA is unlikely. One way to achieve coordinate gene expression would be to transcribe both genes from identical promoters. We therefore determined the DNA sequences around the positions at which the 5'-end of the mature large rRNA and the 5'-end of the precursor for 15S rRNA mapped [1,2]. DNA sequence comparison revealed an identical nonanucleotide sequence (ATATAAGTA) preceding both genes and this turned out to be conserved at the same positions in front of the rRNA genes of the distantly related yeast *Kluyveromyces lactis* [3]. The application of the guanylyl transferase test

---

Abbreviations: bp, base pair(s); kb, kilo base pair(s); mRNA, messenger RNA; rRNA, ribosomal RNA.

Mitochondria 1983

revealed further that several yeast mtRNAs contained 5' di- or triphosphate ends as earmark of de novo initiation [4], among them the transcripts of the rRNA genes. The 5' terminal sequence deduced from sequence analysis of [^{32}]GTP-capped RNAs [3-5] established that in the four rRNA genes of Saccharomyces cerevisiae and K. lactis initiation of transcription occurred at the last A of the box.

The box occurs at several positions in the neighbourhood of genes and in some putative origins of replication [3]. Here we discuss that in all cases - rRNA genes, tRNA genes, protein-encoding genes - as well as origins of replication, initiation of transcription takes place at the last conserved nucleotide of this DNA sequence motif. We have introduced mutations in and around the boxes preceding the rRNA genes using in vitro site-directed mutagenesis with specific DNA fragments in an attempt to define their role in initiation of transcription. Several mutated genes have been constructed and tested for correct initiation in an in vitro reconstitution assay using mtRNA polymerase purified according to Levens et al. [6]. Surprisingly, mutations in the box have no influence on the specificity of initiation, but mutations downstream of the box do.

Transcription of mitochondrial genes

rRNA genes - We have determined the DNA sequence around positions at which rRNA synthesis starts in S. cerevisiae and K. lactis mtDNA and have found a conserved nonanucleotide sequence (ATATAAGTA) preceding the initiation sites (Fig. 1). To study whether this sequence motif contributes to the specificity of initiation observed in vivo, we make use of the in vitro transcription system developed by Edwards et al. [7]. As an example of the specificity of initiation of RNA synthesis in vitro we show in Fig. 2 the run-off transcripts obtained when the cloned 21S rRNA gene is provided as template. The success of this reconstituted, homologous transcription system allows

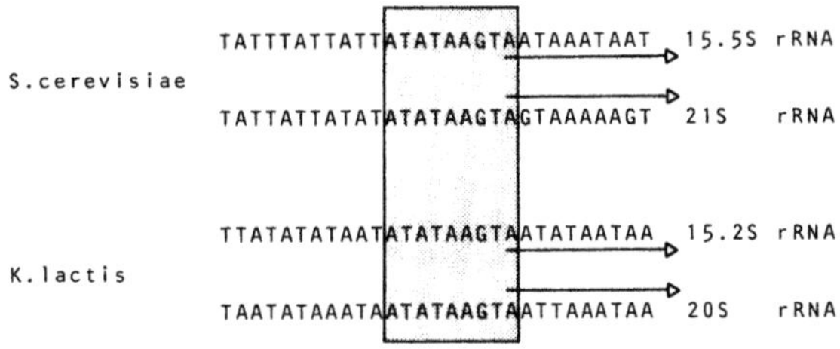

Fig. 1. Sequence comparison of 5'-ends of rRNA genes in S. cerevisiae and K. lactis.

us to use this test for two separate purposes:

1. Search for new RNA initiation sites.

2. Delineation of the DNA sequences essential for faithful *in vitro* transcription initiation.

When we apply this approach to the transcriptional analysis on cloned rRNA genes of K. lactis, we find no evidence for correct initiation with the S. cerevisiae mtRNA polymerase, not even when the rRNA genes of S. cerevisiae and K. lactis

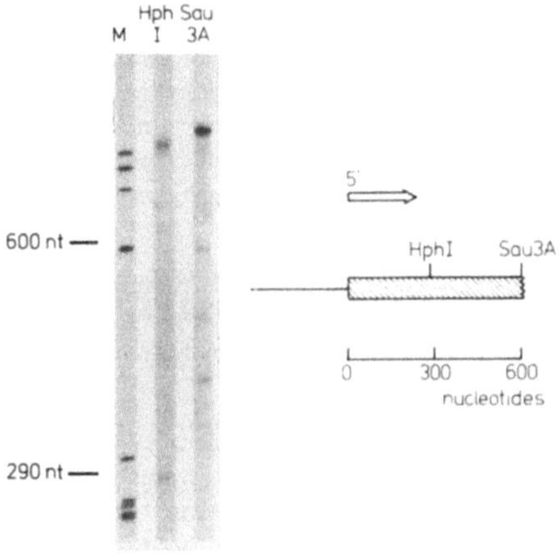

Fig. 2. *In vitro* transcription of the cloned 21S rRNA gene of S. cerevisiae. DNA was cut with Hph1 or Sau3A and the run-off transcript analysed by polyacrylamide gel electrophoresis.

are combined on one plasmid (not shown). In that case we see accurate initiation of the *S. cerevisiae* gene, but not on the *K. lactis* one. We conclude that the presence of the box in front of a gene *per se* does not suffice to specify correct initiation and that other regulatory elements, such as (species) specific DNA sequences and/or protein factors may exist.

Protein encoding genes - The nonanucleotide sequence is also present (Fig. 3) in front of the cytochrome *c* oxidase subunit I and ATPase subunit 9 genes [3]. In the latter case, two boxes are present (Fig. 4) and in collaboration with the group of Dr. M.Rabinowitz (University of Chicago) we have shown that RNA initiation takes place within each box *in vivo* and *in vitro*, although the first box is far more efficient than the second [8]. We have determined the DNA sequence upstream of

rRNA genes			
TATTATTATTATAT	ATATAAGTA	GTAAAAAGTAGAATAA	21S rRNA
ATTTATTTATTATT	ATATAAGTA	ATAAATAATAGTTTTA	15S rRNA
tRNA genes			
AAAGATATAAATAA	TTATAAGTA	TATAAAGTAGTAAAGG	tRNA-cys
ATATATATATATTT	TTATAAGTA	GTATATTATATTTTAT	tRNA-thr
ATATATATATTATA	TTATAAGTA	ATAATAAGTATTATAT	tRNA-phe
ATATATATTAAATT	TTATAAGTA	ATATAATATAAGTATT	tRNA-fMet box 1
TTATAAGTAATATA	ATATAAGTA	TTAATTATATAAATGC	tRNA-fMet box 2
protein encoding genes			
TTTTATTAATATTA	ATATAAGTA	ATATATATAGTTTATG	ATPase 9 box 1
TCTTATACACATTT	ATATAAGTA	TATATATATATTATTA	ATPase 9 box 2
AATAATAAGTATTT	ATATAAGTA	ATAGATATAATAATAA	CO I
replication origins			
TTTTATTTCTAGAT	ATATAAGTA	ATAGGGGGAGGGGGTG	rep-1
TTTATATTTAAAAT	ATATAAGTA	ATGGGGGAGGGGGTGG	rep-2
ATAAAT	ATATAAGTA	ATAAATTAAGTTTTAT	ori-2
TTATATTTATAAAT	ATATAAGTA	ATAAACTAAGTTTTAT	ori-1

Fig. 3. DNA sequence comparison of initiation sites for transcription. rRNA genes [3]; $tRNA_{Cys}$ [25]; $tRNA_{Thr1}$ [26]; $tRNA_{Phe}$ [27]; and $tRNA_{fMet}$ (Martin, N.C., personal communication); ATPase subunit 9 [28]; rep [11]; and ori [12].

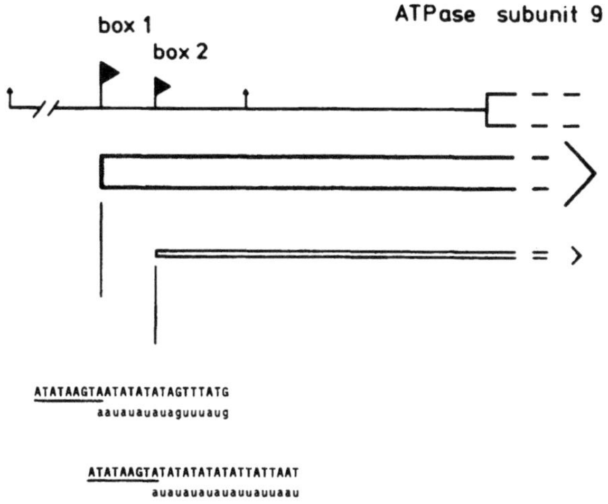

Fig. 4. Schematic presentation of the initiation of transcription of the gene coding for ATPase subunit 9. After ref. 8.

the gene coding for cytochrome c oxidase subunit I and found the box at 540 bp before exon 1 in a position at which the 5'-end of the messenger RNA (mRNA) is located. When the cloned mtDNA segment containing this box is used as a template, specific initiation in the *in vitro* transcription test is observed (not shown). These results imply that also mitochondrial protein genes share this common DNA sequence element.

tRNA genes - The conserved DNA sequence ATATAAGTA or a closely-related sequence TTATAAGTA is found in front of several tRNA genes. For instance, $tRNA_{Cys}$, $tRNA_{Thr1}$ (opposite strand), $tRNA_{Phe}$ and twice in front of the $tRNA_{fMet}$ (the DNA sequence around $tRNA_{fMet}$ was made available by Dr. N.C.Martin) as illustrated in Fig. 3. Edwards *et al.* (this Volume) have shown that initiation of $tRNA_{fMet}$ synthesis initiates within the ATATAAGTA as well as within the TTATAAGTA box both located in front of this gene.

It is interesting to note that most tRNA genes have a T

instead of an A in the first position of the box, which may distinguish them as a group from the rest of the mitochondrial genes in this respect. Moreover, the box occurs at some distance upstream of the 5'-end of the gene encoding the mature tRNA sequence. This is in line with the observed precessing of tRNA precursors [9,10] and may explain the absence of this box in front of other tRNAs, which are processed from common tRNA precursors. Taken together, the results suggest that transcription of all mitochondrial genes is initiated at the ATATAAGTA sequence (rRNA, mRNA) or at a very close variant TTATAAGTA of it (tRNA).

Replication origins - Hypersuppressive petite mutants share a common DNA sequence motif of approximately 300 bp consisting of three GC-rich clusters interlaced with AT-rich DNA. This could constitute a replication origin which would nicely account for the hypersuppressive character of these petite mutants [11,12]. To our surprise we also found the nonanucleotide box in such putative replication origins (Fig. 3). One such a cloned replication origin (rep-2 in clone pSCM128: kindly provided by Drs H.Blanc and B.Dujon) is proficient in RNA synthesis and initiation takes place within the nonanucleotide sequence (Fig. 5). This correlates well with the situation observed in vivo. The RNA 5' terminal sequence determined from a RNA species capped with guanylyl transferase fits exactly with the DNA sequence of rep-2, assuming a start at the last conserved nucleotide of the box [3]. S_1 nuclease protection experiments show that 5'-ends of RNA isolated from some hypersuppressive petite mutants map at the position of the box [13]. We therefore believe that RNA synthesis from replication origins functions in priming of DNA synthesis and that this RNA is also initiated at the box.

Transcription of mutated templates

We have used in vitro, site-directed mutagenesis with synthetic

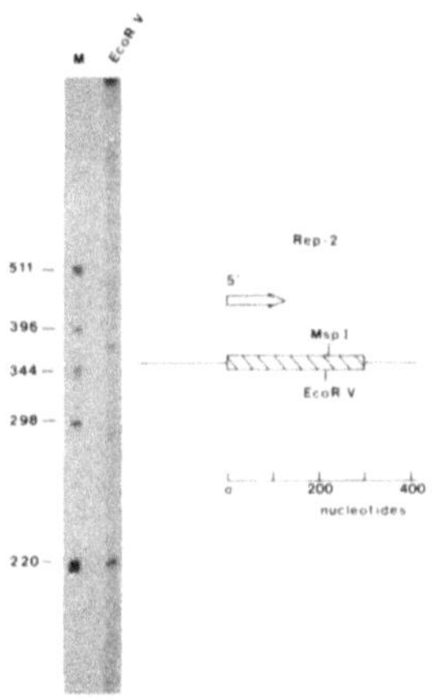

Fig. 5. _In vitro_ transcription of a cloned replication origin. DNA was cut with EcoRV and the run-off transcript analysed by polyacrylamide gel electrophoresis.

DNA fragments to introduce mutations in and around the nonanucleotide box [14]. Apart from the selected alterations aimed at, we have also found unexpected mutations, mainly large deletions due to pairing of the DNA fragments with non-adjacent DNA sequences of the mtDNA (also _vide infra_). When we used these mutated templates (Fig. 6) for _in vitro_ transcription with purified mtRNA polymerase, the results were unexpected and one mutant may suffice to illustrate our surprise (Fig. 7): Substitution of the unique G in the nonanucleotide sequence for a T has no effect on the specificity of initiation of transcription. Also the efficiency of transcription does not seem to be affected (not shown). Only with a template containing two alterations (a T to A substitution at +10 and a T to C at +13) we find in preliminary experiments that the expected run-off transcript is missing. It may be the first mutant deficient in _in vitro_ initiation of transcription and it suggests that DNA sequences downstream of the box rather than the box itself are important for the _in vitro_ initiation of transcription.

Mutants in the 21S rRNA gene

```
TATTATTATTATAT ATATAAGTA GTAAAAAGTAGAATAA     wildtype
TTATTATTATTATT ATATAAGTA GTAAAAAGTAGAATAA     del. A at -10
              *
TATTTATTATTATT ATATAAGTA GTAAAAAGTAGAATAA     del. TATA at -13
              *
TATTATTATTATAT ATATAATTA GTAAAAAGTAGAATAA     sub. G/T at -2
                      *
TTATTATTATTATT ATCTAAGTA GTAAAAAGTAGAATAA     sub. A/C at -6
              *    *                          del. A at -10
```

Mutants in 15S rRNA gene

```
ATTTATTTATTATT ATATAAGTA ATAAATAATAGTTTTA     wildtype
ATTTATTTATTATT ATATAAGTA ATAAATAAAAGCTTAA     sub.T/A at +10
                                 *  *         sub.T/C at +13
```

Fig. 6. Mutants obtained by in vitro, site-directed mutagenesis with synthetic oligonucleotides.

Self-splicing of large rRNA?

The nuclear-encoded large rRNA of *Tetrahymena* can be spliced non-enzymatically *in vitro* [15]. In view of the features common to the introns of the *Tetrahymena* gene and the mitochondrial large rRNA gene [16], it was of interest to investigate whether self-splicing could also occur in the case of mitochon-

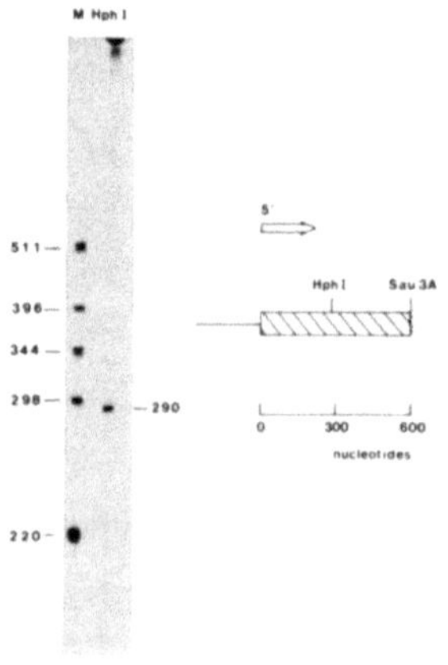

Fig. 7. *In vitro* transcription of the 21S rRNA gene containing a G to T substitution within the box. DNA was cut with HphI and the run-off RNA was analysed by polyacrylamide gel electrophoresis.

drial large rRNA. We have cloned part of the intron containing 21S rRNA gene (SalIxHaeIII fragment) behind the phage lambda promoter present in the vector pHUB 2 (Fig. 8) and used this clone for in vitro transcription with Escherichia coli RNA polymerase. Cech et al. [15] have shown that in such conditions efficient self-splicing takes place. To demonstrate the presence of spliced RNA we have used a synthetic DNA fragment with a sequence complementary to the RNA exon boundaries (Fig. 9). This DNA fragment, labelled with ^{32}P at its 5'-end, is protected against S_1 nuclease degradation when hybridized to the in vitro-synthesized RNA (Fig. 10). That this is indeed due to protection and not to escape from degradation was shown by the predicted shortening of the fragment size due to the deliberate addition of a non-base paired extension (Fig. 9).

However, in view of our results with DNA fragments in site-directed mutagenesis (resulting not only in the desired mutations but also to the introduction of deletions up to 1 kb), this result could be explained by hybridization of the fragment to RNA sections derived from the exons with the folded intron RNA still looping out from this short hybrid. When a DNA fragment complementary to the RNA corresponding to the 3' exon is used as primer for reverse transcriptase on in vitro-synthesized RNA as template in the presence of ddNTPs, we find a unique DNA sequence ladder leading from 3' exon into the intron (not shown). We conclude that thus far we have found no indications for self-splicing of large rRNA. Moreover, the results show that the proposed splicing test [18], which

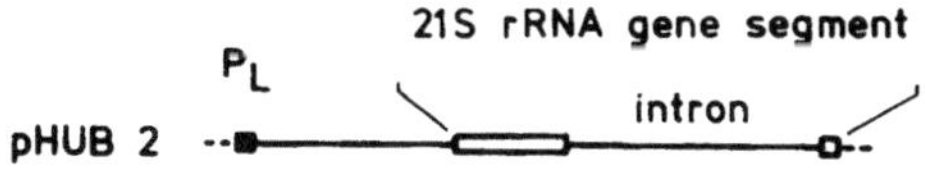

Fig. 8. Schematic presentation of the cloned 21S rRNA gene segment in the phage lambda promoter containing vector pHUB-2. The HaeIIIxSalI 21S rRNA gene segment was cloned in the EcoRI site of the vector pHUB-2 [17].

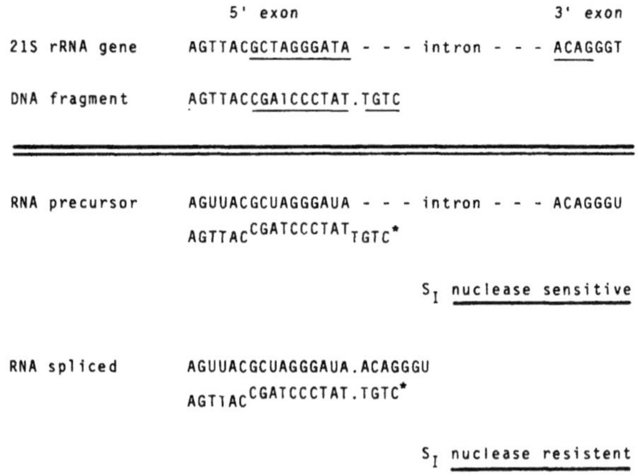

Fig. 9. Design of the test to demonstrate spliced 21S rRNA using a synthetic DNA fragment. The underlined sequence of the fragment is complementary to the RNA flanking the intron. The six nucleotides at the 3'-end are identical to the RNA sequence and cannot base pair with the rRNA. The asterisks symbolize the ^{32}P at the 5' terminal C.

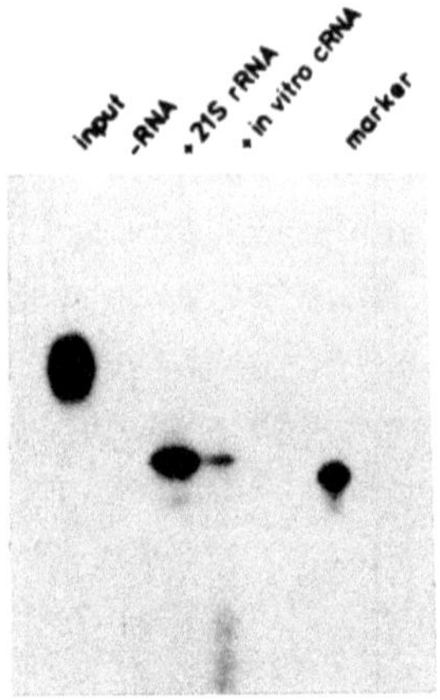

Fig. 10. S_1 nuclease protection analysis of the synthetic DNA fragment hybridized with various RNAs. The marker is a fragment of 14 nucleotides.

is based on the principles outlined in Fig. 9, can only be used with caution.

Discussion

We have focussed our attention on the initiation of transcription of yeast mtDNA with particular interest in the role of the conserved nonanucleotide sequence (ATATAAGTA) in this process. We first identified this box in front of the rRNA genes of S. cerevisiae and K. lactis and show here that it is an element common to many genes, including those for rRNA, some tRNAs and at least two proteins. It is also present in origins of replication (Fig. 11). In some cases this box is conspicuous by its absence. Some tRNAs ($tRNA_{Thr2}$ and $tRNA_{Ser}$) lack the box; the DNA sequence between the gene for cytochrome c oxidase subunit I and the gene for ATPase subunit 6 (DNA

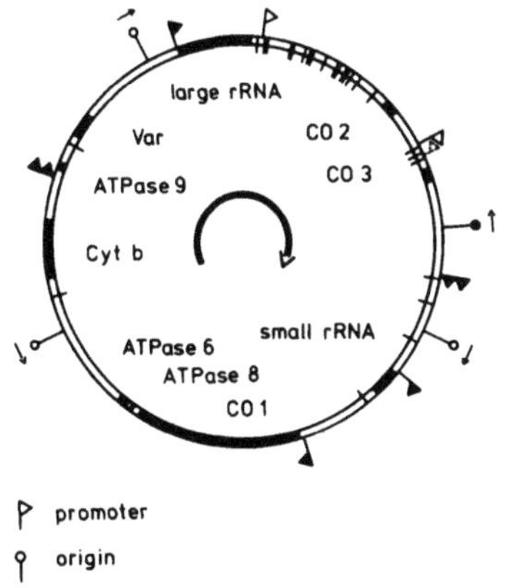

Fig. 11.

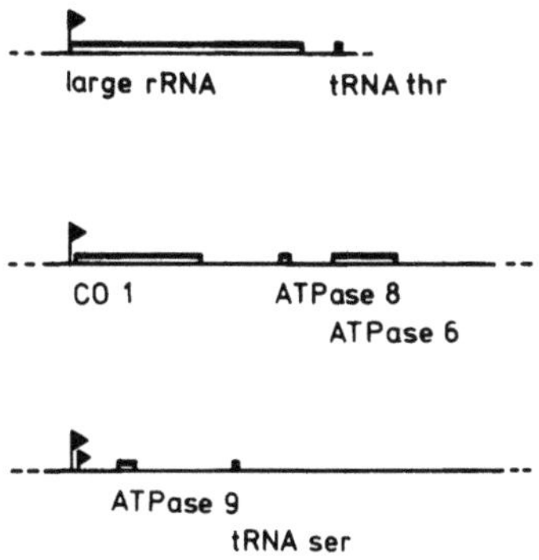

Fig. 12.

Fig. 11. Physical map of yeast mtDNA with initiation sites of transcription and replication origins. Filled in flags are proven sites of initiation. Open flags are positions containing a box. Circles indicate replication origins with the arrow showing the direction of replication. For results on $tRNA_{fMet}$ transcription see Edwards et al. (this volume); for replication origins, see refs 11 and 12.

Fig. 12. Possible multi-gene transcription units. For symbols, see Fig. 11.

sequence kindly made available by Dr. A.W.Linnane) does not contain a box. Such genes are probably transcribed as common precursors initiated at the box of the upstream gene (Fig. 12). The $tRNA_{Thr2}$ is part of the transcription unit with the 21S rRNA gene [19]. The localization of RNAs transcribed from the region containing the ATPase subunit 6 gene [20] shows that the 5'-ends of some of these RNAs map very close to the position at which the mRNA for cytochrome c oxidase subunit I ends. This is most easily explained by assuming processing from a common precursor RNA in which case the genes for cytochrome c oxidase subunit I, ATPase subunit 6 and the newly discovered gene for ATPase subunit 8 in between these two (Linnane, A.W., personal communication) would also belong to one transcription unit. Analysis of RNAs revealed common transcripts hybridizing with both ATPase subunit 9 and $tRNA_{Ser}$ DNA probes [21]. In this case a common transcription unit encoding several genes would be even more interesting in view of the two independent possibilities for initiation of transcription before the ATPase subunit 9 gene both coinciding with boxes. It may allow differential transcription in a way analogous to differential synthesis of rRNA and mRNA in human mitochondria from two closely-located promoters [22,23]. We have shown that the highly purified mtRNA polymerase preparation is capable of correctly transcribing representatives of all classes of mitochondrial genes. It may well be, therefore, that one enzyme is responsible for all the transcriptional events on mtDNA and alludes to the pivotal role that RNA polymerase plays in expressoon of this genome.

The consistency with which the nonanucleotide motif is found at the start of transcription units indicates that it plays some regulatory role in mitochondrial gene expression. In all cases, transcription initiates at the last A of this sequence,it is exclusively present near 5'-ends of genes and where absent this can be reasonably explained by the existence of multi-gene transcription units. We therefore suggested earlier that it could constitute part of a promoter [3].

However, this is not supported by our in vitro mutational and comparative studies: The K. lactis rRNA gene is not specifically transcribed in vitro, although it has the box at the correct position; identical boxes in front of the ATPase subunit 9 gene are very different in the efficiency of transcription; we have not yet found mutations in the box that are detrimental to the initiation of transcription. In this respect our study poses more questions than it answers. The first mutant defective in in vitro initiation has alterations in the DNA sequence immediately downstream of the box. We therefore will focuss our attention also to site-directed mutagenesis of gene-internal regions.

It may be that the box forms part of a negative control circuit and formally can be better compared to the bacterial operator. A comparison with the SV40 enhancer sequence also springs to mind. This DNA element, although much longer, increases transcription, but its action cannot be demonstrated in vitro [24]. The box could be an important element for initiation of transcription, but it may be difficult to demonstrate its function in an in vitro transcription system. We hope to obtain further information from a transcriptional analysis of our collection of mutated templates.

Acknowledgements

We thank Prof. J.H.Van Boom and Mr. G.H.Veeneman (Organic Chemistry, State University Leiden) for synthesis of the various DNA fragments used in this study; Drs B.Dujon and H.Blanc (Centre de Génétique Moléculaire du C.N.R.S., Gif-sur-Yvette, France) for the clone containing the replication origin; Dr. N.C.Martin (The University of Texas Health Science Center, Dallas, Texas, USA) and Prof. A.W.Linnane (Laboratory of Biochemistry, Monash University, Clayton, Victoria, Australia) for providing unpublished DNA sequence data. This work was supported in part by a grant to L.A.Grivell from The Netherlands Foundation for Chemical Research (SON) with financial aid from The Netherlands Organization for the Advancement of Pure Research (ZWO).

References

1. Osinga, K.A., Evers, R.F., Van der Laan, J.C., Tabak, H.F.: Nucl. Acids Res. 9, 1351-1364 (1981).

2. Osinga, K.A., Tabak, H.F.: Nucl.Acids Res. 10, 3617-3626 (1982).

3. Osinga, K.A., De Haan, M., Christianson, T., Tabak, H.F.: Nucl.Acids Res. 10, 7993-8006 (1982).

4. Levens, D., Ticho, B., Ackerman, E., Rabinowitz, M.: J.Biol.Chem. 256, 5226-5232 (1981).

5. Christianson, T., Edwards, J., Levens, D., Locker, J., Rabinowitz, M.: J.Biol.Chem. 257, 6494-6500 (1982).

6. Levens, D., Lustig, A., Rabinowitz, M.: J.Biol.Chem. 256, 1474-1481 (1981).

7. Edwards, J.C., Levens, D., Rabinowitz, M.: Cell 31, 337-346 (1982).

8. Edwards, J.C., Osinga, K.A., Christianson, T., Hensgens, L.A.M., Janssens, P.M., Rabinowitz, M., Tabak, H.F.: Cell (1983) in press.

9. Martin, N.C., Underbrink-Lyon, K.: Proc.Natl.Acad.Sci.U.S. 78, 4743-4747 (1981).

10. Frontali, L., Palleschi, C., Francisci, S.: Nucl.Acids Res. 10, 7283-7293 (1982).

11. Blanc, H., Dujon, B.: in: Mitochondrial Genes (Slonimski, P., Borst, P., Attardi, G., Eds), Cold Spring Harbor Laboratory, Cold Spring Harbor, 1982, pp. 279-294.

12. Bernardi, G.: TIBS 7, 404-408 (1982).

13. Baldacci, G., Bernardi, G.: EMBO J. 1, 987-994 (1982).

14. Zoller, M.J., Smith, M.: Nucl.Acids Res. 10, 6487-6500 (1982).

15. Cech, T.R., Zaug, A.J., Grabowski, P.J.: Cell 27, 487-496 (1981).

16. Lewin, R.: Science 218, 1293-1295 (1982).

17. Bernard, H..U., Remaut, E., Vickers Hershfield, M., Das, H.K., Helinski, D.R., Yanofsky, C., Franklin, N.: Gene 5, 59-76 (1979).

18. Tabak, H.F., Van der Laan, J., Osinga, K.A., Schouten, J.P., Van Boom, J.H., Veeneman, G.H.: Nucl.Acids Res. 9, 4475-4483 (1981).

19. Locker, J., Rabinowitz, M.: Plasmid 6, 302-314 (1981).

20. Cobon, G.S., Beilharz, M.W., Linnane, A.W., Nagley, P.: Current Genet. 5, 97-107 (1982).

21. Martin, N.C., Miller, D.L., Najarian, D.R., Underbrink-Lyon, K.: Abstr. Cold Spring Harbor Meeting on RNA Processing, Cold Spring Harbor, 1983, p. 117.

22. Montoya, J., Christianson, T., Levens, D., Rabinowitz, M., Attardi, G.: Proc.Natl.Acad.Sci.U.S. 79, 7195-7199 (1982).

23. Tabak, H.F., Grivell, L.A., Borst, P.: in: CRC Critical Reviews in Biochemistry (Fasman, G.D., Ed.), CRC Press, Boca Raton, 1983, in press.

24. Dynan, W., Tjian, R.: TIBS 7, 124-125 (1982).

25. Bos, J.L., Osinga, K.A., Van der Horst, G., Borst, P.: Nucl.Acids Res. 6, 3255-3266 (1979).

26. Li, M., Tzagoloff, A.: Cell 18, 47-53 (1979).

27. Miller, D.L., Martin, N.C., Pham, H.D., Donelson, J.E.: J.Biol.Chem. 254, 11735-11740 (1979).

28. Tzagoloff, A., Nobrega, M., Akai, A., Macino, G.: Current Genet. 2, 149-157 (1980).

FUNCTIONS OF NON-GENIC DNA IN YEAST MITOCHONDRIA

Hans Peter Zassenhaus and Ronald A. Butow
Department of Biochemistry
Division of Molecular Biology
University of Texas Health Sciences Center at Dallas
Dallas, Texas 75240

Introduction

The mitochondrial (mt) genome of Saccharomyces cerevisiae is some 75,000 bp in length yet appears to contain about the same number of genes as the mitochondrial genome of animal cells, which is only one fifth as large. Two major features, unique to the yeast mt genome, account for the disparity in genome size: (1) a few genes are interrupted by long intervening sequences some of which contain open reading frames that encode splicing functions (1-3), and (2) the genes themselves are spaced some distance apart, separated by "non-genic" DNA. Clearly, the densely packed configuration of genes on the mt genome of animal cells, with few or no nucleotides separating the end of one gene from the start of the next (4), argues that gene expression does not require long intergenic regions. Our studies concerning the structure and expression of the varl gene on the yeast mt genome have led to unexpected findings that may provide insight into the role non-genic sequences play in the genetic function of mt DNA.

Results and Discussion

In many respects the structure of the varl gene is characteristic of intergenic regions. The contrast between such regions, termed spacer DNA, and gene sequences was first pointed out several years ago by

Mitochondria 1983

Bernardi and co-workers (5). Their model proposes that the spacer DNA is highly AT-rich (about 95%) whereas the genes are of modest G+C content (20-30%). Interspersed in the A+T spacer sequences are short (25-100 bases), highly G+C rich (50-90%) regions, termed GC clusters, many of which contain multiple HpaII and HaeIII sites. Subsequent DNA sequencing of large portions of the yeast mitochondrial genome has largely upheld this picture of the genome organization. However, our recent sequence analysis of the var1 gene (6) has revealed a surprising contrast to this general view. The var1 gene is represented by an uninterrupted reading frame 1188+ nucleotides long that is 90% A+T. It contains one or two GC clusters (depending upon the particular allele) and is separated from its nearest neighboring gene on the 5' side by some 600 bp and on the 3' side by more than 5000 bp. The base composition of the gene and the clustering of G+C residues within it are essentially indistinguishable from the flanking, non-genic sequences; yet, the var1 gene encodes a "proper" protein that is a stoichiometric component of the 38S mitochondrial ribosomal subunit (7,8). Of the genes on mtDNA, var1 is unique not only because of its extremely high A+T content and different codon usage, but also because it is the only major protein gene harboring GC clusters that are both transcribed and translated (9). The var1 gene is not unique, however, when compared to the sequence organization of spacer DNA and, as discussed below, this resemblance extends as well to its transcription.

We suggest that the similarity in sequence organization between var1 and spacer regions is not simply coincidental but rather reflects the evolutionary origin of the var1 gene. As diagrammed in Fig. 1, we propose that the var1 gene was constructed over time from short, potentially functional domains embedded in otherwise non-genic AT-rich spacer DNA. The recruitment of additional domains to graft onto the var1 proto-gene occurred by recombination between homologous segments of the spacer DNA. These recombinants, which led to the establishment of the gene, included either pre-existing GC clusters or those acquired later in the evolution of var1.

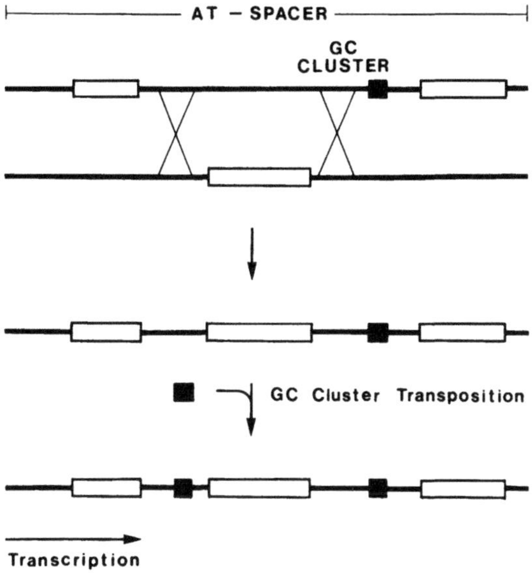

Figure 1. Hypothetical Origin of the Var1 Gene. The open boxes represent coding sequences for the potentially functional domains; the solid boxes represent GC clusters. Diagrammed is one of the products (middle) of recombination between homologous segments of AT-spacer DNA.

This model bears some similarity to the exon shuffling model of Gilbert (10). In the case of var1, however, the recombination joints are not excised as introns but rather are read through as coding information. Support for our model comes from an analysis of the var1 sequences and its comparison with other mtDNA sequences. A computer search of the available sequences of yeast mtDNA reveals many stretches greater than 16 bp long with perfect homology to var1 (L.I. Grossman, personal communication). Furthermore, these segments are scattered about the genome and most are located within the AT-rich intergenic regions.

The intergenic sequences, like those in var1, also show strain dependent length polymorphism. In many portions of the genome, length variation over short distances has been shown by DNA sequencing to be due to a variable number of short, direct repeats composed of pure A+T sequences (11). Such a phenomenon is particularly evident in the var1 gene where alternate alleles contain variable length insertions (termed b1 and b2) composed of multiple AAT repeats embedded in a consecutive run of AAT triplets. These variable regions within var1 share the most sequence homology with the intergenic DNA and may be part of the recombination joints described above. The consequence to var1 is an expansion or contraction in the number of consecutive asparagine residues in the protein.

Some var1 alleles contain an additional insertion, termed the a element, which is an exact, but inverted copy of a 46 bp GC cluster found in all var1 alleles. A cluster of identical sequence, also present in only some yeast strains, is found in the intergenic region between the oli1 and $tRNA^{SerII}$ genes (12), as well as in the AT-rich tRNA processing locus (D. Miller and N.C. Martin, personal communication; here the sequence differs in 5 positions). Thus, GC clusters, which themselves appear to be highly polymorphic among different yeast strains, may also have been recruited as coding information within the evolving var1 gene.

How can the function of the var1 protein tolerate the additional amino acid residues contributed by the a, b1 and b2 elements? In the search for var1 sequences with homology to AT-rich sequences located elsewhere on the genome, we found that most of these homologies were concentrated in the portion of the var1 gene in which the variable elements are located. Since we have been unable to detect any obvious phenotypic difference among strains with alternate var1 alleles, the protein domain encoded by these regions may well be functionally indifferent to their length polymorphism, consistent with the idea that the recombination joints and GC clusters within var1 are coding information.

In this connection, a recent comparison of a given gene product in different species reveals that frequently the product shows length polymorphism associated with regions of the protein corresponding to intron-exon boundaries (13). This observation has been interpreted to suggest that the length polymorphism might be the consequence of a "ragged" elimination at the genomic level of a pre-existing intron or, alternatively, the insertion of an intron that creates (or eliminates) exon sequences as a result of the subsequent splicing reactions. Might those variable regions of the var1 gene containing the a, b1 and b2 elements be evolving toward elimination at the genomic or RNA level, and in effect be considered as incipient introns? Alternatively, we might speculate that these variable regions in the var1 gene are recent recruits, and thus represent new material grafted onto a pre-existing, shorter gene. These sequences still share many properties with the intergenic spacer DNA but, as in the above, serve no required coding function, i.e. they are derelect exon sequences.

The transcription of intergenic sequences is also very similar to that we have observed in our studies of the var1 gene. Our transcriptional analysis of var1 has revealed that the var1 mRNA is made from a much longer, polycistronic transcript (14). The primary transcript initiates several hundred bp 5' to the oli1 gene, located some 2200 bp in front of var1, and includes oli1, $tRNA^{SerII}$ and var1 sequences, in the order 5' to 3', respectively. RNA processing in the intergenic sequences releases the mature mRNAs with correct trimming of the 5' end of the $tRNA^{SerII}$ gene requiring an additional activity encoded in the tRNA processing locus (15). Within the var1 gene sequences, however, we see further RNA processing events and, surprisingly, a pair of "optional" promoters.

By S1-protection experiments, sizing of RNAs on gels, and cDNA sequencing, we have identified where the ends of the var1 RNAs map on the DNA sequence . In this way the processing site in the intergenic region between $tRNA^{SerII}$ and var1 that gives rise to the 16S var1 mRNA has been mapped to the perfect palindromic sequence, AATATAA. This sequence, however, is repeated an additional 15 times in the var1 region (Fig. 2); yet we detect substantial processing activity at only two of these sites, giving rise to the 14S and 13S species (14). The latter RNA, as will be discussed below, also originates as a primary transcript in some varl alleles; neither of these RNAs contain the entire varl gene and so their coding function is doubtful, although we have not ruled out the possibility that they encode some truncated, minor protein species. More likely, it seems that for the varl RNAs, the processing machinery does not discriminate precisely between sites located in intergenic versus gene sequences, although some sites must be preferred over others. Thus, the steady state level of the 16S mRNA probably depends on other factors, e.g. RNA secondary structure or ribosome loading, which determines the extent of cleavage at these potential processing sites.

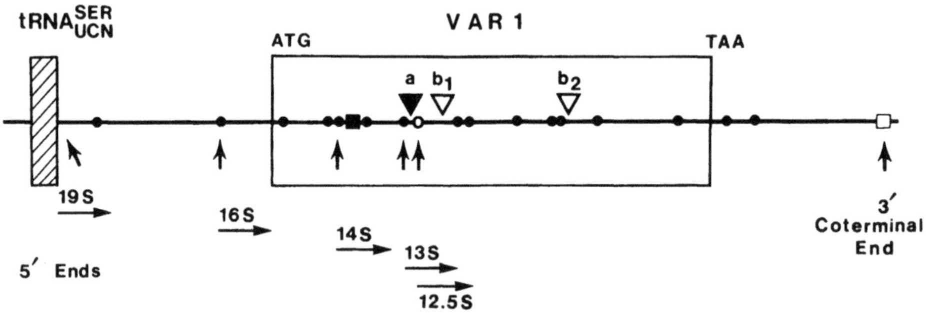

Figure 2. Identified and Potential RNA Processing Sites in the VarI Region. The varl coding region is indicated by the large box. The 5' ends of the varl RNAs are indicated by the horizontal arrows, and the experimentally identified processing and/or initiation sites indicated by the vertical arrows. ●, AATATAA; ○, TTATAAGTA; □, ATTCTTA.

A variety of genetic and biochemical data suggest that intergenic RNA processing is essential for translation. For example, in a varl mutant that accumulates the larger 19S RNA, but contains little 16S mRNA, synthesis of the varl protein is greatly decreased compared to a wild-type where the levels of these RNAs are reversed (16,17). Furthermore, RNAs with the proper 5' end and the entire var1 gene, but which lack the 550 base 3' non-translated region of the 16S RNA, are translationally inactive (18). Thus, the flanking intergenic sequences present on the var1 mRNA, as well as on all other mitochondrial mRNAs, appear critical for translation.

Of the mtDNA promoters identified thus far, all have mapped either immediately preceeding the start of genes or in the upstream intergenic sequences. Within the var1 gene we also find sites active in the initiation of transcription. Surprisingly, these sites appear to be active only in the a^- allele. This is shown in Fig. 3 where, using guanylyltransferase and [α-^{32}P]GTP, we have tested for cappable var1 RNAs isolated from two petite strains, one carrying the var1[40] allele (a^-, A17-10) and the other carrying the var1[44] allele (a^+, 2-33S). The petite 2-33S, in contrast to A17-10, does not retain the entire var1 region and lacks those sequences 3' to the end of the gene; it synthesizes all the equivalent RNAs found in A17-10, but each is shortened at the 3' end. In A17-10, the 13S RNA and a smaller, minor species, the 12.5S RNA, caps but their analogs in 2-33S fail to cap. Indeed, all of the discrete capped RNAs in 2-33S RNA preparations appear to be cytoplasmic RNA contaminants. Fig. 3 also shows our estimates of where the 5' ends of these RNAs map to the var1 DNA sequence. The 5' end of the 12.5S RNA maps in the sequence TTATAAGTA which has been shown, both *in vivo* (19,20) and *in vitro* (20), to be an initiation site for mitochondrial RNA polymerase. The 5' site of the 13S RNA, however, bears no significant resemblence to that sequence, but rather falls within the same sequence, AATATAA, in which RNA processing can occur. Thus, it appears that for both sites, the insertion of the GC cluster abolishes their activity in transcriptional initiation. This is the first demonstration that GC clusters, found scattered throughout the AT-rich regions of the genome, can modulate transcription. These "optional" transcriptional initiation sites, which are unnecessary for generation of the 16S var1 mRNA, flank that region of the var1 gene in which the a element is situated.

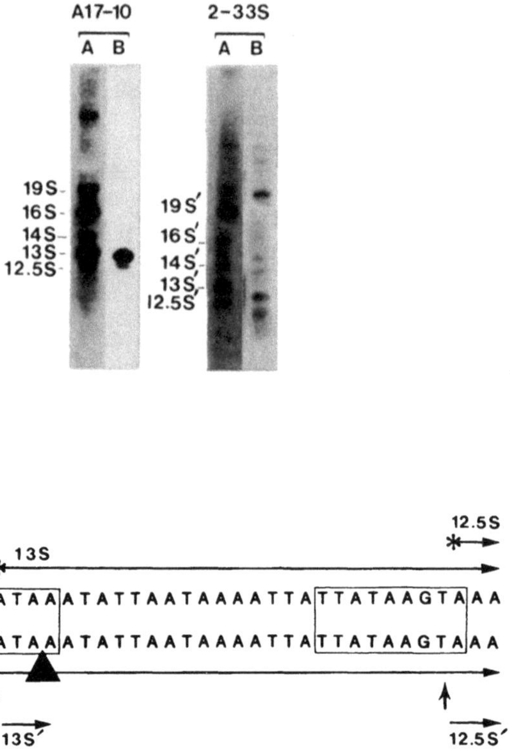

Figure 3. Capping of Var1 RNAs. Top Mt RNAs from petites A17-10 and 2-33S were reacted with [α-^{32}P]GTP in the presence of guanylyltransferase. The products were separated by electrophoresis through 1.5% agarose-6M urea gels and blotted onto DBM paper (lanes B). Unreacted mtRNAs were also fractionated and blotted in parallel (lanes A) and hybridized with cloned var1 DNA The var1 RNAs in 2-33S are each shortened relative to A17-10 (see text).
Bottom Shown is the DNA sequence in A17-10 and 2-33S spanning the 5' ends of the 13S and 12.5S var1 RNAs. ▲, 46 bp GC cluster (the a element); *⟶ RNA with a capped 5' end; ⟶ 5'-uncapped RNA. Vertical arrows indicate RNA processing sites.

Clearly, the promoter sequences and RNA processing sites in the intergenic sequences are important for gene expression. The petite 2-33s, however, does not appear to have a high efficiency promoter in its genome since we detect no cappable RNA corresponding to any of the major mtRNAs present in this petite. Nevertheless, 2-33S synthesizes abundant amounts of mtRNAs that are processed identically to their counterparts found in wild-types and petites retaining the entire varl region (14). How do these RNAs arise in 2-33S? We have shown that cleavage of RNA precursors at processing sites defined in wild-types can and does occur even if the sequences 5' or 3' to the site are non-wild-type, as will occur at the junctions of the petite repeating unit (21). Essentially, whenever an RNA contains such a site, processing can take place. Implicit in this mechanism for the generation of mature RNAs is the absence of a requirement for RNAs initiated and terminated at defined sites. Indeed, in the absence of a wild-type promoter, the mtRNA polymerase may gain entry to gene sequences via the AT-rich intergenic sequences. All promoter elements identified to date, in both procaryotes and eucaryotes, as well as in yeast mitochondria, are composed of short AT-rich sequences. It is not unreasonable to assume, therefore, that the mtRNA polymerase has some affinity for such AT-rich regions. Once initiated, transcription of the circular genome in 2-33S would yield large primary transcripts that could be processed to the discrete species we detect by Northern blot hybridization. Whether such a mechanism also operates in wild-type mitochondria is not known.

Our hypothesis that the varl gene evolved by recruitment of sequences from AT-rich, non-genic DNA of an earlier version of the yeast mt genome raises the question of the presence and origin of a varl cognate in other species. One other well studied example of a mitochondrially translated ribosomal protein is the S-5 small subunit protein of Neurospora crassa (22). Recently, the structural gene for this protein has been tentatively assigned to an open reading frame within the intron of the large mitochondrial rRNA (23). Although varl and S-5 are basic proteins, they share little sequence homology at the DNA or protein level. The absence of any significant homology between the varl DNA sequence and sequences on animal mtDNA (6) suggests further that mitochondrial ribosomal proteins, at least in one case, may have evolved in different organisms in a parallel and independent manner. In other words, although the var1 protein might be unique to yeast, its functional cognate in other organisms could be either a structurally unrelated protein (whether of mitochondrial or nuclear origin) or not exist at all. This apparent flexibility in the assembly of the mitochondrial translation apparatus may be related to the degree of its translational fidelity. Though little data are available on the error rate of mitochondrial translation, we can envision that proofreading might not be a stringent function for mitochondrial protein synthesis. It has already been observed that animal mtDNA has evolved about ten times as fast than nuclear DNA (24). Thus, on an evolutionary time scale rapid changes are tolerated and this may well extend to toleration of a high rate of amino acid substitutions in the mitochondrial translation products themselves. Finally, a variety of observations have led to the idea that mitochondrial codon-anticodon recognition uses primarily a two out of three reading mechanism (25,26). One would imagine that such a mechanism is intrinsically more error-prone than the traditional mechanism postulated by wobble.

Concluding Remarks

We have attempted to show that regions of the yeast mitochondrial genome separating established genes fulfull a number of functions concerned with mitochondrial gene organization and expression. These include (1) transcriptional initiation; (2) sites for post-transcriptional processing; (3) translational efficiency and (4) a potential reservoir of new mitochondrial genes. All of these properties are exemplified by the small ribosomal subunit protein gene, <u>var1</u>.

Acknowledgments

This work was supported by grants from the National Institutes of Health and the Robert A. Welch Foundation.

References

1. Halbreich, A., Pajot, P., Foucher, M., Grandchamp, C., and Slonimski, P.: Cell 19, 321-329 (1980).
2. Schmelzer, C., Haid, A., Grosch, G., Schweyen, R.J., and Kaudewitz, F.: J. Biol. Chem. 256, 7610-7619 (1981).
3. Lamb, M.R., Anziano, P.Q., Glaus, K.R., Hanson, D.K., Klapper, H.J., Perlman, P.S., and Mahler, H.: J. Biol. Chem. 258, 1991-1999 (1983).
4. Attardi, G., Cantatore, P., Chomyn, A., Crews, S., Gelfand, R., Merkel, C., Montoya, J., and Ojala, D.: Mitochondrial Genes (Sonimski, P., Borst, P., and Attardi, G., eds.) Cold Spring Harbor Laboratory, Cold Spring Harbor 1982.
5. Prunell, A., and Bernardi, G.: J. Molec. Biol. 110, 53-74 (1977).
6. Hudspeth, G.S.P., Ainley, W.M., Shumard, D.S., Butow, R.A., and Grossman, L.I.: Cell 30, 617-626 (1982).
7. Terpstra, P., Zanders, E., and Butow, R.A.: J. Biol. Chem. 254, 12653-12661 (1979).
8. Groot, G.S.P., Mason, T.L., and Van Harten-Loosbroek: Mol. Gen. Genet. 174, 339 (1979).
9. Ainley, W., Zassenhaus, H.P., and Butow, R.A.: manuscript in preparation.
10. Gilbert, W.: Nature 271, 501 (1978).
11. Miller, D., Folse, J.R., Benson, P.J., and Martin, N.C.: Biochemistry 22, 1709-1714 (1983).
12. Tzagoloff, A., Nobrega, M., Akai, A., and Macino, G.: Current Genetics 2, 149-157 (1980).

13. Craik, C.S., Rutter, W.J., and Fletterick, R.: Science 220, 1125-1129 (1983).
14. Zassenhaus, H.P., Martin, N.C., and Butow, R.A.: manuscript in preparation.
15. Underbrink-Lyon, K., Miller, D.L., Ross, N., Fukuhara, H., and Martin, N.C.: Mol. Gen. Genet., in press (1983).
16. Farrelly, F., Zassenhaus, H.P., and Butow, R.A.: J. Biol. Chem. 257, 6581-6587 (1982).
17. Zassenhaus, H.P., and Perlman, P.S.: Current Genetics 6, 179-188 (1982).
18. Lopez, I.C., Farrelly, F., and Butow, R.A.: J. Biol. Chem. 256, 6496 (1981).
19. Osinga, K.A., deHaan, M., Christianson, T., and Tabak, H.F.: Nucl. Acids Res. 10, 7993-8006 (1982).
20. Edwards, J.C., Levens, D., and Rabinowitz, M.: Cell 31, 337-346 (1983).
21. Sor, F., and Fukuhara, H.: Cell 32, 391-396 (1983).
22. Lambowitz, A.M., Chua, N.-H., and Luck, D.J.L.: J. Mol. Biol. 107, 223-253 (1976).
23. Burke, J.M., and RajBhandary, U.L.: Cell 31, 509-520 (1983)
24. Brown, W.M., George, M., and Wilson, A.C.: Proc. Nat. Acad. Sci. USA 76, 1967-1971 (1979).
25. Barrell, B.G., Anderson, S., Banker, A.T., DeBruijn, M.H.L., Chen, E., Coulson, A.R., Drouin, J., Eperon, I.C., Nierlich, D.P., Roe, B.A., Sanger, F., Schreier, P.H., Smith, A.J.H., Staden, R., and Young, I.G.: Proc. Nat. Acad. Sci. USA 77, 3164-3166 (1980).
26. Heckman, J.E., Sarnoff, J., Alzner-DeWeerd, B., Yin, S., and RajBhandary, U.L.: Proc. Nat. Acad. Sci. USA 77, 3159-3163 (1980).

REGULATION OF RIBOSOMAL RNA AND MESSENGER RNA SYNTHESIS IN HUMAN MITOCHONDRIA

Giuseppe Attardi, George Gaines, Julio Montoya

Division of Biology, California Institute of Technology, Pasadena, California 91125

Introduction

The recent determination of the nucleotide sequence of human and other mammalian mitochondrial DNAs (mtDNAs) (1-6) and the detailed description of the structural, mapping and metabolic properties of the mtDNA transcripts from HeLa cells (7-10) have provided the framework for posing specific questions concerning the mechanisms and regulation of expression of this genome. The evidence that the sequences coding for the discrete transcripts of the heavy (H) strand of mtDNA form a continuum extending over nearly its entire length (8-10) and the observations indicating that H-strand transcription starts near the origin of replication (11, 12) (Fig. 1) have led us to propose a model of transcription of the H-strand in the form of a single polycistronic molecule destined to be processed by endonucleolytic cleavages to mature or near-mature products (8, 10).

The polycistronic transcript model has raised the question of how a differential control of expression of the different genes encoded in the H-strand is achieved in this system; in particular, one would like to know how is the higher rate of synthesis of the rRNAs relative to that of the individual mRNAs (10- to 30-fold), that is observed in the mitochondrial system (13), as in other systems, produced. In this paper, we will describe experiments aimed at answering this and related questions.

The Dual Mode of Transcription of the rDNA Region of mtDNA

A possible solution to the apparent paradox discussed above was provided by the suggestion (14, 15) that the single polycistronic transcripts normally have a very high probability of premature termination at or near

Mitochondria 1983

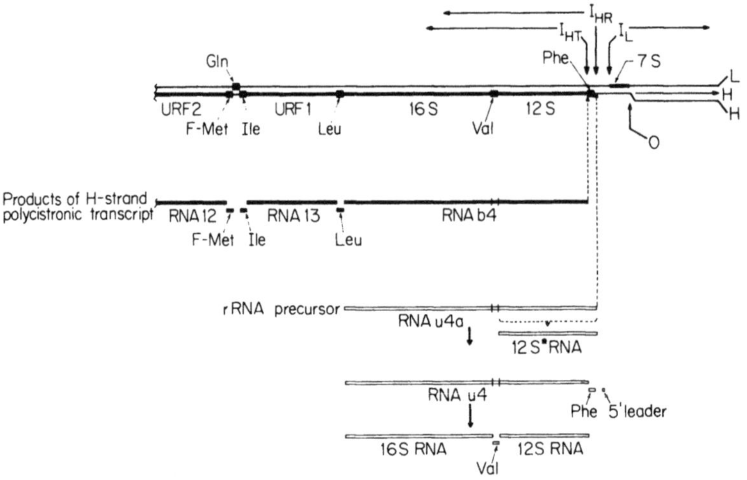

Figure 1. Portion of the HeLa cell mtDNA genetic and transcription maps illustrating the rRNA gene region and the adjacent regions (1, 8). In the upper portion of the diagram, the leftward and rightward arrows indicate the direction of H- and light (L)-strand transcription, respectively, and the downward arrows, the initiation sites for H-strand (I_{HT} and I_{HR}) and L-strand transcription (I_L) (12); the upward arrow indicates the location of the origin of H-strand synthesis (O). The lower portion of the diagram illustrates the proposed pathways of transcription and RNA processing in the rDNA region and adjacent region of the H-strand.

the 3' end of the 16S rRNA gene. The observation (16, 17) that the 3' terminal region of the 16S RNA gene of several mammalian mtDNAs, including human, can be folded in a structure resembling, in a rudimentary form, the hairpin-oligo(U) signal postulated for bacterial termination-attenuation (18) lent credence to this attenuation mechanism. Additional support for a termination of transcription near the 3' end of the 16S rRNA gene came from the discovery that the 3' ends of the mammalian 16S rRNA are ragged (15, 17), an observation which points to a certain imprecision of the process leading to their formation, in contrast to the absolute precision of the processing event leading to the release of the 3' ends of the mRNAs.

Further work, however, revealed a more complicated situation. Two types of observations were crucial in elucidating the process involved in the regulation of rRNA and mRNA synthesis. The first observation concerned the nature of the oligo(dT)-cellulose bound RNA b4 originally identified by Amalric et al. (7). This RNA, because of its mapping position, corresponding to that of the 12S and 16S rRNA species (8), and because of its relatively short half-life (40 min), as determined after cordycepin block of RNA synthesis (13), seemed to be a good candidate for being an rRNA precursor. However, surprisingly, this functional identification was not confirmed by direct experimental evidence. In fact, kinetics of labeling experiments utilizing short (5-^{3}H) uridine pulses (1 to 10 minutes) in exponentially growing cells showed that the incorporation of the precursor into RNA b4, at all time points analyzed, was much lower than the labeling of 12S rRNA, and in addition, similar to the incorporation into most of the other RNA species encoded in the H-strand regions downstream of the rRNA genes (16 19). These observations thus excluded the possibility that RNA b4 is the kinetic precursor of the bulk of mitochondrial rRNAs, and, furthermore, suggested that this RNA is a part of the polycistronic transcript spanning nearly the entire length of the H-strand.

The important implication of the experiments described above was that the rDNA region of human mtDNA is the site of two distinct transcription events, one responsible for the synthesis of the rRNA and the other for the synthesis of the mRNAs and the majority of the tRNAs encoded in the H-strand. Support for this model has recently come from an analysis of the initiation sites for transcription in HeLa cell mtDNA (12). In fact, mapping experiments using mitochondrial RNA molecules "capped" in vitro with (α-^{32}P)GTP and guanylyltransferase, or using nascent RNA chains isolated from mtDNA transcription complexes, have identified two initiation sites for H-strand transcription. Of these, one is situated very near the 5' terminus of the 12S rRNA gene, and the other 90 to 110 nt upstream of this site, i.e., 20 to 40 nt upstream of the tRNA^{Phe} gene (Fig. 1). Although the functional significance of the two initiation sites could not be established, it was a plausible idea that they are implicated in the two different transcription events.

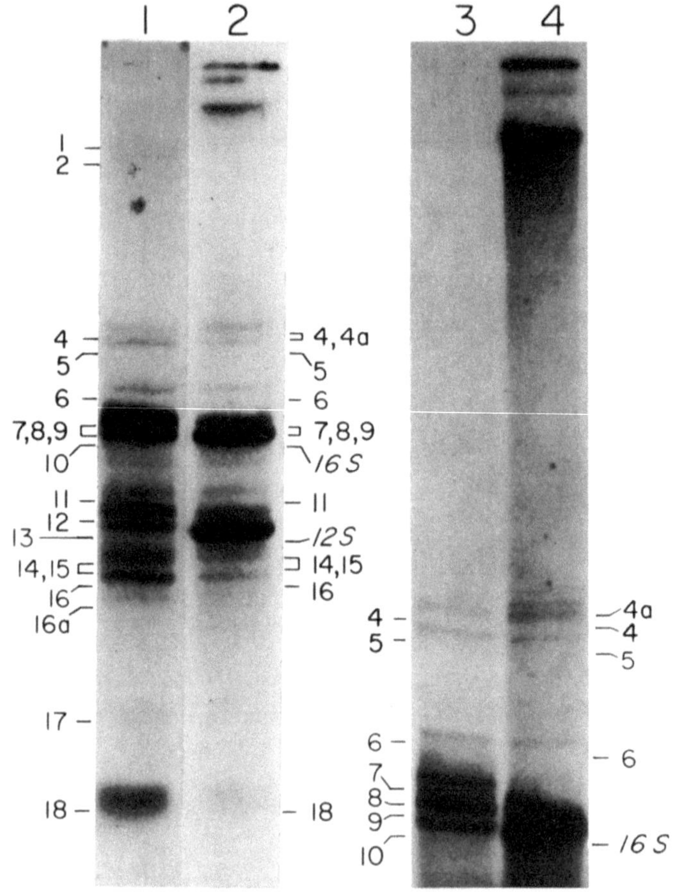

Figure 2. Electrophoretic analysis of oligo(dT)-cellulose bound (lanes 1 and 3) and unbound (lanes 2 and 4) mitochondrial RNA from HeLa cells labeled for 4 hr with ^{32}P-orthophosphate in the presence of 0.1 µg/ml actinomycin D. The samples were run through an agarose-CH^3HgOH slab gel for 6 hr (lanes 1 and 2) or for 9 hr (lanes 3 and 4). (Modified from ref. 19.)

The Search for an rRNA Precursor

With the aim of determining which one of the two initiation sites for H-strand transcription detected in human mtDNA is utilized for rRNA synthesis, a search of the primary rDNA transcripts destined to be processed to mature rRNA species was carried out in the oligo(dT)-cellulose unbound RNA fraction.

Figure 2 shows the autoradiogram, after electrophoresis through an agarose-CH_3HgOH slab gel, of samples of the oligo(dT)-cellulose bound (lane 1) and unbound (lane 2) mitochondrial RNA fractions from cells labeled for 4 hr with ^{32}P-orthophosphate in medium containing 10^{-5} M phosphate in the presence of 0.1 μ/ml actinomycin D. The bound fraction shows the characteristic pattern with 18 discrete poly(A)-containing RNA components previously described (7). The unbound fraction exhibits, besides the prominent 16S and 12S rRNA bands, a series of minor bands which, in their majority, align well with the bands of the poly(A)-containing RNA, and which very probably represent RNA species differing from the latter for the absence or shorter length of the poly(A) tail. In particular, one can see in the pattern of the unbound RNA fraction a fairly pronounced band (u4) with the same apparent mobility as that of RNA b4; furthermore, there is a fainter, somewhat more slowly moving band (u4a), which is not well resolved from u4. The bands u4 and u4a are better resolved after a longer electrophoretic run (lane 4).

In order to verify whether the two RNA species b4 and u4 were related to each other, and in order to test whether RNA u4a mapped in the same mtDNA region as the other two components, S1 protection experiments were carried out to map the 5' end and 3' end of each of these RNA species. The results of these experiments, illustrated schematically in Fig. 3, showed that the 5' ends of RNA b4 and RNA u4 map at the same or very similar position, close to or at the 5' terminus of the 12S rRNA gene, while the 5' end of RNA u4a maps about 95 nt upstream of that position (Fig. 1). It is significant that these two sites correspond precisely to the two H-strand transcription initiation sites mentioned above. The results further showed that, at the 3' end, the RNA species b4, u4 and u4a map in correspondence of the two nucleotides immediately preceding the $tRNA^{Leu}$ gene, exactly as the mature 16S rRNA (Fig. 1).

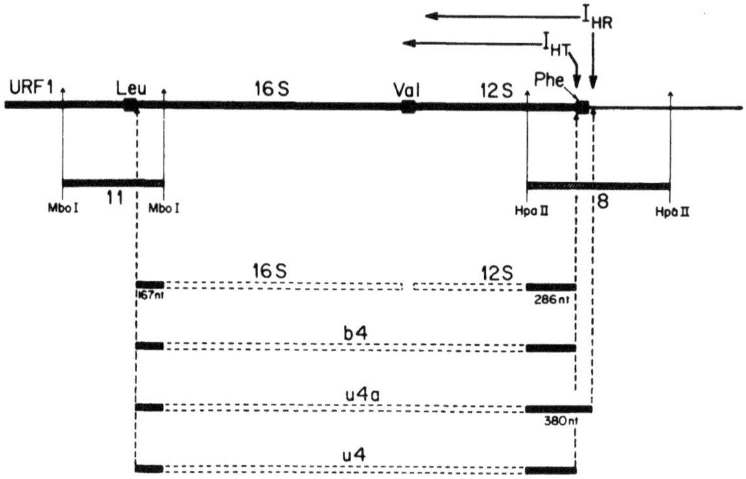

Figure 3. Nuclease S1 mapping of 5'-ends and 3'-ends of mitochondrial rDNA transcripts. The individual RNA species labeled in vivo with ^{32}P-orthophosphate in the presence of 0.1 µg/ml actinomycin D, or unlabeled, were separated in an agarose-CH_3HgOH gel, eluted from slices of the gel, and then hybridized under high formamide conditions with the H-strands of HpaII fragment 8 (unlabeled or 5'-end labeled, respectively) or with the H-strands of MboI fragment 11 (unlabeled or 3'-end labeled, respectively). After S1 digestion, the S1-resistant fragments were analyzed in polyacrylamide gels under native or denaturing conditions. The black bars indicate the protected fragments.

The results described above clearly pointed to the rDNA transcripts u4a and u4 as possible rRNA precursors, prompting an analysis of their metabolic properties. HeLa cells, labeled for 4 hr with ^{32}P-orthophosphate in the presence of 10^{-3} M phosphate and 0.1 µg/ml actinomycin D, were exposed for very short periods to (5-^{3}H) uridine, always in the presence of ^{32}P-orthophosphate. Figure 4 shows the labeling pattern, after a 10 min (5-^{3}H) uridine pulse, of the oligo(dT)-cellulose unbound (panel A) and bound (panel B) rDNA transcripts and of other mitochondrial RNA species migrating in the adjacent regions of the gel tracks, as determined by serial gel slicing and counting after long electrophoretic runs. The RNA species u4 and u4a are both labeled with (5-^{3}H) uridine to a much greater extent than RNA b4. Furthermore, the

$^3H/^{32}P$ ratio in the RNAs u4 and u4a is more than four times higher than in RNA b4, and five times as high as in 12S rRNA. Figure 4C represents the labeling of the various rDNA transcripts in several experiments of this kind involving cell exposure to different pulses of the precursor. One sees that there is no hint of saturation of ^3H radioactivity in RNAs u4 and u4a, as expected if these species were obligatory kinetic precursors of the rRNAs. However, it should be noted that the rapid increase in specific activity of the intramitochondrial UTP and CTP pools during the (5-^3H) uridine pulses (13) contributes to a considerable extent to the observed kinetics of labeling of the various RNA species and would mask a saturation of the putative precursor rRNA pools. Furthermore, the rapid appearance of label in 12S rRNA (Figure 4C), already observed in exponentially growing cells, indicates that a major part of the 12S rRNA is probably directly processed from nascent chains. In any case, the rapid flow of (5-^3H) uridine through RNAs u4 and u4a is clearly consistent with a role of rRNA precursors for these species, in sharp contrast to the labeling behavior of RNA b4.

Regulation of rRNA and mRNA Synthesis

The mapping analysis of the rDNA transcripts has thus established a correlation between the two initiation sites for transcription which have been mapped on the H-strand and the two transcription events of the rDNA region whose existence was indicated by the kinetic properties of these transcripts. The evidence, therefore, strongly suggests that one of the transcription events starts 25 bp upstream of the $tRNA^{Phe}$ gene and terminates at or near the 3' end of the 16S rRNA gene, being responsible for the synthesis of the bulk of rRNA (Fig. 1). Two large oligo(dT)-cellulose unbound RNA species spanning the rDNA region, u4a and u4, are, respectively, a putative primary transcript of this pathway and an intermediate lacking the $tRNA^{Phe}$ and the 5' leader. The other transcription event starts near the 5' end of the 12S rRNA gene, proceeds beyond the 3' end of the 16S rRNA gene and results in the synthesis of a polycistronic molecule corresponding to almost the entire H-strand (Fig. 1). RNA b4 may have no physiological role other than that of joining the promoter near the origin of replication to the main portion of the transcription unit lying downstream with respect to the rRNA genes.

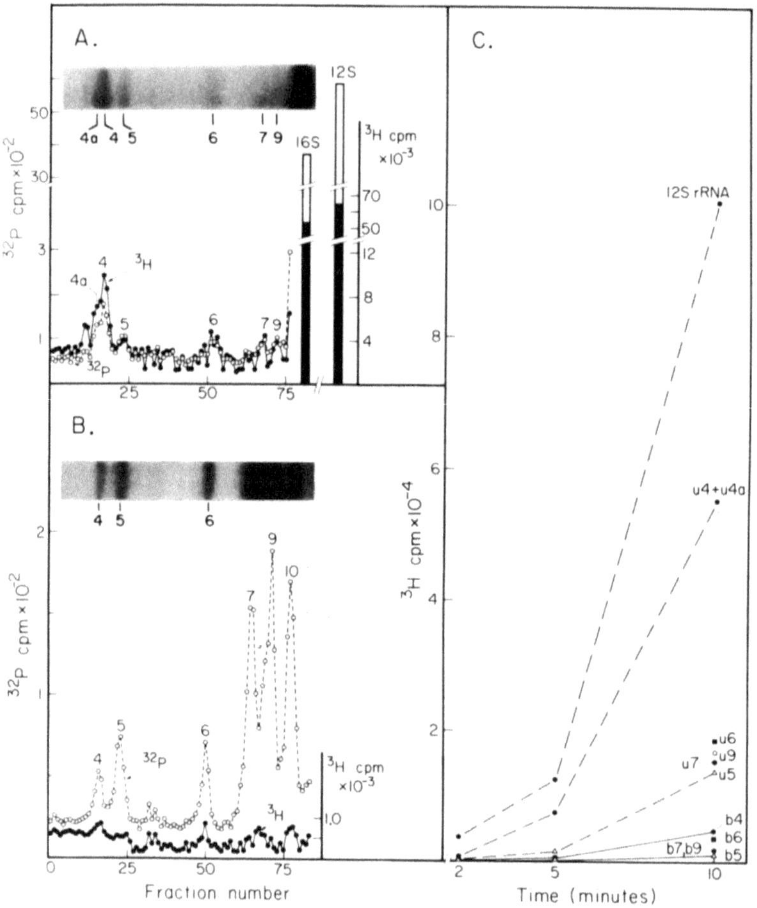

Figure 4. Flow of 5-^{3}H-uridine into oligo(dT)-cellulose bound and unbound rDNA transcripts and other H-strand transcripts in cells treated with 0.1 µg/ml actinomycin D. (A, B) Analysis by serial gel slicing of the labeling of various oligo(dT)-cellulose unbound (A) and bound (B) RNA species separated on agarose-CH_3HgOH slab gels. The RNA was isolated from cells labeled for 4 hr with ^{32}P-orthophosphate in the presence of 10^{-3} M unlabeled phosphate and 0.1 µg/ml actinomycin D, and then labeled for 10 min with 5-^{3}H-uridine in the presence of the antibiotic. (C) Kinetics of labeling of the various RNA species in the experiment illustrated in panels A and B and in other experiments involving a 2 min or a 5 min 5-^{3}H-uridine pulse. The total amount of radioactivity incorporated into each RNA species at the various time points has been normalized to a constant amount of cells and corrected for variations in ^{32}P labeling of that species (from ref. 19).

The proposed existence in the H-strand of human mtDNA of two transcription units with distinct promoters, one used for the synthesis of rRNAs, and the other, for the synthesis of the mRNAs and most of the tRNAs encoded in the H-strand is presumably justified by the need for the cell to maintain a separate control of the rate of synthesis of the rRNAs and of the other two mitochondrial RNA classes. However, the unusual organization of the two transcription units, which overlap in such a way that the two H-strand promoters are brought into close proximity, must satisfy other functional requirements. It seems in fact possible that the close proximity of the two H-strand promoters to the L-strand promoter (12) and to the origin of replication (20) reflects the advantage for the cell to have all controlling elements for mtDNA transcription and replication within the same region of the DNA.

A testable implication of the model shown in Figure 1 is that the $tRNA^{Phe}$ and $tRNA^{Val}$ genes are transcribed at a much higher rate than the majority of the other H-strand tRNA genes, leading to a larger accumulation or turnover of the corresponding tRNA species. Preliminary experiments have indeed indicated that the steady state amounts of the $tRNA^{Phe}$ and $tRNA^{Val}$ in exponentially growing cells are two to four times higher than those of several other H-strand coded tRNAs, including the $tRNA^{Leu}_{UUR}$ encoded on the 3' end side of the 16S rRNA gene. These higher steady state amounts have no relationship to the expected codon usage for mitochondrial translation (King and Attardi, unpublished observation).

Processing and Termination of rDNA Transcripts

A scanning of the DNA sequence does not reveal any significant repeated sequence near the two initiator sites for H-strand transcription. A difference in the efficiency of the two promoters is probably the main factor controlling the different rates of rRNA and mRNA synthesis in human mitochondria. An unusual feature of the mitochondrial system discussed here is that, apart from the control of the rate of transcription, each of the two kinds of transcripts of the rDNA region must have its fate determined by having started at a specific site. The involvement of two different polymerases and/or the extra nucleotide stretch at the 5' end of the RNA u4a may be responsible for this difference in fate. The situation differs from the transcription attenuation described in other systems (21, 22) where alternative secondary structures of the same nascent RNA molecule determine the termination or continuation of transcription. Furthermore, in those systems, the secondary structure of the 5'-leader is assumed to produce the formation of the termination signal in the region immediately downstream. In the mitochondrial system discussed here, if the 5'-end of the rDNA transcript originating at the upstream promoter plays a direct role in termination, it would be required to transmit the signal for transcription termination to a site 2675 nt downstream.

The observation that in exponentially growing cells mature 12S rRNA is labeled to a substantial extent already after a 1 min [5-^{3}H] uridine pulse suggests that most of the processing of the rRNA precursor to mature species under normal conditions occurs while this precursor is being synthesized. The situation may be analogous to that occurring in *E. coli*, where the large rRNA precursor (30S) is not normally observed *in vivo* due to the rapid cleavage by RNase III of the sensitive sites in the nascent chains, but does accumulate in RNase III-deficient strains (23, 24). If processing of the mitochondrial rRNA precursor under normal conditions occurs while this is being synthesized, as is strongly suggested by the kinetic data presented here, any long range effect of the 5'-end of this precursor must involve some form of polar transmission of the signal, as by sequential formation of secondary structures and sequential binding of ribosomal proteins.

MtDNA Transcription in Isolated Mitochondria

Support for the model discussed here for the differential control of expression of the rRNA and structural genes in human mtDNA has come from the results of mtDNA transcription and RNA processing experiments utilizing isolated organelles.

Isolated HeLa cell mitochondria, when incubated in the presence of α-^{32}P-UTP in an appropriate buffer, support transcription of their DNA in a way closely resembling the *in vivo* process. Figure 5 shows a comparison between the *in vitro* labeled mitochondrial RNA species and *in vivo* transcription products. Polyadenylation appears to be less extensive

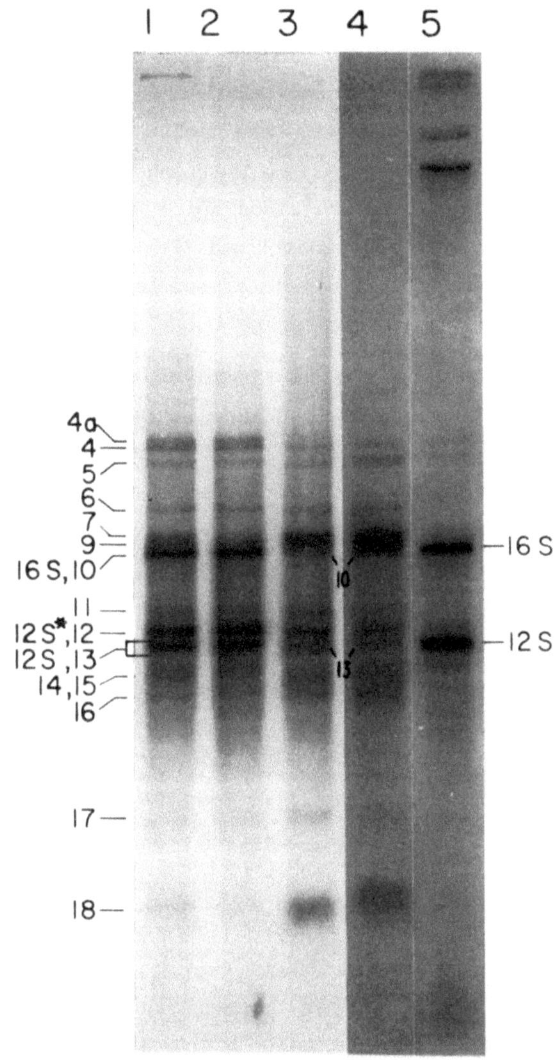

Figure 5. Comparison of mtDNA transcription in isolated mitochondria and _in vivo_. Mitochondrial RNA from organelles incubated _in vitro_ in an appropriate buffer in the presence of (α-^{32}P) UTP (lanes 1-3) and _in vivo_ labeled RNA (lanes 4, 5) were fractionated into oligo(dT)-cellulose bound and unbound RNA, and the various fractions were run through an agarose-CH^3HgOH slab gel. Lanes 1-3: total (lane 1), oligo(dT)-cellulose unbound (lane 2) and bound (lane 3) RNA labeled _in vitro_. Lanes 4, 5: oligo(dT)-cellulose bound (lane 4) and unbound (lane 5) RNA labeled _in vivo_.

in vitro, as shown by the comparison of the oligo(dT)-cellulose bound and unbound fractions in the _in vitro_ (lanes 2 and 3) and the _in vivo_ labeled

RNA (lanes 4 and 5). Furthermore, the processing of the rRNA species is considerably slowed down in isolated organelles. This is shown by the accumulation of RNAs u4a and u4 and the relatively strong labeling of RNA u4a relative to RNA u4. However, the most striking evidence of inefficient processing is represented by the appearance of a processing intermediate which is termed here 12S*, and which migrates in the gel as the polyadenylated RNA 12 (Fig. 5, lanes 1 and 2). As shown by S1 mapping experiments (Fig. 6), this processing intermediate consists of 12S RNA still carrying the $tRNA^{Phe}$ and the leader at its 5' end.

The efficiency of processing of the mitochondrial rRNA appears to depend on the temperature of incubation of isolated mitochondria (Fig. 7). Although there is a temperature dependence of mitochondrial RNA labeling _in vitro_ which affects all RNA species, there are marked differences in the relative labeling of the various rRNA components and their precursors at different temperatures (Fig. 7a and 7c). Thus, at 37°C, the labeling of 12S* RNA is decreased relative to that of 12S RNA, whereas the opposite is seen at 26°C, and even more at 21°C. Furthermore, at the lower temperatures, the labeling of RNA u4a is more pronounced than that of RNA u4. These results suggest that the step involved in the removal of the $tRNA^{Phe}$ or $tRNA^{Phe}$ + leader from the rRNA precursor is strongly inhibited at low temperatures. Also the formation of 16S RNA appears to be strongly affected at suboptimal temperatures, possibly due to pauses of the RNA polymerase or premature termination of the transcripts.

Effects of Intercalating Drugs on mtDNA Transcription in Isolated Organelles

The availability of an _in vitro_ mtDNA transcription system utilizing isolated organelles offered the opportunity to investigate the direct effects of intercalating drugs on this process without the constraints and interference produced by the cellular environment. Four drugs which are known to intercalate into DNA (actinomycin D, proflavine, ellipticine and ethidium bromide) (25) were tested on the _in vitro_ system described above. All four drugs had an overall inhibitory effect upon mtDNA transcription. However, marked differences in the response to each drug of rRNA and mRNA synthesis and of rRNA processing were observed. Three of

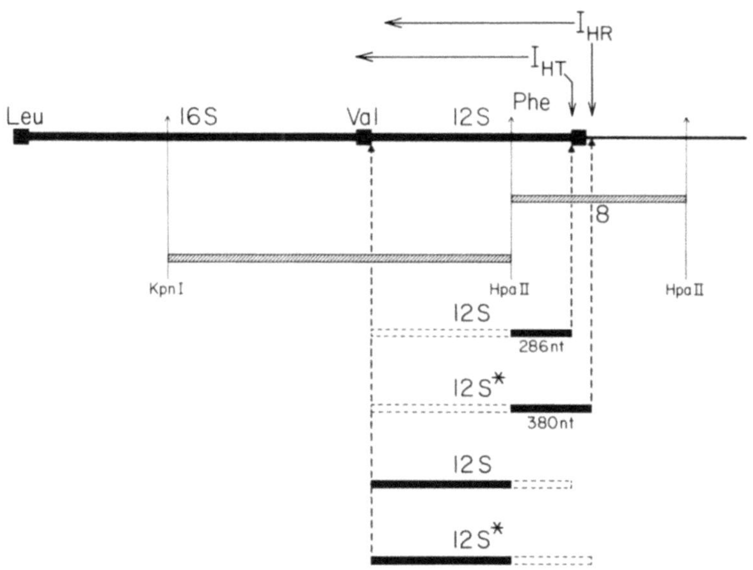

Figure 6. Nuclease S1 mapping of 5'-ends and 3'-ends of RNAs 12S and 12S*. The two RNA species labeled in isolated mitochondria with (α-^{32}P) UTP were separated in an agarose-CH^3HgOH gel, eluted from slices of the gel, and then hybridized under high formamide conditions with unlabeled HpaII fragment 8 or KpnI-HpaII fragment (shown in Fig. 1). After S1 digestion, the resistant fragments were analyzed in a polyacrylamide gel under native conditions. The black bars indicate the protected fragments.

the drugs, i.e., actinomycin D, proflavine and ethidium bromide, had a preferential inhibitory action on rRNA labeling. Particularly striking was the inhibition of labeling of 16S rRNA. These effects are shown in Figure 8b for proflavine. Ellipticine, which is an intercalating drug of similar molecular size, shape, charge and helix-unwinding angle to proflavine (25), in contrast to the latter, caused a parallel inhibition of labeling of the rRNA and mRNA species over the whole range of drug concentrations tested (4 to 100 µM) (Fig. 8a).

The observations described above indicae that suppression of mitochondrial rRNA synthesis by intercalating drugs can be dissociated from the effect on mRNA synthesis. These results, therefore, support the existence of an independent control of rRNA and mRNA synthesis, which is implicit in the model of overlapping transcription units with distinct promoters. The in vitro results also confirm the conclusion, previously reached on the basis of kinetic arguments, that processing of the rRNA precursor at the

$tRNA^{Val}$ site is an early event, and that formation of 12S RNA does not require completion of the 16S RNA.

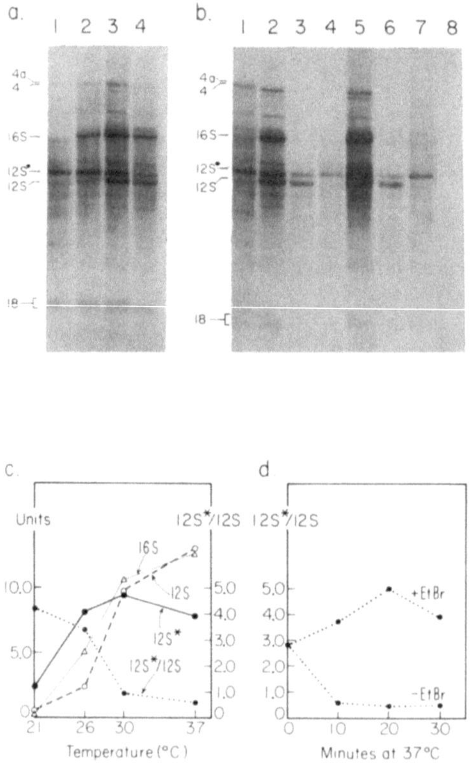

Figure 7. a) Effects of temperature on mtDNA transcription in isolated organelles. Lane 1: 21°C; lane 2: 26°C; lane 3: 30°C; lane 4: 37°C. The relative amounts of material electrophoresed were 10, 5, 2.5 and 1, respectively, in lanes 1 to 4. b) Temperature shift-chase experiment. Mitochondrial samples were incubated at 21°C for 30 min, and then either immediately processed (lane 1), or transferred to 37°C and further incubated for 10 min (lanes 2, 3, 4), 20 min (not shown) or 30 min (lanes 5, 6, 7) in the absence of drugs (lanes 2 and 5), or in the presence of 5.1 µM actinomycin D (lanes 3 and 6) or of 5.1 µM actinomycin D and 16.2 µM ethidium bromide (lanes 4 and 7). Lane 8: a mitochondrial sample was incubated at 21°C for 30 min in the absence of labeled precursor, and then transferred to 37°C and incubated for 30 min in the presence of (α-^{32}P) UTP and 5.1 µM actinomycin D. c) and d) Quantitation of the effects of temperature carried out by densitometric scanning of the autoradiograms.

A Critical rRNA Processing Site

Another specific effect of the intercalating drugs tested was the apparent inhibition of the 5' end processing of the rRNA precursors by proflavine and ethidium bromide. This apparent inhibition, which resembled the effect of low temperatures, resulted in an increase in the labeling of RNA u4a relative to that of RNA u4 with increasing concentrations of the drug, and, correspondingly, an increase in the labeling of $12S^*$ RNA relative to that of the 12S and 16S RNA components (Fig. 8b).

In order to provide evidence that the effect of a low temperature of incubation or of intercalating drugs on the relative labeling of the $12S^*$ and 12S RNA species, and of RNAs u4a and u4, was indeed due to a direct inhibition of the 5' end processing of the rRNA precursors, a temperature shift-chase experiment was carried out. To discriminate between RNA synthesis and processing events, advantage was taken of the observation that actinomycin D inhibits mtDNA transcription without grossly affecting the relative proportions of $12S^*$ and 12S RNAs. Mitochondria were incubated at 30°C in the presence of (α-^{32}P) UTP for 30 min, and then shifted to 37°C in the presence of 5.1 µM actinomycin D or in its absence.

As shown in Figure 7b (lane 8), actinomycin D effectively blocked further incorporation of (α-^{32}P) UTP. At the time of the shift-up, the $12S^*$ RNA band was very prominent, and there was little or no labeled 12S RNA (most of the label in the 12S band being presumably associated with the polyadenylated RNA 13 which migrates to nearly the same position as 12S RNA) (lane 1). After 10 min at 37°C in the presence of actinomycin D, the $12S^*$ band had markedly decreased in intensity, while the 12S band had correspondingly increased so as to become the prominent band (Fig. 7b, lane 3). The effect was more pronounced after a 20 min or 30 min chase (Fig. 6b, lane 6). At the same time, there was an almost complete disappearance of labeled RNA u4a, with only faint u4a and u4 bands being visible. As shown in Figure 7b (lanes 4 and 7), additon of ethidium bromide at 16.2µM at the time of the temperature shift together with actinomycin D blocked completely the processing of $12S^*$ RNA to 12S RNA and, partially, that of RNA u4a to RNA u4. These results indicated that the intercalating drug acted, like the low temperature, by preventing the removal of the 5'-terminal $tRNA^{Phe}$ or $tRNA^{Phe}$ + leader from the rRNA precursors. In the absence of actinomycin D, the same phenomena described above occurred, but they were in part masked by the continuation of RNA synthesis at an appreciably higher rate after the temperature shift-up (Fig. 7b, lanes 2 and 5).

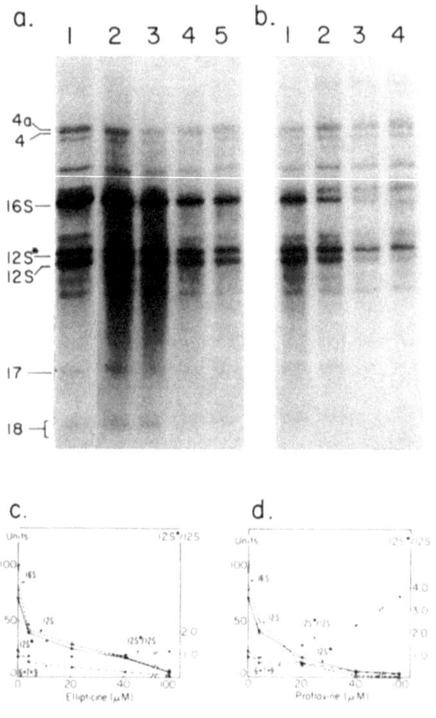

Figure 8. Effects of ellipticine and proflavine on mtDNA transcription in isolated organelles. a) Electrophoretic patterns in an agarose-CH^3HgOH gel of RNA from mitochondria incubated at 30°C in the absence of drugs (lane 1), or in the presence of 4.1 µM (lane 2), 20.3 µM (lane 3), 40.6 µM (lane 4) and 101.5 µM (lane 5) ellipticine. Relative amounts of material electrophoresed were 1, 2, 2, 2, and 5, respectively, in lanes 1 to 5. b) Electrophoretic patterns of RNA from mitochondria incubated at 30°C in the presence of 4.1 µM (lane 1), 20.4 µM (lane 2), 40.7 µM (lane 3) and 101.8 µM (lane 4) proflavine. Relative amounts of material electrophoresed were 2, 5, 10 and 20, respectively, in lanes 1 to 4. c) and d) Quantitation of the effects of ellipticine and proflavine, respectively, carried out as in Figure 7.

The temperature shift experiment described above showed that the specific inhibition of the 5' end processing of the rRNA precursors $12S^*$ and u4a by low temperatures and two of the intercalating drugs tested, proflavine and ethidium bromide, can be dissociated from RNA synthesis; furthermore, this inhibition of 5' end processing appears to occur on molecules already completed and presumably no longer attached to the template. Although one cannot rule out a direct action of the low temperature or intercalating drugs on the processing enzyme or membrane perturbations, it is tempting to speculate that a change in secondary or tertiary structure of the rRNA precursor, affecting the recognition or activity of the processing enzyme, underlies both the effects of a low temperature and those of proflavine and ethidium bromide. These two intercalating agents have been previously shown to affect the processing of nuclear rRNA and mRNA precursors (26-28). A strong binding of ethidium bromide and proflavine to tRNA has also been described (29-31). In view of the postulated role of the tRNA sequences as recognition signals in the processing of the H-strand transcripts, it is quite possible that binding of intercalating drugs to the $tRNA^{Phe}$ segment of the precursor is involved in the block of the 5' end processing of the rRNA precursors by these agents.

Whatever the mechanism may be in the inhibition of this 5' end processing, it is clear that there is a strong specificity of the intercalating drug involved, since neither actinomycin D nor ellipticine have this effect. Furthermore it appears that the processing site affected by the low temperature or by proflavine or ethidium bromide is uniquely sensitive to these agents among the many sites where tRNA processing occurs in the mtDNA transcripts (8, 10). This unique sensitivity may reflect very strong structural constraints in terms of secondary structure and/or protein binding, for processing at this site.

Acknowledgements

This work was supported by a research grant from the U.S.P.H.S. grants GM-11726 and T32GM-07616. We are grateful to Dr. Peter Dervan for a gift of ellipticine. The technical assistance of Ms. Arger Drew is gratefully acknowledged.

References

1. Anderson, S., Bankier, A. T., Barrell, B. G., de Bruijn, M. H. L., Coulson, A. R., Drouin, J., Eperon, I. C., Nierlich, D. P., Roe, B. A., Sanger, F., Schreier, P. H., Smith, A. J. H., Staden, R., Young, I. G.: Nature 290, 457-465 (1981).
2. Bibb, M. J., Van Etten, R. A., Wright, C. T., Walberg, M. W., Clayton, D. A.: Cell 26, 167-180 (1981).
3. Sekiya, T., Kobayashi, M., Seki, T., Koike, K.: Gene 11, 53-62 (1980).
4. Grosskopf, R., Feldmann, H.: Current Genetics 4, 151-158 (1981).
5. Pepe, G., Holtrop, M., Gadaleta, G., Kroon, A. M., Cantatore, P., Gallerani, R., De Benedetto, C., Quagliariello, C., Sbisa, E., Saccone, C.: Biochemistry International 6, 553-563 (1983).
6. Anderson, S., de Bruijn, M. H. L., Coulson, A. R., Eperon, I. C., Sanger, F., Young, I. G.: J. Mol. Biol. 156, 683-717 (1982).
7. Amalric, F., Merkel, C., Gelfand, R., Attardi, G.: J. Mol. Biol. 118, 1-25 (1978).
8. Ojala, D., Merkel, C., Gelfand, R., Attardi, G.; Cell 22, 393-403 (1980).
9. Montoya, J., Ojala, D., Attardi, G.: Nature 290, 465-470 (1981).
10. Ojala, D., Montoya, J., Attardi, G.: Nature 290, 470-474 (1981).
11. Cantatore, P., Attardi, G.: Nucleic Acids Res. 8, 2605-2624 (1980).
12. Montoya, J., Christianson, T., Levens, D., Rabinowitz, M., Attardi, G.: Proc. Natl. Acad. Sci. U.S.A. 79, 7195-7199 (1982).
13. Gelfand, R., Attardi, G.: Molec. Cell. Biol. 1, 497-511 (1981).
14. Attardi, G., Cantatore, P., Ching, E., Crews, S., Gelfand, R., Merkel, C., Ojala, D.: In The Organization and Expression of the Mitochondrial Genome (Kroon, A. M., Saccone, C., eds.) North-Holland, Amsterdam, 103-119 (1980).
15. Dubin, D. T., Timko, K. D., Baer, R. J.: Cell 23, 271-278 (1981).
16. Attardi, G., Cantatore, P., Chomyn, A., Crews, S., Gelfand, R., Merkel, C., Montoya, J., Ojala, D.: In Mitochondrial Genes (P. Sloninski, P. Borst and G. Attardi, eds.) Cold Spring Harbor Laboratory, New York, 51-71 (1982).

17. Dubin, D. T., Montoya, J., Timko, K. D., Attardi, G.: J. Mol. Biol. 157, 1-19 (1982).
18. Rosenberg, M., Court, D.: Ann. Rev. Genet. 13, 319-353 (1979).
19. Montoya, J., Gaines, G. L., Attardi, G. Cell, in press.
20. Crews, S., Ojala, D., Posakony, J., Nishiguchi, J., Attardi, G.: Nature 277, 192-198 (1979).
21. Yanofsky, C.: Nature 289, 751-758 (1981).
22. Hay, N., Skolnik-David, H., Aloni, Y.: Cell 29, 183-193 (1982).
23. Dunn, J. J., Studier, F. W.: Proc. Natl. Acad. Sci. U.S.A. 70, 3296-3300 (1973).
24. Nikolaev, N., Silengo, L., Schlessinger, D.: Proc. Natl. Acad. Sci. U.S.A. 70, 3361-3365 (1973).
25. Gale, E. F., Cundliffe, E., Reynolds, P. E., Richmond, M. H., Waring, M. J. : In The Molecular Basis of Antibiotic Action, Wiley, New York, 258-401 (1981).
26. Snyder, A. L., Kann, H. E., Jr., Kohn, K. W.: J. Mol. Biol. 58, 555-565 (1971).
27. Yannarell, A., Niemann, M., Schumm, D. E., Webb, T. E.: Nucleic Acids Res. 4, 503-511 (1977).
28. Chiu, N. H., Bruszewski, W. B., Salzman, N. P.: Nucleic Acids Res. 8, 153-168 (1980).
29. Bittman, R.: J. Mol. Biol. 46, 251-268 (1969).
30. Wells, B. D., Cantor, C. R.: Nucleic Acids Res. 4, 1667-1680 (1977).
31. Liebman, M., Rubin, J., Sundaralingam, M.: Proc. Natl. Acad. Sci. U.S.A. 74. 4821-4825. (1977).

REGULATION OF MITOCHONDRIAL TRANSCRIPTION IN RAT LIVER: DETERMINATION OF MITOCHONDRIAL TRANSCRIPTS CONCENTRATION IN ADULT AND DEVELOPING RAT.

Palmiro Cantatore, Zina Flagella, Graziano Pesole, Maria Nicola Gadaleta and Cecilia Saccone
Institute of Biological Chemistry, University of Bari
70126 Bari, Italy

Introduction

The determination of the base sequences of several animal mitochondrial (mt) DNAs has revealed the peculiar features of this eukaryotes genome (1-3). Animal mtDNA molecule contains information on both the H-strand, which is completely saturated with genes coding for the two ribosomal rRNAs species, 14 tRNAs and the majority of mRNAs, and the L-strand, which codifies for only 8 tRNAs and perhaps one or two mRNAs.

In contrast to the detailed information about the structure of animal mt genome, the data concerning the regulation of mt gene expression are still inadequate. From studies carried out almost exclusively with HeLa cell, a mechanism of mt gene transcription has been proposed which implies complete transcription of both strands and further processing of the transcripts (4). Two initiation points for the H-strand transcription have been recently found by Montoja et al. (5). One initiation point should be used in the transcription of the two rRNA species, the other for the synthesis of a polycistronic transcript containing tRNA and mRNA species.

The regulation of these events is however little understood: in particular not much is known about the mechanisms by which the rRNA species are preferentially transcribed respect to the messengers (6). Moreover, there is little information on the

Mitochondria 1983

variations of mitochondrial transcription during animal development or in different physiological conditions. The studies on mt biogenesis during development demonstrate that at critical stages an increase in the number of mitochondria and/or mitochondrial differentiation can take place (7,8). Both processes could involve a regulation at mitochondrial gene expression level.
The reasons reported above seem sufficient to justify our interest in determining the steady-state concentration of mt transcripts in animal cell. In this paper we describe the method and report the results found for seven mtRNA species in adult rat liver and an analysis of the ratios between ribosomal (16S rRNA) and CoI mRNA during development.

Results and Discussion

1. Measure of several mtRNA species concentration in adult rat.
The determination of the mtRNA species concentration in adult rat liver has been done by titrating a know quantity of mtRNA with an excess of a radioactive DNA fragment coding for the species to be measured. The amount of hybridized DNA has been used to calculate the steady state concentration of individual RNA species. This has been possible since the gene organization, and a large part of rat mtDNA sequence is known (Fig. 1). The experiments reported in this paper concern two rRNAs and several mRNAs, so that a representative pattern of all the mt transcripts has been obtained.
Radioactive DNA probes have been isolated starting from the cloned rat Eco RI fragments contained in the plasmid pSF2124. The 16S probe is a fragment of 1304 bp, originated from the digestion of Eco RI D with Hha I, whereas the 12S probe (934 bp) comes from the digestion of Eco RI A with Hpa II (Fig. 2). The mRNA species analyzed are URF 1, CoI, ATPase, CoIII and URF 4. The respective probes have been prepared through cross digestions of plasmid DNA with several restriction enzymes,

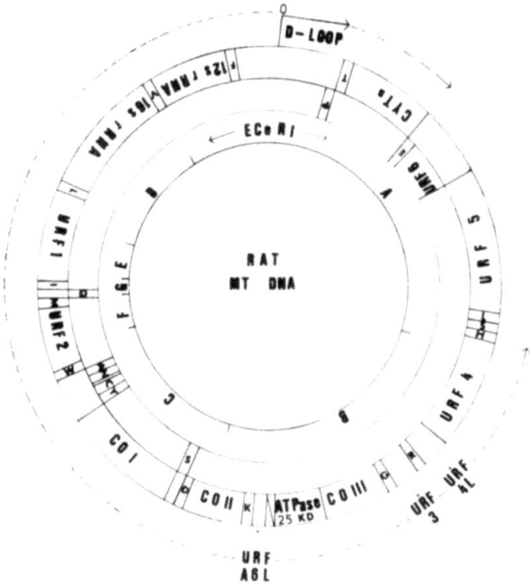

Fig. 1. Gene organization and restriction map of rat liver mtDNA. In the first inner circle the Eco RI restriction map of rat mtDNA is reported. The second and the third circles represent the products codified by the L strand and the H strand respectively. The area comprised between the arrows is the amount of the genome that has been sequenced by ourselves (9, 11) or others (12,13).

as shown in Fig. 3. To accurately determine the concentration of the fragment to be labelled, E. coli cells, containing plasmids, were grown in the presence of methyl (^{3}H)thymidine (0.5 μCi/ml) and plasmid DNA specific activity was used to calculate the concentration of DNA fragments. They were then labelled by nick translation (14) at a specific activity of 1.10^7 cpm/μg and used for the hybridizations.

The RNA was extracted from adult rat mitochondria through several phenol-chloroform extractions. During the preparation of mitochondria, care was taken to reduce the nuclear contamination by centrifuging the nuclear supernatant several times. The mtRNA was analyzed through methyl mercuric agarose gel electro-

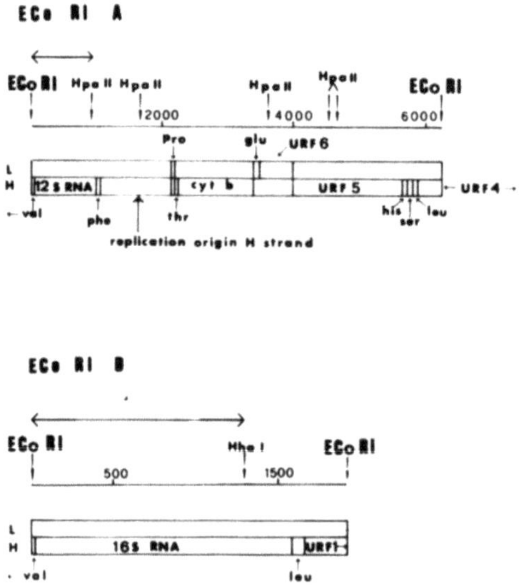

Fig. 2. Physical and transcription map of rat liver Eco RI fragments A and D. Only the restriction sites used for probe preparation are shown. A two headed arrow indicates the fragments that were labelled.

phoresis as shown in Fig. 4. On the gel the mitochondrial transcripts appear well evident (see 12S and 16S bands) even if some cytoplasmic contamination is present. After electrophoresis the mtRNA (about 30 µg) was transferred on nitrocellulose filters according to Thomas (15) and hybridized with the labelled probes for CoI, URF 1 and URF 4. In lanes 3 to 5 of Fig. 4 the higher molecular weight bands correspond to the expected size of the mRNAs. The partial transfer efficiency of the RNA, estimated around 50% by the blotting of radioactive RNA, does not allow however to make quantitative estimates of the single RNA concentration. Therefore, solution hybridization experiments were carried out.
Mitochondrial RNA and increasing concentrations of radioactive DNA fragments were put together in conditions that maximize

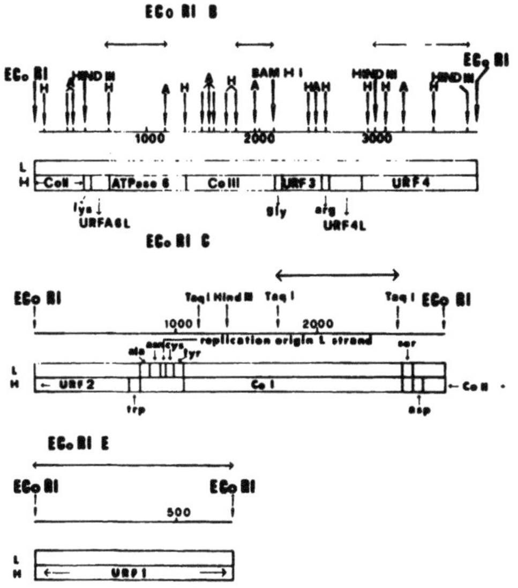

.Fig. 3. Physical and transcription map of rat liver Eco RI fragments B, C and E. The CoI probe is a fragment of 855 bp resulting from the digestion of Eco RI C with Taq I. The URF 4 probe is 800 bp long and comes from the digestion of Eco RI B with Hind III. The probes for ATPase and CoIII have been prepared digesting the Eco RI B fragment with Hind III and Bam HI. Four fragments were obtained. The fragment of 1676 bp was digested with Hinf I and Hae III. A fragment of 508 bp is the probe for the ATPase and two fragments 189 and 195 bp long are the CoIII probe. A and H are the Hae III and Hinf I sites.

the DNA-RNA hybrid formation (16). The mtRNA preparation used for the hybridization also contained DNA coextracted from mitochondria with RNA. Control experiments (not shown) demonstrated that the DNA did not interfere with the hybridization results, provided that corrections were made for the specific activity of the probe and that an equivalent quantity of cold DNA was present in the blanks. The results of such experiments for all the RNA species examined are reported in Fig. 5. Saturation value was reached in all the cases; the pg of hybrid-

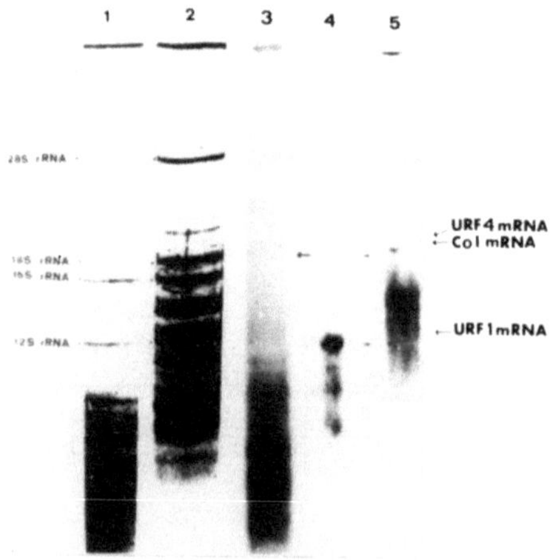

Fig. 4. Characterization of rat liver mtRNA. The electrophoretic pattern of RNA extracted by mitochondria treated and untreated with micrococcal nuclease is reported in slots 1 and 2. Slots 3 - 5 show the results of Northern hybridization experiments between mtRNA (untreated) and probes for CoI, URF 1 and URF 4 mRNA respectively. The treatment of mitochondria with micrococcal nuclease was performed as reported in (6).

ized RNA per µg of total mtRNA were used for the determination of each mtRNA species concentration. This was made by relating them to the amount of mtDNA coextracted from the mitochondrial fraction with the RNA and assuming a mean number of 5 DNA molecules per mitochondrion (7). The results obtained from this calculation are shown in Table I. There are 81 molecules of 12S rRNA and 36 molecules of 16S rRNA per mitochondrion. The mRNAs are instead present at lower levels ranging from 16 molecules of CoI mRNA to 6 molecules of URF 4 mRNA.
This mode of calculation resulted the most direct assuming that DNA and RNA are coextracted with the same yield from mitochondria and that the DNA present in the mitochondrial preparation be of mitochondrial origin only. The first assumption was verified determining the DNA/RNA ratio during the nucleic

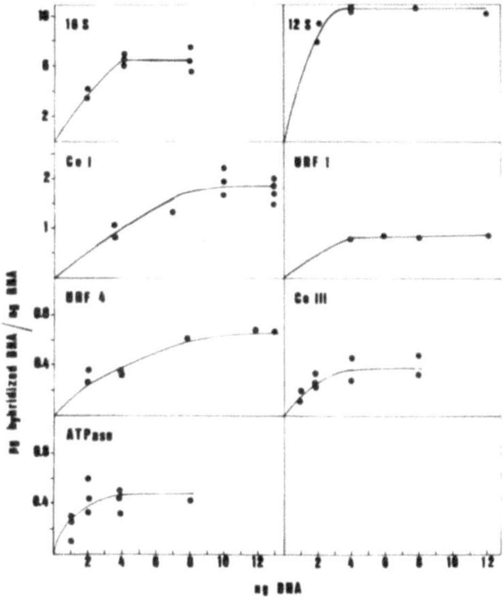

Fig. 5. Solution hybridization between mtRNA from adult rat and probes for 7 mtRNA. The hybridization was done according to the conditions described by Casey and Davidson (16). After incubation (20 h at 48°C) the samples were digested with 100 U/ml of S_1 nuclease and incubated for 30' at 45°C. The amount of hybrids formed was determined by trichloroacetic acid precipitations. Each experimental point was performed in triplicate. The background to be subtracted (1-2% of the input) was determined by measuring the S_1 resistant radioactivity of samples incubated without RNA.

acid extraction from mitochondria. Furthermore the DNA extracted from the mitochondria was 1% of that extracted from the homogenate of the same preparation (7) and the nuclear marker enzyme RNA-polymerase II was absent in mitochondrial preparations. There are two other ways to estimate mtRNA concentrations by refering the pmoles of hybridized RNA either to the mt proteins or to the total cellular DNA. The values obtained, normalized for the losses of mitochondria and RNA occurring

Table I. Mitochondrial concentrations of some mtRNA species in adult rat liver. The pg of the DNA hybridized represent the saturation values.

mtRNA	Hybridization (pg DNA/ng RNA)	RNA molecules per mitochondrion
12S	10.8 ± 0.66	81.2 ± 3.3
16S	6.5 ± 0.69	36.3 ± 4.2
URF1	0.82 ± 0.04	8.5 ± 0.5
CoI	1.84 ± 0.24	16.1 ± 2.1
ATPase	0.48 ± 0.07	7.1 ± 1.2
CoIII	0.38 ± 0.09	7.3 ± 1.6
URF4	0.66 ± 0.04	6.4 ± 0.4

during the extraction are similar to those reported in Table I. These data may be compared to those found by Gelfand and Attardi in HeLa cells mitochondria (6)by using methods different from our. These Authors measured the concentration of the 12S rRNA and of different mRNAs. The concentration of 12S rRNA is comparable in the two systems whereas the concentration of the mRNAs is very different. It is about 30 times higher in rat liver. This difference can be related to the different metabolism of the two systems. The rat liver cells, having a higher respiratory rate respect to the HeLa, may fully utilize the mt mRNA species that are known to code for chain respiratory components (the three subunits of the cytochrome oxidase, the subunit 6 of the ATPase, the apocytochrome b). It is interesting to note that the concentrations of CoI mRNA in the two systems correlate perfectly with their respective cytochrome content (17,18).

In rat liver the 12S rRNA concentration is about twice that of 16S rRNA. This could be explained with the gene order in mtDNA. The 12S rRNA is in fact immediately adjacent to the transcription initiation point on the H strand so that we can hypothesize an attenuation mechanism along the DNA molecule. The concentra

tion of mRNAs however does not seem to be related to the gene order. In fact, except for the CoI mRNA all the species studied are present more or less at the same concentration, indicating that the mRNAs are synthesized in a cohordinate fashion.
Gelfand and Attardi have demonstrated also that both rRNA and mRNA are short living in HeLa cell mitochondria, half lives of rRNA being however about one order higher than mRNA (6). They furthermore suggested that the rate of mtRNA decay is under feedback control by some mechanism linked to RNA synthesis or processing. Several studies have in fact demonstrated that when mtRNA synthesis is inhibited (19-21) mt protein synthesis proceeds for a longer time than could be expected on the basis of their half lives.
Studies carried out in sea urchin have demonstrated that in spite of mtDNA replication absence, mtRNA synthesis takes place during early development (22). However, the newly synthesized mRNA is not found on mitochondrial polysomes until the gastrula stage (23). In the cerebellum, a late developing organ in the rat, mtDNA and mt protein synthesis peak at 10 days of age, whereas mtRNA synthesis remains quite constant during the first three weeks of postnatal life (24,25). If we assume that the RNA decays with a kinetic of the first order (26) its concentration at steady state is given by the ratio between the rate of synthesis and that of degradation. It could be hypothesized that mtRNA synthesis proceeds in a constitutive way and that the concentrations of the mRNAs are regulated on the basis of cell necessity. Therefore, in condition of elevated respiratory rate, as in rat liver, the increased concentration of mtmRNA species could depend on an increase of their stability.
A regulation operating at mRNA synthesis level cannot however be ruled out and only experiments specifically directed towards the measurements of the mRNA stability under different physiol ogical conditions may solve the problem.
The model described previously does not seem to explain the fact that rat and HeLa concentrations of mtrRNA species are

similar, thus suggesting that the mtrRNA concentrations in different cell type are independent of their metabolism. Quantitation of mtrRNAs in other systems will be useful to clarify this point.

2. Variations of 16S and CoI levels in developing rats. It is well known that in rat liver at birth there is a sudden increase of oxidative phosphorilation accompained by an increase in the number and in cytoplasmic volume occupied by mitochondria. Mitochondrial maturation immediately after birth has also been reported. According to some Authors this rapid mitochondrial maturation requires the expression of the mitochondrial genes, according to others transcription and translation of mitochondrial genes are not implicated (8, 27-29). We think that the problem might be solved by a quantitative study on the single rat liver mtRNA species level variations during

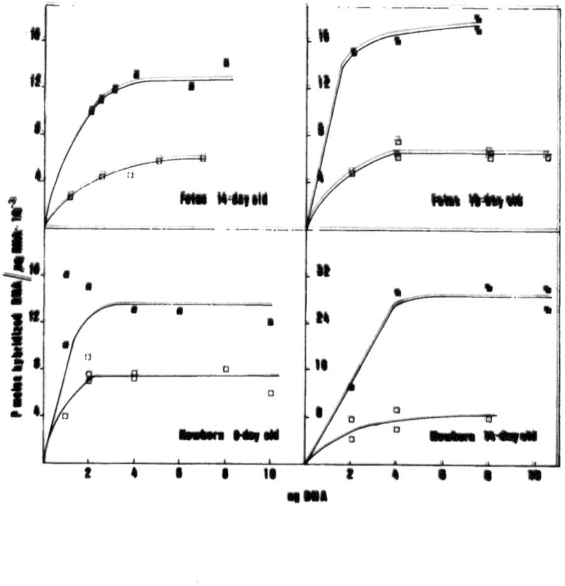

Fig. 6. Hybridization of mtRNA from developing rats with 16S (black triangles) and CoI (white triangles) DNA probes.

development.
In order to do this we isolated mitochondria from fetuses at eight and three days before birth and from rats at four hours after birth. MtRNA was extracted through lysis of mitochondria with guanidine thiocyanate and centrifugation of lysate on CsCl cushion (30). The resulting pellet consisted of RNA, the DNA being trapped in the CsCl layer as results from gel electrophoresis and cholorimetric analysis. Hybridization experiments between RNA and increasing amounts of DNA probes coding from the 16S rRNA and CoI mRNA were performed as described previously. The results are summarized in Fig. 6.
As a first approach we chose to measure the ratio between the 16S rRNA and CoI mRNA at various ages withouth making any estimate of the absolute concentration of the individual RNA species. This makes the results independent of the recovery and of contamination of the single RNA preparation and may give indications about the variations of the levels of one species respect to another. The ratio between the pmoles hybridized at saturation per µg of total mtRNA at the four development stages examined are reported in Table 2.
We can observe a slight decrease of the ratio at -3 days of age and at birth. This can be explained by assuming a constant level of mtrRNAs during development and an increase in the concentration of CoI mRNA, at the time of fast mitochondrio-

Table 2. Ratio between the pmoles of 16S and CoI RNAs per µg of mtRNA hybridized at saturation.

Animals (age)	16S rRNA (pmol/µg RNA.10^{-3})	CoI mRNA (pmol/µg RNA.10^{-3})	Ratio
Fetus (14-day old)	12.7 ± 1.0	5.0 ± 0.5	2.54
Fetus (19-day old)	14.0 ± 1.5	7.7 ± 0.7	1.82
Newborn (0-day old)	14.0 ± 2.0	7.4 ± 0.1	1.90
Newborn (14-day old)	29.1 ± 2.0	8.3 ± 0.5	3.52
Adult	15.1 ± 1.6	6.6 ± 0.8	2.28

genesis. It is interesting to note that the cytochrome oxidase is just one of the mitochondrial enzymes that upsurges in the neonatal enzyme cluster (31). The determination of the exact number of rRNAs molecules per mitochondrion will allow to clarify these data.

Acknowledgements

We thank F. Fracasso for the skilfull technical assistance and M. Badini for the preparation of the manuscript.
This work was partially supported by a grant from the Progetto Finalizzato Ingegneria Genetica e Basi Molecolari delle Malattie Ereditarie, CNR and partially by a grant from M.P.I. (40%), Italy.

Non standard abbreviations: mt: mitochondrial; bp: basepair(s); nt: nucleotide(s).

References

1. Anderson, S., Bankier, A.T., Barrel, B.G., de Bruijn, M.H.L., Coulson, A.R., Drouin, J., Eperon, I.C., Nierlich, D.P., Roe, B.A., Sanger, F., Schreier, P.H., Smith, A.J.H., Staden, R., Young, I.G.: Nature 290, 457-465 (1981).
2. Bibb, M.J., Van Etten, R.A., Wright, C.T., Walberg, M.W., Clayton, D.A.: Cell 26, 167-180 (1981).
3. Anderson, S., de Bruijn, M.H.L., Coulson, A.R., Eperon, I.C., Sanger, F., Young, I.G.: J. Mol. Biol. 156, 683-717 (1981).
4. Attardi, G., Cantatore, P., Chomyn, A., Crews, S., Gelfand, R., Merkel, C., Montoya, J., Ojala, D. In "Mitochondrial genes" (Slonimski, P., Borst, P. and Attardi, G. eds) pp. 51-71 (1982). Cold Spring Harbor, New York.
5. Montoja, J., Christianson, T., Levens, D., Rabinowitz, M., Attardi, G.: Proc. Natl. Acad. Sci. USA 79, 7195-7199 (1982).
6. Gelfand, R. and Attardi, G.: Mol. Cell Biol. 1, 497-511 (1981).
7. Neubert, D., Gregg, C.T., Bass, R., Marker, H.J. In "Biochemistry of animal development" (Weber, R. ed) vol. III. Academic Press, New York, 387-464 (1975)

8. Pollack, J.K., Sutton, R.: Trends in Biol. Sci. 1, 23-25 (1980).

9. Saccone, C., Cantatore, P., Gadaleta, G., Gallerani, R., Lanave, C., Pepe, G., Kroon, A.M.: Nucl. Ac. Res. 9, 4139-4147 (1981).

10. Cantatore, P., De Benedetto, C., Gadaleta, G., Gallerani, R., Kroon, A.M., Holtrop, M., Lanave, C., Pepe, G., Quagliariello, C., Saccone, C., Sbisà, E.: Nucl. Ac. Res. 10, 3279-3289 (1982).

11. Pepe, G., Holtrop, M., Gadaleta, G., Kroon, A.M., Cantatore, P., Gallerani, R., De Benedetto, C., Quagliariello, C., Sbisà, E., Saccone, C.: Biochem. Int. 6, 553-563 (1983).

12. Kobayashi, M., Seki, T., Yaginuma, K., Koike, K.: Gene 16, 297-307 (1981).

13. Koike, K., Kobayashi, M., Yaginuma, K., Taira, M., Yoshida, E., Imai, M.: Gene 20, 177-185 (1982).

14. Rigby, P.W.J., Dieckmann, M., Rhodes, C., Berg, P.: J. Mol. Biol.: 113, 237-251 (1977).

15. Thomas, P.S.: Proc. Natl. Acad. Sci. USA 77, 5201-5205 (1980).

16. Casey, J., Davidson, N.: Nucl. Ac. Res.: 4, 1539-1552 (1977).

17. Williams, J.N., Jr: Biochem. Biophys. Acta 162, 175-181 (1968).

18. Hare, J., Ching, E., Attardi, G.: Biochem. 19, 2023-2030 (1980).

19. England, J., Costantino, P., Attardi, G.: J. Mol. Biol. 119, 445-462 (1978).

20. Lansman, R.A., Clayton, D.A.: J. Mol. Biol. 99, 777-793 (1975).

21. Gadaleta, M.N., Di Reda, N., Bove, G., Saccone, C.: Eur. J. Biochem. 51, 495-501 (1975).

22. Saccone, C., Nicotra, A.: Bull. Mol. Biol. Med. 3, 23s-31s (1978).

23. Devlin, R.: Dev. Biol. 50, 443-456 (1976).

24. Gadaleta, M.N., Giuffrida, A.M., Renis, M., Serra, I., Del Prete, G., Geremia, E., Saccone, C.: Neurochem. Res. 4, 25-35 (1979).

25. Giuffrida, A.M., Gadaleta, M.N., Serra, I., Renis, M., Geremia, E., Del Prete, G., Saccone, C.: Neurochem. Res. 4, 37-52 (1979).

26. Galau, G.A., Lipson, E.D., Britten, R.J., Davidson, E.H.: Cell 10, 415-432 (1977).

27. Herzfeld, A., Federman, M., Greengard, O.: J. Cell Biol.

57, 475-483 (1973).

28. Jakovcic, S., Haddock, J., Getz, G.S., Rabinowitz, M., Swift, H.: Biochem. J. 121, 341-347 (1971).

29. Hallman, M.: Biochem. Biophys. Acta 253, 360-372 (1971).

30. Chirgwin, J.M., Przybyla, A.E., Mac Donald, R.J., Rutter, W.J.: Biochemistry 18, 5294-5299 (1979).

31. Greengard, O. In Biochemistry (Campbell,P.N. and Diekens, F. eds) vol. 7, Academic Press, New York, 159-205 (1971).

HETEROLOGOUS IN VITRO TRANSCRIPTION OF MITOCHONDRIAL DNA SEGMENTS

Gabriele Görtz, Rüdiger Grosskopf, Horst Feldmann
Institut für Physiologische Chemie, Physikalische Biochemie und Zellbiologie der Universität München
D 8000 München 2

Introduction

In vitro studies of RNA processing in mammalian mitochondria need appropriate substrates. The isolation of RNA presursors from natural sources, however, is still difficult, and homologous in vitro transcription systems yielding large RNA precursors have not been described. Therefore, we have tested the possibility to obtain in vitro transcripts from defined mitochondrial DNA segments in heterologous systems. Two different approaches for the construction of recombinant plasmids and their subsequent transcription in HeLa cell free extracts, or with E.coli polymerase, respectively, will be reported briefly.

Results and Discussion

1. Integration of the cytochrome b gene region from rat liver mitochondria into an expression plasmid and transcription with E.coli polymerase.

Recently, we determined the nucleotide sequence of a 1.76 Kb DNA fragment from rat liver mitochondria spanning the cytochrome b gene and three flanking tRNA genes (1). The 1.92 Kb HpaII fragment C containing this region was isolated from a recombinant plasmid (pBR322-HindIII fragment A from rat liver mitochondria (1)) and integrated into the HpaI site of

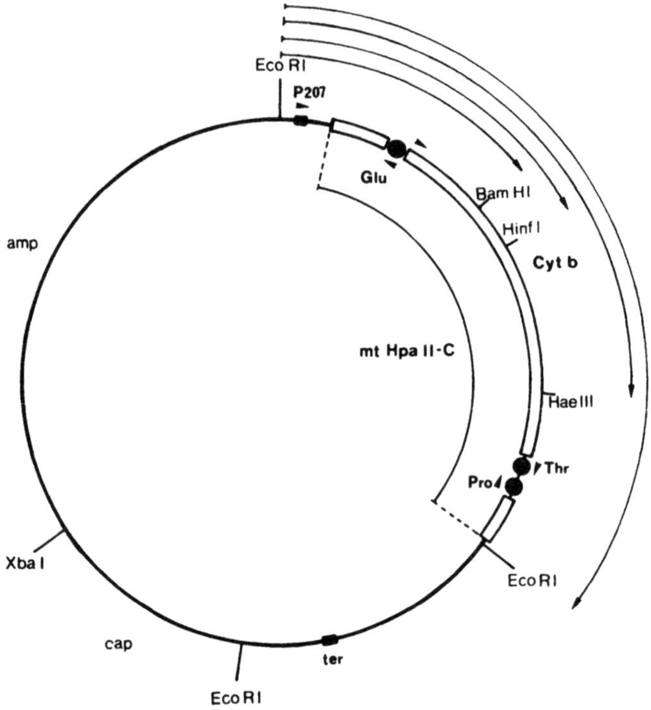

Figure 1: Construction of expression plasmid pmGG3.

The 1.92 Kb fragment C (open block) was integrated in the HpaI site of p2/5 811-1*13 (2)(solid line). Amp, ampicilline resistance gene; cap, chloramphenicol resistance gene; P207, strong T5 promoter; ter, phage fd termination signal; cyt b , cytochrome b gene; the tRNA genes are symbolized by the dots and the three letters of the corresponding amino acids. The small arrows indicate the direction of transcription; the large arrows refer to the restriction fragments used for "run off" transcription (see Figure 2).

p2/5 811-1*13 (2) as a vector (Figure 1). For this purpose, the protruding ends of the HpaII fragment were filled in with DNA polymeraseI from E.coli, and this segment blunt-end ligated to the vector DNA that had been linearized by cleavage with HpaI. This procedure places the mitochondrial DNA between the strong promoter and the transcription terminator in the expression vector (2), with the promoter

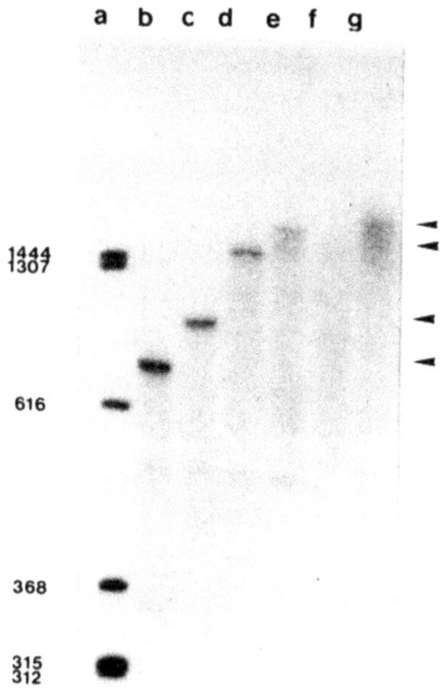

Figure 2: Autoradiogramme of RNA transcripts resolved by gel electrophoresis.
Length standard (a); transcripts from restriction fragments of Fig.1: EcoRI/BamH1 (b), EcoRI/HinfI (c), EcoRI/HaeIII (d), EcoRI/EcoRI (e).

being located 125 bp upstream from the inserted DNA fragment. E.coli cells were transformed with the recombinant DNA, and clones were selected that gave signals in colony hybridization experiments with the nick-translated HpaII fragment. Further restriction analyses together with filter hybridizations demonstrated that three of the clones had integrated the HpaII fragment with the cytochrome b gene in proper orientation for its transcription (Figure 1). In vitro transcription was explored under a variety of conditions, using commercial E.coli RNA polymerase as enzyme, ^{32}P UTP as

label, and full length plasmid DNA or shorter restriction fragments as templates. The transcripts were separated on polyacrylamide gels. Figure 2 shows the results of one such experiment. The radioactively labeled products were then isolated and rehybridized to appropriate restriction fragments in order to assure that the transcripts originated from the DNA fragments used as templates (data not shown).
With these experiments we demonstrate that mitochondrial DNA can be stably integrated in a prokaryotic expression vector and is faithfully transcribed in an in vitro system by E.coli polymerase.

2. Transcription of mitochondrial DNA segments by HeLa cell free extracts.

The majority of the tRNA in mammalian mitochondria (e.g. 3-6) have lost the characteristic features of the D-and the T-loops of prototype tRNA. Only three of them, namely $tRNA^{Leu}_{UUR}$, $tRNA^{Gln}$, and $tRNA^{Ser}_{UCN}$, have retained the consensus sequences (Figure 3) that in eukaryotic systems serve as the bipartite promoter elements for transcription by RNA polymerase III (e.g.:7). Therefore, we have tested the possibility, whether HeLa cell extracts containing RNA polymerase III activity would yield transcripts from appropriate mitochondrial DNA segments. Figure 4 shows portions of the rat liver mitochondrial genome which were cloned in pBR322 (3). Either the plasmid DNAs or the excised restriction fragments were used as templates; the transcription products were resolved by gel electrophoresis (Figure 5). No signals were obtained with the linear HindIII/EcoRI fragment (Fig.4c), with the linear EcoRI fragment, or with the plasmid DNA containing the EcoRI fragment in opposite orientation to that indicated in Fig. 4b. The plasmid DNA containing the EcoRI fragment in the orientation of Fig. 4b, yielded several products of different

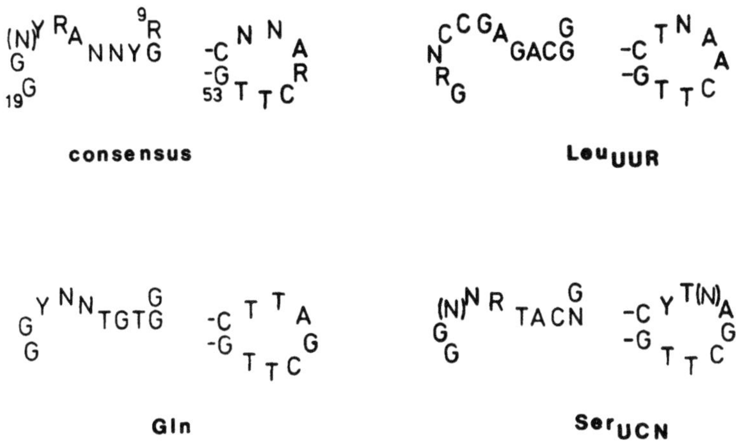

Figure 3: Comparison of the D-and T-loop consensus sequences from mammalian mitochondrial tRNAs (3-6).

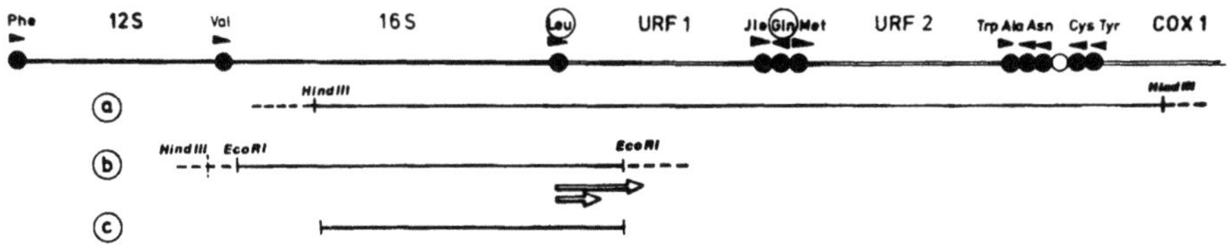

Figure 4: In vitro transcription of cloned rat mitochondrial DNA segments with HeLa cell free extracts.

(a) HindIII fragment D (ref.3)
(b) EcoRI fragment D (refs.3 and 8)
(c) HindIII/EcoRI fragment

length (Figure 5 d-g).The largest (band 1) is 380 nucleotides long, the smallest (band 5) has the size of the mature $tRNA^{Leu}$. Mapping of these products by rehybridization to smaller

Figure 5: Autoradiogramme of in vitro transcription products from Figure 4.

a: control DNA, 160 nucleotides (pH_{III} ß*-KpnI); b: no DNA; c: HindIII/EcoRI fragment (Fig. 4c); d - g: different concentrations of EcoRI plasmid DNA (Fig. 4b). The arrows indicate bands produced as a background.

restriction fragments showed that all of them originate from the region of the $tRNA^{Leu}$ gene, and that therefore they must have a precursor-product relationship. An explanation that the fragment is active in transcription only in the one orientation is offered by the fact that in this case 5 T's from the pBR322 sequence are placed ca. 380 nucleotides distal from the 5'-end of the tRNA gene in the non-coding strand, and thus can serve as a transcription termination signal (e.g.:8). Transcription experiments using restriction fragments as templates that extend beyond the EcoRI site (e.g.:Fig.4a) were unsuccessful so far. It may be that a T cluster is missing or is too far downstream in the mitochondrial DNA. An answer has to await data on nucleotide sequences from the URF1/URF2 region of rat liver.

With these experiments we have shown that eukaryotic RNA polymerase III can recognize a mitochondrial tRNA gene provided it has conserved promoter elements, and that obviously the primary transcript(s) is processed by this in vitro system to yield a stable product of tRNA size. On this basis, it would be interesting to know, why explicitely the three above mentioned tRNA structures have been conserved in mammalian mitochondria, and whether the cytosolic processing enzymes could also interact with larger transcripts from the H-or L-strands.

Materials and Methods

[α^{32}P]dATP, [α^{32}P]dCTP, [α^{32}P]UTP, and HeLa cell free extracts (9) were purchased from New England Nulear Corp. E.coli RNA polymerase was bought from P-L Biochemicals Inc. Restriction enzymes were obtained from the indicated sources and used as suggested by the suppliers: EcoRI, BamH1, HpaI, Sau3A, TaqYI, XbaI (Boehringer); RsaI, MnlI, DdeI (Biolabs); HinfI, HaeIII (R.E.Streeck,this laboratory); HindIII, HpaII (H.Feldmann). T4 DNA ligase and E.coli DNA polymerase were from Boehringer GmbH.,Mannheim. SeaKem and SeaPlaque agarose were products of Marine Colloids, Rockland. p2/5 811-1*13 was kindly given by H.Bujard,Heidelberg.

Ligation (10) of DNA and transformation (11) of E.coli 490A were performed as detailed in the references. Nick-translation, colony hybridization , and filter hybridization followed the protocols of references 12, 13, and 14, respectively. Minilysate assays were performed according to

Birnboim and Doly (15). Plasmid DNA was prepared by scaling up this procedure; the RNase step was omitted, instead the DNA was purified by repeated centrifugations in a CsCl-EtBr gradient. Defined restriction fragments were prepared by gel electrophoreses as described earlier (3).

Transcription with E.coli polymerase: 0.2 pMole DNA was incubated with 0.3 mM GTP and ATP,each, 0.2 mM CTP and UTP, each, 0.015 µM labeled UTP (3000 Ci/mMole) and 1 pMole RNA polymerase in 50 µl transcription buffer (120 mM KCl, 10 mM $MgCl_2$, 20 mM Tris-HCl, 1 mM EDTA, 5% glycerol, pH 7.9) for 2 to 5 min at 37°C. The reaction was stopped by adding 2 µl of 25mM UTP. A 3µl aliquot was added to 6 µl of a solution of 0.1% bromophenol blue , 45 mM Tri-borate buffer, 95% formamide, the mixture heated for 45 sec to 95°C, chilled on ice and immediately electrophorized on a polyacrylamide gel 0.5 x 40 cm, 5% acrylamide, 1:37.5 methylene bisacrylamide, 7 M urea, Tris-borate buffer).

Transcription with HeLa cell extracts followed the procedure of ref. 9. The total volume was 25 µl, incubation was for 60 to 9o min at 30°C. The reaction was stopped by adding 1µl 10% SDS, 1µl proteinase K (2 mg/ml) ,and incubating 15 min at 37°C. 10 µg yeast tRNA was added as carrier and the solution made 0.3 M in NaOAc. Deproteinization was achieved by 3 fold extraction with phenol/chloroforme , and the RNA was precipitated with ethanol. The pellet was dissolved in 0.1 M NaOAc buffer, pH 5.2, heated 5 sec to 65°C for denaturation and electrophorized on a 1 x 40 cm gel (5% acrylamide, 1:20 methylene bisacrylamide, 8 M urea, Tris-borate buffer). Autoradiography was performed at -70°C with Kodak X-ray film.

Acknowledgements

We wish to thank H.Bujard and his collaborators for the gift of plasmid p2/5 811-1*13 and their help with the transcription system. The Deutsche Forschungsgemeinschaft and the Fond der Chemischen Industrie have supported this work.

References

1. Görtz, G., Feldmann, H.: Current Genet. 5, 221-225 (1982)
2. Gentz, R., Langner, A., Chang, A.C.Y., Cohen, S.N., Bujard, H.: Proc.Natl.Acad.Sci.USA 78, 4936-4940 (1981)
3. Grosskopf, R., Feldmann, H.: Current Genet. 4, 151-158; 191-196 (1981)
4. Anderson, S., Bankier, A.T., Barrell, B.G., de Bruijn, M.H.L., Coulson, A.R., Drouin, J., Eperon, I.C., Nierlich, D.C., Roe, B.A., Sanger, F., Schreier, P.H., Smith, A.J.H., Staden, R., Young, I.G.: Nature 290, 457-464 (1981)
5. Anderson, S., de Bruijn, M.H.L., Coulson, A.R., Eperon, I.C., Sanger, F., Young, I.G.: J.Mol.Biol. 156, 683-717 (1982)
6. Bibb, M.J., van Etten, R.A., Wright, C.T., Walberg, M.W. Clayton, D.A.: Cell 26, 167-180 (1981)
7. Hofstetter, H., Kressmann, A., Birnstiel, M.L.: Cell 24, 573-585 (1981)
8. Saccone, C., Cantatore, P., Gadaleta, G., Gallerani,R., Lanave, C., Pepe, G., Kroon, A.M.: Nucleic Acids Res. 9, 4139-4148 (1981)
9. Manley, J.L., Fire, A., Cano, A., Sharp, P.A., Gefter, M.L.: Proc.Natl.Acad.Sci.USA 77, 3855-3859 (1980)
10. Pech, M., Streeck, R.E., Zachau, H.G.: Cell 18, 883-893 (1979)
11. Lederberg, E.M., Cohen, S.: J.bacteriol. 119, 1072-1074 (1974)
12. Rigby, P.W.J., Dieckmann, M., Rhodes, C., Berg, P.: J.Mol.Biol. 113, 237-251 (1977)
13. Grunstein, M, Hogness, D.S.: Proc.Natl.Acad.Sci.USA 72, 3961-3965 (1975)
14. Southern, E.M.: J.Mol.Biol. 98, 503-517 (1975)
15. Birnboim, H.C., Doly, J.: Nucleic Acids Res. 7, 1513-1523 (1979)

TRANSCRIPTION OF YEAST MITOCHONDRIAL tRNA GENES AND PROCESSING OF tRNA GENE TRANSCRIPTS

D.L. Miller, K. Underbrink-Lyon, D.R. Najarian, J. Krupp, N.C. Martin

Department of Biochemistry
The University of Texas Health Science Center at Dallas
Dallas, Texas

Introduction

Yeast mitochondrial DNA codes for the complete set of transfer RNAs (tRNAs) used in mitochondrial protein synthesis. Petite deletion mapping, restriction enzyme mapping and DNA sequence analysis has provided detailed information about the organization of tRNA genes on yeast mitochondrial DNA (reviewed in 1). This information has, in turn, enabled studies on the expression of mitochondrial tRNA genes.

Our laboratory has been using a combined genetic and biochemical approach to achieve an understanding of tRNA biosynthesis in yeast mitochondria. We find that tRNA sequences are transcribed as higher molecular weight RNAs. The structure of these transcripts suggests that the production of mature tRNAs requires nucleolytic RNA processing steps at both the 5' and 3' ends as well as nucleotide modification and the addition of a 3' CCA end. Most of the genetic information necessary for mitochondrial tRNA biosynthesis is encoded by nuclear genes, but one mitochondrial locus other than the tRNA genes is also required. This mitochondrial tRNA synthesis locus codes for an RNA molecule that appears to be essential for the maturation of the 5' end of mitochondrial tRNAs.

Mitochondria 1983

Results and Discussion

tRNA gene transcripts have been identified in wild type strains as well as in strains carrying deletions of mitochondrial DNA (petites) or point mutations in tRNA genes (syn^{-}). Mitochondrial RNA is isolated, separated by gel electrophoresis and transferred to DBM paper prior to hybridization with radiolabeled genes. Transcripts which appear to contain ribosomal RNA plus tRNA sequences, messenger RNA plus tRNA sequences and tRNA sequences alone have been identified, and an example of each type of transcript is shown below.

Figure 1 shows that the large ribosomal RNA (21S) and tRNA$^{Thr}_{ACN}$ genes may be cotranscribed. Although we have not detected the combined transcript in wild type cells, a transcript which seems to contain both the 21S rRNA and tRNA$^{Thr}_{ACN}$ is clearly present in the petite strain F11.

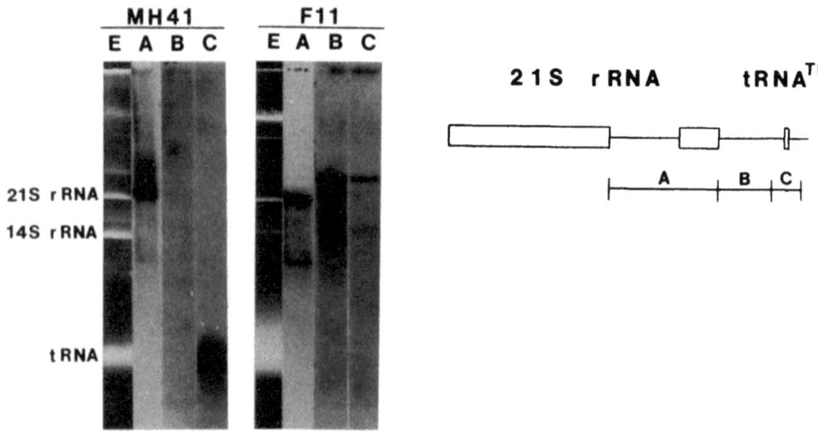

FIGURE 1. Northern hybridization experiments using labeled restriction fragments from the region of the 21S rRNA and tRNA^{Thr} genes. RNAs from isolated mitochondria of wild type strain, MH41-7B, and petite strain F11, were separated on 1.5% agarose-urea gels, stained with ethidium bromide and photographed (E). Restriction fragments corresponding to the regions marked A, B and C were labeled and hybridized with the RNA transferred in situ to DBM paper. The exon-intron probe, A, hybridized with 21S rRNA and a precursor-21S rRNA retaining the intron in wild type and with 21S rRNA and the excised intron in F11. The 3' extension probe, B, does not hybridize with wild type RNA but hybridizes with a 21S rRNA precursor retaining a 3' extention in F11. The tRNA^{Thr} probe, C, hybridizes with the mature tRNA^{Thr} in wild type and with the 21S rRNA precursor retaining the 3' extension in F11.

This result is consistent with previous electron microscopic (2) and hybridization (3) studies.

Messenger RNA and tRNA sequences are also cotranscribed. Figure 2 shows that the $tRNA^{Ser}_{UCN}$ and the ATPase subunit 9 genes are cotranscribed. The predominant transcript in wild type cells detected by the tRNA probe is the $tRNA^{Ser}_{UCN}$ but at least one larger RNA which contains this sequence can also be seen (➡). In contrast, the petite RNA shows a number of high abundance, large transcripts which contain tRNA sequences as well as ATPase subunit 9 mRNA sequences but no mature tRNA is detected. Mature mRNA does appear to be made (⧫).

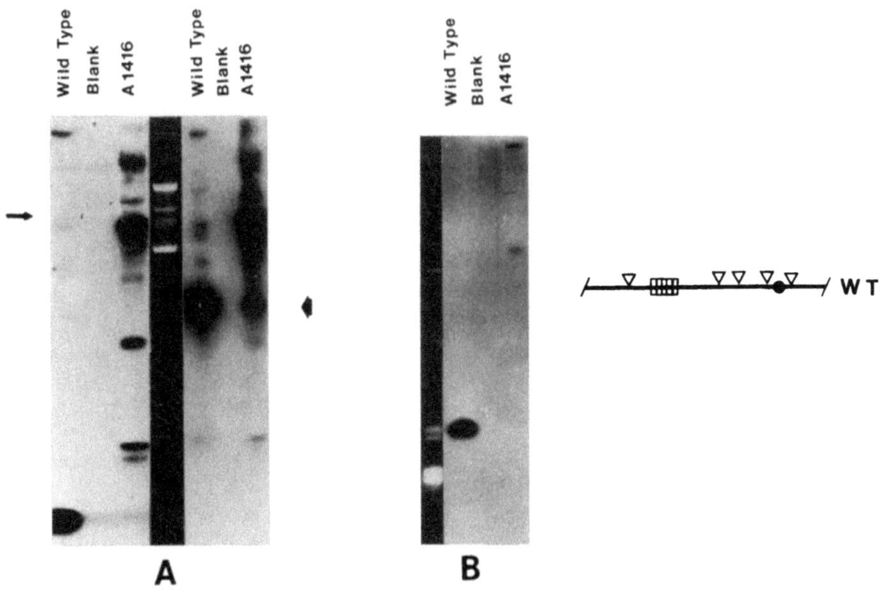

FIGURE 2. Identification of $tRNA^{Ser}_{UCN}$ gene transcripts in wild type and mutant cells. Mitochondrial RNA was isolated from the wild type strain COP41-161 and a petite strain A14-16, separated on 1.5% agarose urea gels (Panel A) or 10% acrylamide urea gels (Panel B), transferred to DBM paper and hybridized with the Hpa II fragments that contain the $tRNA^{Ser}$ gene (left three lanes of panel A and all lanes of panel B) or the ATPase subunit 9 gene (right three lanes of panel A). The map on the right shows the relative location of the ATPase subunit 9 gene (▥) and $tRNA^{Ser}_{UCN}$ gene (●) as well as the HpaII sites (▽) in the region.

Some petites are able to make mature tRNA. Figure 3 shows hybridization of a $tRNA_f^{Met}$ gene probe to RNA from wild type cells and a petite mutant able to make tRNA (MB81). The mature $tRNA_f^{Met}$ and a higher molecular weight RNA of about 100 nucleotides are clearly evident in the petite MB81 and we assume it is an intermediate in $tRNA_f^{Met}$ synthesis. This 100 base transcript accumulates in mutants unable to make tRNAs (ND40). Frontali et al. have also found RNA molecules that are larger than mature tRNAs and some of these transcripts contain several tRNA sequences (4). These data are consistent with the idea that large tRNA gene transcripts are produced and must be processed by nucleolytic events.

The two methods used in the yeast mitochondrial system to determine whether transcripts detected by Northern analysis are actually primary transcripts have been pioneered by the Rabinowitz laboratory. They are labeling di- or triphosphate termini with [α^{32}P]GTP and the enzyme guanylyl transferase and in vitro transcription. Transcription of the 21S rRNA is initiated at the 5' end of the rRNA gene (5), so the transcript

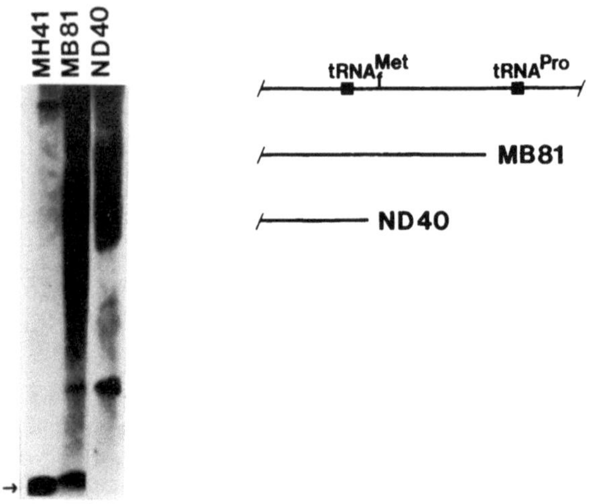

FIGURE 3. $tRNA_f^{Met}$ transcripts in wild type and mutant cells. Mitochondrial RNA from the wild type MH41-7B and petite strains MB81 and ND40 was isolated, transferred to DBM paper and hybridized to $tRNA_f^{Met}$ gene probes. The mature tRNA (→) can be seen in the wild type and MB81 RNA but not ND40 while a 100 base transcript that accumulates in ND40 can be seen in MB81. The map at the right represents the wild type DNA and the DNA retained in each petite.

that contains the 21S RNA and the $tRNA^{Thr}$ sequences must be a primary transcript.

Some tRNA and mRNA transcripts that are cotranscribed also appear to be primary transcripts. The transcript that contains the ATPase subunit 9 mRNA and $tRNA^{Ser}_{UCN}$ sequences (Figure 2) can be capped (Zassenhaus, Martin and Butow, in preparation). Also, only one site for the initiation of transcription in vitro is found on a clone that contains the ATPase subunit 9 gene, the $tRNA^{Ser}_{UCN}$ gene and the Var I gene suggesting that all of these genes are cotranscribed (Edwards, Martin and Rabinowitz, unpublished). There is a transcription initiation site 391 base pairs upstream from the $tRNA^{Glu}$ gene and this transcript appears to contain the $tRNA^{Glu}$ and the mRNA for cytochrome b (7). Finally, petite deletion mapping studies indicate the presence of a promotor upstream of the $tRNA^{Val}$ gene which is responsible for the transcription of the $tRNA^{Val}$ as well as the mRNA for cytochrome oxidase subunit III (8).

There is a consensus sequence which is located at the site of initiation of transcription for the ribosomal RNAs (9,10) as well as for a number of other primary transcripts (11). This sequence, 5'A/T-TATA-A/G-GTA3', must be an important element of the mitochondrial promoter. It is found twice in the DNA sequence reported in Figure 4 which contains the $tRNA^{Met}_{f}$ gene. Sequence analysis of the 100 base RNA which accumulates in ND40 (Figure 3) demonstrates that its 5' end falls at nucleotide 967 in the DNA sequence shown in Figure 4. Nucleotide 967 is the last nucleotide in one of the consensus sequences. This RNA is a primary transcript because it labels in the presence of [α^{32}P] GTP and the enzyme guanylyl transferase (See Figure 9, lane 7). Further, we have also observed initiation of transcription from this site in vitro (Edwards, Miller, Martin and Rabinowitz, unpublished). Although both consensus sequences are 5' to the $tRNA^{Met}_{f}$ gene, only one of them is used in vivo under the growth conditions we have tested to date. Factors other than the consensus sequence which are necessary for the initiation of transcription of yeast mitochondrial DNA remain to be determined. Since the consensus sequence used to initiate the 100 nucleotide transcript is the only functional consensus sequence in this region, it is likely that this whole region is expressed as a single transcript which includes $tRNA_f^{Met}$, 9S RNA (see below) and $tRNA^{Pro}$.

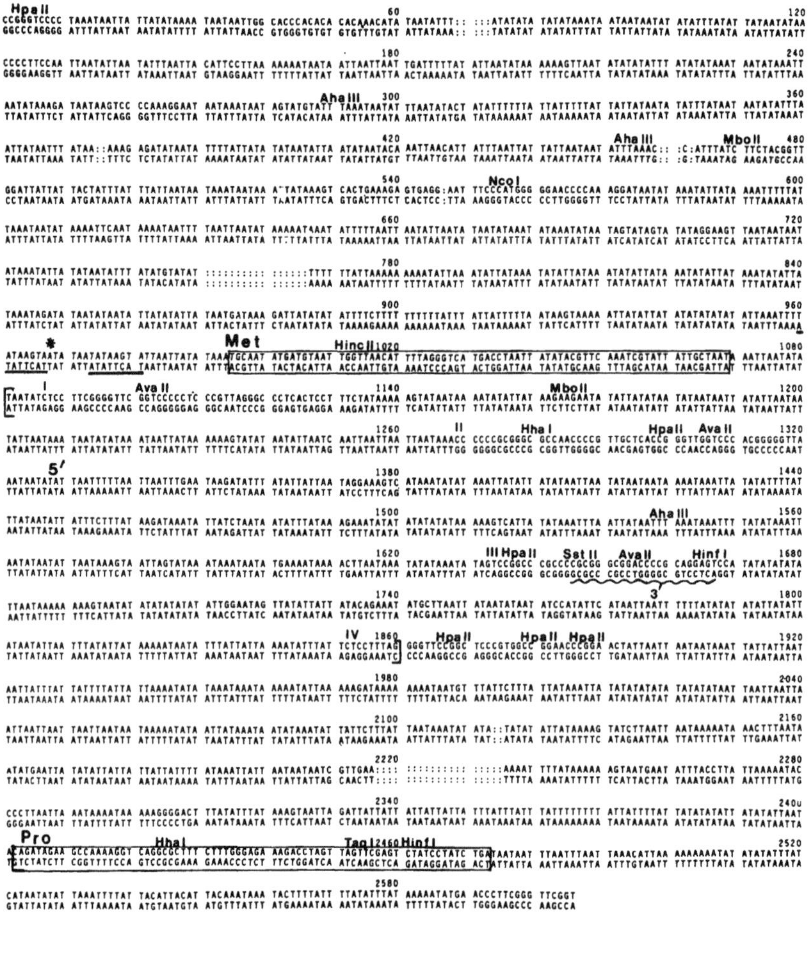

FIGURE 4. DNA sequence of a portion of the yeast mitochondrial genome from wild type strain, MH41-7B containing the $tRNA_f^{Met}$ gene, tRNA synthesis locus and the $tRNA^{Pro}$ gene. The $tRNA_f^{Met}$ and $tRNA^{Pro}$ gene sequences are boxed. Brackets mark the boundary of the tRNA synthesis locus. Consensus sequences found at the sites of initiation of transcription are underlined. The site of initiation of transcription of the $tRNA_f^{Met}$ gene is designated with an asterisk (*). 5' and 3' mark the 5' and 3' ends of the 9S RNA. I, II, III and IV indicate the various G+C rich elements located in this sequence. The wavy line denotes the primer used in primer extension experiments. Various restriction enzyme recognition sites are shown.

All of these results on the initiation of transcription of mitochondrial genes show that gene expression in yeast mitochondria resembles that found in prokaryotes such as E. coli, in that promotion occurs external to tRNA genes and probably does not depend on sequences internal to the genes as has been found for eukaryotic tRNA genes. Our examination of tRNA gene transcripts from point mutants in yeast mitochondrial tRNA genes is also consistent with this conclusion. We reasoned that such mutations might disrupt either transcription or processing of tRNA transcripts. Transcription in these mutants does occur, since each syn^- mutant we have examined does make mature tRNA from the mutant gene although the steady state level of transcripts often varies from wild type and from other mutants. Figure 5 shows an experiment with RNA from FF1210-170, a strain with a mutation in the $tRNA^{Asp}$ gene (13,14). An RNA that migrates on 10% acrylamide gels as if it were a larger RNA than the mature tRNA can be seen and is indicated by an arrow in Figure 5. One other syn^- mutant we have studied also accumulates an RNA larger

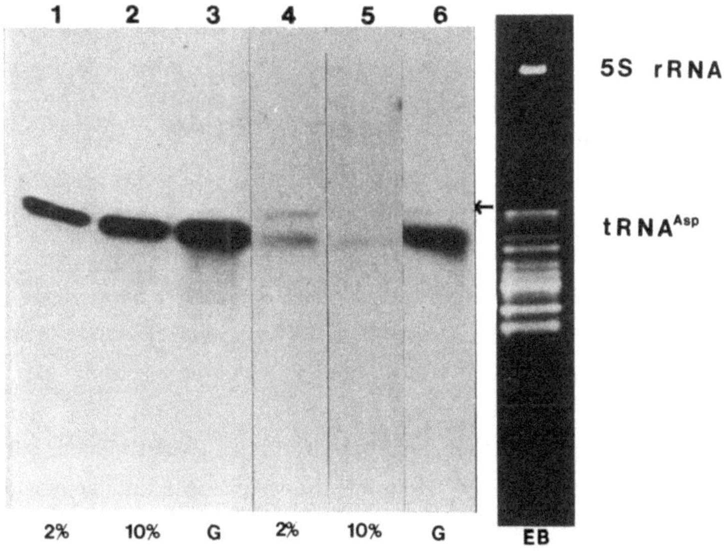

FIGURE 5. Transcripts of the $tRNA^{Asp}$ gene in wild type and syn^- mutant FF1210-170. Whole cell RNA was isolated, separated on a 10% acrylamide-urea gel and transferred to DBM paper prior to hybridization with a $tRNA^{Asp}$ gene probe. Lanes 1-3 are wild type RNA, Lanes 4-6 are mutant RNAs isolated from cells grown in 2% glucose (2%), 10% glucose (10%) or 2% galactose (G) media. Equal amounts of whole cell RNA (EB) were added in each lane. The arrow indicates the larger RNA containing the $tRNA^{Asp}$ sequences.

than the mature tRNA and in very long exposures such RNAs can also be seen in wild type cells. The structure of these RNAs is unknown. There are numerous examples of point mutations in tRNA genes which alter the processing of tRNA gene transcripts in other systems. An analysis of the structure of these mutant transcripts may be useful in understanding the tRNA biosynthetic process in mitochondria.

Some petite mutants transcribe their tRNA genes and make mature tRNAs and some accumulate larger RNAs containing tRNA sequences and do not make mature tRNAs. An analysis of the transcripts in petite mutants defective in tRNA biosynthesis has given insights into the mechanism of action of the tRNA synthesis locus. Further, deletion mapping of petites with and without tRNA synthetic ability has enabled the location of the tRNA synthesis locus and the discovery of its product. This information is required for further work towards understanding how the gene is expressed and how the product functions.

The sequence of the $tRNA_f^{Met}$ gene transcript seen in Figure 3 has been completed (Underbrink-Lyon, K., Miller, D.L., Ming, X., and Martin, N.C., in preparation). The transcript has a mature 3' end and can be charged with methionine. These results are consistent with three previous lines of experimental evidence suggesting that petites which do not make mature tRNAs can have transcripts with mature 3' ends. First, RNAs from petites such as A17-10 can be labeled with the E. coli enzyme nucleotidyl transferase which adds the 3' CCA sequence to tRNAs (Martin, unpublished). Second, Wesolowski and Fukuhara (15) reported that some transcripts from tRNA genes in petities that do not make mature tRNAs can still be charged with the appropriate amino acids. Third, Frontali et al. (4) and Zassenhaus, Martin and Butow (in preparation) have shown by S1 nuclease mapping that some tRNA gene transcripts in petites without the synthesis locus have correctly processed 3' ends but are extended on their 5' ends. These results lead to the conclusion that petites unable to make mature tRNAs are defective in 5' end processing and that information necessary for 5' end processing is encoded by mitochondrial DNA.

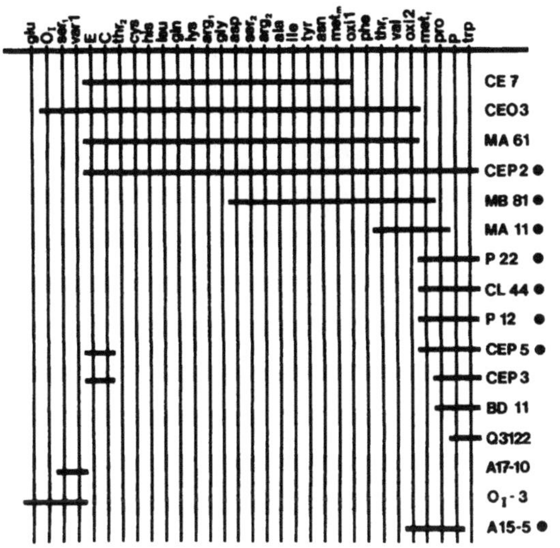

FIGURE 6. Deletion map of the tRNA synthesis locus. Each petite listed on the right side of this figure was tested for its ability to make tRNA and those that make tRNA are indicated by a dot.

We have shown that the function necessary for tRNA biosynthesis acts in trans (16). A deletion mapping study limited the tRNA synthesis locus to an area of the genome near the $tRNA_f^{Met}$ gene (Figure 6). The isolation of very small petites from this region of the genome and restriction mapping of these petites limited the synthesis locus to a 780 base pair sequence of DNA located between the $tRNA_f^{Met}$ and $tRNA^{Pro}$ genes on yeast mitochondrial DNA (16). The DNA sequence of the region is found in Figure 4 and extends from nucleotides 1081 to 1860. In the wild-type strain MH41-7B this region contains four G+C rich sequence elements and A+T rich DNA. Hybridization of restriction fragments from this region to mitochondrial RNA from wild type and petite strains identified a 9S RNA (Figure 7). Its presence is correlated with the ability of petites to make tRNA. Larger RNA species which could be precursors to this RNA can also be seen in Northern blot analyses (Figure 7). Other methods used to confirm that an RNA molecule is made from this region include primer extension and S1 nuclease digestion experiments (17). The primer that was extended in cDNA synthesis using mitochondrial RNA from MH41-7B as a template was the sequence underscored with a wavy line in Figure 4 and the sequence of the DNA made in this reaction matched the sequence of DNA found 5' to the primer through nucleotide 1347. This result confirms that the region is transcribed, that the gene coding for the RNA does not have introns and further has given an apparent 5' end at

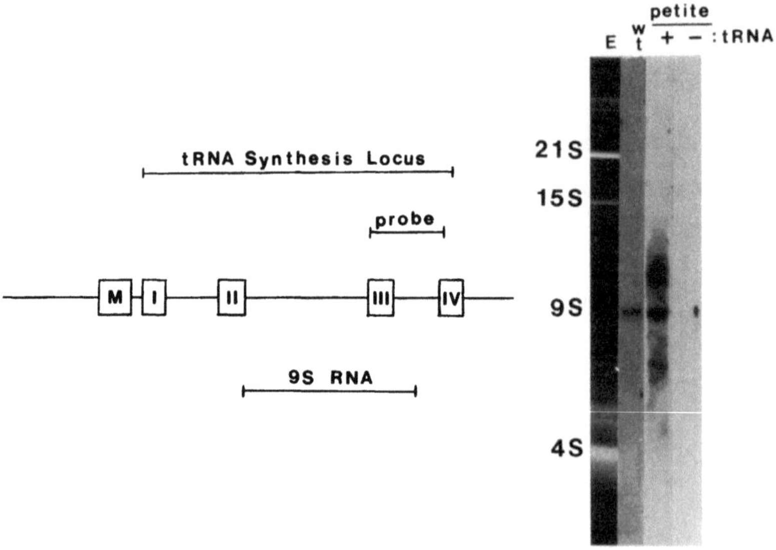

FIGURE 7. Northern hybridization of a cloned mitochondrial DNA restriction fragment with mitochondrial RNAs from various yeast strains. At left is the region of the DNA containing the $tRNA_f^{Met}$ gene (M) and the tRNA synthesis locus. The region labeled "probe" shows the extent of the restriction fragment used in the Northern hybridizations shown at right. I, II, III, and IV indicate the G+C rich sequence elements in this region. RNA obtained from isolated mitochondria were separated on 1.5% agarose-urea gels, stained with ethidium bromide and photographed (E). The RNA was then transferred to DBM paper in situ and hybridized with the radiolabeled restriction fragment indicated at left. WT indicates mitochondrial RNA from a wild type strain, + indicates mitochondrial RNA from a petite that makes tRNA, - indicates mitochondrial RNA from a petite which does not make tRNA. The extent of the 9S RNA as determined by primer extension and S1 nuclease experiments is shown at left.

nucleotide 1347 in Figure 4. S1 nuclease mapping fixed the 3' end of the RNA near nucleotide 1778 (Figure 4). These ends indicate that the RNA in MH41-7B is about 450 bases, consistent with the size of the RNA detected by Northern analysis. We have named this newly identified transcript the 9S RNA.

The structure of the tRNA synthesis locus in another wild type strain, D273-10B, has also been examined because D273-10B does not have the Hha I site at position 1279 (Figure 4) and DNA sequence analysis of this region from D273-10B demonstrated that it does not contain the entire

G+C rich sequence element at that position. Other differences between MH41-7B and D273-10B DNA include a 27 base A+T rich insertion located between nucleotides 1479-80, a 6 base A+T rich deletion from 1502 to 1507, and a GC to AT base pair substitution at nucleotide 1511 (See Figure 4). The sequence of a cDNA copy from D273-10B RNA matches these sequence changes indicating that the structure of the 9S RNA in D273-10B is not identical in size to the 9S RNA in MH41-7B.

We have used the information from these studies to construct a "minimal structure" of the 9S RNA (431 bases) and it is shown in Figure 8. Certain features of the structure are similar to an RNA molecule found in the E. coli enzyme RNase P. This enzyme is necessary for the

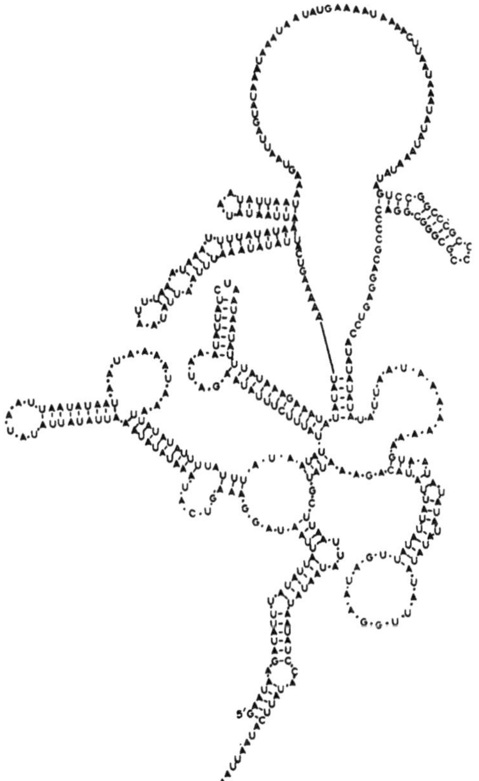

FIGURE 8. A possible secondary structure for the minimum sequence of the 9S RNA. This structure was produced by searching for regions of intrastrand base pairing which generated a thermodynamically optimal structure. Short range interactions are emphasized (19).

processing of the 5' end of tRNA gene transcripts and requires a specific RNA for activity (20). Since transcripts with unprocessed 5' ends accumulate in petites without the 9S RNA, the 9S RNA may play an analogous role in tRNA processing in mitochondria. Further support for this suggestion comes from experiments where the 100 base transcript that accumulates in ND40 (Figure 3) has been processed with *E. coli* RNase P (Figure 9). If this hypothesis is correct then the analogous protein portion of the mitochondrial enzyme must come from the nucleus. None of the mitochondrial tRNA gene transcripts described here or reported by others have been processed in homologous cell free systems in a 9S dependent fashion or shown to have a precursor-product relationship to mature tRNAs in *in vivo* pulse-chase experiments. Nonetheless, it is likely these transcripts are precursors which must be processed.

Other tRNA processing functions that must be carried out by mitochondria include 3' nucleolytic processing, CCA addition and base modification. All of these clearly can be done in the absence of mitochondrial protein synthesis since tRNAs isolated from petites can be charged and have been shown to have modified bases (Pham, H., and Martin, N.C.,

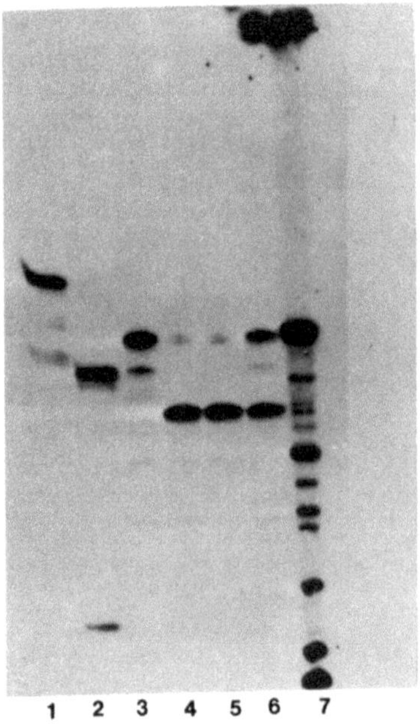

FIGURE 9. Processing of tRNA gene transcripts with *E. coli* RNaseP. *E. coli* RNA and the 100 base mitochondrial RNA from ND40 were incubated with *E. coli* RNaseP and the products analyzed by polyacrylamide gel electrophoresis. Lane 1, uncut *E. coli* $tRNA^{Tyr}$ precursor; Lane 2 *E. coli* $tRNA^{Tyr}$ precursor cut with *E. coli* RNAse P; Lane 3, ND40 $tRNA_f^{Met}$ transcript; Lanes 4-6 ND40 $tRNA_f^{Met}$ transcript cut with decreasing amounts of *E. coli* RNaseP. Lane 7 is the ND40 transcript labeled with $\alpha^{32}P$ GTP and guanylyl transferase.

unpublished). Nuclear genes that affect base modification in yeast mitochondria have been identified (21,22).

As this short review of results from many laboratories indicates, great progress has been made in understanding the expression of yeast mitochondrial tRNA genes. Nevertheless, many nucleo-mitochondrial interactions important for the biosynthesis of these tRNA remain to be elucidated.

Acknowledgements

We would like to thank Raquel Voss for preparing the manuscript and Carl Sidle for preparing the figures. This work was supported by grants to NCM from the National Institutes of Health and The Robert A. Welch Foundation.

References

1) Grivell, L.A. (1983) Scientific American 63, 78-89.

2) Merten, S., Synenki, R., Locker, J., Christianson, T. and Rabinowitz, M. (1980) Proc. Nat. Acad. Sci. 77, 1417-1421.

3) Locker, J., and Rabinowitz, M. (1981) Plasmid 6, 302-314.

4) Frontali, L., Palleschi, C., and Francisci, S. (1982) Nucleic Acid Res 10, 7283-7293.

5) Levens, D., Ticho, B., Ackerman, E., and Rabinowitz, M. (1981b) J. Biol. Chem. 256, 5226-5232.

6) Edwards, J.C., Levens, D., and Rabinowitz, M. (1982) Cell 31, 337-346.

7) Christianson, T., Edwards, J., Mueller, D., and Rabinowitz, M. (1983) submitted for publication.

8) Thalenfeld, B., Hill, J., and Tzagaloff, A. (1983) J. Biol. Chem. 258, 610-615.

9) Christianson, T., Edwards, J., Levens, D., Locker, J. and Rabinowitz, M. (1982) J. Biol. Chem. 257, 6494-6500.

10) Osinga, K., and Tabak, H.F. (1982) Nucleic Acids Res. 10, 3617-3626.

11) Christianson, T., and Rabinowitz, M. (1983) submitted for publication.

12) Miller, D.L., Najarian, D.R., Folse, J.R. and Martin, N.C. (1981) Biol. Chem. 256, 9774-9777.

13) Faye, G., Bolotin-Fukuhara, M., and Fukuhara, H. (1976) in The Genetics and Biogenesis of Chloroplasts and Mitochondria (Bucher, C.T., Neupert, W., Sebalt, W., and Werner, S., eds.) p. 547-555, North Holland Publishing Co., Amsterdam.

14) Wesolowski, M., and Fukuhara, H. (1979) Mol. Gen. Genet. 170, 261-275.

15) Martin, N.C., and Underbrink-Lyon, K. (1981) Proc. Nat. Acad. Sci. 78, 4743-4747.

16) Underbrink-Lyon, K., Miller, D.L., Ross, N., Fukuhara, H. and Martin, N.C. (1983) Mol. Gen. Genet., in press.

17) Miller, D.L. and Martin, N.C. (1983) submitted for publication.

18) Auron, P.E., Rindome, W.P., Vary, C.P.H., Celentano, J.J., and Vournokis, J.N. (1982) Nucl. Acids Res. 10, 403-419.

19) Reed. R.E., Baer, M.F., Guerrier-Takada, C., Donis-Keller, H., and Altman, S. (1982) Cell 30, 627-636.

20) Hopper, A., Furukawa, A.H., Pham, H.D., and Martin, N.C., (1982) Cell 28, 543-550.

21) Martin, N.C., and Hopper, A.K. (1982) J. Biol. Chem. 257, 10562-10565.

EXPRESSION OF MITOCHONDRIAL GENES IN SACCHAROMYCES CEREVISIAE: RELEASE FROM GLUCOSE REPRESSION AND TRANSCRIPTION OF THE tRNA GENES.

Giuseppe Baldacci, Silvia Francisci, Claudio Palleschi, Elisabetta Zennaro, Laura Frontali.

Department of Cell and Developmental Biology - University of Rome I - Rome - Italy.

Introduction

In S.cerevisiae the mitochondrial function is highly regulated, but the control mechanisms and nucleo-cytoplasmic interactions that constitute this regulation are still poorly understood. Variations in the overall level of transcription of the mitochondrial (mit) genome, observed in anaerobic or glucose-repressed cells, are probably not sufficient to explain the patterns of induction of mitochondrially coded subunits of respiratory enzymes. Our previous work (1) demonstrated different kinetics of induction for the various products of mitochondrial protein synthesis during release from glucose repression. Moreover a comparison of mitochondrial transcripts from repressed and derepressed cells has shown a complex pattern in which the level of only certain transcripts exhibited strong differences under these two conditions (2). This suggests an important role of splicing and processing events in modulating the expression of the mitochondrial genome. In particular much variation had been observed in the levels of mit tRNAs in cells grown under different physiological conditions. In this paper we report an analysis of mitochondrial transcripts in the course of respiratory induction as well as a study of the transcripts of mitochondrial tRNA genes.

Results

I. Mitochondrial Transcription During Respiratory Induction.

We have used gene specific probes (see Fig. 1) to investigate the trans-

Mitochondria 1983

cripts from different regions of the mit genome in cells grown to mid-exponential phase on 15% glucose (repressed) or 2% galactose (derepressed) and in cells undergoing release from glucose repression in resting conditions. The latter procedure was adopted to eliminate the possibility that the observed variations might be affected by differences in the growth rate. Results obtained from cells collected at different stages of respiratory induction have confirmed and extended those obtained from growing repressed and derepressed cells.

Examples of transcripts exhibiting a differential dependence on the physiological conditions are reported in Fig. 2. The oxi2 transcripts exhibited a very similar pattern in the different stages of respiratory induction, and no important changes in the level of the 3.2 kb messenger RNA were observed. In contrast the oli1 messenger RNA was much more abundant in the mitochondria of derepressed cells and exhibited a strong increase in the course of respiratory induction. The 1.5 kb oli1 precursor did not show a

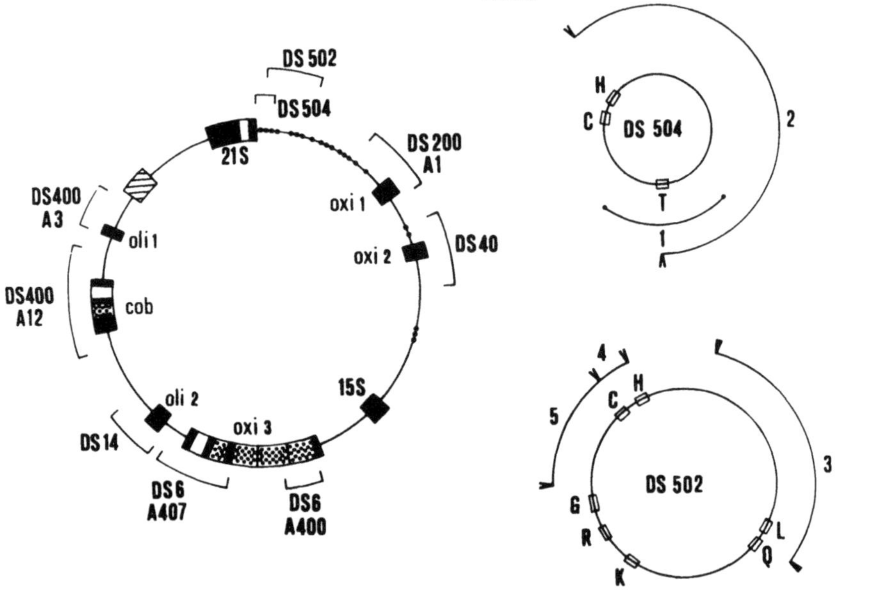

Fig. 1 - Deletion structure of the rho^- strains used in this work and structure of the probes obtained by restriction endonuclease digestion from DNA of strain DS502 and DS504 (3,4). All rho^- strains were derived from the w.t. D273-10B. ▼ HinfI site; ∨ TaqI site; ● AvaII site. The one letter aminoacid code is used to indicate the corresponding tRNA genes.

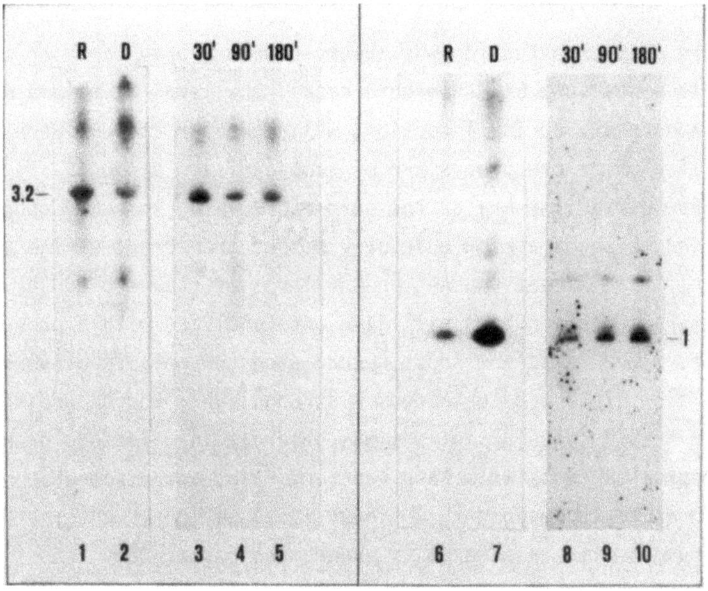

Fig. 2. Hybridization of oxi2 and oli1 gene specific probes to mit RNA extracted from repressed (slots 1,6) and derepressed (slots 2,7) cells and from cells undergoing release from glucose repression (slots 3-5 and 8-10 respectively). RNAs (10 μg in each slot) were electrophoresed on 1.5% agarose 6M urea gels as reported by Locker (5), blotted to nitrocellulose filters by the procedure of Thomas (6), and hybridized with nick translated DNAs from rho⁻ strains containing the oxi2 gene (slots 1-5) or the oli1 gene (slots 6-10). Nick translation and hybridization procedures were as previously reported (2). Release from glucose repression was obtained in the following conditions. Cells were grown in a medium containing 1% yeast extract, 1% peptone and 15% glucose, collected in mid-exponential phase, washed several times and resuspended in 0.05M phosphate-buffer pH 6, containing 0.1% glucose and 2% lactate. At the indicated times aliquots were withdrawn and mit RNA prepared.

parallel increase, suggesting that both transcription and processing are involved in the control of oli1 gene expression. Since equal amounts of the same RNA preparations were used for hybridization with the two probes, the observed constant steady state level of the oxi2 transcripts can be taken as an internal standard to evaluate the variations in the concentrations of other transcripts.

A more complex situation was observed when probes carrying the split cob and oxi3 genes were used. In both cases, the level of mature messenger increased moderately during induction, while greater changes were observed in the level of some precursors and of some excision products. Major examples of these variations are a) the strong decrease, in early phases of respiratory induction, of a high molecular weight transcript of the oxi3 gene b) the highly significant changes in the level of stable circular transcripts.

When transcripts of the oxi3 gene were analyzed with a probe carrying only introns 1 and 2 and the short second exon (36 bp), results reported in Fig. 3 (slots 1,2) were obtained: a strong signal corresponding to the 2.4 and 2.5 stable circular introns was observed in galactose grown cells, while in repressed cells this band was very faint and a somewhat higher band (which could correspond to a linear form) was observed. The same experiment performed with resting cells undergoing respiratory induction showed a gradual increase of stable circular transcripts during the first 180' (see Fig. 3 slots 3-5). The lack of stable circular transcripts has also been observed in sporulating cells (7).

The variations in steady-state concentrations of mitochondrial transcripts observed in these experiments might be due to variations in stability or to differences in the rate of synthesis or processing of transcripts. The second hypothesis seems to be favored by the similarity of results obtained in growing and in resting cells, and by the observation that the levels of only some transcripts were variable.

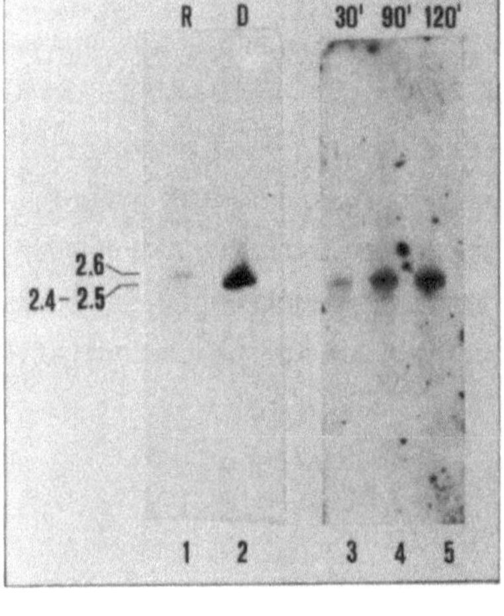

Fig. 3. Hybridization of nick translated DNA from strain DS6/A400 carrying the first and second intron of the oxi3 gene on mtRNA extracted from growing repressed (slot 1) and derepressed (slot 2) cells and from cells undergoing release from glucose repression (slots 3-5). All other procedures were as described in Fig. 2.

Highly specific and differentiated nucleolytic activities would be necessary to explain, on the basis of differential stability, the complex pattern of variations in transcripts observed in the course of respiratory induction. On the other hand the absence of stable circular transcripts in repressed as well as in sporulating (7) cells, might be explained by the deficiency in these cells of a circularizing activity which in respiring cells could preserve circular RNAs from degradation. As a whole these results are in favor of a specific rather than a generalized control of gene expression in yeast mitochondria.

II. The Expression of Mitochondrial tRNA Genes Is Strongly Affected by Glucose Repression.

Previous work has shown significant differences between repressed and derepressed cells in the concentration of tRNAs (2). However a complete analysis of the effect of respiratory induction on the expression of tRNA genes required the use of specific probes. To this purpose, we have used nick-translated DNA from strain DS502, which carries a region of the mitochondrial genome spanning the first 6 map units (containing only tRNA genes) as well as 5' end labelled single-stranded fragments containing single genes. Hybridization of mit RNA from repressed and derepressed cells with nick-translated DNA from strain DS502 (Fig. 4a) showed that the amount of mature tRNA from this region of the genome was much higher in derepressed than in repressed cells. Some high molecular weight transcripts (the main ones approximately 2200, 1500 and 900 bases long) were also observed in both conditions. An example of the transcript pattern obtained with a gene-specific probe from the same region (probe 1 containing the gene for $tRNA_{cys}$; see Fig. 1) is reported in Fig. 4b. A similar pattern of high molecular weight transcripts has been obtained with a probe containing the adjacent gene for $tRNA_{his}$. The high molecular weight transcripts could be detected in the w.t. cells only by the use of very highly radioactive probes (specific activity greater than 10^7 c.p.m./µg DNA).

These data agree with previous observations showing the existence in w.t. cells of high molecular weight transcripts carrying more than one tRNA sequence (10) and they suggest that tRNA genes are transcribed in polygenic precursor RNAs which are then processed to yield mature tRNAs. Therefore at least in this region of the mitochondrial genome, expression of tRNA ge

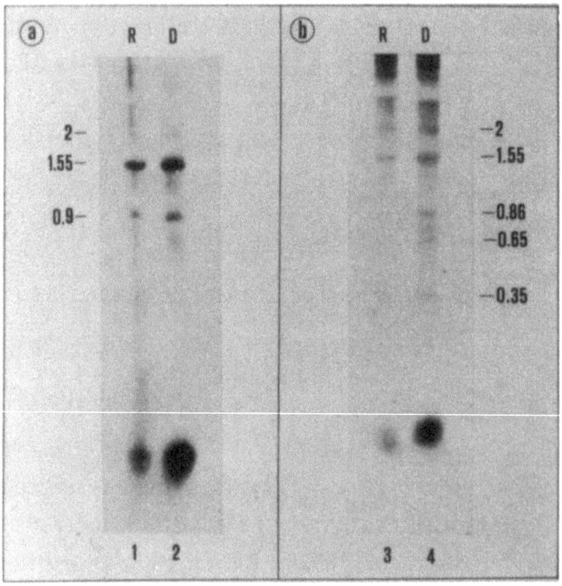

Fig. 4. Hybridization of probes carrying tRNA genes to mit RNA extracted from repressed (slots 1 and 3) and derepressed (slots 2 and 4) cells. RNAs were electrophoresed on 1.5% agarose, 6M urea gels, blotted to DBM paper by the procedure of Alwine (8) as modified by Wahl (9) and hybridized with P^{32} labelled DNA from petite mutant DS502 in a) and with 5' end labelled probe 5 in b). For the preparation of probe 5, DNA from strain DS502 was digested with TaqI restriction endonuclease, and 5' end labelled with polynucleotide kinase in the presence of P^{32} ATP as previously reported (10). Strand separation and idenfitication by sequencing was performed by the Maxam and Gilbert procedure (11).

nes would not require internal promotors (as is generally the case for eukaryotic tRNA genes), but would require extensive processing events (as with tRNA genes from bacteria and from mammalian mitochondria.ref. 12).

Both the level of mature tRNAs and, to a lesser extent, the level of high molecular weight transcripts decreased in glucose repressed cells. Moreover some transcripts were present in the derepressed condition and not visible in repressed cells (Fig. 4b). This might indicate differences in the complex sequence of processing events in the two conditions.

III. Processing of Transcripts Carrying tRNA Genes

A further study of the control of tRNA levels in yeast mitochondria requires a better understanding of the transcription of tRNA genes and of the mechanisms that leads to the appearance of mature tRNA species; for this reason, we have further investigated the transcripts of the region spanning the first 6 map units in w.t. cells and in some petites, which, due to the absence of a region of the mitochondrial genome which is necessary for the synthesis of mature tRNA (13,14), accumulate transcripts of different sizes carrying tRNA sequences. These species are probaly intermediates in the processing of tRNAs from polygenic transcripts and therefore they should be very useful for the analysis of the maturation process. This analysis was carried out both by hyrbdizing 5' end labelled single-stranded probes to RNA from w.t. and <u>rho</u>$^-$ strains blotted onto DBM paper, and by S1 mapping of 5' ends of transcripts. Hybridization of gene-specific probes on northern blots revealed a striking accumulation of high and low molecular weight transcripts in the mutants DS502 and DS504. Several transcripts of the same size hybridize with more than one gene-specific probe (Fig. 5).

Some results of the S1 mapping experiments are reported in Fig. 6 and 7. We will consider three aspects of these results.

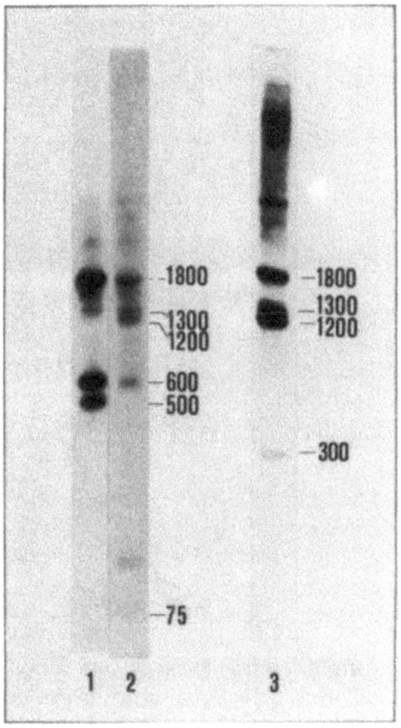

Fig. 5. Hybridization of 5' end labelled single stranded probes carrying the gene for $tRNA_{his}$ (slot 1), the gene for $tRNA_{cys}$ (slot 2) or the gene for $tRNA_{thr2}$ (slot 3) to mit RNA from strain DS504 blotted onto DBM paper.

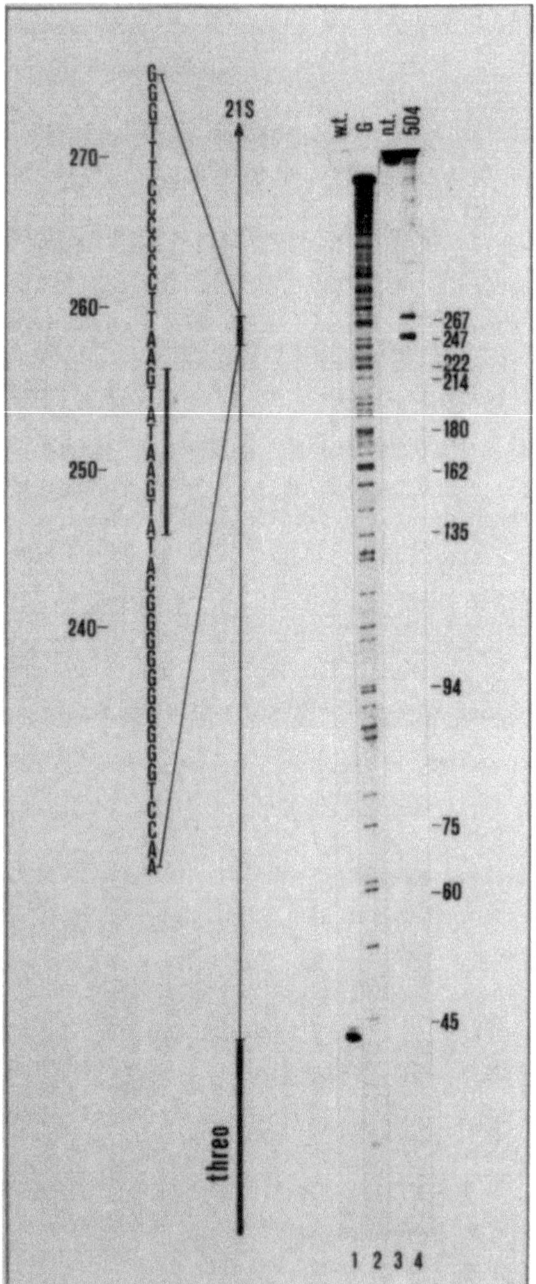

Fig. 6. S1 mapping of 5' ends of transcripts containing $tRNA_{thr2}$. The procedure by Berk and Sharp (15) as modified by Weaver and Weissmann

(16) was followed. Fragments protected from S1 digestion after hybridization to probe 2 were run on 10% acrylamide 8M urea sequencing gels for 2h at 2000 volts. Lanes are as follows: (1) probe 2 after hybridization to mit RNA from the w.t. strain D273-10B and subsequent treatment with S1 nuclease; (2) reference ladder obtained by a G-specific reaction of a known DNA fragment; (3) probe 2 hybridized to mit RNA from strain DS504, untreated; (4) probe 2 treated with S1 nuclease after hybridization to mit RNA from DS504. The numbers on the right refer to the length of fragments in nucleotides calculated using the reference ladder. On the left side, a schematic drawing shows the position of the gene for $tRNA_{thr2}$ and the sequence in which the 5' ends of the major transcripts map. The sequence is that of the sense non transcribed strand; the distance in nucleotides from the TaqI site inside the tRNA gene at which labelling was introduced is reported.

a) Transcripts of the gene for $tRNA_{thr2}$.

To map the transcripts from this gene, which is the first one downstream from the 21S ribosomal RNA gene, we made use of probe 2 (see Fig. 1). When this probe was hybridized to mit RNA from strain DS504 and then digested with S1 nuclease, a strong signal was obtained (Fig. 6 slot 4) corresponding to a very peculiar sequence which is shown in Fig. 6. This sequence contains a nonanucleotide (indicated by a bar) nearly identical to the one that has been identified as an initiation site of mitochondrial transcription (17, 18), and it is flanked by a G cluster and a pyrimidine cluster. A second signal of lesser intensity was found to correspond to the pyrimidine cluster. When the same probe was hybridized with mit RNA from w.t. cells, only the signal corresponding to mature tRNA was observed (Fig. 6 slot 1); correspondingly, no high molecular weight transcripts were detected on northern blots, probably because the specific activity of this probe was never sufficient. Unique signals corresponding to mature tRNAs were obtained in these experiments in which Seikaguki S1 nuclease was used; in previous work (10), performed with a different enzyme product, even at very high concentration a set of bands had been obtained, and this had suggested a possible heterogeneity in the 5' termini of mature tRNAs.

The localization of the 5' end of the transcripts of strain DS502 in the proximity of a promotor-like sequence suggests that these species might be primary transcripts; the nonanucleotide sequence might serve as a site of initiation for the transcription of tRNA genes. However, Locker and Rabinowitz (19) have reported that precursors of the 21S ribosomal RNA carrying a 3' extension hybridize to the region which contains the gene for

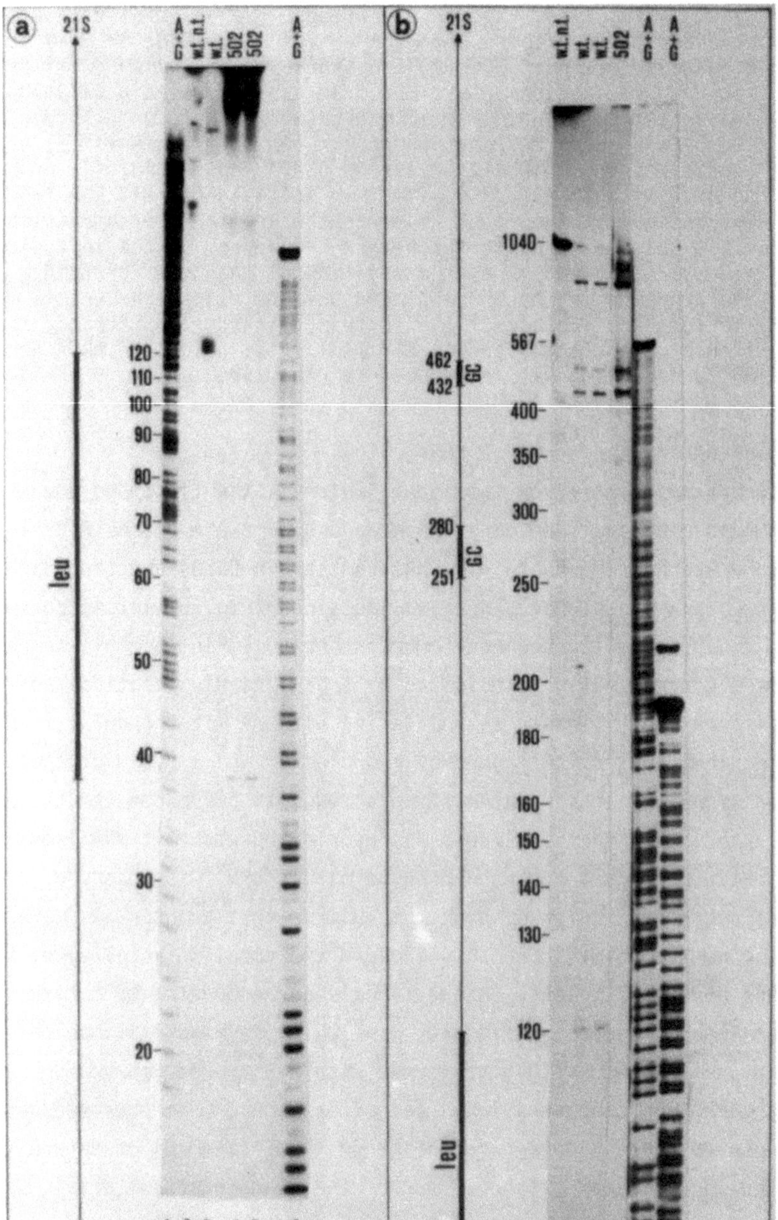

Fig. 7. S1 mapping of 5' ends of transcripts after hybridization to probe 3. Fragments protected from S1 digestion were run on 10% acrylami-

de 8M urea sequencing gels for 2h at 2000 volts (panel a) or for 3h at 2000 volts (panel b). Lanes are as follows: 1) probe 3 cleaved by a purine-specific reaction; 2 and 7) probe 3 hybridized to mit RNA from w.t. cells untreated; 3 and 9) the same treated with S1 nuclease for 2h; 8) the same but S1 treatment for 1h; 4) probe 3 hybridized to mit RNA from strain DS502 and treated with S1 nuclease for 1h; 5 and 10) the same but S1 treatment for 2h; 6, 11 and 12) purine-specific reactions on other fragments used as reference ladders.
In both panels, the distances in nucleotides from the HinfI cleavage site are reported, and a schematic drawing shows the position of the $tRNA_{leu}$ gene and of the GC-rich clusters.

$tRNA_{thr2}$. In this case one might postulate either that the mapped 5' ends of transcripts result from cleavage of longer transcripts, or that the above-mentioned sequence is a secondary initiation site for transcription, which comes into play in a petite lacking the promoter of the 21S RNA gene. Determination (by capping experiments) of the presence or absence of di- or triphosphates at the 5' end of the transcripts of the $tRNA_{thr2}$ gene in mutant and w.t. strains will be necessary to allow a definite conclusion.

b) Small molecular weight transcripts.

Our previous results (10) had demonstrated that mutants lacking the region of the mitochondrial genome necessary for tRNA maturation contained small molecular weight transcripts carrying tRNA sequences and the upstream region up to the 3' end of the preceding tRNA gene. This could be explained by the absence, in these petites, of a mitochondrial function that is necessary for the processing of tRNAs at their 5' ends. The accumulation of longer transcripts could be a consequence of this primary defect. The mitochondrial region necessary for the synthesis of mature tRNA has been precisely mapped (20) and an RNA species transcribed from this region has been detected (21); this RNA could, as is the case for RNAse P from bacteria, represent an essential moiety of a nucleolytic enzyme necessary for processing the 5' ends of tRNAs.

Further information on low molecular weight transcripts has been obtained by S1 mapping. Results reported in Fig. 7 a) (slots 4 and 5) confirmed that the 5' extension of the low molecular weight transcript of the $tRNA_{gln}$ gene corresponds to the 3' end of the upstream $tRNA_{leu}$ gene.

A low molecular weight protected fragment has also been detected in RNA from w.t. cells hybridized with probe 3 and treated with S1 nuclease (Fig. 7, slot 3). This fragment extends from the HinfI site inside the gene for $tRNA_{gln}$ to the 5' of the $tRNA_{leu}$ gene, indicating the presence in w.t. cells of a transcript carrying two tRNA sequences. This also shows that in w.t. cells a cleavage event must occur at the 5' end of the tRNA sequences; this cleavage does not occur in the mutants.

Further data on the low molecular weight transcripts have been obtained for another group of tRNA genes present on the genome of strain DS502 (genes for $tRNA_{hys}$, $tRNA_{arg}$ and $tRNA_{gly}$). A detailed map of the protected fragments is shown in Fig. 8.

c) High molecular weight transcripts.

Some new results have been obtained by the use of probe 3 which carries 5 bases of the gene for $tRNA_{gln}$, 33 bases of intergenic region, the gene for $tRNA_{leu}$ and about 1000 nucleotides of the upstream region. We used this long probe to map the 5' termini of transcripts of the approximate length of 500-700 nucleotides that had been detected on northernblots with probes specific for $tRNA_{gln}$ and $tRNA_{leu}$ (10).

Three signals have been observed after hybridization of mtRNA from w.t. cells or from strain DS502 with probe 3 followed by S1 treatment (Fig. 7 b, slots 8,9, 10). The two shorter transcripts have been mapped and found to correspond to protected fragments of the length of 425±20 and 450±20 nucleotides. Their termini could correspond to the 5' and 3' sides respectively of a G+C cluster that is located between 432 and 462 bases from the HinfI site used to prepare the probe.

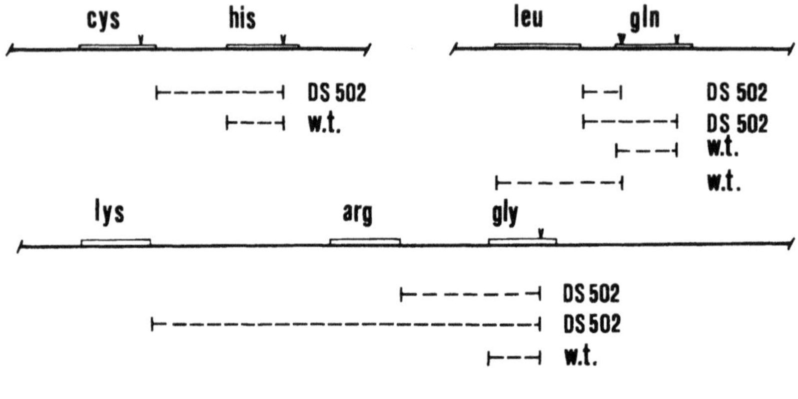

Fig. 8. S1 mapping of the low molecular weight transcripts. Hatched segments indicate the fragments protected from S1 digestion after annealing with RNA from the indicated strains.
ᵛ TaqI site; ▼ HinfI site.

As a whole, our results completely support the previous data indicating the existence of polygenic transcripts carrying multiple tRNA sequences and suggest a complex series of processing events, some of which are lacking in the mutants defective in tRNA maturation.

The present results seems to indicate that cleavage of longer precursors could occur: i) at the 3' end of tRNA sequences by a cleavage that occurs correctly also in rho$^-$ mutants, ii) at the 5' end of tRNA sequences by a different nucleolytic activity that requires the presence of a mitochondrial locus, iii) in the long intergenic regions in the proximity of G+C clusters by a processing event that seems identical in w.t. cells and in the mutants. It is worthwhile noticing that several G+C clusters, including the one mentioned above, are repeated in inverted orientation in different regions of the genome; this has suggested that they may be involved in processing events (3, 22).

Acknowledgements

We are deeply indebted to Prof. A.Tzagoloff for the gift of the rho$^-$strains, to Prof. H. Fukuhara for information about unpublished sequences and to Dr. G. Macino for helpful discussions. This work was supported by CNR grants n. 82.01142.95, 8202363/51 and partially by Istituto Pasteur-Fondazione Cenci-Bolognetti.

References

1. Falcone, C., Agostinelli, M., Frontali, L.: J.Bacteriol. 153, 1125-1132 (1983)
2. Baldacci, G., Zennaro, E.: Eur.J.Biochem. 127, 411-416 (1982)
3. Berlani, R.E., Bonitz, S.G., Coruzzi, G., Nobrega, M. and Tzagoloff, A.: Nucleic Acid Research 8, 5017-5030 (1980)
4. Berlani, R.E., Pentella, C., Macino, G., Tzagoloff, A.: J.Bacteriol. 141, 1086-1097 (1980)
5. Locker, J.: Anal.Biochem. 98, 358-367 (1979)
6. Thomas, P.: Proc.Natl.Acad.Sci. USA 77, 5201-5205 (1980)
7. Schroeder, R., Breitenbach,M., Schweyen, R.J.: Nucleic Acid Research

11, 1735-1746 (1983).

8. Alwine, J.C., Keny, J.D., Stark, G.R.: Proc.Natl.Acad.Sci.USA 74, 5350-5354 (1977).

9. Wahl, G.M., Stern, M., Stark, G.R.: Proc.Natl.Acad.Sci.USA 76, 3683-3687 (1979)

10. Frontali,L., Palleschi, C., Francisci, S.: Nucleic Acid Research 10, 7283-7293 (1982).

11. Maxam, A.M., Gilbert, W.: Methods in Enzymology 65, 499-560 (1980)

12. Ojala, D., Montoya,J., Attardi, G.: Nature 290, 470-474 (1981)

13. Morimoto,R., Locker, J., Synenki, R.M., Rabinowitz, M.: J.Biol.Chem. 254, 12461-12470 (1979)

14. Martin,N.C., Underbrink-Lyon, K.: Proc.Natl.Acad.Sci.USA 78, 4743-4747 (1981)

15. Berk,A.J., Sharp,P.A.: Cell 12, 721-732 (1977)

16. Weaver, F., Weissmann, C.: Nucleic Acid Research 7, 1175-1193 (1979)

17. Osinga, K.A. and Tabak, H.F.: Nucleic Acid Research 10, 3617-3625 (1982)

18. Baldacci, G. and Bernardi, G.: EMBO Journal 8, 987-994 (1982)

19. Locker, J., Rabinowitz, M.: Plasmid 6, 302-314 (1981)

20. Underbrink-Lyon, K., Miller, D.L., Ross, N.A., Fukuhara, H., Martin, N.C.: Mol.Gen.Gen. in the press

21. Miller, D.L., Martin, N.C.: personal communication

22. Tzagoloff, A., Nobrega, M., Akai, A., Macino, G.: Current Genetics 2, 149-157 (1980)

IMPLICATIONS OF FUNGAL MITOCHONDRIAL INTRON RNA STRUCTURE FOR THE MECHANISM OF RNA SPLICING

R. Wayne Davies, Richard B. Waring and Terence A. Brown
UMIST, Department of Biochemistry and Applied Molecular Biology, P.O. Box 88, Manchester M60 1QD, U.K.

Claudio Scazzocchio
University of Essex, Department of Biology, Wivenhoe Park, Colchester CO4 3SQ, U.K.

Introduction

Introns occur in nuclear, chloroplast and mitochondrial genes of eukaryotes, and their removal from RNA precursors is one of the major steps in RNA processing. The accessibility of fungal mitochondrial genomes to genetic as well as molecular approaches has allowed a very precise analysis of the process of RNA splicing to be made in these systems. All fungal mitochondrial introns are among those introns that do not obey the GT ... AG rule (1) of nuclear introns in protein-coding genes. Apart from the small introns of nuclear tRNAs of *Saccharomyces cerevisiae*, which have their own well-studied unique splicing mechanism, this class contains introns in protozoal ribosomal rRNA genes and in plant chloroplast rRNA and tRNA genes in addition to mitochondrial introns. Genetic analysis of RNA splicing defects in *S.cerevisiae* (2) showed that some mitochondrial introns coded for trans-acting functions essential for RNA splicing, and that regions of these introns were required intact in *cis*. DNA sequencing of some of these *cis*-acting mutants (3,4) showed them to be very closely linked within short sequence which we recognised as being among the sequences within introns that are strongly conserved between distant species (5). Four short (10 or 12 bp) sequences are conserved, which we have called P, Q, R and S (6), and the occurrence of partial inverted repeats within these sequences suggested that they pair P with Q and R with S in the RNA secondary structure. If these pairings are taken as nucleation points for RNA secondary structure formation then many (21 to date) introns take up a particular conserved RNA structure (6). In fact, mitochondrial introns can be divided into two classes on this basis (7); class 1 the majority class, is that with P, Q, R and S; class 2 is a group of introns with some, less striking, sequence conservation and a conserved secondary structure that is different from that of class 1. We will consider what roles the conserved intron RNA structures may play in RNA splicing, concentrating on the class 1 introns since all our own data is from Aspergillus nidulans, which has only class 1 introns in its mitochondria. In the context of this meeting, the discovery of nuclear mutations affecting mitochondrial RNA splicing in a specific manner, suggests that RNA splicing could be one of the key points where control is exercised in nuclear-mitochondrial interactions.

Introns of Aspergillus nidulans

Four introns occur in the mitochondrial genome of A.nidulans, one in the gene for apocytochrome b (cobA, 8,5), one in the large rRNA gene (9), and three in the gene for cytochrome oxidase subunit 1 (oxiA, 10). The DNA sequence is almost complete, and no other obvious intron has been discovered. The cobA intron (NC), and the second and third oxiA introns (NOX2 and NOX3) all have open reading frames contiguous and in frame with the upstream exon, and could thus code for mRNA maturases (II), though the open reading frame of NOX3 is interrupted by a GC-rich sequence which may be related to transposable elements (10). There is an open reading frame within the rRNA intron which is not contiguous with the exons, and NOXI has no significant open reading frame. This latter fact makes the involvement of ribosomes in splicing of NOX1 very unlikely. The conserved decapeptides which have been proposed to be typical of maturase coding sequences are found in the NC, NOX2 and NOX3 introns (Table 1), and a possible vestigial first decapeptide in NOX1.

The NC and rRNA introns are at the same sites within the genes as related (cobB3 and rRNA respectively) introns of S.cerevisiae, but the quite different distribution of introns in Neurospora crassa (G. Macino personal communication) and the variation of numbers of introns in other species of Aspergillus (12, 13) makes it probable that the presence of introns is not related to gene function. It is possible that introns can transpose or be transposed from time to time, and this process may have a preference for some target sequences.

Table 1. Conserved decapeptide sequences in putative maturases coded for by introns of Aspergillus nidulans

Intron	First decapeptide		Second decapeptide
NC	YLVGLFEGDG	- 122 -	WLVGFIEAEG
NOX2	YLAGLIEGDG	- 109 -	WLSGFIEADG
NOX3	YLAGLIDGDG	- 101 -	WLAGFSDADA
NOX1	YLAGLF*GNI	- ? -	?

The number shown between the two decapeptides is the separation in amino acids (codons). The asterisk indicates a nonsense codon. Intron terminology is described in the text.

RNA Secondary Structure of Class I Introns

A proposed secondary structure for NOX3 RNA based on the model of Davies et al. 1982 (6), is shown in Fig. 1 by way of illustration. We have built such secondary structures for the 21 class 1 introns whose consensus conserved sequences are given in Table 2.

Table 2. Conserved sequences in class 1 introns

P

13	19	14	18	10	14	15	14	19	12
A	U	G	C	U	G	G	A	A	A
U			G	A	A	A	G		U
7			3	8	5	6	6		5

Q

16	21	17	12	9	17	13	19	18	11
A	A	U	C	A	G	C	A	G	G
U		C	U	U	C				C
5		3	6	8	3				7

R

17	18	20	12	21	21	19	21	17		19
U	C	A	GA	G	A	C	U	A	X	A
	U		AC							
	3		6							

S

21	18	19	18	14	20	17	21	20	21	20	13
A	A	G	A	U	A	U	A	G	U	C	C
	U		G								U
	3		3								6

The consensus sequences shown are based on 21 introns, which are the 19 introns analysed in ref. 14 plus the maize chloroplast tRNA ^{leu} intron (7, 14) and the second intron of N. crassa URF5 (G. Macino, per. comm.). Alternative bases to the consensus sequence that occur more than once are given underneath. Pairing regions are underlined.

NOX3 is typical of the major maturase - coding subclass, except that it (together with NOX2 and one S.cerevisiae intron, AB (15)) lacks the stem-loop structure called P2-L2 (6) which is usually present between the internal guide sequence (boxed in Fig. 1) and the E sequence. The introns in mitochondrial 21S rRNA genes, together with the fifth intron of the cob gene of S.cerevisiae and the second intron of the cob gene N.crassa, form another subclass with minor but characteristic differences (14), most notably the insertion of a 32 to 35 bp sequence that forms an extra stem-loop structure between R and E'. Although these are such minor difference, the structures formed by the RNA of all 21 introns are strikingly similar. The following structures are always present:

(1) a pairing of parts of R and S (P7);

(2) a pairing (P3) of two unconserved 5 to 7 bp sequences, E and E', where E' is usually 1 bp downstream of R, E 3 to 5 bp upstream of P, (the significance of both of the above is supported by mutational evidence [6,3,16,17]

Fig. 1. Proposed secondary structure of the RNA of the <u>A.nidulans</u> intron NOX3. The boxed sequence at the top is the internal guide sequence, that just 5' to S is the UAA stop codon of the maturase-coding open reading frame. The structure has been arbitrarily cut for display purposes at the site marked by asterisks.

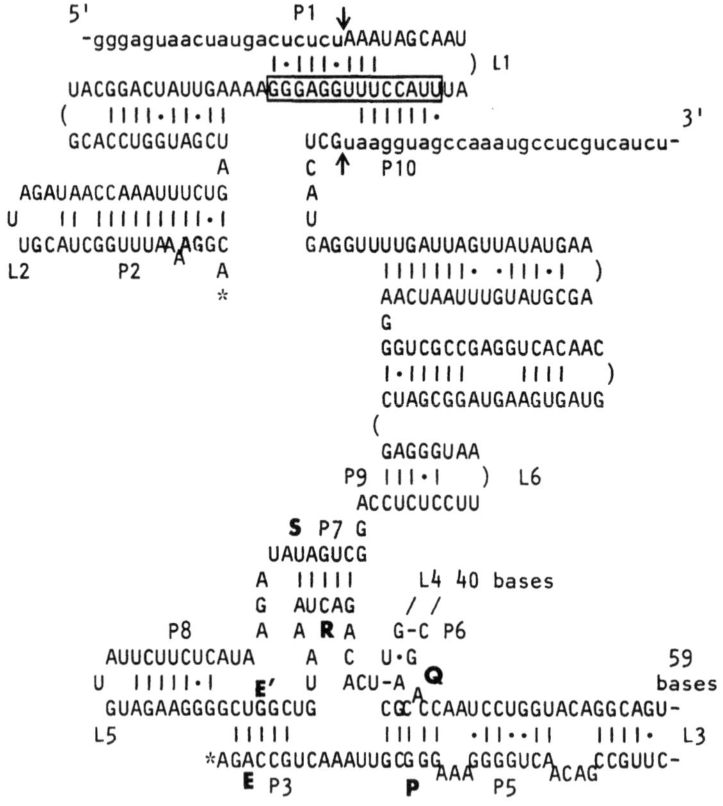

Figure 2. The proposed secondary structure of the intron of the large nuclear ribosomal RNA of *Tetrahymena thermophila*, which has been shown to excise itself from precursor RNA in the absence of proteins by Cech and his coworkers (Kruger et al., 1982). The structure shown here differs slightly from that published by Waring et al., (1983), in that the P2 stem has been extended to end closer to E, as suggested to us by Dr.T.R. Cech. The internal guide sequence is boxed, and the splice junctions are indicated by arrows. Intron bases are in capital letters, exon bases in lower case letters. Stems and loops are named by the convention of Davies et al., (1982) as in Figure 1.

(3) a pairing of P and Q (P4);

(4) the P5 pairing in the L3 loop adjacent to P4 (P - Q), separated from P4 by a characteristic small asymmetrical loop containing the highly conserved ninth base of P;

(5) the P8 pairing adjacent to the P7 (R-S) pairing, forming the stem of the long maturase-coding loop in many introns;

(6) the P9 pairing, forming a strong stem with a very short loop, is always situated just 3' to S.

The significance of all the above is supported by examples of conservation of pairing despite sequence variation (6, 14, 18). Several other structures are very frequent:

(1) 18/21 introns have the pairing P2 forming a stem-loop just upstream (2-7 bases) of sequence E;

(2) 20/21 introns have a stem-loop structure (P6) between Q and R;

(3) 15/21 (and possibly as many as 20 (14)) have a pairing, P1, between upstream exon sequences adjacent to the splice site and intron RNA immediately next to P2;

(4) all introns that can definitely form P1 also form, adjacent to P1, a somewhat more variable pairing, P10, with downstream exon sequences next to the splice site.

The last two pairings make up the components of a possible alignment structure for the splice junctions (6), while the remainder make up the "core" structure of these introns. We find this structure-function division useful, since the evidence from available structures indicates that the core can occur separately from the alignment structure (14, 18).

Functions of Components of Intron RNA Secondary Structure

(a) The core structure

It was clear from the outset that intron folding *per se* must be thermodyuamically required for splicing. In addition the "core" structure brings the 5' and 3' splice junctions reasonably close together, and this presumably is a prerequisite for the recognition of the splice junction and the subsequent splicing reaction. There is no obvious reason, however, why any particular structure should be conserved for this purpose. Moreover, the "core" structure is conserved in rRNA introns and particularly a chloroplast intron, where the splice junctions can be apposed by the exon tertiary structure. There are two extreme explanations for the strong conservation of sequence and structure within intron RNA: (a) the RNA provides a complex of specific recognition sites for a number of proteins, so that RNA structure underlies the assembly of a ribonucleoprotein splicing particle; (b) the RNA structures function directly in the splicing reaction. A compromise between these extremes probably approximates the truth. The former view is supported by genetic evidence that maturase function (11) and/or nuclear gene product function (19) is required for efficient splicing, while the latter view receives support from the fact that the RNA of the intron of the Tetrahymena nuclear large rRNA can fold up into an essentially identical structure to the mitochondrial introns around the same conserved sequences (18, 7). This Tetrahymena intron has been convincingly shown to be a self-splicing RNA molecule, or "ribozyme" (20).

Which part(s) of the structure are critically involved in the cutting and ligation reactions is not yet known. The cleft formed by folding the structure about an axis between R-S and P-Q is a tempting candidate for an active site, since it would allow one to propose functions for the unpaired conserved bases in interacting with the substrate RNA. Table 2 shows that a number of unpaired bases in P, Q, R and S are very highly conserved, including some well away from the pairing regions. These bases are clearly available for binding nucleotide cofactors, pairing with bases in and around the splice junctions to position the structure correctly for splicing and to hold intermediates in place. However, even in the Tetrahymena case proteins may be involved _in vivo_ (20), and whether one can detect significant _in vitro_ self-splicing of mitochondrial introns depends on the degree of dependence of the structure on protein components, whether or not these actually carry out the reaction. We have not yet succeeded in demonstrating such a reaction with the NC intron. Nevertheless, the structures are strikingly similar, so that either the Tetrahymena intron has acquired "ribozyme" function secondairly (which seems very unlikely), or the mitochondrial introns were or are "ribozymes" that are now in various stages of increasing dependence on protein factors.

(b) The internal guide sequence

The contiguous stretches of intron RNA that can take part in the P1 and P10 pairings (Figs. 1 and 2) make up the internal guide sequence (IGS). All 15 introns that have a short distance between E and the 5' splice site possess a sequence in this short region that can form P1 and P10 pairings of varying strengths with exon sequences. With as few as 20 bases in this region, 8 to 10 or more of which could be involved in pairing, the probability of this occurring in 15 introns is negligible. This biological argument strengthens the case for these pairings being of real importance for splicing, despite the flimsy nature of some of the examples (6, 14). We note that pairings that in isolation would be very unstable could become stabilised within a tertiary RNA structure even without the intervention of proteins. Another interesting feature of the alignment structure proposed by Davies et al. (6) is that the last base of the upstream exon$^{(U)}$ always forms a weak UG base pair with a G within the IGS, while the end of P10 is always at or one base away from the last intron base, a G. This has allowed us to predict the most probable splice site among alternatives (14), and in the three cases where the precise site is known the prediction proved correct.

It is possible that the whole IGS acts as an alignment sequence as suggested previously (6), or the P1 and P10 components may form at different times during the splicing process, the P1 providing the initial structure for recognition of the 5' splice junction, the P10 forming subsequently within the ribozyme cleft to position the 3' splice junction near the free 5' end and perhaps to provide further recognition signals. The fact that possible P1 pairings can be found in introns where no P10 is discernable, and several examples of complete overlap of these pairings, suggests that they may indeed play non-simultaneous roles, which might be taken over separately during evolution by proteins or other parts of the RNA molecule. This seems to be the case for six other introns which we have looked at carefully, five of which have long sequences between the 5' splice site and the P2-E region (14). Perhaps once again we are sampling a population of structures based upon common principles that are evolving separate detailed mechanisms of RNA splicing.

Experimental investigation of the model

Genetic data for S.cerevisiae has already substantiated the importance of parts of the secondary structure, and this work will continue. We have obtained a collection of cold and heat-sensitive mitochondrial mutations in A.nidulans some of which may prove relevant. Until the development of an in vitro splicing system for mitochondrial introns, the Tetrahymena nuclear rRNA intron provides a homologous system in which structure-function relationships can be tested. We and others are using oligonucleotide-directed site-specific mutagenesis to achieve this.

References

1. Breathnach., R. and Chambon, P.: Ann. Rev. Biochem. 50, 49-83 (1981)

2. Jacq, C., Pajot, P., Lazowska, J., Dujardin, G., Claisse, M., Groudinskey, O., De La Salle, H., Grandchamp, C.,

Labouesse, M., Gargouri, A., Guiard, B., Spyridakis, A., Dreyfus, M. and Slonimski, P.P.: in Mitochondrial Genes (Slonimski, P.P., Borst, P. and Attardi, G. eds), Cold Spring Harbour Press, pp 155-183 (1982).

3. De La Salle, H., Jacq, C., and Slonimski, P.P.: Cell 28, 721-732 (1982).

4. Netter, P., Jacq, C., Carignani, G. and Slonimski, P.P.: Cell 28, 733-738 (1982).

5. Waring, R.B., Davies, R.W., Scazzocchio, C. and Brown, T.A.: Proc. Natu. Acad. Sci. U.S.A. 79, 6332-6336 (1982).

6. Davies, R.W., Waring, R.B., Ray, J.A., Brown, T.A. and Scazzocchio, C.: Nature 300, 719-724 (1982).

7. Michel, F. and Dujon, B.: EMBO J. 2, 33-38 (1983).

8. Waring, R.B., Davies, R.W., Lee, S., Grisi, E., McPhail Berks, M. and Scazzocchio, C.: Cell 27, 4-11 (1981).

9. Netzker, R., Köchel, H.G., Basak, N. and Küntzel, H.: Nucleic Acids Res. 10, 4783-4794 (1982).

10. Waring, R.B., Brown, T.A., Ray, J.A., Scazzocchio, C. and Davies, R.W.: submitted for publication.

11. Lazowska, J., Jacq, C. and Slonimski, P.P.: Cell 22, 333-348 (1980).

12. Earl, A.J., Turner, G., Croft, J.H., Dales, R.B.G., Lazarus, C.M., Lunsdorf, H. and Küntzel, M.: Curr. Genet. 3, 221-228 (1981).

13. Turner, G., Earl, A.J., and Greaves, D.R.: in Mitochondrial Genes (Slonimski, P.P., Borst, P. and Attardi, G. eds), Cold Spring Harbour Press, pp 411-414 (1982).

14. Waring, R.B., Davies, R.W., Brown, T.A. and Scazzocchio, C.: submitted for publication.

15. Hensgens, L.A.M., Bonen, L., de Haan, M., vander Horst, G. and Grivell, L.A.: Cell 32, 379-389 (1983).

16. Weiss-Brummer, B., Rodel, G., Schweyen, R.J. and Kaudewitz, F.: Cell 29, 527-536 (1982).

17. Anziano, P.Q., Hanson, D.K., Mahler, H.R. and Perlman, P.S.: Cell 30, 925-932 (1983).

18. Waring, R.B., Scazzocchio, C., Brown, T.A. and Davies, R.W.: J. Molec. Biol. 167, 595-606 (1983).

19. Dieckmann, C.L., Pape, L.K. and Tzagoloff, A.: Proc. Natu. Acad. Sci. U.S.A. 79, 1805-1809 (1982).

20. Kruger, K., Grabowski, P.J., Zaug, A.J., Sands, J., Gottschling, D.E., and Cech, T.R.: Cell 31, 147-157 (1982).

COB INTRON 1 AND 4: STUDIES ON MUTANTS AND REVERTANTS UNCOVER FUNCTIONAL INTRON DOMAINS AND TEST THE VALIDITY OF PREDICTED RNA-SECONDARY STRUCTURES

Gerhard Rödel, Jürgen Holl, Carlo Schmelzer, Cornelia Schmidt, Rudolf J. Schweyen, Brigitte Weiss-Brummer, Fritz Kaudewitz

Institut für Genetik und Mikrobiologie, Universität München, Maria-Ward-Str. 1a, D-8000 München 19, FRG

Introduction

The removal of intron sequences interrupting the coding sequences of many viral and nuclear as well as of several mitochondrial and chloroplast gene transcripts is a critical step in the formation of messenger RNA. At least three types of introns which basically differ in their splicing mechanisms are described so far: a) Intron-containing nuclear tRNAs of yeast: In this case, a unique subset of enzymes with endonucleolytic and ligating activities is required (1,2). b) Intron-containing nuclear and viral protein-coding genes: They obey (with one exception (3)) the GT/AG rule: at the 5' and 3' end of the intron, the invariable dinucleotides GT and AG, respectively, are an absolute requirement for splicing (4). Besides that a less striking conservation of sequences in the immediate vicinity of the splice junction is observed (5). In the case of yeast introns, a heptanucleotide, 20 to 55 nucleotides upstream to the 3' splice site may be relevant for the recognition of the 3' splice site (6,7). So far, there is no experimental indication of intronic secondary structures as prerequisites for splicing of these introns. c) Mitochondrial intron-containing genes in fungi, chloroplast rRNA and tRNA genes of some protozoa: They do not obey the GT/AG rule, but rather have short sequences in common which allow the construction of characteristic RNA secondary structures. According to this criterion,

Mitochondria 1983

two classes of introns (class I and II) have been distinguished (8,9). For some of the well-studied mitochondrial introns of Saccharomyces cerevisiae the involvement of either nuclear or nuclear plus mitochondrial components in the splicing reaction has been clearly established (10-13). In the case of Tetrahymena rRNA transcript containing a class I intron a self-splicing capacity of the RNA has been demonstrated, which does not, however, exclude the involvement of other components in vivo (14). In all other cases, a 'ribozyme'-like function so far has not been observed.

A major advantage of the yeast mitochondrial splicing system is the availability of various mutations interfering with the correct splicing reaction. Analysis of such mutations and of reversions restoring the capacity to remove the introns correctly is an excellent system to elaborate structures and/or components essential for splicing. Especially the relevance and validity of RNA secondary structures deduced by computer analysis on the basis of conservation of sequences and stability of base-paired regions can be tested. In this paper we concentrate on introns 1 and 4 of the mitochondrial cob gene (bI1 and bI4, respectively) of S. cerevisiae, which represent class II and I introns. We will sum up all mutations so far analyzed within these two introns and discuss their alterations and locations with respect to the proposed secondary structures. Moreover, the study of revertants will be presented as an important tool to substantiate conclusions drawn from studies on mutations and from computer analysis.

Results and Discussion

Cob Intron 1

Besides the oxi3 introns aI1, aI2 and aI5γ, bI1 is one of the mitochondrial-type introns of class II (8). A prominent feature of this class is that the corresponding RNAs can be folded up to secondary structures with almost identical cores of six base-paired stems and five rather conserved stem intervals (8,15). Of the few mutations which affect the removal of bI1 and which have been allocated to the 765 bp long intron 1, three have been sequenced so far (15,16). Fig.1 shows the positions of these mutations within bI1. Mutant M1301 was found to comprise a - 1 bp deletion some 30 bp upstream to the 3' splice site. Mutant G5049 has been shown to carry three mutational alterations one of which (G5049-3) is irrelevant for splicing as deduced from crosses with definite rho^- strains. The two relevant mutations (G5049-1 and G5049-2) are both located in the 5' part of the intron, 10 and 59 bp apart from the 5' splice site. Mutant M1301 has been shown to be a - 1 bp deletion about 500 bp distal to the 5' splice junction within the intron. What are the positions of these mutations in a secondary structure model of bI1?

Based on the criteria of conservation and stability as outlined in (8), one can construct a secondary structure of bI1 exhibiting the typical core structure of class II introns (Fig.2). Comparison with intron 5γ of the oxi3 gene reveals an excellent conservation of the primary sequences of the central wheel. As can be seen from Fig.2, the mutations M4873, M1301 and G5049-1 all affect stems of the central core structure either by interfering with internal base pairings or by the alteration of characteristic small loops within these stems. Mutation 1301 comprises a deletion of one A in a run of five A's (position 503 - 507) which is conserved in aI5γ and which takes part in the formation of a stem structure and of one of its internal loops. Similarly, mutation M4873 increases the size of a bulge loop in another central stem by the deletion of a G in a run of five G's. Mutation G5049-1, a G→A transition, affects the stability of a central base-paired region.

Does the study of revertants substantiate the conclusions concerning the primary effect of the mutations?

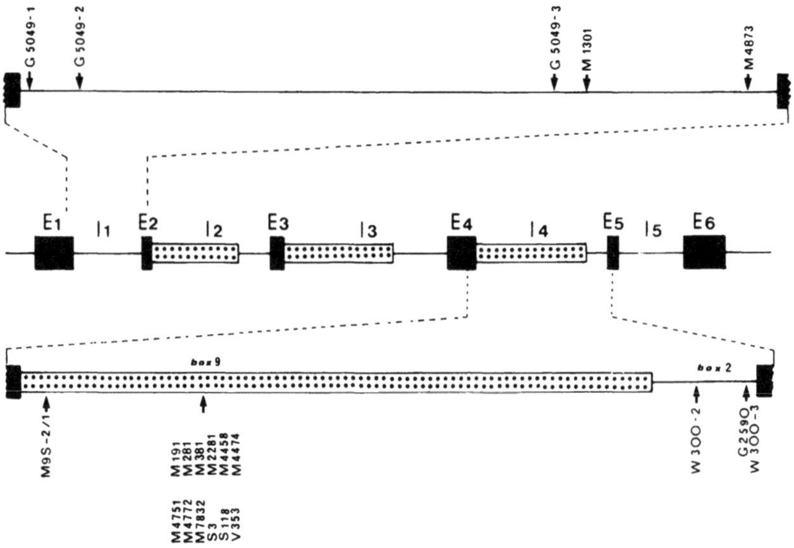

Fig.1: Organization of the cob gene in strain 777-3A and positions of sequenced cis-acting mutations in bI1 and bI4

black bars: exon sequences; dotted bars: open reading frames in introns; black lines: closed reading frames
upper part: enlarged drawing of bI1; positions of the mutations according to (15,16)
lower part: enlarged drawing of bI4; positions of the mutations according to (20-22, 24)

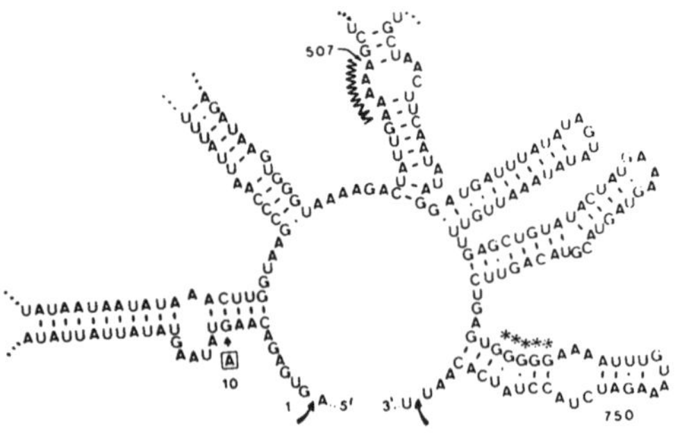

Fig.2: Central core of the secondary structure of bI1 RNA according to (15)
Large arrows: proposed splice sites
Mutational alterations: sequences harboring the mutational alterations of mutant M1301 (ʌʌʌʌ) and of M4873 (✱✱) as well as of G5049-1 (□) are indicated.

Two types of revertants have been found for mutant M4873: (i) those restoring the wild type sequence by insertion of a G, and (ii) those conserving the - 1 bp deletion but creating a novel sequence by a transition of an A next to the run of G's in the 3' direction. By that the reversion restores a conserved bulge loop of the original size without affecting the stem formation. Thus the primary effect of the mutation indeed might be the creation of an unusual bulge loop rather than the destabilization of the stem structure. In the case of mutant M1301, out of six mitochondrial revertants studied three are true wild-type revertants, while the other three replace the deleted A by a G, creating the novel sequence AAAAAG. Again, the size of the original bulge loop is restored without any significant destabilization of the stem structure. The failing to find other types of mitochondrial revertants with the same effect may be an indication for a functional constraint on the sequence per se. Hence, combined with the finding that (i) this sequence is perfectly conserved in aI5γ by means of sequence and relative position to the central core, and (ii) a nuclear allele specific suppressor of this mutation has been identified, one may speculate that the sequence is bifunctional: on one hand, it is necessary to form the typical stem structure with a definite bulge loop, while on the other hand it represents a primary sequence domain, possibly directly interacting with splicing components.

Cob intron 4

bI4 is one of the three introns in the long form cob gene, which exhibits an open reading frame (orf) in phase with the preceding exon (17-19). As for bI2 (and possibly for bI3) the intron orf, together with the preceding exon sequences, function as template for a protein (a 'maturase') necessary for splicing of the corresponding intron (18,20-22). Most probably, a 27 kd polypeptide comprising the C-terminal part of the (precursor) protein is equivalent to the bI4 maturase or TAF

(Trans-Acting Factor), which is necessary for splicing of bI4 as well as of aI4 (20,23). Besides mutations affecting this TAF (which correspond to the 'box7' cluster), three other functional domains have been revealed by the analysis of a variety of bI4 mutants (20-22,24): In contrast to box7 mutants, these mutants cannot be complemented in trans, i.e. the sequences affected act in cis. Fig.1 sums up all mutations in bI4 which have so far been sequenced. Two of the cis-acting domains are found within the orf, while the third is located in the 3' closed reading frame. Location within the orf includes the interesting possibility that a mutation affects at the same time a signal structure acting in cis and the trans-acting function by preventing the expression of a functional TAF. This has been found to occur in the case of mutant M191, deleting the cis-acting function by a -2 bp deletion and at the same time the trans-acting function by alteration of the reading frame with a subsequent TAA stop codon (21).

What do secondary structure models of bI4 tell us concerning these sequences?

As outlined in the introduction, bI4 is typical for the majority class of mitochondrial-type introns (= class I). Several authors noticed (8,9,25) that a striking feature of this intron class is the conservation of short sequences in the order given

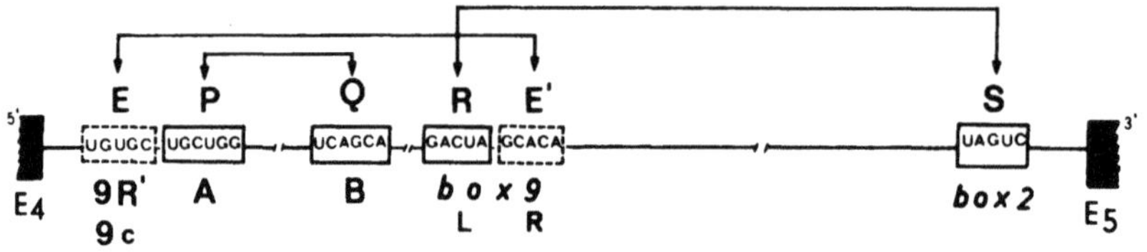

Fig.3: Conserved sequences of class I introns within bI4

Sequences conserved by means of the primary sequence (solid boxes) and of pairing potential (dashed boxes) are given. Potential base pairings are indicated by brackets. The nomenclature given in the upper part is according to (9) and in the lower part according to (26) and (28).

in Fig.3. These sequences have the potential to base pair with one another and indeed some of them appear as paired stem regions in RNA secondary structure models. As a consensus, six regions crystallize which potentially can interact to form three paired stretches. Most strikingly, three out of the six regions are exactly those which have been revealed by the analysis of bI4 mutants as cis-acting domains. Thus the clustering of mutations within bI4 is far from being fortuitous but rather is a very sensitive indicator of those regions which are either essential in building up RNA structures as prerequisites for splicing or which are recognized by the splicing machinery on the basis of their sequence.

The contribution of genetic studies to prove secondary structure models

Although the coincidence of the location of mutational alterations in prominent parts of RNA core structures is striking, it has to be demonstrated that those structures are indeed relevant in vivo. A promising experimental assay has been applied in the case of the Tetrahymena rRNA intron, which shares all the features of class I introns and for which autocatalytic splicing has been demonstrated (14). With the aid of nucleases specific for either single or double stranded RNAs, regions of paired RNA sequences could be deduced, most of which are in good agreement with the predicted secondary structure (26). But again, the result merely reflects the in vitro situation. Moreover, in the case of an equilibrium of alternative secondary structures in vivo, preferentially the major component which must not be the functionally important one, is detected. In addition, an application of this method to a structural analysis of bI4 RNA so far fails because of the limited quantity and stability of complete pure intron RNA. Other biochemical methods like cross-linking of paired RNAs harbor the same difficulties. A distinct assay to probe the in vivo situation is a genetical approach, the study of second site revertants.

While this method is restricted in that it reveals spot-like interactions rather than overall structures, it has the major advantage of avoiding in vitro artifacts and hence reflecting the in vivo situation.

Primary mutations affecting RNA base-pairing and hence the formation of stems in secondary structures, in principle should be revertible (i) by direct reversion to the wild type sequence; (ii) by mutations which circumvent the necessity for the formation of correct RNA structures, e.g. deletion of complete intron sequences (27) or mutations altering the components which recognize the RNA structures; (iii) by second site mutations which restore functional RNA stem structures. One possibility is the restoration of the stem structure by a compensatory bp exchange leaving the primary mutation unaffected but replacing the unpaired nucleotide of the opposite strand by a base with pairing capacity. The sequencing of such second site suppressors thus is a powerful method in detecting sequences involved in base-pairing with the region around the primary mutation. This method has been successfully used in revealing an interaction between the box9R and 9R' sequence as predicted by secondary structure models (for nomenclature, see Fig.3). Systematic search for second site suppressors of the cold-sensitive box9 mutants M4751 and M4772 resulted in the detection of two strains, M4751/R2 and M4772/R1, which compensated the original box9 mutation, a G→A exchange, by a compensatory C→T transition about 300 bp further upstream within bI4 (28). Thus an interaction between box9R/box9R' in vivo is evident, independent of whether this hybrid is one of the prominent stems of the bI4 core (8) or rather an interaction on the level of tertiary structure as a consequence of a built-up RNA structure (as proposed by 26).

From crosses between the box9R$^-$/box9R'$^-$ double mutant with wild-type or definite rho$^-$ clones, the single mutants box9R$^-$ (which is identical with the primary mutation) and box9R'$^-$ should arise by recombination. Indeed, crossing with the wild type results in the appearance of cold-sensitive respiratory deficient colonies. The same holds true for crosses between the double mutant with rho^- clones, covering box9R but not box9R'. Again conditional respiratory deficient colonies appear the cs phenotype of which is not as stringent as in the original mutant (V.Forsbach, personal communication). A plausible explanation for this may be that the weak G·U pairing in case of $box9R'^-$ mutant is sufficient for a partial restoration of the hybrid's function at 20°C, whereas this is not possible for the A/C pair in the original mutant. Alternatively, one might think of the possibility that the RNA sequence is not only involved in a hybrid formation, but at the same time part of a primary recognition sequence with a more or less pronounced tolerance for certain nucleotides.

In conclusion, the results presented prove genetic analyses of splicing deficient mutants and their revertants to be powerful sensors for sequences which interact in vivo on the level of RNA to build up characteristic secondary structures. Thus they are fruitful complementations to the theoretical work based upon the systematic analysis of available sequence data and to in vitro studies.

Acknowledgements

We thank Mrs. K. Hohlenburger for excellent technical assistance and Mrs. I. Nerkorn for typing the manuscript. This work was supported by the Deutsche Forschungsgemeinschaft.

References

1. Peebles, C.L., Gegenheimer, P., Abelson, J.: Cell 32, 525-536 (1983)

2. Greer, C.L., Peebles, C.L., Gegenheimer, P., Abelson, J.: Cell 32, 537-546 (1983).

3. Wieringa, B., Meyer, F., Reiser, J., Weissmann, C.: Nature 301, 38-43 (1983).

4. Breathnach, R., Benoist, C., O'Hare, K., Gannon, F., Chambon, P.: Proc.Natl.Acad.Sci.USA 75, 4853-4857 (1978).

5. Mount, S.M.: Nucl.Acids Res. 10, 459-472 (1982).

6. Langford, C.J., Gallwitz, D.: Cell 33, 519-527 (1983).

7. Mount, S.M.: Nature 304, 309-310 (1983).

8. Michel, F., Jacquier, A., Dujon, B.: Biochimie 64, 867-881 (1982).

9. Davies, R.W., Waring, R.B., Ray, J.A., Brown, T.A., Scazzocchio, C.: Nature 300, 719-724 (1982).

10. Dieckmann, C.L., Pape, L.K., Tzagoloff, A.: Proc.Natl.Acad. Sci. USA 79, 1805-1809 (1982).

11. Faye, G., Simon, M.: Cell 32, 77-87 (1983).

12. Pillar, T., Lang, B.F., Steinberger, I., Vogt, B., Kaudewitz, F.: J.Biol.Chem. 258, 7954-7959 (1983).

13. Borst, P., Grivell, L.A.: Nature 289, 439-440 (1981).

14. Kruger, K., Grabowski, P.J., Zaug, A.J., Sands, J., Gottschling, D.E., Cech, T.R.: Cell 31, 147-157 (1982).

15. Schmelzer, C., Schmidt, C., May, K., Schweyen, R.J.: EMBO J., in press

16. Schmelzer, C., Schmidt, C., Schweyen, R.J.: Nucl.Acids Res. 10, 6797-6808 (1982).

17. Nobrega, F.G., Tzagoloff, A.: J.Biol.Chem. 255, 9828-9837 (1980).

18. Lazowska, J., Jacq, C., Slonimski, P.P.: Cell 22, 333-348 (1980).

19. Lazowska, J., Jacq, C., Slonimski, P.P.: Cell 27, 12-14 (1981).

20. De la Salle, H., Jacq, C., Slonimski, P.P.: Cell 28, 721-732 (1982).

21. Weiss-Brummer, B., Rödel, G., Schweyen, R.J., Kaudewitz, F: Cell 29, 527-536 (1982).

22. Anziano, P.Q., Hanson, D.K., Mahler, H.R., Perlman, P.S.: Cell 30, 925-932 (1982).

23. Dhawale, S., Hanson, D.K., Alexander, N.J., Perlman, P.S., Mahler, H.R.: Proc.Natl.Acad.Sci.USA 78, 1778-1782 (1981)

24. Lamb, M.R., Anziano, P.Q., Glaus, K.R., Hanson, D.K., Klapper, H.J., Perlman, P.S., Mahler, H.R.: J.Biol.Chem. 258, 1991-1999 (1983).

25. Michel, F., Dujon, B.: EMBO J. 1, 33-38 (1983).

26. Cech, T.R., Tanner, N.K., Tinoco, I. Jr., Weir, B.R., Zuker, M., Perlman, P.S.: Proc.Natl.Acad.Sci.USA 80, 3903-3907 (1983).

27. Gargouri, A., Lazowska, J., Slonimski, P.P.: this volume

28. Weiss-Brummer, B., Holl, J., Schweyen, R.J., Rödel, G., Kaudewitz, F.: Cell 33, 195-202 (1983.)

SPLICING OF CYTOCHROME b mRNA BY IN VITRO COMPLEMENTATION OF MITOCHONDRIAL EXTRACTS FROM SPLICING DEFICIENT YEAST MUTANTS

Renée Schroeder[1] and Piotr P. Slonimski[2]
[1]Institut für Allgemeine Biochemie und Ludwig Boltzmann Forschungsstelle der Universität Wien, Austria and [2] Centre de Génétique Moléculaire du C.N.R.S., 91190 Gif/Yvette, France

Many mitochondrial genes are split and their expression requires a complex RNA processing of successive intron excisions and exon ligations. A considerable understanding of various elements involved has been gained by molecular genetics. Mainly, three classes of macromolecules are essential for splicing: i) critical RNA sequences necessary for correct folding of the substrate and/or its recognition by the splicing machinery, ii) specific proteins, mRNA-maturases, encoded by certain mitochondrial introns, iii) nuclear-gene products, presumably proteins, which interfere with or modulate the overall process. Each class of these molecules can be identified by specific mutations located either in the mitochondrial or in the nuclear genome (1-6, 16, 17).
It is clear, however, that a different approach based essentially on in vitro biochemical techniques is required in order to get further insight. In this article we shall describe the approach we have used.

1) Rationale of the in vitro splicing assay

A priori one could think of two ways in developping an in vitro splicing system: i) one can start from a wild type cell which obviously contains all the elements involved because it is able to carry on the whole splicing process, ii) one can start from two cell types, each one lacking a different ele-

Mitochondria 1983

ment and restore the overall activity by mixing. Instead of using the wild type extracts we have attempted to reconstruct in vitro the activity from two splicing deficient mutants that complement in vivo (7) : a) A trans-recessive box7-G1659 mutant which is deficient in the bI4 (= cytochrome b intron 4) maturase was used as a source of the unspliced precursor RNA. This RNA has a correct three dimensional structure and accumulates in the mutant because of the lack of maturase (1). The accumulation of the precursor RNA containing the unexcised intron 4 (referred to as 23S) has been shown by DNA-RNA hybridisations (8,9). b) a cis-dominant box2-G2590 mutant which is deficient in the three dimensional structure of the bI4 RNA (1) was used as a source of the active maturase, which accumulates in the mutant because of the incorrect folding of the precursor. The presence of the maturase has been shown by protein SDS gels demonstrating that the translation product of the bI4 ORF (p27) accumulates in large amounts in this mutant (2, 10, 11). In order to avoid interferences from the intron aI4 (cytochrome c oxidase subunit one intron 4) a double mutant CO66: box2-G2590 plus a cis dominant "box9" like aI4 mutation G192 (12) was used. This mutant was constructed and kindly provided by P. Netter. c) As a control, mitochondrial extracts from a mutant, G625 (13), which carries a complete deletion of the cytochrome b gene were used.

2) Preparation of extracts and assay conditions

Mitochondria were isolated by differential centrifugation after lysis of spheroplasts in 0.6M mannitol, washed twice in 0.6M mannitol and shocked in 0.15M mannitol (14). After centrifugation the pellet was resuspended in the assay lysis buffer: 40mM phosphate buffer pH 6.5, 200mM NH_4Cl, 20mM $MgCl_2$, 0.2 mM EDTA, 2mN dithiothreitol, 16mM creatine phosphate, 0.4mg/ml creatine kinase, 1mM ATP, 1mM GTP, 30% glycerol, PMSF, pepstatin, chymostatin and TLCK all 2mM, 0.2% Triton-X-100 and 10mM vanadyl-rinucleoside. Crude mitochondrial extracts were incubated at 30° either separatly or mixed. Samples were harvested at different times and RNA isolated by phenol extraction, ethanol precipitation and DNase I digestion. The RNA was identified by the northern-technique the hybridization probe being the plasmid p14 containing the whole intron free cytochrome b gene (15). After autoradiography the films were scanned with a Beckman DU8 photometer to quantify the amounts of hybrids. The hybrid at 23S corresponds to the precursor RNA (intron 4 containing preRNA), while the hybrid around 18S corresponds to the presumably intronfree mature mRNA (9).

3) Results

Figures 1a and 1b and table I show that:

a) during the first 15 minutes of incubation of the mixture of the mutants there is a general degradation of RNAs hybridizable with the cytochrome b exon probe (30 to 70% decrease depending on the experiment). The inital degradation in total RNA may be due to RNase activity and the following stabilization after 30 minutes may result from a protection of the RNA by proteins or membrane vesicles.

b) after 1 to 3 hours incubation a RNA band corresponding in size and hybridization properties to the mature cytochrome b mRNA can be seen. Depending on the experiment this band amounts from 10 to 20% of the total (bI4 containing 23S precursor + mature 18S messenger) RNA.

c) this band is absent or very low (1 %) in vivo or in vitro at 0 time in unmixed extracts.

d) this "mRNA" is much less abundant (3%) when G1659 is incubated alone or supplemented with G625.

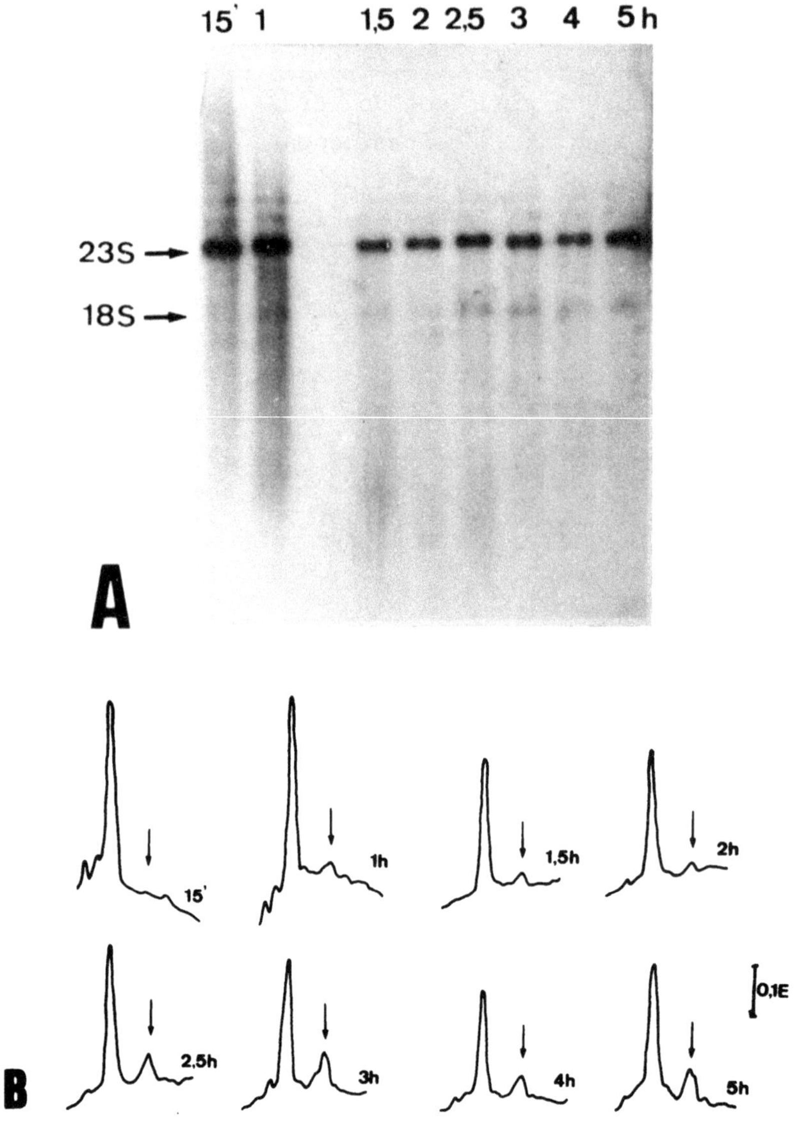

Mitochondrial extracts from strains G1659 and CO66 were mixed and incubated as described in the text. Samples were harvested at different times, RNA was isolated and identified by hybridization. After autoradiography (fig. 1A), the films were scanned (fig. 1B) to quantify the abundancy of the precursor (intron 4 containing, 23 S) and mature (18 S) messenger RNAs

Mitochondria from			Abundancy			
Mutation	Maturase	Unspliced Precursor	Incub. (hrs)	Precursor 23S	mRNA 18S	mRNA %
G1659	deficient	active	0			1
CO66	active	deficient	0			1
Mixture:			0.25	1.25	0.01	1
G1659	active	active	1	0.9	0.08	8
+			2.5	0.74	0.19	20
CO66			5	0.97	0.16	14

Table 1:The hybrid amounts of precursor (23S) and messenger (18S) RNAs were quantified by integration of the peaks shown in figure 1b. The percentage 18S RNA is related to total (bI4 containing precursor + mature messenger) RNA.

When increasing amounts of extract from strain CO66 were added to a constant amount of G1659 extract also increasing amounts of the putative 18S RNA could be detected.
We have tried several other conditions which all gave negative results. Among them were incubations in protein free extracts and addition of purified RNA from strain G1659 to mitochondrial extracts from strain CO66.

4) Conclusions

It is clear that an RNA processing occurs in our <u>in vitro</u> system. The fact that it takes place essentially in the mixture of extracts and <u>not</u> in individual extracts from splicing deficient mutants strongly suggests that this <u>in vitro</u> RNA processing is related to the in <u>vivo</u> complementation which leads to the correct splicing and maturation of cytochrome b mRNA. Furthermore the size and hybridization properties of this 18S RNA are in

agreement with the expected values for the mature mRNA. These two results constitute, in our opinion, evidence in favor of a bona fide in vitro splicing system.
We consider, however, our results as preliminary. A more extensive characterization of the in vitro made 18S RNA, by S1 mapping and sequence analysis, are necessary to demonstrate that its primary structure is identical to the true mRNA and does not correspond to a product of an aberrant processing. Fractionation and characterization of various components of the mitochondrial extracts necessary for RNA processing to occur should be also of interest.
Whatever the issue would be we believe, that this first attempt to realize an in vitro splicing of mitochondrial mosaic gene transcripts, has fullfilled its initial aim: it has provided evidence that the in vitro complementation approach is promising and worth to be exploited more thoroughly.

Acknowledgements

We thank Dr. P. Netter for constructing the strain CO66 and Dr. M. Labouesse for the kindly gift of plasmid DNA. The excellent technical assistance of Christian Bernhart was highly appreciate. R. Schroeder acknowledges the receipt of an EMBO long-term fellowship held at Gif in 1982. This work was supported by grants from the CNRS, DGRST, INSERM and Ligus Nationale Francaise contre le Cancer.

References

1. De La Salle, H., Jacq, C., Slonimski, P.P.: Cell 28, 721 731 (1982)
2. Weiss-Brummer, B., Rödel, G., Schweyen R.J., Kaudewitz, F.: Cell 29, 527-536 (1982)
3. Michel, F., Jacquier, A., Dujon, B.: Biochimie 64, 867-881 (1982)

4. Dieckmann, C., Pape, L., Tzagoloff, A.: Proc. Natl. Acad. Sci., USA 79, 1805-1809 (1982)

5. Faye, G., Simon, M.: Cell 32, 77-87 (1983)

6. Davies, W.R., Waring, R.B., Ray, J.A., Brown T.A., Scazzochio C.: Nature 300, 719-724 (1982)

7. Lamouroux, A., Pajot, P., Kochko, A., Halbreich, A., Slonimski, P.P.: The organization and Expression of the Mitochondrial Genome, A.M.Kroon and C.Saccone eds. Elsevier North Holland, Amsterdam, pp. 153-156 (1980)

8. Church, G., Slonimski, P.P., Gilbert, W.: Cell 18, 1209-1215, (1979)

9. Labouesse, M., Netter, P., Schroeder, R.: in preparation

10. Claisse, M.C., Slonimski, P.P., Johnson, J., Mahler, H.R. Mol. Gen. Genet. 177, 375-387 (1980)

11. Anziano, P.Q., Hanson, D.K., Mahler H.R., Perlman, P.S.: Cell 30, 925-932 (1982)

12. Netter, P., Jacq, C., Carignani, G., Slonimski, P.P.: Cell 28, 733-738 (1982)

13. Slonimski, P.P., Claisse, M.C., Foucher, M., Jacq, C., Kochko, A., Lamouroux, A., Pajot, P., Perrodin, G., Spyridakis, A.: Biochemistry and Genetics of Yeast, Bacila M. et. als. eds. (Academic Press, Sao Paulo) pp.391-401 (1978)

14. Boehni, P.C., Daum, G., Schatz, G.: J. Biol. Chem. 258 4937-4943 (1983)

15. Labouesse, M., Slonimski, P.P.: EMBO J. 2, 269-276 (1983)

16. Lazowska, J., Jacq, C., Slonimski, P.P.: Cell 22, 333-348 (1980)

17. Groudinsky, O., Dujardin, G., Slonimski, P.P.: Mol. Gen. Genet. 184, 493-503 (1981)

MITOCHONDRIAL CONTRIBUTIONS TO HYBRID ENZYME COMPLEXES FOR MITOCHONDRIAL RNA PROCESSING AND ELECTRON TRANSPORT

Warren C. MacKellar, Deborah K. Hanson*, Philip M. Sass, C. David Zarley and Henry R. Mahler

Department of Chemistry, and the Molecular, Cellular and Developmental Biology Program, Indiana University, Bloomington, IN 47405

Philip S. Perlman

Department of Genetics, and the Molecular, Cellular and Developmental Biology Program, Ohio State University, Columbus, OH 43210

Introduction

Our laboratory has been dealing with two aspects of nucleomitochondrial interactions in *Saccharomyces cerevisiae*. 1) The as yet hypothetical splicing complex(es) required for the processing of the primary transcripts of the mitochondrial gene for cytochrome *b* (*cob*) and 2) the effect of the cytochrome *b* gene on the biogenesis of the bc_1 complex (Complex III) of the respiratory chain.

I. Processing of *cob* transcripts

The mitochondrial *cob* gene exists in either of two pleiomorphic forms, the long form with six exons and five introns or the short form with only three exons and two introns (1). This communication will deal entirely with the long form of the *cob* gene. Three of the introns (I2, I3 and I4) contain long open reading frames contiguous with their upstream exons (2-5). Recent genetic and biochemical evidence strongly suggest that these introns, at least in part, encode proteins essential for splicing of primary transcripts (2,6,7). These hypothetical intron-encoded proteins have been termed maturases (2) or spligases (8). The maturase of the

*Present address: Section of Genetics & Development, Bradfield Hall, Cornell University, Ithaca, New York 14853

penultimate intron of *cob* (I4) is also required for the splicing of at least one intron of the mitochondrial gene *oxi3*, which encodes subunit one of cytochrome oxidase (9-13).

There are two classes of splicing defect mutants for I2 and I4. They are: i) *trans*-recessive, with mutations located in the intron open reading frame (loci *box3* in I2 and *box7* in I4), which fail to splice because they affect the synthesis or activity of the intron product, but whose *cis*-acting sequences are functional since the *trans*-acting mutations can be complemented by exon mutations, and ii) *cis*-acting mutations which produce a normal active intron product, but are defective in the splicing sites. Examples of *cis*-acting mutations in I4 are found in the closed reading frame (locus *box2*) and in the 5' portion of the I4 open reading frame (ORF) (locus *box9*) (9-17). Current information suggest that all mutants of I2 and I4, whether *cis* or *trans*-acting, accumulate a single major, intron-specific, partially spliced transcript (3,5,7,11,12,16,18-20), while novel proteins, presumably translated from these RNA species, are more diverse, and include allele-specific species (6,9,14-17).

Intron 4 mutants accumulate from three to eight novel proteins, three of which, P35, P27 and P23, are *cob* related (6,8,11,15,20-25). While P35, a hybrid protein analogous to P43 of I2, does not appear to be affected by I4 nonsense mutations, P27, which appears to be encoded wholly within the intron, is shortened by these mutations (22,23). Three research groups have recently provided evidence for P27 being the maturase for intron 4 (10,11,14,15).

Two different mechanisms have been proposed for the biosynthesis of P27. In the first, protein synthesis is initiated within the intron reading frame, so that P27 is made as a primary translation product (10). In the second, a hybrid precursor protein of calculated mass (from the DNA sequence) equal to 74,000 is translated from B1 through B4 and the contiguous intron ORF of I4, very similar to the situation in I2. This hybrid precursor is then processed by proteases to yield P27 as the C-terminal and P35 as the N-terminal fragments, respectively (11,14,15,17). In this communication we will present evidence that a minor protein band, accumulated by selected I4 mutants and with a relative molecular mass of 65,000 (P65), has the properties of the postulated hybrid protein.

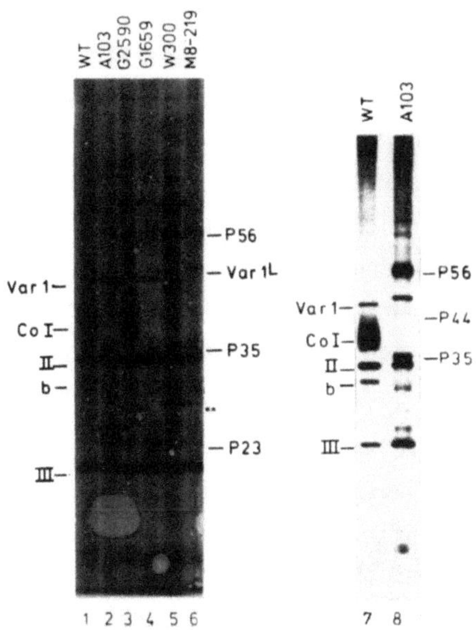

Figure 1: Mitochondrial translation products described in this study. Var1 (high and low mobility forms), cytochrome *b* (cyt *b*) and cytochrome oxidase (Co) subunits I, II and III are included for comparison. The allele specific proteins in the high (>40,000) and low (<30,000) M range are indicated by single and double asterisks, respectively; they include P65 in W300 and G2590; P46 in G1659; P28 in W300; P27 in G2590 and M8-219 and P25 in A103. Two additional species P62 in A103 and P44 in M8-219 are prominent in some analyses, but not the one shown here. P46, although exhibiting the same mobility as var1, is not the same protein: G1659 contains the high mobility form of var1 and the two proteins generate different fingerprint patterns.

Results

Figure 1 depicts *trans* and *cis*-acting mutants used in this study. *Trans*-acting mutants (lanes 2,4,5 and 6) all lack apocytochrome *b* (cyt *b*, P30) and subunit 1 of cytochrome oxidase (Cox I) and accumulate five additional *oxi3*-related proteins (P56, P44 and a triplet of bands around P29; refs. 22,24) and three *cob*-related bands (P35, P27 and P23) (15,17,21-23). Most *cis*-acting mutants retain P35, P27 and P23 but, because they express *oxi3*, lack P54, P43 and the bands at P29.

W300 (Fig. 1; lane 5) is a triple mutant (10,14): it contains two single base changes in the closed reading frame of I4, one of which causes a *cis*-acting splicing defect at the *box2* locus and the other a G-A transition at bp 1665 in the ORF (changing Gly to Asp) that results in a slightly lower mobility of P27 to P28. However, W300 is fully competent in complementation studies, indicating that this change does not affect the function of the *trans*-acting product. When the mitochondrial proteins of W300 are labeled to very high specific radioactivity, an additional larger protein (P65, marked with an asterisk) can be seen (Fig. 1, lane 5).

P65, P28 and P35 were excised from a preparative gel of W300 and partially digested with *S. aureus* V8 protease yielding the patterns shown in Fig. 2A. These data confirm the earlier demonstration that P35 and P28 are unrelated (16,22) and that P35 is related to cyt *b* (P30). They also reveal that P65 is related to both P35 and P28. For example, two prominent large fragments (bands A and B) are present in both P35 and P65, but are absent or represent very minor species, in P28. Two smaller fragments (bands C and D) are present in P28 and P65, but not in P35. If P65 were only an aggregate of P35 and P28 the partial digest of P65 (Fig. 2A, lane 7) should be identical with the combined partial digest of P28 and P65 (Fig. 2A, lane 2). However, this is not the case. P65 contains at least one extract band (marked with an asterisk) not seen in the combined digest of P28 and P35. Also, the combined digest of P28 and P35 contains at least four additional bands not seen in the P65 digest. These results rule out the possibility that P65 is an aggregate of P35 and P28, and support the hypothesis that P65 may be a hybrid precursor of P35 and P28.

In order to determine if the fragments labeled A, B, C, and D in Fig. 2A migrate at the same location on the primary digest gel due to extensive sequence homologies or to happenstance, we carried out sequential digestion with a second protease. Bands were excised from the primary gel and their fingerprints were examined after digestion with Nagarse. The patterns obtained are shown in Fig. 2B. It is evident that bands A and B are identical for P65 and P35, while they appear to be different from the minor species found at these positions in P28. The reciprocal result

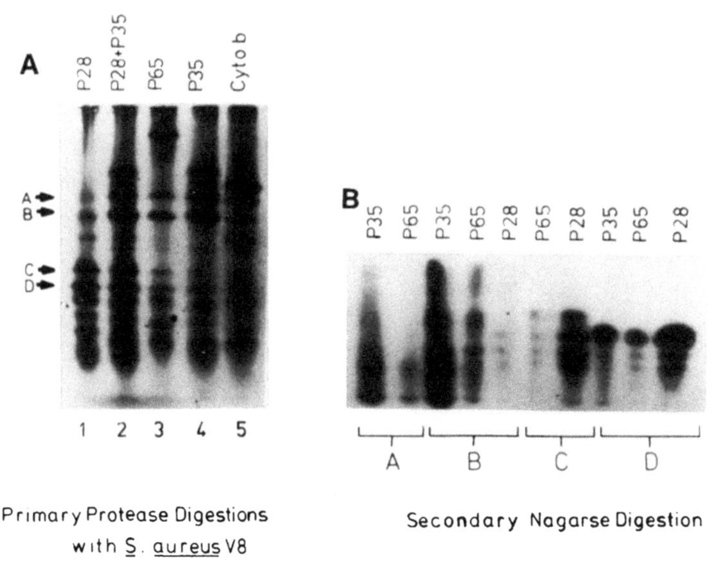

Figure 2: A. Partial digestion of novel polypeptides derived from W300 with *S. aureus* V8 protease (15 µg/lane for 2 h at 30°C). Asterisk in lane 3 denotes a band present in P65 digests and absent from those of P35, P28 and the mixture. B. Nagarse digestions (10 µg/lane for 2 h at 30°C) of polypeptides A, B, C and D from Fig. 2A.

is shown for bands C and D, they are identical for P65 and P28, but differ for P35.

If the hybrid precursor model is correct we would predict that P65 should be present in other *cis*-acting mutants and, more importantly, that *trans*-acting mutants with shortened intron reading frames should produce shortened forms of the hybrid precursor protein. In Fig. 1 candidates for the expected species are indicated by a single asterisk. These species meet the criterion of being *cob*-related (24).

G2590 is a *cis*-acting mutant with a splicing defect in the *box2* locus (14). Like W300, it produces a P65 (Fig. 1, lane 3). G1659 and M8-219 both have a truncated ORF in I4 due to the introduction of chain termination codons at bp 1510 (15,16) and bp 1482-1484 (17), respectively. Both should make truncated forms of the hypothetical hybrid protein. G1659 makes a candidate at P46 (Fig. 1, lane 4), and M8-219 at P44.

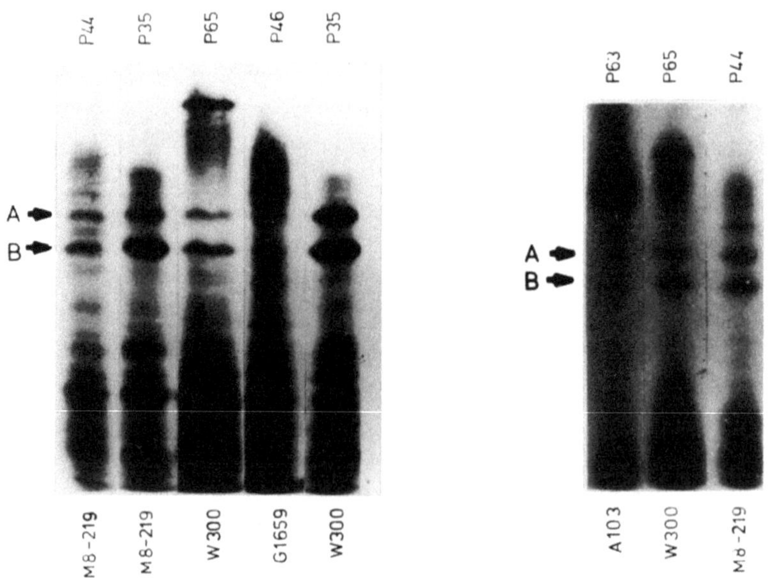

Figure 3: *S. aureus* V8 protease digestion (15 μg/lane) of novel polypeptides derived from selected I4 mutant strains. Arrows A and B indicate comigrating polypeptides produced by this protease in the various mutants.

Finally, mutant A103 contains a chain termination codon around bp 1820, near the end of the I4 ORF (17). This mutant produces a P25 instead of a P27 and also a P62 instead of a P65 (Fig. 1).

When the various proposed precursor bands were excised from preparative gels and partial digests were done with *S. aureus* V8 protease, the diagnostic bands A and B were seen in all the candidates (Fig. 3). These results indicate that the P65 from other *cis*-acting mutants along with the truncated forms from various *trans*-acting mutants are all related to P65 and P35 from W300.

The hypothetical origins and pathways for the formation of P27 are depicted in Fig. 4. In this scheme the primary transcript comes from the fusion of B1-B4 with the contiguous ORF of I4. The locations of the various mutants used in this study have been placed on the primary transcript. In the wild type this primary transcript is translated to produce the P65 which can undergo proteolytic cleavage to produce the P35 and P27. The P35 can then undergo further proteolytic cleavage to produce the P23. The truncated forms of the P65 from the mutants A103, G1659 and M8-219 are also indicated

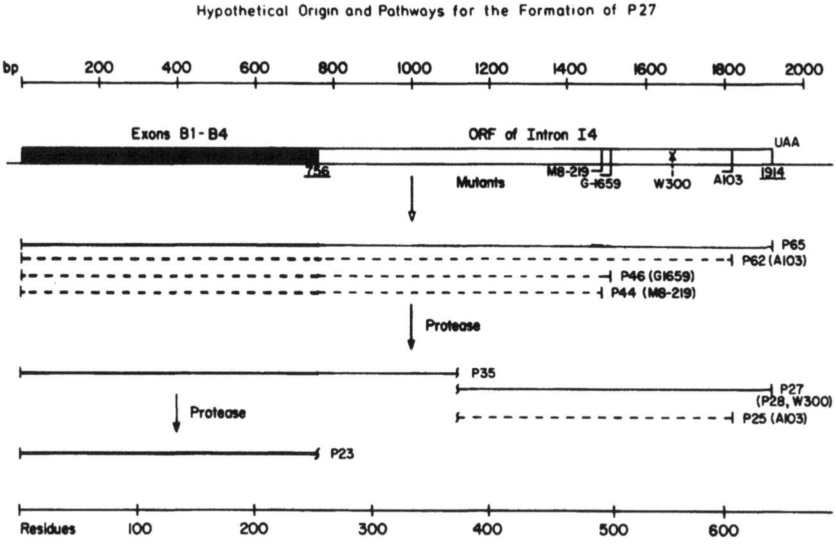

Figure 4: Hypothetical origin and pathways for the formation of P27, the proposed maturase of I4, and related products. The coordinates are those given by Nobrega and Tzagoloff (3). Shown also are the locations of the mutants described in the text.

on this scheme. G1654 and M8-219 should also produce truncated forms of the P27, like A103. However, these fragments are too small to be resolved by our gel system, and they are not indicated in the scheme.

II. Effect of *cob* gene on bc_1

The bc_1 complex (Complex III) is a portion of the mitochondrial electron transport chain, which carries reducing equivalents from reduced coenzyme Q to oxidized cytochrome *c*. It is also an energy conservation site and has thus been shown to pump protons across the inner mitochondrial membrane (26-28). Complex III is composed of 8-10 proteins (29-37) of which only cytochrome *b* is encoded by mitochondrial DNA (36-41). The nuclear encoded subunits are transported into the mitochondria as larger precursor proteins, which are then processed to their mature forms (34-36,40,41). Only cytochromes *b* and c_1 and the iron-sulfur protein have been shown to carry electrons in the complex; the other subunits are without known function, but they may be important in proton translocation (28,40).

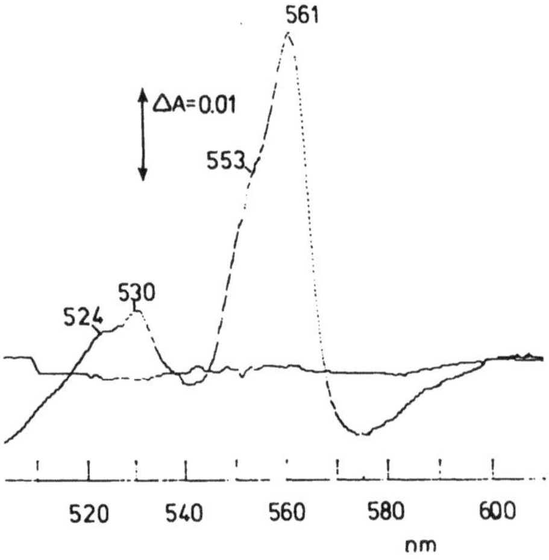

Figure 5: Difference spectrum (dithionite reduced-ferricyanide oxidized) of Complex III, taken at room temperature.

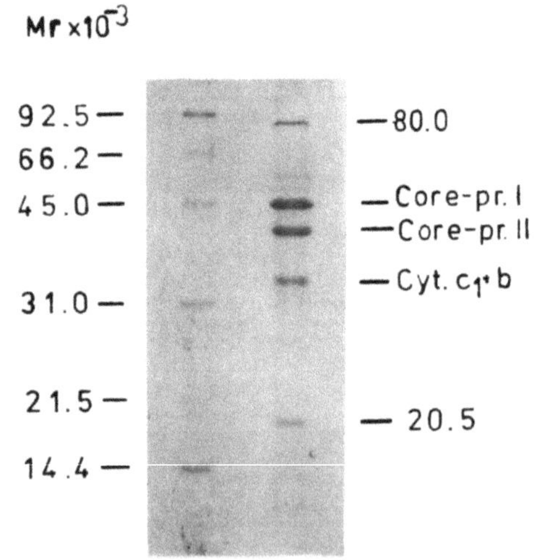

Figure 6: Subunit composition of the antimycin Complex III. The sample was electrophoresed on 10-15% linear gradient SDS-polyacrylamide gel.

Complex III has been isolated from commercial yeast (29-31,37), but to date there has been no report of the isolation of Complex III from genetically characterized laboratory strains of *S. cerevisiae*. The initial part of this communication will deal with the isolation and characterization of Complex III from a wild type strain ID41-6/161 (cob^L) and the production of an antibody to the complex. The last part will deal with immunoprecipitation of the complex from wild type and selected *cob* missense mutants to determine the effect of these mutants on the biogenesis of Complex III.

Results

Using modifications of a published procedure (38,42), we were able to isolate Complex III from a genetically defined strain (ID41-6/161, cob^L) grown under laboratory conditions. The difference spectrum (dithionite reduced-ferricyanide oxidized) of the complex, taken at room temperature, has an α-maximum absorbance peak at 561 nm and a shoulder at 556 nm characteristic of cytochrome *b* and c_1, respectively (Fig. 5) (29-32). It was determined that there were 6 nmoles cytochrome *b*/mg protein and 3.3 nmoles cytochrome c_1/mg protein yielding a b/c_1 ratio of 1.8, which is very close to the theoretical ratio of 2. The lack of absorbance at 603 nm indicates the complete absence of cytochrome a-a_3 in this preparation. When a sample of Complex III was electrophoresed on a 10-15% linear gradient SDS-polyacrylamide gel the pattern in Fig. 6 is observed. There are three major bands in the 45,000-31,000 M_R range. We have labeled these core-pr 1, core-pr 2 and bc_1, which have M_R of 45,000, 39,000 and 34,000, respectively. There are three higher M_R bands at 80,000, 55,000 and 49,000, which are probably contaminants, along with three lower M_R bands. The band at about 25,000 corresponding to the Reiske Fe-S protein is missing, but the Reiske protein is absent in most antimycin bc_1 preparations (29,42,43). This preparation was used as an immunogen and an antibody was produced to the bc_1 complex. Using the "Western" blot transfer-immunoperoxidase technique of Towbin *et al.* (25) we determined that the antibody was heterospecific; it reacted strongly with core-pr 1, cytochrome c_1 and slightly with the 55,000 M_R contaminant (data not shown).

Employing the Complex III antibody we immunoprecipitated a bc_1 complex from Triton X-100 extracts of submitochondrial particles (SMP), which had been prelabeled *in vivo* overnight with $^{35}SO_4^=$, but without cycloheximide (11) (Fig. 7, lane 1). The labeled pattern was identical with the Coomassie Blue stained pattern of the original immunogen with the exception that the lower M_R components did not label well. Poor labeling of the smaller subunits with ^{35}S has also been observed in rat liver (44). However, if the cells were not labeled overnight, but rather pulse labeled for one hour, there was a decrease in label of all the components of Complex III except core-pr 1 and cytochrome c_1 (Fig. 7, lane 2). This differential labeling may indicate one of several things: i) that during the one hour pulse, core-pr 1 and cytochrome c_1 may be synthesized and processed to their mature forms, but not assembled into a bc_1 complex. They appear on the gel only because they react with the antibody; ii) that there is a large unlabeled pool of the other components of the bc_1 complex, so that

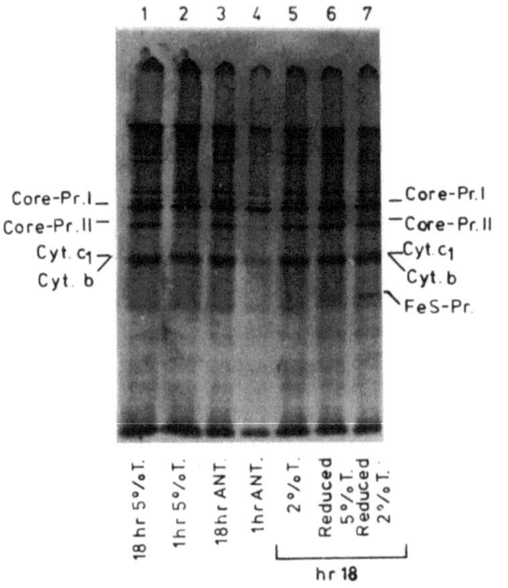

Figure 7: Immunoprecipitation of Complex III from wild type yeast under reducing vs non-reducing conditions. Mitochondria were labeled with $^{35}SO_4^=$ and immunoprecipitated. The samples were run on a 10-15% linear gradient SDS-polyacrylamide gel. The gel was dried and exposed to film for 12 hr.

we do not see the label appearing in them as quickly as in core-pr 1 and cytochrome c_1. There are different pool sizes for the subunits of *Neurospora crassa* cytochrome oxidase (45). iii) Subunits of Complex III do not turnover as a unit, rather, some components are more rapidly replaced in the membrane than others. In lane 3 and 4 (Fig. 7) a repeat of the experiments seen in lanes 1 and 2 are shown with the exception that the mitochondria were loaded with antimycin before the immunoprecipitation, indicating that the bc_1 complex can be precipitated plus or minus antimycin.

There have been reports that the cytochrome bc_1 complex changes its configuration depending upon its redox state (33,46) and also that the Triton X-100 concentration used to solubilize the SMP affects the binding of the Reiske Fe-S protein (M_R = 24,500-25,000) (43). Since we are interested in a complex that is as close to its physiological state as possible, we decided to determine whether a complex which contained the

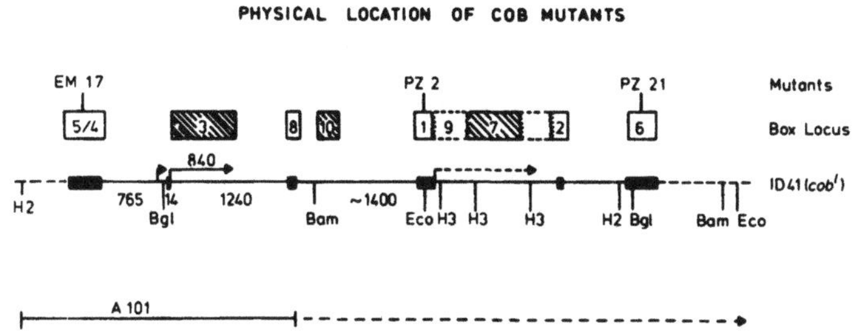

Figure 8: Physical location of the *cob* mutants used in this study. Exons are represented as filled rectangles, introns as solid lines.

Reiske protein could be isolated. When the SMP were solubilized in 2% Triton X-100 instead of 5% a faint band was seen with M_R = 24,500 (Fig. 7, lane 5). If the SMP were first reduced with succinate in the presence of NaN_3 then solubilized with 5% Triton X-100 the band at M_R = 24,500 was more pronounced. Finally, when the SMP were first reduced then solubilized in 2% Triton X-100 the band at M_R = 24,500 was very prominant (Fig. 7, lane 7).

Once the procedure for isolation of Complex III had been established, we could analyze the affect of *cob* mutations on Complex III assembly. We began with missense mutants which produced a protein with M_R = 30,000, but had no cytochrome *b* spectrum and no Co-Q—cytochrome *c* reductase activity. The missense mutants chosen were: EM-17, which has a mutation in box locus 5/4 in the first exon of *cob*, PZ2 in box locus 1 in the 4th exon of *cob* and PZ21 in box locus 6 in the last exon of *cob* (Fig. 8). The large *cob* deletion A101, which does not produce a P30, and CAD245, which has an *oxi3* deletion but has a normal *cob* gene were also used.

Wild type cells were labeled overnight (Fig. 9, lane 1) or for 4 hours (Fig. 9, lane 6) with $^{35}SO_4^{=}$. When Complex III was immunoprecipitated, all the major proteins observed in the isolated complex (except subunit V), were recovered (compare with Fig. 7, lanes 1 and 2). The labeling of all the subunits except core-pr 1 and cytochrome c_1 was greatly decreased in the 4 hour pulse. Since the mutants were all labeled using a 4 hour pulse, we will compare the mutant patterns to the pattern observed in lane 6 (Fig. 9). It can be seen that the deletion of *oxi3* (mutant CAD245) had no

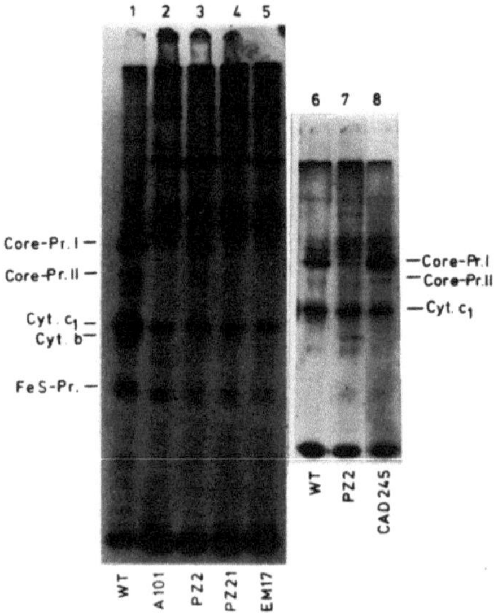

Figure 9: Immunoprecipitation of Complex III from the various *cob* mutants described in the text. The samples were run on a 10-15% linear gradient SDS-polyacrylamide gel. However, the 15% solution contained 10% glycerol. This modification gave good separation of cytochrome c_1 and b. Lanes 1-5 were run on a gel 20.5 cm long and lanes 6-8 on a gel 14 cm long.

effect on the formation of Complex III since the patterns in lane 6 and 8 are identical. The immunoprecipitates of the various missense mutants along with the partial *cob* deletion mutant (A101) all displayed a band migrating at the same position as cytochrome c_1 (Fig. 9, lanes 2-3). The missense mutations at both the N-terminus (EM-17) and the C-terminus (PZ21) showed no other bands corresponding to subunits of Complex III (lanes 4 and 5, respectively). However, a missense mutation in the middle of the *cob* gene (PZ2) had a band comigrating with core-pr 1. The partial *cob* deletion (A101) had the same pattern as the N + C-terminal mutants (lane 2).

It is not surprising that cytochrome c_1 was observed in all the *cob* mutants tested since it is observed even in petite mutants (47) and our antibody reacted towards cytochrome c_1. The lack of the Reiske Fe-S protein is also not surprising. Even though the immunoprecipitation was done under reducing conditions, it has been shown that there is no E.P.R. signal corresponding to this protein in cytochrome b mutants (48). The fact that mutations at the N + C termini, but not in the middle of the *cob* gene, do not contain a mature core-pr 1 indicates that there is something important about these locations which is somehow involved in either the assembly of Complex III or the import and processing of precursors of Complex III subunits. This involvement will be more extensively investigated using a battery of *cob* mutants. Since only cytochrome c_1 and core-pr 1 are observed in the 4 hour labeling experiments, little can be said about the formation of an intact complex, because the antibody is heterospecific and reactive towards both of these subunits. Further studies to investigate complex formation are underway utilizing antibodies produced against each of the individual subunits and also utilizing a procedure which we have developed for the co-purification of unlabeled wild type Complex III, as a carrier with $^{35}SO_4^=$ labeled mutants.

Conclusion

We have presented data that indicates that the hypothetical maturase (P27), encoded within the ORF of intron 4 of *cob*, which has been proposed to combine with nuclear encoded proteins to splice the primary transcript of *cob*, comes from a higher M_R precursor (P65) as depicted in Fig. 4. This higher M_R precursor is a hybrid protein encoded from the fusion of Exons B1-B4 with the contiguous ORF of I4. Proposed precursors react to mutations in a allele-specific fashion as does the proposed mature maturase protein.

While our new findings are consistent with the precursor model, they raise several important questions: First, even though P65 and its derivatives are capable of functioning as precursor for P27, is their relationship an obligatory one? That is, are all P27 molecules derived from P65, or are some due to internal initiation of translation? Second, is the P65 itself the maturase rather than one of its proteolytic fragments? If so, then the original maturase model postulated by Lazowska *et al.* (2) for the case of *cob* intron 2 would apply directly to the case of *cob* intron 4, without the modifications required by the precursor model. These questions will be investigated further.

With regard to the role of cytochrome *b* on the biogenesis of Complex III, it appears that the N + C termini are required in order to produce a mature core-pr 1. Further experiments are presently underway to investigate this question more thoroughly.

References

1. Borst, P., Grivell, L.A.: Nature 289, 439-440 (1981).
2. Lazowska, J., Jacq, C., Slonimski, P.P.: Cell 22, 333-348 (1980).
3. Nobrega, F.G., Tzagoloff, A.: J. Biol. Chem. 255, 9828-9837 (1980).
4. Lazowska, J., Jacq, C., Slonimski, P.P.: Cell 27, 12-14 (1981).
5. Bonitz, S.G., Homison, G., Thalenfeld, B.E., Tzagoloff, A., Nobrega, F.G.: J. Biol. Chem. 257, 6267-6274 (1982).
6. Alexander, N.J., Perlman, P.S., Hanson, D.K., Mahler, H.R.: Cell 20, 199-206 (1980).
7. Schmelzer, C., Haid, A., Grosch, G., Schweyen, R.J., Kaudewitz, F.: J. Biol. Chem. 256, 7610-7619 (1981).
8. Church, G.M., Gilbert, M.: in Mobilization and Reassembly of Genetic Information (Joseph, D.R., Schultz, J., Scott, W.A., Werner, R., eds.) Academic Press, New York, pp. 379-395 (1980).
9. Dhawale, S., Hanson, D.K., Alexander, N.J., Perlman, P.S., Mahler, H.R: Proc. Natl. Acad. Sci. USA 78, 1778-1782 (1981).
10. Jacq, C., Pajot, P., Lazowska, J., Dujardin, M., Claisse, M., Groudinsky, O., de la Salle, H., Grandchamp, C., Labouesse, M., Gargouri, A., Guiard, B., Spyridakis, A., Dreyfus, M., Slonimski, P.P.: in Mitochondrial Genes (Slonimski, P.P., Borst, P., Attardi, G., eds.), Cold Spring Harbor Laboratory, New York, pp. 155-185 (1982).
11. Mahler, H.R., Hanson, D.K., Lamb, M.R., Perlman, P.S., Anziano, P.Q., Glaus, R., Haldi, M.L.: in Mitochondrial Genes (Slonimski, P.P., Borst, P., Attardi, G., eds.), Cold Spring Harbor Laboratory, New York, pp. 185-199 (1982).
12. Grivell, L.A., Hensgens, L.A.M., Osinga, K.A., Tabak, H.F., Arnberg, A.C., Boer, P.H., Crusius, J.B.A., van der Laan, J.C., de Haan, M., van der Hoarst, G., Evers, A.F.: in Mitochondrial Genes (Slonimski, P.P., Borst, P., Attardi, G., eds.), Cold Spring Harbor Laboratory, New York, pp. 225-240 (1982).
13. Dujardin, G., Jacq, C, Slonimski, P.P.: Nature 298, 628-632 (1982).
14. de la Salle, H., Jacq, C., Slonimski, P.P.: Cell 28, 721-732 (1982).
15. Weiss-Brummer, B., Rodel, G., Schweyen, R.J., Kaudewitz, F.: Cell 29, 527-536 (1982).

16. Lamb, M.R., Anziano, P.Q., Glaus, K.R., Hanson, D.K., Klapper, H.J., Perlman, P.S., Mahler, H.R.: J. Biol. Chem. 258, 1991-1999 (1983).

17. Anziano, P.Q., Hanson, D.K., Mahler, H.R., Perlman, P.S.: Cell 30, 925-932 (1982).

18. Halbreich, A., Pajot, P., Foucher, M., Grandchamp, C., Slonimski, P.P.: Cell 19, 321-329 (1980).

19. van Ommen, G.-J.B., Boer, P.H., Groot, G.S.P., de Haan, M., Roosendall, E., Grivell, L.A., Haid, A., Schweyen, R.J.: Cell 20, 173-183 (1980).

20. Claisse, M.L., Spyridakis, A., Wambier-Kluppel, M.L., Pajot, P., Slonimski, P.P.: in Biochemistry and Genetics of Yeast (Bacila, M., Horecker, B.L., Stoppani, A.O.M., eds.), Academic Press, New York, pp. 369-390 (1978).

21. Solioz, M., Schatz, G.: J. Biol. Chem. 254, 9331-9334 (1979).

22. Claisse, M.L., Slonimski, P.P., Johnson, J., Mahler, H.R.: Molec. gen. Genet. 177, 375-387 (1980).

23. Bechmann, H., Haid, A., Schweyen, R.J., Matthews, S., Kaudewitz, F.: J. Biol. Chem. 256, 3525-3531 (1981).

24. Hanson, D.K., Lamb, M.R., Mahler, H.R., Perlman, P.S.: J. Biol. Chem. 257, 3218-3224 (1982).

25. Towbin, H., Staehelin, T., Gordon, J.: Proc. Natl. Acad. Sci. USA 76, 4350-4354 (1979).

26. Rieske, J.S.: Biochem. Biophys. Acta 456, 195-247 (1976).

27. Boyer, P.D., Chance, B., Ernster, L., Mitchell, P., Racker, E., Slater, E.C.: Annu. Rev. Biochem. 46, 955-1026 (1977).

28. Beattie, D.S., Villalobo, A.: J. Biol. Chem. 257, 14745-14752 (1982).

29. Katan, M.B., Pool, L., Groot, G.S.P.: Eur. J. Biochem. 65, 95-105 (1976).

30. Reed, J., Hess, B.: H.-S. Z. Physiol. Chem. 358, 1119-1124 (1977).

31. Siedow, J.N., Power, S., de la Rosa, F.F., Palmer, G.: J. Biol. Chem. 253, 2392-2399 (1978).

32. Beattie, D.S., Battie, C.A., Weiss, R.A.: J. Biol. Chem. 253, 2392-2399 (1980).

33. Sidhu, A., Beattie, D.S.: J. Biol. Chem. 257, 7879-7886 (1982).

34. Ohashi, A., Gilson, J., Gregor, I., Schatz, G.: J. Biol. Chem. 257, 13042-13047 (1982).

35. Côté, C., Solioz, M., Schatz, G.: J. Biol. Chem. 254, 1437-1439 (1979).

36. Katan, M.B., van Harten-Loosbroek, N., Groot, G.S.P.: Eur. J. Biochem. 70, 409-417 (1976).

37. Lin, L.-F.H., Clejan, L., Beattie, D.S.: Eur. J. Biochem. 87, 171-179 (1978).

38. Gellerfors, P., Johansson, T., Nelson, B.D.: Eur. J. Biochem. 115, 275-279 (1981).

39. Kreike, J., Bechmann, H., van Hemert, F.J., Schweyen, R.J., Boer, P.H., Kaudewitz, F., Groot, G.S.P.: Eur. J. Biochem. 101, 607-617 (1979).

40. Teintze, M., Slaughter, M., Weiss, H., Neupert, W.: J. Biol. Chem. 257, 10364-10371 (1982).

41. van Loon, A.P.G.M., de Groot, R.J., van Eyk, E., van der Horst, G.T.J., Grivell, L.A.: Gene, in press (1983).

42. Riccio, P., Schägger, H., Engel, W.D., von Jagow, G.: Biochim. Biophys. Acta 459, 250-262 (1977).

43. Engel, W.D., Schägger, H., von Jagow, G.: Biochim. Biophys. Acta 592, 211-222 (1980).

44. Gellerfors, P., Nelson, B.D.: Biochem. Biophys. Res. Comm. 99, 170-175 (1981).

45. Beattie, D.S., Clejan, L.: J. Bioenerget. Biomembr. 12, 33-45 (1980).

46. Engel, W.D., Michalski, C., von Jagow, G.: Eur. J. Biochem. 132, 395-402 (1983).

47. Claisse, M.L., Pajot, P.F.: Eur. J. Biochem. 49, 49-59 (1974).

48. Capeillere-Blandin, C., Ohnishi, T.: Eur. J. Biochem. 122, 403-413 (1982).

THE FIRST INTRON OF THE GENE OXI3 IN YEAST MITOCHONDRIA ENCODES A mRNA-MATURASE

Olga Groudinsky, Giovanna Carignani*, Emma Schiavon*, Domenico Frezza*, Elisabetta Bergantino* and Piotr P. Slonimski

Centre de Génétique Moléculaire, 91190 Gif-sur-Yvette, Laboratoire propre du C.N.R.S. associé à l'Université Pierre et Marie Curie and *Istituto di Chimica Biologica dell'Università di Padova, Via Marzolo 3, Padova

Introduction

The yeast mitochondrial gene oxi3, which specifies the cytochrome oxidase subunit I (COX I), is split, as are many eukaryotic genes, in several exons and introns. The number of these varies with the yeast strain. Nine to ten exons are separated by eight to nine introns in the "long" strains KL14-4A and 777-3A. The first five introns contain long ORFs in phase with their upstream exons and are virtually able to specify polypeptides of 30 to 80 kd. (1, 2). The overall organization of this gene reminds that of the cytochrome b gene, suggesting that the expression of the two genes could be regulated by a similar mechanism. As far as the cytochrome b gene is concerned it has been demonstrated that RNA splicing is dependant on the activity of proteins (mRNA-maturases) encoded by introns. Direct evidence comes from the study of two introns in this gene: intron bI2 (3) and intron bI4 (4, 5, 6).

Until now such a role could not be assigned to any of the wild type oxi3 introns. This was essentially due to the impossibility of distinguishing different functional domains in the oxi3 gene: all the mutations isolated, some of which are certainly located in introns, (1, 7) constituted a single complementation group (7, 8; P. Pajot, G. Church and A. Lamouroux, unpublished results). Moreover, it was shown that the splicing of the fourth intron of the gene (aI4), which displays a strong sequence homology with the fourth intron of cytochrome b gene (bI4) was controlled

Mitochondria 1983

by the activity of a mRNA-maturase encoded by the intron bI4 (4, 9, 10). Only when mutated the intron aI4 acquiers the capacity to specify an active maturase (11). These observations have raised the possibility that, contrary to the cytochrome b gene, the wild type oxi3 introns would not encode any mRNA-maturase.

Here we show that the first, aI1, intron codes for a protein involved in splicing of the intron which specifies it. The protein has all the characteristics of a typical mRNA-maturase. Analysis of a new collection of mutants, isolated by Putrament et al. (12), has allowed us to identify a set of $oxi3^-$ mutations which constitute a separate complementation group from the exon ones. These mutations are located in the first aI1 intron of the gene and are phenotypically suppressed by the antibiotic paromomycin (13), which is known to enhance ribosomal misreading. They are also genetically suppressed by the informational suppressors nam 3-1 and mim 3-1 (14) which have been shown to suppress nonsense mutations located in the second bI2 intron of the cytochrome b gene (Gargouri and Lazowska, in preparation). Taken together these results suggest that a protein, active in trans and endowed with a different function from the one specified by exons, is encoded by the intron. We present direct evidence in favor of this idea based on a comprehensive analysis of the structure of novel polypeptides synthesized by the mutants and on the analysis of RNA transcripts.

Results and Discussion

1) A set of $oxi3^-$ mutations has been located inside intron aI1. Ten $oxi3^-$ mutations (isolated from the strain 777-3A), all preventing the synthesis of COXI, have been positioned, both genetically and physically, along the intron by petite deletion and restriction mapping (see figure 1). Most of them are located inside the aI1 ORF (= Open Reading Frame), while two are located in the vicinity of exon A2, probably in the blocked frame at the end of the intron.
2) All mutations prevent RNA splicing of intron aI1. We have analyzed the oxi3 transcripts present in mutants by Nothern-blot hybridiza-

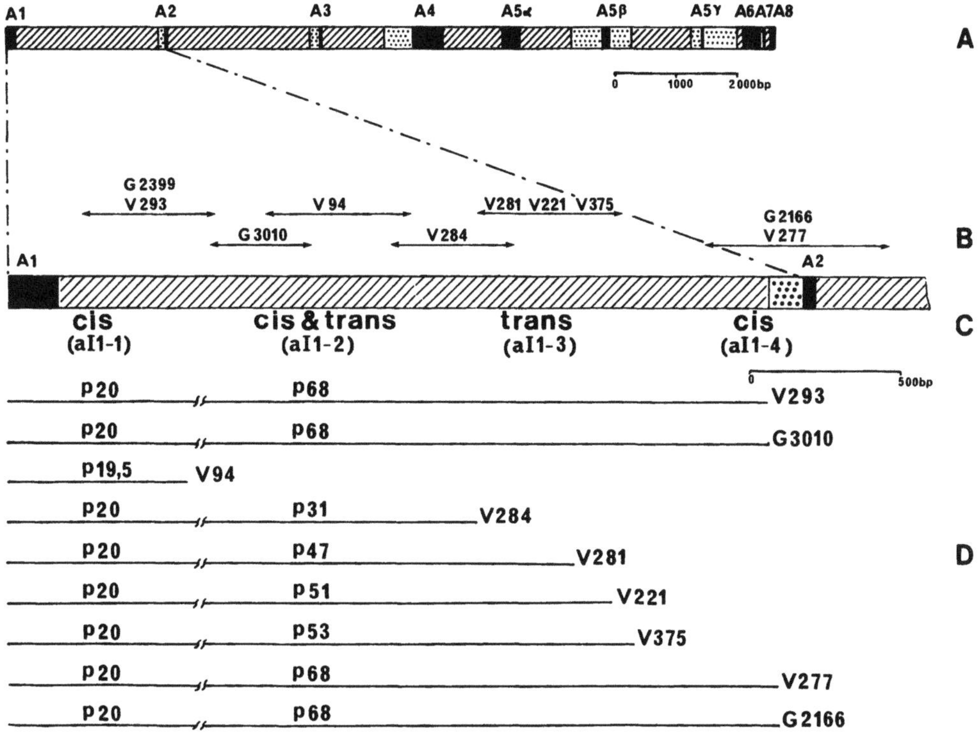

Figure 1: Synopsis of physical, genetic and complementation data. Nature of mitochondrially synthesized polypeptides in RNA-splicing deficient mutants of the first, aI1, intron of the gene oxi3.

Part A: overall organization of the $oxi3^+$ gene in the long strains of S. cerevisiae (after 1, 2, 8). Black bars indicate exons, hatched areas indicate intron ORF's while dotted areas correspond to blocked reading frames.

Part B: localization of $oxi3^-$ mutations on the physical map of intron aI1, as deduced from crosses with discriminating rho^- petites.

Part C and D: the four functional domains and the translation products. Cis-acting mutants in the domains aI1-1 and aI1-4 are splicing deficient because the structure of the substrate i.e. intronic RNA is changed in such a way that it cannot be excised, albeit the mutants synthesize and accumulate an active mRNA-maturase (presumably p68) translated from the full length of the intron ORF. Trans-acting mutants in the domain aI1-3 result probably from chain termination mutations in the intron ORF; they synthesize and accumulate an inactive maturase whose length is truncated in accordance with the mutant's position; these mutations do not impair the structure of the pre-mRNA so they can be complemented by the active maturase from mutants in aI1-1 and aI1-4. Mutations in the domains aI1-2 affect simultaneously the activity of the maturase and the structure of the substrate; they can not be complemented (and can not complement) any other class of $oxi3^-$ mutants.

tions with molecular probes containing either an internal fragment of intron aI1 or the exon A4. We have shown that the lack of synthesis of COXI is due to a deficiency in RNA processing: the intron aI1 is not excised from the primary RNA transcript, the mature mRNA is not synthetised and large molecular weight precursors accumulate instead.

3) Complementation analysis uncovers different domains in intron aI1. Pairwise complementation tests have been performed by the technique of Lamouroux (15) among the aI1 mutants, on the one hand, and between the aI1 mutants and other oxi3⁻ exon and intron mutants, on the other hand. The results disclosed four functional domains in the intron aI1:
   - The most 5'-proximal (aI1-1) and 3'-proximal (aI1-1) domains harbour cis-dominant mutations (see figure 1). It is probable, by analogy with the organization of the fourth intron of the cytochrome b gene (4, 5, 6), that these mutations alter some critical sequences whose integrity is essentially for RNA folding of the intron.
   - The three trans-recessive mutations (domain aI1-3) are located in the second half of the intron's ORF.
   - Three adjacent mutants (domain aI1-2) are simultaneously trans- and cis-acting as they do complement neither trans- nor cis-acting mutants. These mutations probably affect the intron RNA three dimensional structure and, at the same time, interfere with the activity of the mRNA-maturase.

4) The aI1 mutants accumulate novel polypeptides. All the mutants investigated, which do not synthesize COXI, accumulate novel mitochondrially synthesized polypeptides of various molecular weights. Figure 1 shows that all mutants, with the exception of mutant V94, accumulate a polypeptide of some 20 kd (p20) and another polypeptide whose apparent molecular weight (aMV) varies with the mutant: in the cis-acting mutants located both at the beginning and at the end of intron, this polypeptide has an aMV of 68 kd. (domains aI1-1 and aI1-4). The same polypeptide appears also in one mutant of the aI1-2 domain (mutant G3010). Interestingly, the aI1 mutants which are deficient in the trans-acting factor (domains aI1-2 and aI1-3) accumulate instead a shorter polypeptide whose aMW is directly correlated with the posi-

tion of the mutation in the intron's ORF. Fingerprint analysis by single and double digests showed that the p68 polypeptide and the shorter polypeptides are closely inter-related, while clearly different from COXI. This observation strongly suggests that the shorter polypeptides are truncated forms of p68, probably due to nonsense mutations. Mutant V94 (domain aI1-2) is worth of a particular note: it accumulates only one polypeptide (p19.5) that could correspond to a truncated form of p20, while no accumulation of a second polypeptide is observed.

Colinearity between lenghts of polypeptides and positions of mutations constitutes a strong evidence in favor of mitochondrial translation of intron aI1. Complementation experiments suggest that some of these polypeptides are active in RNA-splicing. As predicted by the maturase model (3), the cis-dominant mutations localized in the domain aI1-4 should synthesize and accumulate a fully active maturase. Which protein is it? We believe that the translation of the exon A1 and of the intron's ORF would lead to a very long polypeptide of some 90 kd. This polypeptide would be then cleaved post-synthetically into a p20 and a p68. The latter could be endowed with maturase activity. It is quite remarkable that it is the second instance of apparent proteolysis which may be related to maturase formation (cf. 4, 5, 6). Whatever the physiological role (if any) of this proteolysis, it is clear that at least one wild type intron (which otherwise belongs to the class II of potential secondary strutures, cf. 16) of a gene different from cob-box encodes a maturase.

Acknowledgements

This work was supported by grants from the CNRS, DGRST, INSERM, Ligue Nationale Française contre le Cancer and CNR (progetto finalizzato ingegneria genetica).

References

1. Bonitz, S., Coruzzi, G., Thalenfeld, B., Tzagoloff, A. and Macino, G. J. Biol. Chem. 255, 11927-11941 (1980).

2. Hensgens, L.A.M., Bonen, L., de Haan, M., Horst, G., Grivell, L.A. Cell 32, 379-389 (1983).

3. Lazowska, J., Jacq, C. and Slonimski, P.P. Cell 22, 333-348 (1980).

4. De La Salle, H., Jacq, C. and Slonimski, P.P. Cell 28, 721-732 (1982).

5. Weiss-brummer, B., Rodel, G., Schweyen, R.J., Kaudewitz, F. Cell 29, 527-536 (1982).

6. Anziano, P.Q., Hanson, D.K., Mahler, H.R., Perlman, P.S. Cell 30, 925-932 (1982).

7. Netter, P., Carignani, G., Jacq, C., Groudinsky, O., Clavillier, L and Slonimski, P.P. Mol. Gen. Genet. 188, 51-59 (1982).

8. Foury, F., Tzagoloff, A. J. Biol. Chem. 253, 3792-3797 (1978).

9. Dhawale, S., Hanson, D.L., Alexander, N.J., Perlman, P.S. and Mahler, H.R. Proc. Natl. Acad. Sci. USA 78, 1778-1782 (1981).

10. Netter, P., Jacq, C., Carignani, G. and Slonimski, P.P. Cell 28, 733-738 (1982).

11. Dujardin, G., Jacq, C. and Slonimski, P.P. Nature 298, 628-632 (1982)

12. Putrament, A., Kruszewska, A., Baranowska, H., Ejchart, A., Polakowska, A. and Szczesniak, B. Curr. Genetics 3, 57-63 (1981).

13. Dujardin, G., Groudinsky, O., Kruszewska, A., Pajot, P., Slonimski, P.P. : The organization and expression of the mitochondrial genome, Kroon, A.M., Saccone, C., eds. Elsevier/North Holland Biomedical Press, Amsterdam. p 157-160 (1980).

14. Kruszewska, A.: Mitochondrial Genes, Slonimski, P.P., Borst, P., Attardi, G., eds. Cold Spring Harbor Laboratory, Cold Spring Harbor. p 323-326 (1982).

15. Lamouroux, A. These de Doctorat de troisième cycle de Génétique, Paris XI (1979).

16. Michel, F., Jacquier, A., Dujon, B. Biochimie 64, 867-881 (1982).

GENETIC AND BIOCHEMICAL STUDIES OF THE NUCLEAR SUPPRESSOR NAM2 : EXTRANEOUS ACTIVATION OF A LATENT PLEIOTROPIC MATURASE.

Geneviève Dujardin, Michel Labouesse, Pierre Netter, Piotr P. Slonimski
Centre de Génétique Moléculaire du C.N.R.S, 91190 Gif-sur-Yvette (France) Laboratoire associé à l'Université Pierre et Marie Curie.

INTRODUCTION

To uncover the functional circuitry between the mitochondrial and the nuclear genome, we have developped, three years ago, a general method of selecting and genetically characterizing nuclear suppressors of mitochondrial mit$^-$ mutations (1). These suppressors were called NAM (for nuclear accomodation of mitochondria). One of these nuclear suppressors, NAM2 has been shown to be gene specific, allele non specific (2). It preferentially compensates pleiotropic box 7$^-$ mutations which are located in the fourth intron of the cytochrome b gene and which lead to a deficiency in both cytochrome b and subunit I of cytochrome oxidase.

Extensive genetic and biochemical studies of the cob-box gene, coding for apocytochrome b and the oxi3 gene, coding for subunit I of cytochrome oxidase (COXI), demonstrated a mosaic organization composed of intron and exon sequences of the two genes (3, 4, 5, 6, 7, 8, 9). Respiratory deficient mit$^-$ mutants have been mapped in both exons and introns of these genes. Intron mutations have been shown to block the mitochondrial RNA processing (10, 11, 12). Complementation analyses(13), DNA sequences analyses of trans-recessive intron mutations and characterization of the mitochondrial translation products have led to the idea of a mRNA maturase, a trans-acting protein involved in splicing and encoded within the long open reading frames present in certain mitochondrial introns (8, 14, 15, 16).

Mitochondria 1983

Furthermore, it has been shown that pleiotropic trans-recessive box7⁻ mutations in the fourth intron (bI4) of the cob-box gene prevent not only the processing of cytochrome b pre-mRNA but also that of COXI pre-mRNA (10) and it has been suggested that the mRNA maturase translated from the intron bI4 controls the coordinated excision in two introns of two distinct and well separated genes : the intron bI4 of cob-box gene and the fourth intron (aI4) of oxi3 gene (14, 17, 18) which display a striking sequence homology (7).

To investigate further the interactions between the two introns bI4 and aI4 and the role of nuclear encoded products in mitochondrial splicing machinery, we have undertaken an extensive genetic and biochemical analysis of the suppressor NAM2 which compensates for a deficiency of the mitochondrial bI4 maturase.

We have now : 1) isolated seven dominant alleles at the NAM2 locus and compared their action spectra on mitochondrial mutations,

2) located the NAM2 locus on the chromosomal map of Saccharomyces cerevisiae,

3) analyzed the action of NAM2 product on cytochromes b and oxidase pre-mRNA splicing,

4) analyzed the effects of a deletion mutation of the cob-box gene and of a deletion or point mutations of the oxi3 gene on the NAM2 action.

We show that the product of gene NAM2 activates a latent maturase encoded by the oxidase intron aI4.

RESULTS

I - Genetic characterization of seven alleles at the NAM2 locus

Seven dominant alleles at the nuclear locus NAM2 have been isolated from three different mitochondrial respiratory deficient mutants or gene-constructs. The NAM2-1 to NAM2-5 derived from the point mutation box7-G1659 (1, 2) which is located in the open reading frame coding for the bI4 mRNA maturase (14). The NAM2-6 was isolated from a constructed intron-free cytochrome b gene (19). The NAM2-7 appeared concomitantly with a spontaneous deletion at the DNA level of introns bI4 and bI5 and was obtained by P. Pajot (20).

We have first shown that these nuclear mutations segregate as a single mendelian gene. Analysis of the recombinant frequencies between

CROSSES	TETRAD ANALYSIS		
	MITOCHONDRIAL GENOME	SEGREGATION OF THE SUPPRESSOR NAM2:nam2$^+$	RECOMBINANTS FREQUENCIES BETWEEN THE SUPPRESSORS
NAM2-1 x NAM2-3	rho$^+$ box7-G1659	4:0 (24)	$\leqslant$1 %
NAM2-1 x NAM2-4	rho$^+$ box7-G1659	4:0 (14)	$\leqslant$2 %
NAM2-1 x NAM2-6	rho$^+$ bI$^\Delta$	4:0 (10)	$\leqslant$3 %
NAM2-1 x NAM2-7	rho$^+$ box7-G1659	4:0 (29)	$\leqslant$1 %

Table 1. Recombinants frequencies between 5 nuclear dominant suppressors of bI4-maturase deficiency..

The suppressor NAM2-1 derived from the point mutation box7-G1659 was crossed with four other nuclear suppressors isolates as described in the text. For each cross, diploids grown on non-fermentable substrate were sporulated and tetrads were microdissected. The presence or absence of the suppressor mutations was determined by the capacity to restore the respiration to the mutant box7-G1659:

1) either by direct replica-plating of the haploid spores on non-fermentable medium (phenotypic test),

2) or by converting the spores to rho^0 and crossing these rho^0 clones with the initial mutant rho$^+$ box7-G1659 tester strain. This genotypic test presents the advantage of eliminating the effects of recessive modificator gene(s) which can exist in a heterozygous nuclear background. Such a modificator gene, presenting an epistatic effect on the NAM2 mutations was observed in the fourth cross. In the two last crosses, a rather high proportion of rho$^-$ spores was obtained. It is important to notice that the suppressors NAM2-1 to NAM2-5, NAM2-6 and NAM2-7 were isolated in three different genetic backgrounds.

bI$^\Delta$: intron-free cob-box gene (19).

five of these alleles (table 1) shows that they do not recombine, but the number of tetrads analyzed was too low to determine whether they are homo- or hetero-alleles.

We have compared the action spectra of the various alleles. Although they all suppress the trans-recessive box7⁻ mutations located in the open reading frame of the intron bI4, their activities are slightly different.

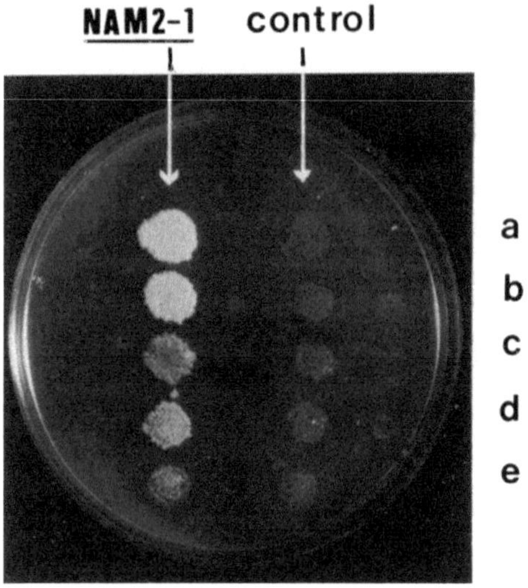

Figure 1. Examples of suppression by NAM2-1

Crosses were performed on minimal medium. After 3 days of incubation, the diploids were replicated on non fermentable medium (glycerol). The plates were incubated 5 days at 28 °C and then photographed. The genotype of the diploids is as follows :

left row : NAM2-1/$nam2^+$ rho^+ boxi ; **right row :** $nam2^+$/$nam2^+$ rho^+ boxi (isogenic control) ; **(a)** i : $bI4^{\Delta}$ $bI5^{\Delta}$ (deletions of introns bI4 and bI5) ; **(b)** i : box7-G1659 (trans-recessive mutation in intron bI4) **(c)** i : box9-M7832 (cis-dominant mutation in intron bI4) ; **(d)** i : box8-V343 (cis-dominant mutation in intron bI2) ; **(e)** i : box4-G5064 (cis-dominant mutation in intron bI1).

(1) The efficiency of suppression varies from one allele to another: for instance the NAM2-1 allele appears stronger than the NAM2-4 which is itself stronger than the NAM2-3 .

(2) Certain alleles suppress differentially the various box7$^-$ trans-recessive mutations. Particularly, the frameshift mutation A103 (15) located at the end of the bI4 maturase is often less suppressed than the upstream box7$^-$ mutations.

(3) A broader range of action is observed with the NAM2-1 allele that suppresses not only trans-recessive box7$^-$ mutations but also some cis-dominant mutations in the introns bI4, bI2, bI1 and aI4 (see Figure 1, and ref. 2).

In conclusion, the seven nuclear dominant suppressor isolates deriving from three different mitochondrial genomes are all allelic. The fact that they present slight differences in their action spectra suggests that they are actually heteroallelic.

II - Genetic localisation of the NAM2 gene on the chromosomal map of Saccharomyces cerevisiae.

We have located the NAM2 gene on the chromosomal map of S. cerevisiae by using two strains (constructed and kindly provided by Dr. D.C. Hawthorne) which carry a large number of auxotrophic markers already mapped on various chromosomes. The strain carrying the NAM2-1 allele , associated with the mitochondrial mutation box7-G1659, was crossed with the rho^0 derivatives of the two strains . The recombinant frequencies between the NAM2 gene and these markers were analyzed (Table 2). This analysis has shown that the NAM2-1 : nam2$^+$ alleles always segregate 2:2 in these new nuclear backgrounds and that the NAM2 gene is not closely linked to the centromere : the frequency of tetratype tetrads between NAM2 and a centromere markers (e.g. trp1) is high. With fourteen couples of NAM2/chromosomal markers , the number of parental ditypes is not significantly higher than the number of non-parental ditypes : the NAM2 gene is independant from these 14 genes. On the contrary with the NAM2/ura4 couple, the frequency of parental ditypes is quite high and no non-parental ditypes was observed : the NAM2 gene is linked to the ura4 marker and the genetic distance between these two genes is about 20 cmorgans.

GENE PAIR (chromosome)	ASCUS TYPE PD	NPD	TT	LINKAGE DISTANCE c morgans
NAM2-gal10 (II)	1	3	17	NL
NAM2-his4 (III)	8	6	17	NL
	6	1	10	NL
NAM2-leu2 (III)	4	1	26	NL
NAM2-a/α (III)	6	3	19	NL
NAM2-trp1 (IV)	3	4	19	NL
NAM2-tyr6 (IV)	4	6	15	NL
NAM2-ura3 (V)	4	1	14	NL
NAM2-ade5-7 (VII)	1	2	12	NL
	7	3	20	NL
NAM2-leu1 (VII)	3	4	11	NL
NAM2-his5 (IX)	5	5	21	NL
NAM2-lys1 (IX)	3	7	20	NL
NAM2-ile3 (X)	5	3	23	NL
NAM2-trp3 (XI)	2	7	10	NL
NAM2-gal2 (XII)	5	1	13	NL
NAM2-ura4 (XII)	18	o	13	21

Table 2. Meiotic linkage of NAM2 to previously mapped chromosomal genes

In conclusion, the gene NAM2 is on the right arm of the chromosome XII of *Saccharomyces cerevisiae*. It is genetically linked to the ura4 marker.

III - Action of NAM2 on mitochondrial RNA splicing.

Trans-recessive mutations in the bI4 intron are known to block the excision of the bI4 intron from cytochrome b pre-messenger and to alter COXI pre-messenger processing at unknown step (10, 12, 21).We have undertaken an analysis of processing intermediates generated after transcription of cob-box and oxi3 genes in mutants suppressed by NAM2 using molecular probes specific for introns bI4 and aI4 as well as for cob-box and oxi3 exons . As shown in Figure 2:

Legend of Table 2 (see facing page)

A rho^+ strain: NAM2-1, mat a, box7_G1659 was crossed with two rho^o strains carrying numerous nuclear auxotrophic markers:

3942-10D/50 matα gal2, trp3,ura3, ade5-7, leu1, arg1, suc, mal $nam2^+$
3133-1A/50 matα gal10, his5, trp1, ura4, rad8, ade5-7, leu2, suc, mal, $nam2^+$.

For each cross, diploids were sporulated and tetrads microdissected. Spores were analyzed (i) for auxotrophic markers by replica plating on media lacking single nutritional requirements and (ii) for the presence or absence of suppressor by replica-plating on non fermentable substrate. To discriminate between the his4 and his5 markers and to determine the mating type, the spores were crossed with tester strains of opposite mating type carrying the his4 or the his5 markers.

Each tetrad was classified either PD (parental ditype), NPD (non-parental ditype) or T (tetratype). If the number of PD tetrads exceeds the number of NPD tetrads by a statistically significant amount, the markers are considered linked. The map distance between linked markers was calculated using the formula derived by Perkins (22)

$$d\ cM = \frac{T + 6\ (NPD)}{2\ (PD + NPD + TT)} \times 100$$

The first column indicates the gene pair and the chromosomal location of the previously mapped auxotrophic markers, the second one the number of different ascus types and the last one the linkage distances in cmorgans. **NL** means not significantly linked.

1) strains with the box7-G1659 mutation or an intron-free cob-box gene lack COXI messenger (panel C) and accumulate a precursor RNA of about 3100 nucleotides which hybridizes with a molecular probe containing intron aI4 DNA (panel D) ; taking into account the length of intron aI4 (1010 nt) (7), we believe that this precursor contains the unspliced intron aI4 in addition to mRNA sequences (2100 nt according to 23) and conclude that the bI4 maturase is necessary for its excision;

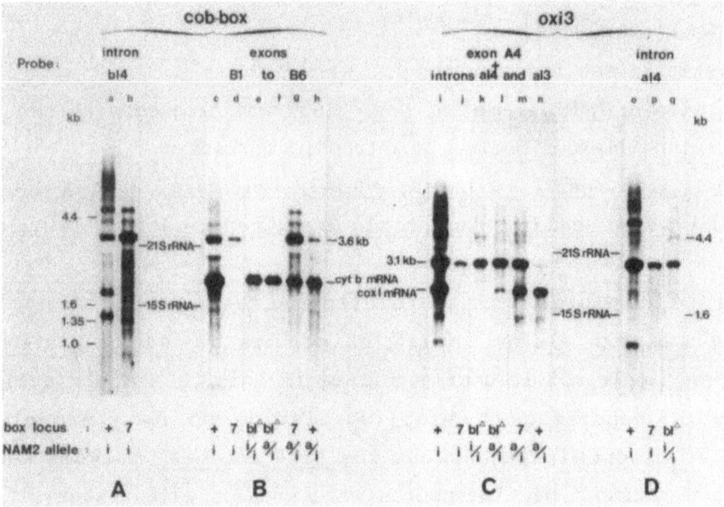

Figure 2. Action of NAM2 on the excision of mitochondrial introns from cytochrome b and oxidase premessenger RNAs.

Mitochondrial RNAs of the strains, which genotype is indicated below each track, were purified (24) and separated by electrophoresis in denaturing agarose/formaldehyde gels (25). RNAs were transfered to nitrocellulose filters (26) and hybridized with ^{32}P-nick-translated probes specific for cob-box intron bI4 (panel A), cob-box exons (panel B), oxi3 exon A4 and introns aI4 and aI3 (panel C) and oxi3 intron aI4 (panel D). Molecular weight markers obtained in parallel tracks are the 21S rRNA, the 15S rRNA and restriction enzyme fragments of plasmid DNA .Allele at the NAM2 locus: **i**= inactive nam2$^+$ allele; **a**= active NAM2-6 allele ; homozygous **i/i** or heterozygous **a/i**. bI$^{\Delta}$ =cob-box intron free gene (19).

2) while the strain carrying the intron-free cob-box gene has quite normal amounts of cytochrome b messenger (panel B), the mutant box7-G1659 is unable to excise intron bI4 from cytochrome b pre-messenger (in agreement with other studies on similar mutants, see 27 and 21). Thus the inability to excise intron aI4 (vide Supra) is not related at all to the presence of a functional cytochrome b messenger and protein;

3) The NAM2 suppressor restores the ability to excise introns aI4 and bI4 (panels B and C). However, it is apparent that the abundancy of oxidase mRNA in these strains is low when compared with an isogenic wild type control, while the concentration of aI4 containing precursor is high. A similar observation can be made for cytochrome b mRNA in strain carrying the box7$^-$ mutation.

These results show that the bI4 maturase controls the excision of introns bI4 and aI4 and that the NAM2 gene product compensates, but imperfectly, for the deficiency of this maturase.

IV -Effects of a deletion of the cob-box gene and of deletion or point mutations of the oxi3 gene on NAM2 action

The NAM2 gene product can, a priori, act in several different ways : either it does not require any mitochondrially encoded product to perform the excision of introns bI4 and aI4, or it does. In the latter alternative, the mitochondrial product could be coded for by the cob-box gene, the oxi3 gene or finally by any other mitochondrial gene.

In order to choose between these different models we have determined whether specific mitochondrial mutations will prevent the NAM2 action or not. Indeed if a mitochondrially encoded product is required, a mutation in the gene coding for this product will not allow NAM2 to compensate for the deficiency. First we have looked on the effect of a complete deletion of the cob-box or the oxi3 gene.

We have already shown (2) that a complete deletion of the cob-box gene, leading (see Figure 3, slot b) as expected to a total deficiency of cytochrome b and of COXI, does not prevent the action of the NAM2 gene which partially restores the synthesis of COXI (Figure 3, slot c).

We have, now, constructed a double-mutant carrying a deletion of the oxi3 gene and a point mutation box7^{-} in the bI4 maturase. In this double mutant, COXI is lacking (because of the oxi3 deletion) and the

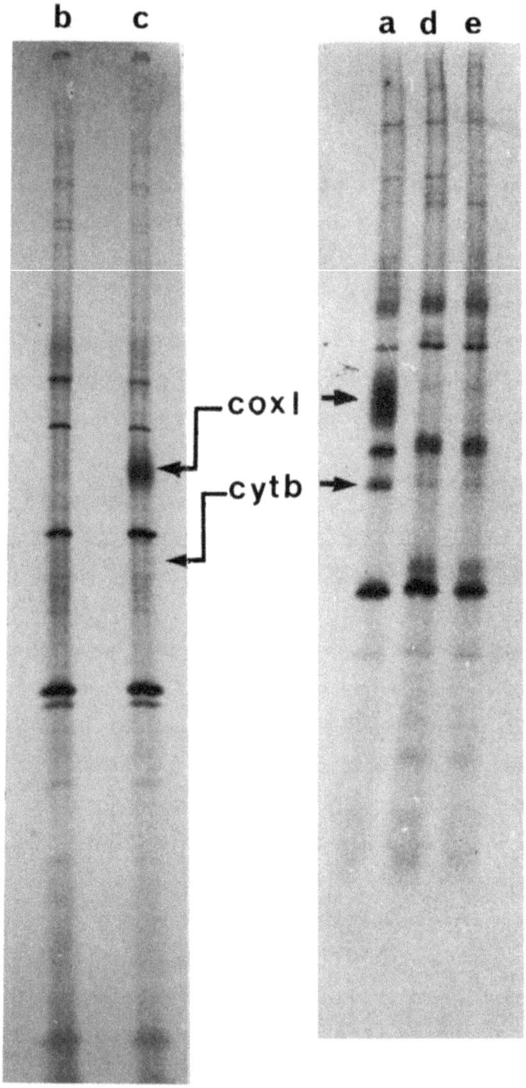

Figure 3. Effect of deletions in the cob-box gene or in the oxi3 gene on the NAM2-1 action.

cytochrome b is also absent (because of the box7⁻ mutation) (see Figure 3,slot d). Figure 3, slot e, shows that the introduction of the NAM2-1 mutation does not allow restoration of cytochrome b synthesis.

We conclude that the presence of the cob-box gene is not necessary for the excision of the intron aI4 in strains carrying the suppressor NAM2-1. On the contrary, the presence of the oxi3 gene *is* necessary for the excision of intron bI4 in strains carrying the NAM2-1 suppressor associated to a mutation in the bI4 maturase.

In order to precise which region of the oxi3 gene is essential for the excision of intron bI4 in strains carrying the NAM2-1 suppressor and a mutation in the bI4 maturase, we have constructed two other double mutants oxi3 and box7. Both oxi3 mutations chosen block the excision of intron aI4 (P. Netter, unpublished results) and lead to a deficiency of COXI (Figure 4, slot a and h). The two double mutants are deficient in cytochrome b and COXI (Figure 4, slot c and i). When the NAM2-1 suppressor is associated with the double mutant oxi3-G1765 and box7-G1659, the synthesis of cytochrome b is not restored (Figure 4, slot j). On the contrary, when the NAM2-1 mutation is associated with the other double mutant oxi3-G192 box7-G1659, the cytochrome b synthesis is partially restored (Figure 4, slot d). The oxi3-G192 mutation is a cis-dominant one and corresponds to a single base substitution in the consensus box9 sequence of the intron aI4 (17) while the oxi3-G1765 mutation corresponds to multiple alterations located in the region spanning the fourth exon (A4) and the intron aI4 (P. Netter, unpublished results).

Legend of Figure 3. (see facing page)

Analysis of mitochondrial translation products of strains which genotype is indicated below, were performed as described in (2).
(a) : $nam2^+/nam2^+$ rho^+ mit^+ (wild type) ; **(b)** : $nam2^+/nam2^+$ rho^+ cob-box^{Δ} (deletion of cob-box) ; **(c)** : NAM2-1/$nam2^+$ rho^+ cob-box^{Δ} (deletion of cob-box) ; **(d)** : $nam2^+/nam2^+$ rho^+ oxi3Δ box7-G1659 (deletion in oxi3 and mutation in bI4) ; **(e)** : NAM2-1/$nam2^+$ rho^+ $oxi3^{\Delta}$ box7-G1659 (deletion in oxi3 and mutation in bI4).
cob-boxΔ = G625 (3) ; $oxi3^{\Delta}$ = M10-150/8 (28)

Therefore, the specific mutation oxi3-G1765 acts in trans, preventing the action of NAM2 on splicing the cytochrome b pre-mRNA ; the cis-acting mutation oxi3-G192 does not. We conclude that a trans-acting product coded for by the region of the intron aI4 is necessary for the excision of the intron bI4 in strains carrying the NAM2-1 suppressor associated to a mutation in the bI4 maturase. On the contrary, the correct folding of the intron aI4 pre-mRNA is not essential for the NAM2-1 action on splicing of the intron bI4.

DISCUSSION

The nuclear locus NAM2 is interesting in many respects. We shall discuss but a few.

It is a major gene and not one of several minor modifier genes since alleles of NAM2 have been found in seven independant isolates, in three different nuclear backgrounds (either haploid or diploid) and in three different types of organization of the cob-box gene (from a nonsense mutation in bI4-maturase, from a deletion of introns bI4 and bI5, as well as from an intron-free gene). Mutations at the locus NAM2 are therefore the most frequent nuclear suppressors of the bI4-maturase deficiency and even possibly the only ones. We can predict that other suppressors, if isolated in a similar way, will turn out to be alleles of NAM2.

Legend of Figure 4. (see facing page)
Conditions are as in figure 3. Strains genotype is as follows :
(a) : $nam2^+/nam2^+$ rho^+ box^+, oxi3-G192 ; **(b)** : $nam2^+/nam2^+$ rho^+ box7-G1659, $oxi3^+$; **(c)** : $nam2^+/nam2^+$ rho^+ box7-G1659, oxi3-G192 ; **(d)** : NAM2-1/$nam2^+$ rho^+ box7-G1659, oxi3-G192 ; **(e)** : NAM2-1/$nam2^+$ rho^+ box7-G1659, $oxi3^+$; **(f)** : NAM2-1/$nam2^+$ rho^+ box^+, oxi3-G192 ; **(g)** : $nam2^+/nam2^+$ rho^+ mit^+ ; **(h)** : $nam2^+/nam2^+$ rho^+ box^+, oxi3-G1765 ; **(i)** : $nam2^+/nam2^+$ rho^+ box7-G1659, oxi3-G1765 ; **(j)** : NAM2-1/$nam2^+$ rho^+ box7-G1659, oxi3-G1765 ; **(k)** : NAM2-1/$nam2^+$ rho^+ box^+, oxi3-G1765.

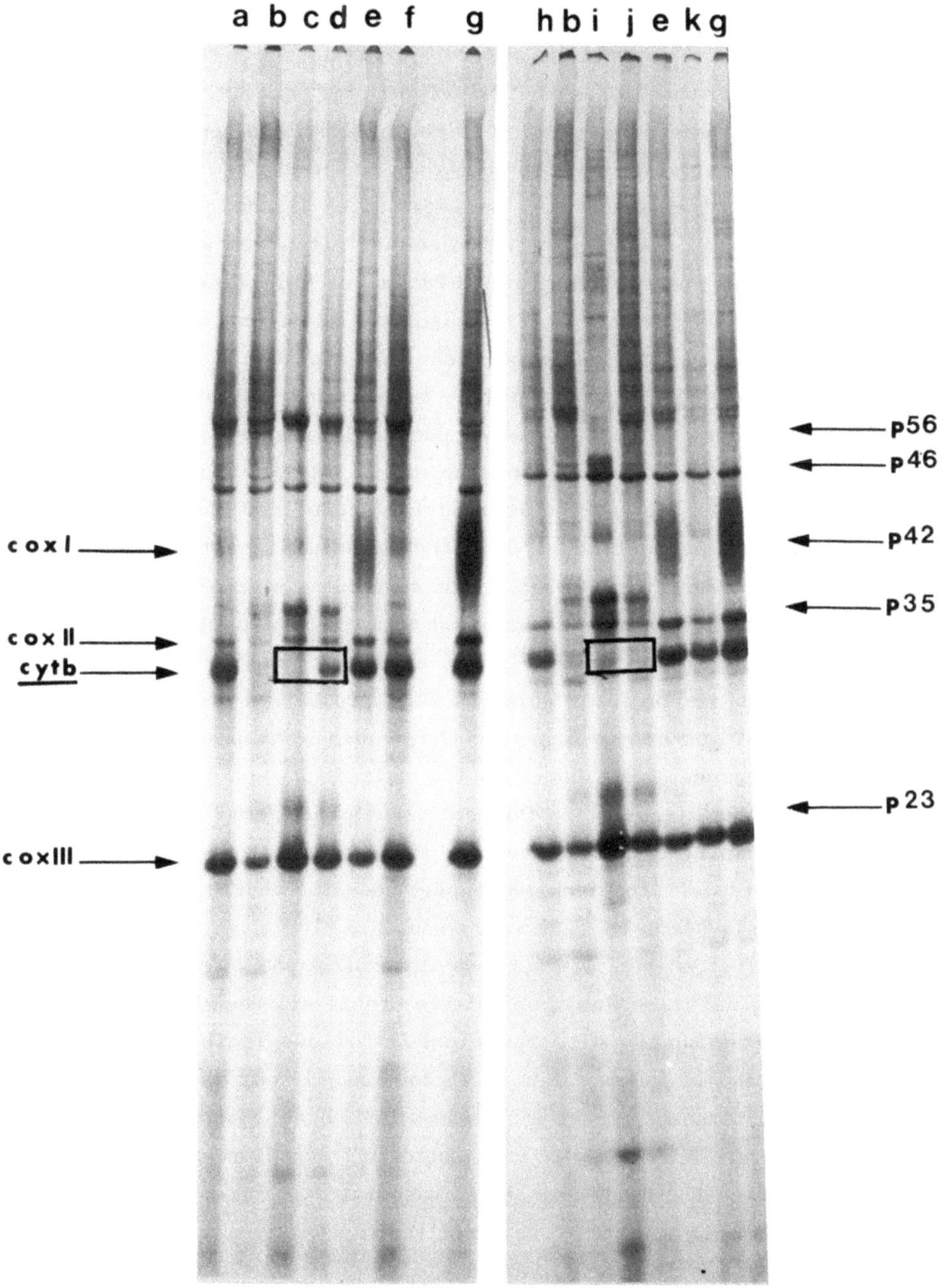

Figure 4. Effect of point mutations in the oxi3 gene on the NAM2-1 action.

The main questions which remain are what is the product of NAM2 and how does it act. Active alleles should operate by a positive mechanism since all are dominant over the wild-type $nam2^+$ allele. It will be of interest to see whether activity is proportional to gene dosage i.e. whether the phenotype of NAM2-a/NAM2-a is different from NAM2-a/$nam2^+$. The fact that in the latter case the excision of introns aI4 and bI4 is less efficient, in comparison to the normal situation when an active bI4-maturase is present (see Figure 2), makes this type of analysis quite feasible. Furthermore, our finding that different mutant alleles of NAM2 display slight, but significant, variations in their action spectrum (e.g. NAM2-1 suppresses the mitochondrial mutation box8-V343 while NAM2-3 does not), raises two possibilities: either different mutations of NAM2 concern the promoter region i.e. increase to different extents the production of the same active product or they change the quality of this product and affect the structural part of the gene. Whatever the answer, these observations rule out a simple on/off switch mechanism. Such a mechanism should lead to one mutant phenotype only.

The gene is in the nucleus but it exerts its action in the mitochondrion. Thus its product, which is most probably a protein (nucleic acids are believed not to traverse the double membrane), has to get inside the organelle, either into the matrix or /and into the inner membrane. Either the mutation that activates $nam2^+$ permits the entrance to the organelle by creating necessary import signals, or the wild-type gene has already all the required import-to-mitochondrion signals. The first hypothesis seems to us not very probable. The mechanisms by which proteins enter organelles are surely complex (29) and it seems unlikely that a single mutation would be able to create all the structures required for entering the mitochondrion. It follows logically that the $nam2^+$ allele should possess them all beforehand. If so, one expects that the gene $nam2^+$ probably makes a protein which normally enters the organelle. Its nature , at present, is a matter of conjecture. We have asked the question: do the suppressor mutations present a phenotype by themselves? The answer appears to be negative, since NAM2-1 or NAM2-6 when combined with a wild type mitochondrial genome grow normally on non fermentable media. Consequently, either the $nam2^+$ allele has no function under the test conditions (which seems to us implausible , vide supra), or the suppressor mutations modify its function in such a way as to preserve the normal one and to permit simultaneously the acquisition of a new one.

Experiments described in section IV give us some clues about the mode of action of NAM2. It does not act directly on RNA splicing, requiring a mitochondrial relay or helper. This relay is the intron aI4 of the oxidase gene. Significantly, it is not the structure of this intron per se that matters but rather a diffusible protein encoded by this intron. We are confronted with a situation similar to the one produced by the mitochondrial mutation mim2-1 (18). In this latter case, a single base substitution (G→A), occuring in the ORF of aI4, creates a lysine codon instead of a glutamic one and activates a previously silent maturase. By analogy, we can hypothesize that the mutated NAM2 protein also activates the same maturase. How this occurs is not known at present. Activation could result either from a post-synthetic alteration of the protein by a proteolytic cleavage or some other covalent modification, or from an increase of its rate of synthesis. Another alternative would be that the protein NAM2-a activates by quaternary interactions the protein translated from intron aI4. The resulting nuclearly activated maturase (= $aI4^N$-maturase) would mimic the action of its normal homologue the $bI4^+$-maturase or of the mitochondrially activated $aI4^M$-maturase. Interestingly, the three maturases are not identical in their action: $aI4^N$- and $aI4^M$- maturases are less efficient in splicing the wild type introns aI4 and bI4 (see paragraph II). However, the former has a broader specificity since it permits splicing of some structurally modified introns (bI1, bI2, bI4 and aI4 see figure 1) which cannot be accomplished by the latter. It is worth stressing that intron bI1 does not code for a protein (8, and this volume) and does not require a maturase (11). Only the correct folding of RNA structure is required for its splicing (30). However, when this structure is changed a modified maturase, but not the normal one, allows some splicing to occur! Such an observation gives substance to the notion (14) of compensating structural modifications which allow altered intron RNA to be spliced by altered maturase. Maturases appear to recognize critical elements of this structure, to stabilize them and thus allow splicing to occur. This mutual adjustement concept has several interesting evolutionary implications which we shall discuss elsewhere.

In order to characterize further the nature of the active NAM2 gene, we have constructed with a shuttle E. coli- S. cerevisiae vector, a gene bank from the strain carrying the allele NAM2-1. An active gene was isolated by complementation of a bI4-maturase deficient yeast mutant. Its analysis is in progress.

Aknowledgments. We are grateful to Drs. O. Groudinsky and P. Pajot for fruitful discussions, friendly collaboration and for communicating unpublished results. We thank B. Poirier for help in preparing the manuscript. Dr. W. Fangman should be thanked for critical reading of the manuscript . This work was supported by grants from the CNRS, DGRST, INSERM and Ligue Nationale contre le Cancer. M.L. acknowledges Ecole Polytechnique and Ligue Nationale contre le Cancer for financial support.

References

1. Dujardin, G., Pajot, P., Groudinsky, O., and Slonimski, P. P.: Molec. gen. Genet. 179, 469-482 (1980).

2. Groudinsky, O., Dujardin, G., and Slonimski, P. P.:Molec. gen. Genet. 184, 493-503 (1981).

3. Slonimski, P. P., Claisse, M.L., Foucher, M., Jacq, C., Kochko, A., Lamouroux, A., Pajot, P., Perrodin, G., Spyridakis, A., Wambier-Kluppel, M.L.: Biochemistry and Genetics of yeast, Bacila, M., et al eds.,Academic Press, Sao Paulo, 391-401 (1978).

4. Grivell, L.A., Arnberg, A.C., Boer, P.H., Borst, P., Bos, J.L., Van Bruggen, E.F.J., Groot, G.S.P., Hecht, N.B., Hensgens, L.A.M., Van Ommen,G.J.B., Tabak, H.F.,: Symposia on molecular and cellular biology XV, Cummings, D., et al eds., Academic Press, New York, 305-324 (1979).

5. Haid, A., Schweyen, R.J., Bechmann, H., Kaudewitz, F., Solioz, M., and Schatz, G.: Eur. J. Biochem. 94, 451-465 (1979)
6. Alexander, N.J., Perlman, P.S., Hanson, D.K., Mahler, H.R.: Cell 20,199-206 (1980)
7. Bonitz, S., Coruzzi, G., Thalenfeld, B., Tzagoloff, A., and Macino, G.: J. Biol. Chem. 255, 11927-11941 (1980)
8. Lazowska, J., Jacq, C., and Slonimski, P.P.: Cell 22, 333-348 (1980)
9. Nobrega, F.G., Tzagoloff, A.: J. Biol. Chem. 255, 9828-9837 (1980)
10. Church, G., Slonimski, P.P., and Gilbert, W.: Cell 18, 1209-1215 (1979)
11. Halbreich, A., Pajot, P., Foucher, M., Grandchamp, C., and Slonimski, P.P.: Cell 19, 321-329 (1980)
12. Van Ommen, G.J.B., Boer, P.H., Groot, G.S.P., De Haan, M., Roosendaal, E., and Grivell, L.A.: Cell 20, 173-183 (1980)
13.Lamouroux, A., Pajot, P., Halbreich, A. and Slonimski, P.P.: The organization and expression of the mitochondrial genome, Kroon, A.M., et al, eds., North Holland Press, Amsterdam, 153-156 (1980)
14. De la Salle, H., Jacq, C., and Slonimski, P.P.: Cell 28, 721-732 (1982)
15. Anziano, P.Q., Hanson, D.K., Mahler, H.R., Perlman, P.S.: Cell 30, 925-932 (1982)
16. Weiss-Brummer, B., Rodel, G., Schweyen, R.J., Kaudewitz, F.: Cell 29 527-536 (1982)
17. Netter, P., Jacq, C., Carignani, G., and Slonimski, P.P.: Cell 28, 733-738 (1982)
18. Dujardin, G., Jacq,C., and Slonimski, P.P.: Nature 298, 628-632 (1982)
19. Labouesse, M., and Slonimski P.P.: The EMBO J. 2, 269-276 (1983)
20. Jacq, C., Pajot, P., Lazowska, J., Dujardin, G., Claisse, M., Groudinsky, O., De la Salle, H., Grandchamp, C., Labouesse, M., Gargouri, A., Guiard, B., Spyridakis, A., Dreyfus, M., and Slonimski P.P.: Mitochondrial genes, Slonimski, P.P., eds., Cold Spring Harbor Laboratory, New York, 155-184 (1982)
21. Lamb, M.R., Anziano, P.Q., Glaus, K.R., Hanson, D.H., Klapper, H.J., Perlman, P.S., Mahler, H.R.: J. Biol. Chem. 258, 1991-1999 (1983)

22. Perkins, D.D: Genetics 34, 607-626 (1949).

23. Hensgens, L.A.M., Bonen, l, De Haan, M., Van der Horst, G., and Grivell, L.A.: Cell 32, 379-389 (1983).

24.Locker, J., Morimoto, R., Synenki, R.M. and Rabinowitz, M.: Current Genetics 1, 163-172 (1979).

25. Goldberg, D.A.,: Proc. Natl. Acad. Sci. 77, 5794-5798 (1980).

26. Thomas, P.S.: Proc. Natl. Acad. Sci. 77, 5201-5205 (1980).

27. Schmelzer, C., Haid, A., Grosch, G., Schweyen, R.J., and Kaudewitz, F.: J. Biol. Chem. 256, 7610-7619 (1981).

28.Slonimski, P.P. and Tzagoloff, A. : Eur. J. Biochem. 73, 276-286 (1976).

29. Schatz, G., and Butow, R.A.: Cell 32, 316-318 (1983)

30.Schmelzer, C., Schmidt, C., Schweyen, R.J.: Nuc. Ac. Res. 10, 6797-6808 (1982).

NUCLEAR-MITOCHONDRIAL INTERACTIONS CAUSE SENESCENCE IN THE FILAMENTOUS FUNGUS PODOSPORA ANSERINA

Karl Esser, Heinz Dieter Osiewacz, Ulf Stahl, Paul Tudzynski
Lehrstuhl für Allgemeine Botanik, Ruhr-Universität Bochum,
D-4630 Bochum 1, FR Germany

Introduction

Senescence or ageing is a phenomenon found in all forms of life. With different levels of biological organization this phenomenon is expressed differently but in all organisms as a result of irreversible alterations in metabolism, these is ultimately senescence and cellular death. Three decades ago a phenomenon of senescence comparable to cellular ageing in mammalian cells (e.g. fibroblasts) was described by Rizet and coworkers (1). They observed that a filamentous fungus, the ascomycete *Podospora anserina*, regularly becomes senescent after prolonged vegetative growth. The peripheral cells of the mycelium stop growing and eventually die. These authors showed further that in senescent cells an infective principle must exist, which when transferred via hyphal fusion to juvenile cells, immediately brings about ageing (2).

Experimental work

1. First genetic approach (chromosomal)

In studying a series of morphological mutants of *P.anserina* it was found that the onset of ageing, which in wild strains occurs after some 25 days (20 cm in length), may be postponed indefinitely by a pairwise action of several nuclear genes (3). Many of the strains seem to have "eternal life"; they have been kept in laboratory culture for more than eight years by now and have grown a distance of about 2,000 cm (4), Table 1.

Conclusion: The onset of senescence is under nuclear control.

Mitochondria 1983

Table 1 : Characteristics of long-lived double mutants of P.anserina

Strain	Life span	Senescence		mt DNA	pl DNA			
		inducible	irreversible		integrated	liberated	replicated	expressed
wild	25 d	+	+	normal	+	+	+	+
i viv	>7 a	−		normal	+	+	(+)	−
gr viv	>2 a	+	−	normal	+	−	+	+

2. Physiological approach

The onset of senescence may also be postponed indefinitely if the fungus is grown on media containing sub-lethal amounts of inhibitors of the mitochondrial (mt) DNA and of mt protein synthesis (5). In contrast to the mutated genes these substances have no permanent effect, since after removal from the medium ageing proceeds. Thus, there is no curing of the infective principle by application of these metabolic inhibitors. These experiments have further revealed the existence of a presenescent phase, morphologicaly indistinguishable from the juvenile phase. During this phase the onset of ageing is only prevented by mtDNA inhibitors (4,6).

Conclusion: The mtDNA must be somehow involved in the onset of senescence.

3. Second genetic approach (extrachromosomal)

Experiments with protoplasts showed that when these were infected with DNA of senescent strains these immediately become senescent. A more detailed analyasis of the DNA was then undertaken. According to biochemical and biophysical experiments ageing cells contain besides nuclear (n) and mtDNA a third DNA species, which was identified as a plasmid (7,8). This plasmid was the first plasmid to be detected in a filamentous fungus. This plasmid (contour length 0.75 µm, molecular weight 2.4 kb) was found to be able to transform juvenile protoplasts to senescence (9). It is thus identical with the infective principle.

Conclusion: The onset of senescence is caused by a plasmid.

In order to find out the origin of this plasmid the mtDNA was also submitted to a detailed biochemical and biophysical analysis. Having a contour length of 30 µm corresponding to a molecular weight of 95 kb, it was found to be the largest chondriome detected so far in filamentous fungi (11). Its physical map (restriction sites) and genetic map (functional sites) revealed that in juvenile cells the plasmid is an integrated part of the mtDNA (11,12).

Conclusion: The senescence plasmid is part of mitochondrial DNA.

To find out whether the plasmid is able to replicate autonomously it was ligated to a bacterial plasmid (pBR 322) which carries the gene for ß-lactamase (13). After transfer of this hybrid DNA into juvenile hyphae senescence and the production of ß-lactamase was induced (14). Thus the plasmid of P.anserina must possess an origin of replication since pBR 322 alone is not able to replicate (or express) in the fungus (15).

Conclusion: In ageing strains the plasmid is excised from the mtDNA, replicates autonomously and expresses its phenotype.

4. Nuclear-mitochondrial correlation

In order to test whether there is a correlation between the two kinds of genetic information which are connected with the senescence phenomenon in P.anserina, the nuclear genes and the mitochondrial plasmid, the long lived strains, described above, were used for transformation experiments (9,14). Two types of double mutants were used. The first (gr viv) when infected with the plasmid became senescent and the second (i viv) did not. Both mutants however show no alterations of their mtDNA's in comparison to the juvenile wild strain. As may be seen from table 1, the mutant gr viv is not able to liberate the plasmid whereas the mutant i viv albeit liberating the plasmid does not allow its expression (16). Consequently gr viv when infected with plasmid shows both replication and expression (senescence) of the plasmid. In contrast, in i viv when infected the plasmid probably replicates but certainly fails to express as it is the case for "own plasmids" (17). According to recent information the plasmid was found to enter the nucleus (18) and there is some circumstantial evidence that it might also enter mitochondria. The elucidation of both phenomena might become the key for understanding of the infectivity of the Podospora plasmid.

Conclusion: The liberation and expression of the senescence plasmid depends on the genetic constitution of the nuclear genome.

5. Molecular approach

In order to obtain information about the function of the plasmid at the molecular level and about its liberation mechanism a sequence analysis was performed (19). The results (Fig. 1) show that the plasmid has distinctive sequences at its 5' and 3' ends in common with certain introns of genes of other eucaryotes, such as Saccharomyces cerevisiae and Zea mays .According to a classification of Michel and Dujon (20) the P.anserina plasmid shares characteristics of introns termed 'group II'. In analyzing adjacent mtDNA sequence on both sides of the plasmid, it becomes obvious that it is enclosed by coding regions for polypeptides. A comparison of the derived amino acid sequence with those of other well defined mitochondrial genes resulted in the identification of this region. According to these data the plasmid is deliminated by exon I and II respectively of the cytochrome c oxidase (subunit I) gene, homologous to corresponding genes of S.cerevisiae and N.crassa . Furthermore it is remarkable that palindromic sequences at both ends of the plasmid were detected which probably are involved in the process of its liberation. In addition, inverted repeats were observed in both adjacent exons a few base pairs apart.

Conclusion: The P.anserina plasmid is an intron of a structural mitochondrial gene (subunit I of cytochrome c oxidase).

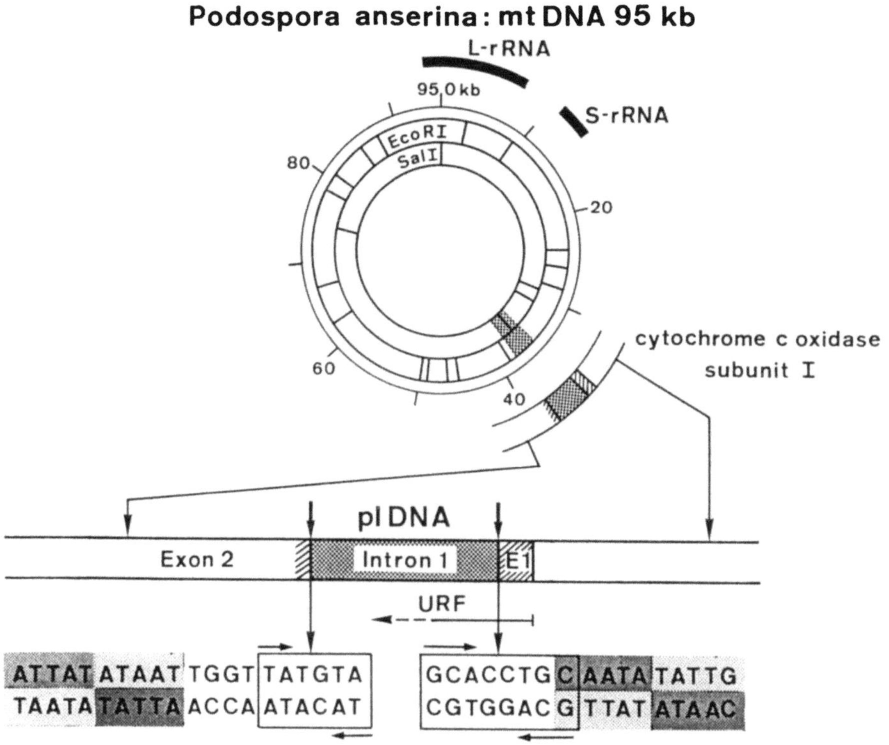

Fig.1 : Structure of the mtDNA and the mt plasmid of P.anserina . The mtDNA is represented by its physical map (Eco R I- and Sal I-sites); as markers serve the genes for the (L) and small (S) rRNA. The integrated plasmid is identical to intron I of the cytochrom c oxidase-gene, subunit I. It may be seen that from exon I an unidentified reading frame (URF) proceeds into the plasmid. Insertion sites are characterized by palindromes each side which are followed by inverted repeats.

Discussion

In the case of the fungus P.anserina the genetic control of senescence is now understood: It is caused by a plasmid originating from mtDNA, which once liberated is able to replicate autonomously and to express its phenotype. (The system itself is comparable to a bacterium infected with a lysogenic bacteriophage). Both the replication and the expression of the plasmid is under the control of the nuclear genome. This plasmid is a bifunctional genetic element. On the one hand it represents an intron of a mitochondrial structural gene, whose sequence might be processed during RNA maturation; on the other hand it behaves like a mobile element, as proposed for mitochondrial introns by Borst and Grivell (21).

In this context it seems worthwile to speculate about a possible relationship between the senescence phenomenon of P.anserina and cancer. The "eternal life" of some mutants may be compared with the capacity of a cancer cell (e.g. HELA cells) to propagate indefinitly. The limited growth of the wild strain, however, is equivalent to the limited growth capacity of normal cells (e.g. fibroblasts). In the recent years it has been shown that cancer is both under nuclear and extranuclear control. As an example of the first, the comprehensive work of Anders and collaborators (22) may be quoted. These authors could unequivocally prove that the onset of melanomic tumors in Xiphophorus depends on the presence of specific nuclear genes. In P.anserina permanent growth depends also on the action of specific nuclear genes. An example of the influence of extrachromosomal elements on cancer is provided by the observations that transformation of cells to cancer cells is caused by carcinogenic substances like Ethionine (23) which is known to provoke mutations of mtDNA. It is further postulated that oncogenes are activated by the insertion of mt sequences into the genome (24). In P.anserina the liberation of mitochondrial plasmids stops the unlimited growth. From this it follows that senescence in P.anserina is the inverse of the phenomenon of cancer. Further studies along these lines could contribute to the current discussion on models of cancer.

References

1. Rizet, G., Marcou, D.: C.R. VII .Congr. intern. Bot. Sect. Vol.10, pp. 121-128, Paris 1954.

2. Marcou, D.: Ann. Sci. Natur. Bot. 11, 653-764 (1961).

3. Esser, K., Keller, W.: Molec. Gen. Genet. 144, 107-110 (1976).

4. Tudzynski, P., Esser, K.: Molec. Gen. Genet. 173, 71-84 (1979).

5. Esser, K., Tudzynski, P.: Nature 265, 454-456 (1977).

6. Tudzynski, P., Esser, K.: Molec. Gen. Genet. 153, 111-113 (1977).

7. Stahl, U., Lemke, P.A., Tudzynski, P., Kück, U., Esser, K.: Molec. Gen. Genet. 162, 341-343 (1978).

8. Cummings, D.J., Belcour, L., Grandchamps, C.: Molec. Gen. Genet. 171, 239-250 (1979).

9. Tudzynski, P., Stahl, U., Esser, K.: Curr. Genet. 2, 181-184 (1980).

10. Jamet-Vierny, C., Begel, O., Belcour, L.: Cell 21, 189-194 (1980).

11. Kück, U., Stahl, U., Esser, K.: Curr. Genet. 3, 151-156 (1981).

12. Kück, U., Esser, K.: Curr. Genet. 5, 143-147 (1982).

13. Stahl, U., Kück, U., Tudzynski, P., Esser, K.: Mol. Gen. Genet. 178, 639-646 (1980).

14. Stahl, U., Tudzynski, P., Kück, U., Esser, K.: Proc. Natl. Acad. Sci. USA 79, 3641-3645 (1982).

15. Esser, K., Tudzynski, P., Stahl, U., Kück, U.: Mol. Gen. Genet. 178, 213-216 (1980).

16. Tudzynski, P., Stahl, U., Esser, K.: Curr. Genet. 6, 219-222 (1982).

17. Tudzynski, P.: Habilitation, Bochum 1983.

18. Wright, R.M., Cummings, D.J.: Nature 302, 86-88 (1983).

19. Osiewacz, H.D., Esser, K.: Curr. Genet. 7, 219-223 (1983).

20. Michel, F., Dujon, B.: The EMBO Journal 2, 33-38 (1983).

21. Borst, P., Grivell, L.A.: Nature 289, 439-440 (1981).

22. Anders, F.: 82. Kolloquium Moosbach (1982).

23. Wilkie, D., Evans, I.: TIBS 7, 39-47 (1982).

24. Galloway, D.A., McDougal, J.K.: Nature 302, 21-24 (1983).

DNA-SPLICING OF INTRONS IN THE GENE: A GENERAL WAY OF REVERTING INTRON MUTATIONS

Ali Gargouri, Jaga Lazowska and Piotr P. Slonimski

Centre de Génétique Moléculaire, 91190 Gif-sur-Yvette, Laboratoire propre du C.N.R.S. associé à l'Université Pierre et Marie-Curie

Introduction

Many eukaryotic genes whether nuclear, mitochondrial or plastidial are discontinuous. In order to get expressed, the primary RNA transcript has to be spliced i.e. intron-RNA has to be excised and exon-RNA has to be ligated. Some but not all sequences of the intron are essential for this process: when they are absent or modified the correct splicing does not take place, the gene is not expressed and a mutant phenotype is displayed.

We have asked the question: can one eliminate a RNA-splicing defect resulting from an intron mutation by deleting from the gene the entire intron that bears the mutated structure? In other terms, can one recover the expression of the gene by transforming a non-functional discontinuous gene into a functional continuous one?

In order to answer this question we have taken advantage of the existence of numerous RNA-splicing deficient mutations located in introns of the mitochondrial cob-box gene in S. cerevisiae. These mutations are of two types: they either inactivate intron encoded maturases or they modify critical sequences necessary for RNA folding (1, 2, 3, 4, 5, 6, 7). All lead to respiratory deficiency since the mutants do not synthesize cytochrome b. We have undertaken a systematic analysis of the molecular organization of this gene in the respiratory sufficient revertants from such mutants.

We show that clean DNA-splicing of introns occurs quite frequently and that the phenomenon is general. Some of its mechanistic and evolutionary implications are discussed.

Mitochondria 1983

Results and Discussion

We begun our hunt of intron-DNA deletions with the mutant box 8-P366 isolated from the long gene present in the strain 777-3A. It carries a cis-dominant, RNA-splicing deficient mutation located in the closed reading frame of the second, bI2, intron of the gene. Determination of its DNA sequence (Gargouri and Lazowska, unpublished results), has shown a single base, G toA, substitution which eliminates a Taq I restriction site present in the wild type. Consequently, the two Taq I restriction fragments of 3.3. kb and 1.9 kb, which span exons B1 to B3 and introns bI1, bI2 and a part of bI3 are replaced in the mutant by a single restriction fragment 5.2 kb long (see Fig. 1 for the structure of the gene). The rationale for the choice of this mutant is quite obvious: i) a true back mutation should restore the wild type sequence and give rise to two Taq I fragments of 3.3 kb and 1.9 kb; ii) a suppressor mutation at another place or in a different gene should leave the 5.2 kb fragment unchanged in length; iii) an intron deletion should replace the 5.2 kb fragment by a shorter one truncated by the length of the eliminated intron.

We have plated ca 10^9 cells from individual subclones (in order to be sure that revertants arise from independant genetic events) of the mutant box8-P366 on a glycerol medium and after 7 days of incubation we have isolated revertant colonies. All large size colonies were collected at random. From each revertant clone we have extracted the mtDNA by a quick procedure (8), restricted it with TagI, hybridized by Southern blotting with the ^{32}P-ATP nick-translated probe pYAG6-2, electrophorezed and autoradiographed (data not shown). This molecular probe permits to characterize the DNA fragments of the cob-box gene (see Fig. 1). In 7 revertants out of 20 analyzed the 5.2 kb fragment present in the mutant was replaced by shorter fragments: some revertants contained a 3.1 kb and a 1.2 kb fragment, while others contained only one hybridizable fragment, some 2.6 kb long. Perusal of the sequence of the gene (Fig. 1) suggests that the first class of revertants has lost two introns, bI1 and bI2, while in the second class three introns, bI1, bI2 and bI3, have been deleted from the DNA.

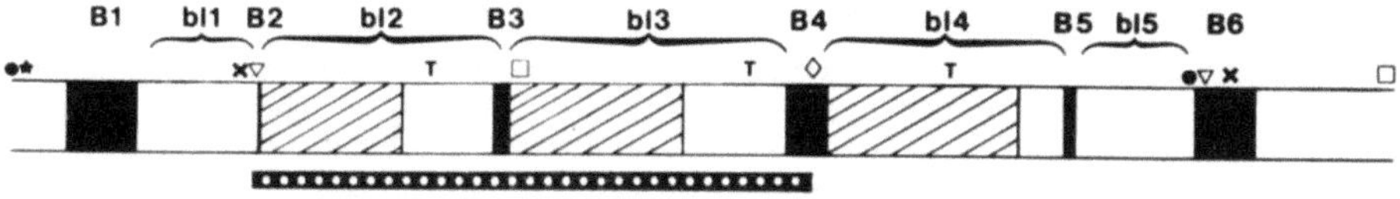
B1
bI1
B2
bI2
B3
bI3
B4
bI4
B5
bI5
B6
T
T
T

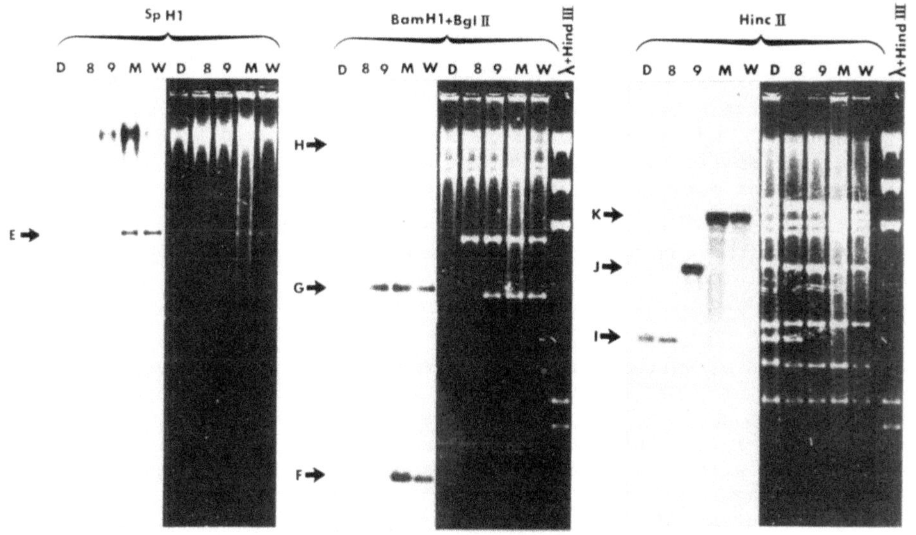
Sp H1
D 8 9 M W D 8 9 M W
E
BamH1+Bgl II
λ+Hind III
D 8 9 M W D 8 9 M W
H
G
F
Hinc II
λ+Hind III
D 8 9 M W D 8 9 M W
K
J
I

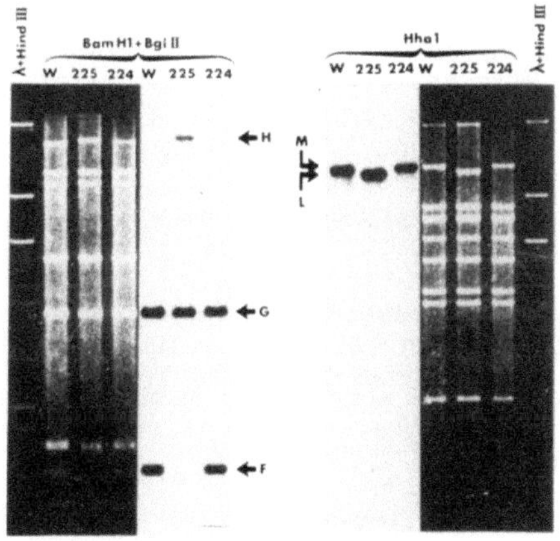
λ+Hind III
BamH1+Bgl II
W 225 224 W 225 224
H
G
F
Hha1
λ+Hind III
W 225 224 W 225 224
M
L

Hinc II
Hha I
Sph I
Bgl II
Taq I
Bam H I
ECoR I
exon
open reading frame
blocked reading frame
intron
hybridization extent with pYAG6-2

Figure 1. Revertants of intron mutants have processed genes.

Upper panel shows the structure of the wild type gene in the strain 777-3A : cytochrome b exons are in black and maturase coding intron ORF's are hatched. Underneath is shown a Bgl II-Eco R1 fragment of 2350bp, cloned in pBR322, which was used as a molecular probe (pYAG6-2) for Southern blots.

Middle panel shows the restriction and hybridization patterns of two respiratory sufficient revertants (lanes 8 and 9) isolated from the splicing deficient mutant box8-P366 localized in intron bI2. Total mtDNA was purified according to (3), restricted with the indicated enzymes and hybridized with the ^{32}P-ATP nick-translated pYAG6-2 plasmid. Ethidium bromide stained, 1% agarose gel, is on the right and the corresponding autoradiogram is on the left. As a control, mtDNA's of the wild type strain 777-3A (lane W), of the original box8-P366 mutant derived from it (lane M) and of a non-isomitochondrial wild type strain D273-10B (lane D), are also shown. Molecular markers are at the extreme right lane.

Bottom panel shows the restriction and hybridization patterns of two respiratory sufficient revertants isolated from the splicing deficient mutant box4-M1301 localized in intron bI1. Other conditions and symbols are like in the middle panel, except that the electrophoresis gel contains 0.7% agarose.

Comparison of various panels demonstrates that in the revertant 8 three introns, bI1, bI2 and bI3, have been deleted from the gene: DNA fragment E (6.0 kb) is lost because the Sph 1 site in intron bI1 is absent; fragment F (1.6 kb) is lost because the Bgl II site in intron bI1 and the Bam H1 site in intron bI3 are absent, the latter change also eliminates the fragment G (4.2 kb); fragment K (7.2 kb) which is present in the original mutant (M) and in its mother strain (W) is replaced by a new fragment I (3.2 kb). The difference in size between K and I corresponds to the sum total of the eliminated introns (3.9 kb) as predicted from DNA sequence. Furthermore, another wild type gene, (lane D), that is naturally devoid of the first three introns, gives identical hybridization patterns to the revertant 8. A similar reasoning demonstrates that the revertant 9 has kept the intron bI3 and has lost the first two introns (compare fragment J, 5 kb, with fragments I and K). Finally, in the revertant 225 only the first intron is eliminated from the gene: fragment F is replaced by a very long fragment H (ca 15 kb) as expected from the elimination of the Bgl II site in intron bI1 and the Hha I fragment M (12 kb) is sligthly, but significantly, shortened to the fragment L (11.2 kb). This shortening corresponds exactly to the length of the first intron, 767 bp (see Lazowska et al., this volume).

Symbols of some restriction sites are shown at the bottom left panel.

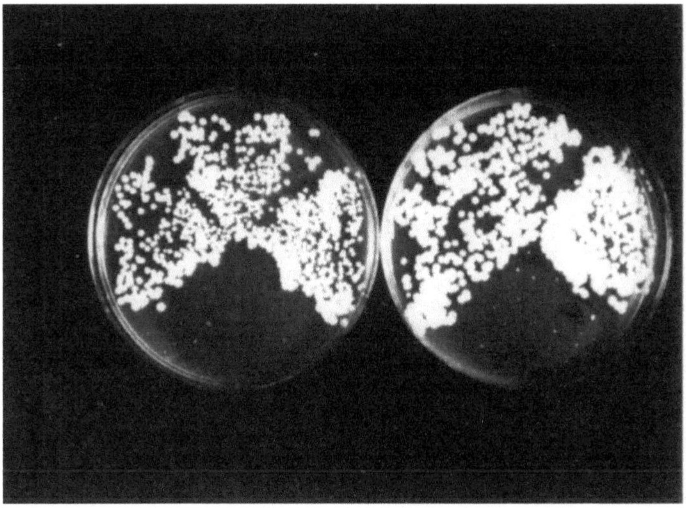

Fig. 2. Revertants with processed genes grow well on non-fermentable substrates. The mutant box8-P366 (lower quarter) and two revertants, N°8 (deletion of introns bI1, bI2, bI3, left quarter) and N°9 (deletion of introns bI1 and bI2, right quarter) as well as the isogenic wild type (top quarter) have been streaked out on a glycerol medium (medium N3) and incubated for 7 days either at 28°C (left Petri dish) or at 36°C (right Petri dish).

A more extensive analysis was undertaken, after purification of their mtDNA, on two revertants: N°9 (from class I) and N°8 (from class II). Restriction and hybridization studies described in Fig. 1 demonstrate that a complete deletion of two introns has occured in the first case and of three introns in the second case. This is confirmed by a comparison with the strain D273-10B which lacks the first three introns (9,10). It should be noted that the revertant 8, which has a structure of the cob-box gene like the latter wild type strain, differs from it in other parts of the mitochondrial genome. The deletion has affected only the cytochrome b gene. Since, as shown in Fig. 2, the two revertants grow very well on non-fermentable media they must synthesize a functional cytochrome b. We have determined, by Maxam and Gilbert procedure (11), the sequence across the exon junctions (data not shown). Revertant mtDNA was 5' labeled at the NdeI site present in the exon B1 (seen Lazowska et al., this

Mutants located in intron	Number of different mutants tested	Number of revertants tested per mutant	Number of mutants reverting by gene processing	Introns deleted from DNA
bI1 : BRF	2	10	2	ΔI1
bI2 : ORF	10	10 to 40	6	ΔI1 ΔI2
bI2 : BRF	3	10 to 20	2	or
bI3 : ORF	1	10	1	ΔI1 ΔI2 ΔI3
Total	16	ca 350	11	ca 130

Table 1. Summary of processed genes recovered in revertants isolated from different RNA-splicing deficient intron mutants of the cob-box gene. ORF: maturase coding open reading frame; BRF: blocked reading frame.

volume) and cut with EcoR1 (revertant 8) or EcoR1 + BamHI (revertant 9). Introns bI1, bI2 and bI3 have been excised exactly where they should be, in agreement with the previous sequence determinations (1, 12 and Lazowska et al, this volume). Thus, intron deletions are "clean" with neither a single base too much nor too little.

Do intron deletions occur only in revertants of this particular mutant box-P366? The results summarized in Table 1 show just the contrary. The phenomenon is quite general and we have found it in 11 mutants out of 16 analyzed. Genes processed by intron deletions have been recovered from splicing deficient mutants localized in the first, second and third intron of cytochrome b gene, from cis-dominant as well as from trans-recessive maturase deficient mutants. We predict that other mosaic genes are likely to be processed too. DNA-intron deletions can be single (like in the case of mutant box4-M1301 (7), where only the intron bI1 which carries the mutation is eliminated, see fig.1) or double and triple. Intrestingly enough, multiple deletions are by far more frequent than the single ones.

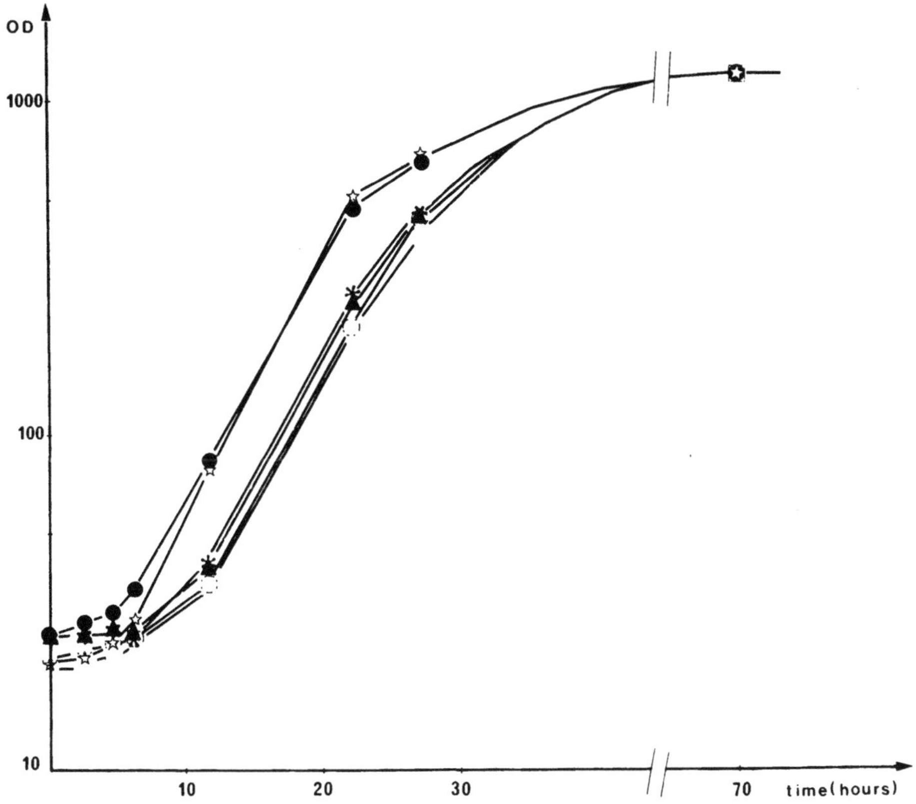

Fig. 3. Some revertants resulting from an intron-DNA deletion grow better than revertants resulting from an extragenic suppression. The cell density (in Klett units) was followed in a glycerol containing medium (medium N3) at 28°C. All revertants were isolated from the mutant box3-M3041 which inactivates the bI2-maturase as a result of a +1 frameshift at position 652 of bI2ORF (cf. 2). (●): a revertant in which the introns bI1 and bI2 were deleted grows as well as an isogenic strain (✫) which has the wild type introns, while other revertants in which the mutated intron has not been deleted display a 3 to 4 hrs longer growth lag.

Obviously, as shown previously by Pajot (cf. 2, 13 and Dujardin et al., this volume), a deletion of the intron bI4, which encodes a pleiotropic maturase, can not be recovered in revertants, unless accompanied by a compensating extra-genic mutation. It should be noted that in several instances a deletion of the mutated intron makes a "better" gene, as judged by respiration and growth, than extragenic suppression. An example given in Fig. 3 illustrates this point. The mutant box3-M3041 can be easily suppressed by extragenic suppressors mim3-1 and nam3-1 (2, 14, 15). However, intron deletions lead to a more performing respiratory phenotype.

The discovery of gene processing by reversion of intron mutations raises a number of interesting questions. We shall discuss but a few. The first set concerns the mechanism of the phenomenon and the second its evolutionary implications. In the majority of cases it occurs spontaneously rather frequently , between 0.2×10^{-8} to 2×10^{-8} revertants with processed genes per cell. This value is of the same order of magnitude as the frequency of some true back mutants or suppressors. It is by far too frequent to explain the phenomenon by a random deletion process followed by intracellular selection, especially in the case where three clean intron deletions have to occur in the same gene. We can envisage essentially three classes of hypotheses: i). The gene would be processed by a DNA-DNA recombination from a cDNA reverse-transcribed from a template of mature mRNA or partially spliced pre-mRNA. Such a mechanism takes place in higher eukaryotes but produces pseudogenes rather than functional wild type genes (cf. 16). If true, it would imply that yeast mitochondria have a reverse transcriptase. ii) The gene would be processed by DNA-RNA recombination from a hetoroduplex involving mRNA or partially spliced pre-mRNA. In addition (or instead) to the DNA-recombination machinery, the RNA-splicing machinery and mRNA-maturases could be involved in the process. iii) The intron DNA would, intrinsically, have a property of excision, transposition etc like other mobile genetic elements. Obviously, no answer can be provided at present. We would like, however, to point out some features of the process that we find quite remarkable.

Let us consider two RNA-splicing deficient mutants located in the second, bI2, intron of the gene: box3-G55 and box3-G1909. We have sequenced them previously (1) and found that each of them results from two non-adjacent single base changes in the middle part of maturase coding ORF. The first one has two ochre mutations and the second two missense mutations. The first accumulates a truncated inactive maturase and the second a full length structurally modified maturase. Both mutants have an extreme negative phenotype. The first one reverts, although with a rather low frequency, the second not at all. We have analyzed recently 42 revertants of box3-G55. In all of them the gene hase been processed by intron deletions! It is easy to understand that box3-G55 can not revert by true back-mutations but the question remains why box3-G1909 does not revert. We believe that it is due to the accumulation of large amounts of structurally modified mRNA-maturase. The accumulation would prevent DNA-splicing. This may be a clue in favor of the second hypothesis.

If introns can be so easily and precisely eliminated from a gene, why they are still there? It is one of the major unanswered questions of to day's molecular evolution (cf. 17). We hope, by extending the experiments initiated in this work and by combining them with intron additions described previously (13), to provide at least some clues for the raison d'être of introns.

Acknowledgments

We thank Gisèle Perrodin for technical assistance and Jacqueline Brizard for help in preparing the manuscript. This work was supported by grants from the CNRS, DGRST, INSERM and Ligue Nationale Française contre le Cancer. A. Gargouri acknowledges the receipt of a Bourse de Cooperation du Gouvernement Français.

References

1. Lazowska, J., Jacq, C., and Slonimski, P.P.: Cell 22, 333-348 (1980)
2. Jacq, C., Pajot, P., Lazowska, J., Dujardin, G., Claisse, M., Groudinsky, O., De la Salle, H., Grandchamp, C., Labouesse, M., Gargouri, A., Guiard, B., Spyridakis, A., Dreyfus, M., and Slonimski, P.P. in Mitochondrial genes, Slonimski, P.P. et al., eds., Cold Spring Harbor Laboratory, New York, pp 155-184 (1982)
3. Dela Salle, H., Jacq, C., and Slonimski, P.P.: Cell 28, 721-732 (1982).
4. Weiss-Brummer, B., Rodel, G., Schweyen, R.J., Kaudewitz, F.: Cell 29, 527-536 (1982)
5. Anziano , P.Q., Hanson, D.K., Mahler, H.R., Perlman, P.S.: Cell 30,
6. Schmelzer, C., Schmidt, C., Schweyen, R.J.: Nuc. Ac. Res. 10, 6797-6808

7. Schmelzer, C., Haid, A., Grosch, G., Schweyen, R.J., and Kaudewitz, F.: J. Biol. Chem. 256, 7610-7619 (1981)

8. Dujon, B., Blanc, H. in The Organization and Expression of the Mitochondrial Genome, Kroon AM et al, eds, North Holland Press, Amsterdam pp 33-36 (1980)

9. Borst, P. in Biological Chemistry of Organelle Formation, eds Bucher T et al, Springer, Berlin pp 27-41 (1980)

10. Nobrega, F.G., Tzagoloff, A.: J. Biol. Chem. 255, 9828-9837 (1980)

11. Maxam, A., Gilbert, W. Meth. Enzymol. 65, 499-560 (1980)

12. Lazowska, J., Jacq, C., Slonimski, P.P.: Cell 27, 12-14 (1981)

13. Labouesse, M., and Slonimski, P.P.: Embo J. 2, 269-276 (1983)

14. Dujardin, G., Groudinsky, O., Kruszewska, A., Pajot, P., Slonimski, P.P. in The organisation and expression of the mitochondrial genome, Kroon A M et al, eds, North Holland Press, Amsterdam, pp 157-160 (1980)

15. Kruszewska, A. in Mitochondrial Genes, Slonimski, P.P. et al., eds, Cold Spring Harbor Laboratory, New York (1982)

16. Wilda, C.D., Growther, C.E., Cripe, T.P., Lee, M.G., Cowan, N.J. Nature 297, 83-84 (1982)

17. Crick, F. Science 204, 264-271 (1979)

MITOCHONDRIAL GENES AND THEIR EXPRESSION IN HIGHER PLANTS

C.J. Leaver, E. Hack, A.J. Dawson, P.G. Isaac and V.P. Jones

Department of Botany
University of Edinburgh
Edinburgh, EH9 3JH, Scotland

Introduction

The mitochondrial genome of higher plants is much larger and more variable in size than the mitochondrial genomes of protists, fungal and mammalian cells. Current estimates of the size of plant mitochondrial DNA (mtDNA), based on restriction enzyme analysis and resaturation kinetics, give values of between ca. 200kb for several Brassica sp (1) to ca. 2400kb for muskmelon, with a seven to eight-fold variation within a single plant family (2). The complexity of mtDNA restriction enzyme patterns, coupled with electron microscopic analysis of native mtDNAs which reveals linear molecules of varying sizes as well as a small proportion of open circular and supercoiled molecules (1-100kb), has led to problems in constructing a complete physical map of the genome. Recent data obtained by Lonsdale and his colleagues working with a cosmid library of maize mtDNA has allowed them to postulate that the genome exists in one of two states; either as a single large circular molecule of ca. 1200kb or as equimolar amounts of three smaller circular molecules (ca. 600, 350 and 250kb), two of which arise by intra-molecular recombination, between paired direct repeats (of ca. 3kb), from the larger of the three circules (3). The complexity of the genome is estimated to be between 500-600kb, with less than 10% being comprised of short (3-14kb), interspersed, direct and inverted repeat sequences. If this proposed structure for the maize genome is confirmed and shown to be representative of the organisation of plant mitochondrial genomes as a whole several major questions will still however remain to be answered. Why for example is the plant mitochondrial so large and what is the functional significance, if any, of the wide

Mitochondria 1983

range of genome size found in different plants?

Sequence analysis has not revealed the existence of long stretches of A+T-rich 'spacer' DNA such as that found in S. cerevisiae and the buoyant density of all plant mtDNAs examined to date is in the range 1.705-1.707 g/cm^3 (47% G+C). The recent finding that mtDNA from a range of plant species shows extensive sequence homology to chloroplast DNA provides a partial answer to the first of these questions (4). However there is no apparent correlation between mitochondrial genome size and the amount of chloroplast DNA homology (5). The regions of homology have been shown to include both the chloroplast 16S rRNA, several tRNAs (4) and in at least one case the chloroplast gene for the large subunit of ribulose bisphosphate carboxylase (6). Thus raising further fundamental questions concerning the extent, origin, significance and function of such 'foreign' DNA sequences in the mitochondrial genome.

One further explanation for the large size of the plant mitochondrial genome could be that it contains structural genes not present in the mtDNA of other organisms. It is also conceivable that there are additional 'regulatory sequences' in green plant mtDNA which are involved in 'inter-organellar communication' with chloroplast and nuclear genomes. Such communication being of particular importance during the coordinated changes in cellular function which occur during plant development and differentiation, e.g. during the transition from a heterotrophic seedling, where the provision of energy is dependent upon oxidative phosphorylation, to the leaves of an autotrophic plant where the major energy requirement is provided by photophosphorylation in the chloroplast.

We are currently attempting to answer some of the questions raised above by two approaches directed towards the identification and analysis of plant mitochondrial genes and translation products (see 7 for review). The first of these approaches exploits the fact that the nucleotide sequence of many mitochondrial genes are evolutionarily conserved, thus allowing the use of cross-hybridisation between plant mtDNA restriction fragments and specific mitochondrial gene probes from fungi and animals, to identify the homologous plant gene. The second approach is designed to establish the identity of the polypeptides synthesised by isolated plant mitochondria and, by extrapolation, encoded by mtDNA.

Identification and Sequence Analysis of Plant Mitochondrial Genes.

The first genes to be isolated from higher plant mtDNA were those for the 26S, 18S and 5S rRNAs (7). They have been localised on the proposed physical map of maize mtDNA (8) and sequence analysis of the 18S and 5S RNA genes has just been completed (9).

To isolate protein coding genes,specific mitochondrial gene probes from yeast and bovine cells were nick translated and hybridised under conditions of low stringency to restriction fragments of maize mtDNA (10). The first plant gene isolated by this approach was subunit II of cytochrome c oxidase (COII). A single 2.4kb EcoRI fragment of maize mtDNA which hybridised with a yeast Oxi-1 probe was cloned into the plasmid pBR322 and sequenced (10). The predicted amino acid sequence of the plant gene revealed a homology of 47% and 40%, respectively with the corresponding yeast and bovine sequences. In contrast to the homologous genes the plant gene contains a single centrally located intervening sequence of 794bp. Northern blot analysis of total maize mitochondrial RNA shows four major transcripts of 3.2, 2.8, 2.45 and 1.95kb which suggests that the smaller, more abundant transcripts are spliced (10) (a full length transcript of the gene containing both exons and the intron would be at least 1619bp, while the minimum length for a spliced mRNA would be 825bp). A comparison of the plant COII nucleotide and amino acid sequences with the predicted amino acid sequence of yeast, Neurospora and animal protein shows that Trp is almost certainly not encoded by TGA,as found in all other mitochondrial systems analysed to date,and in most cases is coded by the standard TGG codon. There is also strong evidence that in addition the codon CGG (normally Arg) codes for Trp in maize mitochondria. This is confirmed by sequence analysis of the COII gene from Oenothera mitochondria by Axel Brennicke from Tubingen (11). Using a maize COII probe he identified and sequenced the homologous gene in this dicotyledonous plant. The predicted amino acid sequence shares 85% homology with the maize gene but does not contain an intron. TGA codons are absent and in common with the maize gene, in regions of high amino acid conservation Trp is encoded on four occasions by TGG and on four occasions by CGG. Of particular interest is the finding that in one conserved Trp position encoded by TGG in maize the Oenothera sequence contains a CGG.

Using a similar approach we have now used cloned genes from yeast and bovine mtDNA to identify the homologous genes for apocytochrome b (COB) and cytochrome c oxidase subunit I (COI) in maize mtDNA. The M13 clone designated B6, containing the amino-terminal region of the bovine apocytochrome b gene (a gift of Dr. S. Anderson, refn. 12), was used to identify 5.1kb EcoRI and 18kb Bam HI, restriction fragments which appeared to contain this gene in maize. Subcloning these fragments into M13 and sequence analysis by the dideoxy-chain termination method, has identified an 1164 base pair open reading frame (see Figure 1) which encodes a polypeptide of molecular weight 42,900, with a predicted amino acid sequence which shows homology of 50% and 51% with the amino acid sequence of the polypeptide from S. cerevisiae (13) and B. taurus (12).

A similar approach, using the pBR322 probe pKLD for the yeast COI gene (a gift from Dr. L. Hensgens (4)), has been adopted to identify and clone a 3.9kb Bam HI-EcoRI fragment of maize mtDNA containing the COI gene. Sequence analysis of this gene is now more than 90% complete and the predicted amino acid sequence shares a homology of 67-72% with the corresponding sequences from yeast (15) and bovine (12) mitochondria.

A number of interesting features have emerged from sequence analysis of the plant COB and COI genes: 1) The plant COB gene in common with the homologous mammalian gene, does not contain intervening DNA sequences; which contrasts with the situation in S. cerevisiae and Aspergillus nidulans. The estimated size of the plant COI gene is ca. 1550bp and the sequence analysis suggests that there are no introns. 2) The codon TGA is not found in either the COB or COI genes whereas TGG (Trp) appears relatively frequently. In the COB gene, twelve TGG codons occur and eight of these are located in the same position as TGA codons in S. cerevisiae and B. taurus. Similarly, in the COI gene, nine TGG codons have been found at positions where TGA codons occur in the homologous gene in S. cerevisiae, Neurospora and man. These data thus confirm the results from sequence analysis of the COII gene (10). The two CGG codons occuring in the COB and COI genes are found in non-conserved regions so we are unable to confirm that this codon also specifies Trp. 3) All the methionine residues would appear to be coded by ATG and not ATA as in mammals; ATA in maize is a codon for isoleucine. 4) Transcript analysis

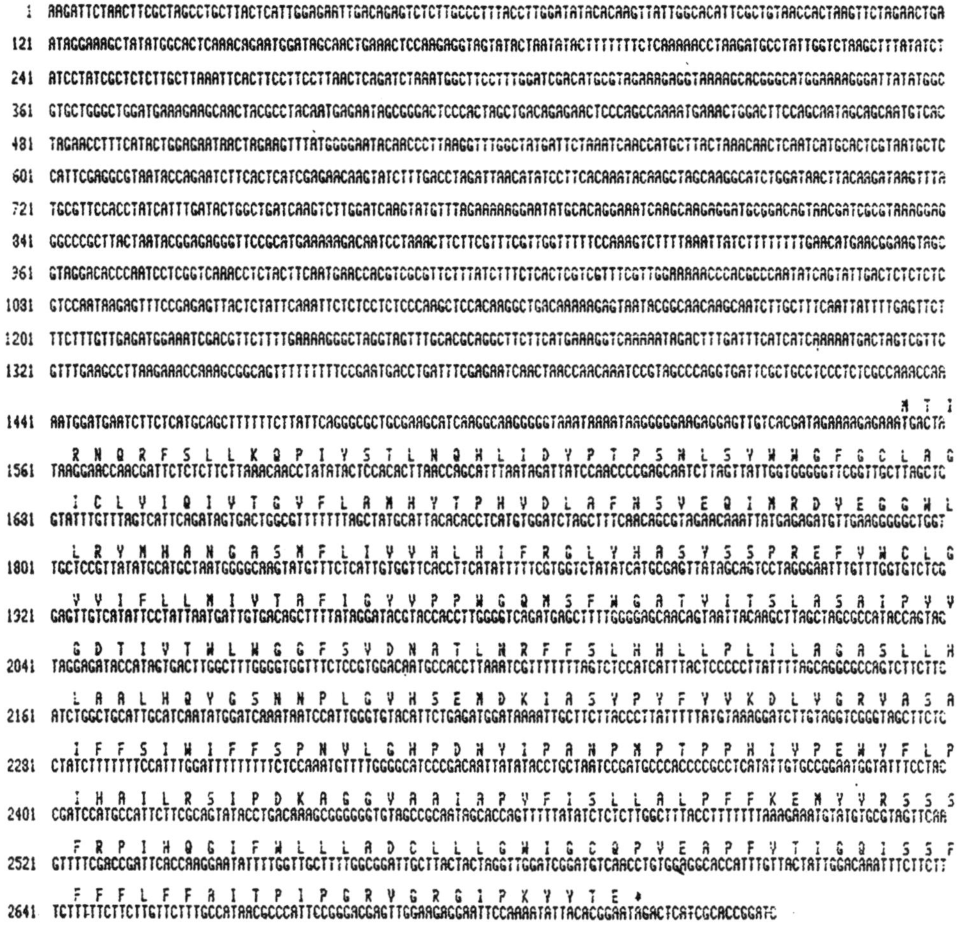

Figure 1. Nucleotide sequence of the apocytochrome b gene in Zea mays. The predicted amino acid sequence is shown using the universal genetic code.

by 'Northern' blotting has shown that the most abundant transcripts hybridising to internal COB gene probes are 2.2kb and 4.2kb, and to COI probes are 2.4kb. These transcripts are considerably larger than the open reading frames specifying these genes (cf. COB, 1.16kb; COI, ca. 1.55kb). Comparison of the nucleotide sequences immediately 5' and 3' to the COI,

COII and COB genes has, so far, revealed no strikingly conserved sequences.

All these observations raise interesting questions concerning the mechanism of gene expression and regulation in higher plant mitochondria.

Identification of Plant Mitochondrial Translation Products.

The use of cross hybridisation to identify structural genes in plant mtDNA is ultimately limited by the availability of homologous gene probes from other organisms. As an alternative approach designed to establish the number and identity of polypeptides synthesised within plant mitochondria, and by extrapolation, encoded by mtDNA, we are currently analysing the proteins synthesised by isolated plant mitochondria (7, 16). One dimensional SDS-polyacrylamide gel electrophoresis of the [^{35}S]-methionine-labelled mitochondrial translation products from a range of plant tissues and species, reveals an essentially similar spectrum of some 18-20 polypeptides with molecular weights ranging from 8,000-58,000 (e.g. Figure 2 and refn. 7, 16, 17, 18). While the absolute number of translation products synthesised by the mitochondria isolated from the different species and tissue sources remains constant, the relative rates of synthesis of several of the major polypeptides varies according to the developmental state of the tissue from which the mitochondria were isolated.

Two-dimensional IEF-SDS polyacrylamide gel electrophoresis (Figure 3) of the [^{35}S]-methionine labelled mitochondrial translation products from maize reveals a reproducible pattern of more than 30 polypeptides, although it is possible that certain labelled 'spots' are products of premature termination, charged modification or proteolytic clearage. Two of the major labelled translation products can be tentatively identified by immunoprecipitation with monospecific antibodies against subunits I and II of yeast cytochrome c oxidase (17, 18). Subunit I has an estimated molecular weight of ca. 38,000, while subunit II has an estimated molecular weight of ca. 34,000,which is larger than the molecular weight of 30,848 predicted from the gene sequence (10). The major low molecular weight translation product has a molecular weight of ca. 8,000, is soluble in butanol or a chloroform-methanol mixture and specifically binds [^{14}C]-

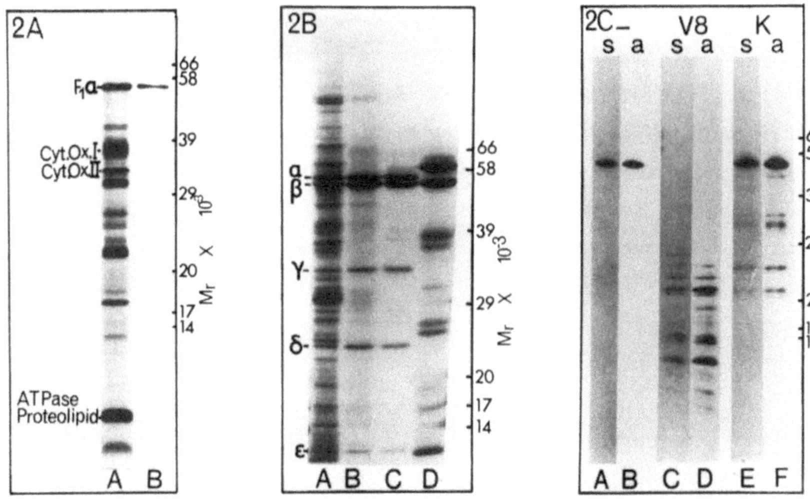

Figure 2A: Total polypeptides synthesised by isolated maize mitochondria (A) and immunoprecipitation of the α-subunit of the F_1-ATPase (B). Mitochondria were labelled with $[^{35}S]$-methionine and immunoprecipitated with antibody raised against α-subunit of yeast F_1-ATPase as described in refn.[19] The labelled polypeptides were analysed on a 15% (w/v) SDS-polyacrylamide gel and autoradiographed.

Figure 2B: One dimensional gel electrophoresis on 10-15% (w/v) SDS-polyacrylamide gradient gels of (A) total maize mitochondrial protein, (B) partially purified F_1-ATPase, (C) sucrose gradient purified F_1-ATPase and (D) chloroform extracted maize chloroplast CF_1-ATPase (19).

Figure 3C: Proteolytic digestion patterns of purified and in vitro labelled F_1 α-subunit. The α-subunit was located by staining two-dimensional gels of total mitochondrial protein labelled in vitro with $[^{35}S]$-methionine, excised, digested with proteases and the peptide products separated on 15-20% (w/v) SDS-acrylamide linear gradient gels. s = stained with Coomassie Brilliant Blue. a = autoradiograph of same track. A, B: undigested; C, D: digested with 0.1µg *S. aureus* protease V8; E, F: digested with 5ng *T. album* proteinase K.

(Figures reproduced from refn. 19 with permission)

dicyclohexyl-carbodiimide, suggesting that it is the proteolipid subunit 9, of the F_o portion of the ATPase complex (7, Hack & Leaver, in preparation).

Unfortunately we have not been able to obtain cross reactivity between antibodies prepared against yeast cytochrome *c* oxidase subunit III, apocytochrome b or subunit 9 of the ATPase complex. We have therefore embarked on a more laborious, but ultimately more useful approach aimed at identifying the polypeptides synthesised by the plant mitochondrion. This involves fractionation of [^{35}S]-methionine labelled mitochondria into outer membrane, inter-membrane space, inner membrane and matrix, followed by localisation of the labelled polypeptides. In common with other organisms most of the labelled polypeptides are associated with the inner membrane and therefore we are initially concentrating on the purification of the major enzyme complexes from this fraction.

Purification and characterisation of the subunit composition of the maize mitochondrial F_1-ATPase.

The F_1-ATPase complex was solubilised from 'osmotically shocked', gradient purified maize mitochondria, by a modification of the chloroform extraction procedure (19) and further purified by ammonium sulphate precipitation and centrifugation on a 5-20% (w/v) sucrose density gradient. SDS-polyacrylamide gel electrophoresis of aliquots of the gradient fractions revealed that five polypeptides co-purified (M_r ca. 3 x 10^5) and were associated with the peak of ATPase activity (19). The five polypeptides were identified as the α, β, γ, δ and ε subunits by comparison with the well characterised five subunit structure of the F_1-ATPase from other organisms. The subunits had estimated molecular weights of 58,000 (α), 56,000 (β), 35,000 (γ), 22,000 (δ) and 8,000 (ε) (see Fig. 2B) and were present in a stoichiometry of approximately 3:3:1:1:1 as judged by Coomassie Blue staining. All the subunits were distinct from the homologous CF_1-ATPase purified from the maize chloroplast, as judged by SDS-gel electrophoresis (Figure 2B) and partial proteolysis (19). The mitochondrial α, β and δ subunits were well resolved on two dimensional, IEF-SDS gels, while in contrast the basic γ and ε subunits did not enter the IEF gel. The α and β subunits were both represented by more than one related stained 'spot', probably due to charge variance of a single polypeptide, and in common with the homologous subunits in other F_1-ATPases, the α subunit is more basic than the β subunit (Figure 3).

Isolated maize mitochondria synthesise the α-subunit of the F_1-ATPase complex.

A comparison of the polypeptides synthesised by isolated mitochondria with the Coomassie Blue stained mitochondrial polypeptides separated by one (Figures 2A and 2B) and two dimensional (Figures 3A and B) gel electrophoresis and autoradiography, revealed that the major labelled polypeptide, with a molecular weight of ca. 58,000, comigrated with the α-subunit of the F_1-ATPase. None of the remaining F_1-ATPase subunits had mobilities corresponding to the other mitochondrial translation products. Further confirmation for the suggestion that the α-subunit of the F_1-ATPase is a mitochondrial translation product was obtained by partial proteolysis of the authentic α-subunit and the [^{35}S]-methionine labelled, putative α-subunit. The appropriate polypeptides were located on stained two-dimensional gels, excised and subjected to partial proteolysis with Staphylococcus aureus, V8 protease or proteinase K from Tritirachium album (see Figure 2C). The stained peptides produced by proteolysis of the authentic α-subunit and the labelled peptides (visualised by autoradiography) obtained from the 'in organello' synthesised putative α-subunit, corresponded precisely in mobility and closely in relative intensity.

Final proof that the α-subunit of the F_1-ATPase is a mitochondrial translation product was obtained by specific immunoprecipitation of the labelled 58,000 molecular weight mitochondrial translation product with an antibody prepared against the α-subunit of yeast F_1-ATPase (Figure 2A) (kindly provided by Professor G. Schatz). In contrast non-immune serum and antisera to the β-subunit did not immunoprecipitate any labelled polypeptides (19).

Using similar approaches to those described above we have also shown that the α-subunit of the F_1-ATPase is a mitochondrial translation product in sorghum, cucumber, pea and field bean.

In the other eucaryotes examined to date all five subunits of the mitochondrial F_1-ATPase are encoded in the nuclear genome, translated on cytosolic ribosomes, usually as higher molecular weight precursors and subsequently imported into the mitochondrion. Our demonstration that the α-subunit is synthesised in the plant mitochondrion and is thus almost

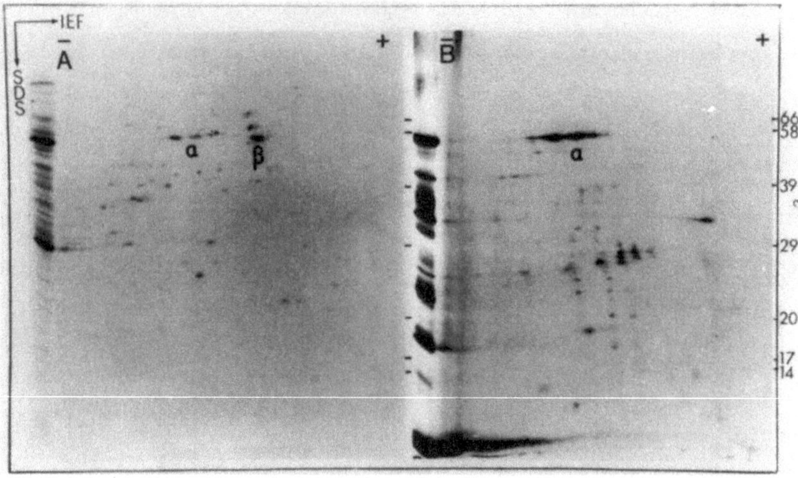

Figure 3. Two-dimensional polyacrylamide gels of polypeptides from mitochondria labelled <u>in vitro</u> with [^{35}S]-methionine. A: gel stained with Coomassie Brilliant Blue. B: autoradiograph of stained gel. The α and β subunits of F_1-ATPase are indicated. Samples were focussed in a gel containing 9.5M urea, 2% (w/v) Triton X-100, 3.2% (w/v) acrylamide, 0.56% (w/v) N,N'diallyltartiamide, 1.6% (w/v) Ampholines, pH 5-7, 0.4% (w/v) Ampholines, pH 3.5-10.0. Polypeptides were fractionated in the second dimension on 16% (w/v) SDS-polyacrylamide gels. A one-dimensional SDS-polyacrylamide gel fractionation in parallel is shown on the left. (Figure reproduced from refn. 19 with permission).

certainly encoded on mtDNA, suggests that the plant mitochondrial genome does indeed contain additional genetic information not found in the mitochondrial genomes of other organisms. Preliminary hybridisation experiments indicate that maize mtDNA contains sequences homologous to the nuclear gene encoding the α-subunit from yeast. Confirmation of this result followed by a comparison of the DNA sequence for the plant and yeast genes may provide some clues as to both the evolutionary origin of the gene and the significance of its genetic location.

Mutations in Plant mtDNA.

One of the major constraints in studying the structure and expression of the mitochondrial genome of higher plants is our inability to isolate and characterise mutations in the mitochondrial and nuclear genomes which affect mitochondrial biogenesis and/or function. The vast majority of mutations would either be 'silent' (because of the large size of the mtDNA most of which probably does not have a coding function), or lethal, because higher plants, unlike yeast, cannot live without functional mitochondria. Non-lethal mutations in plant mtDNA on the other hand might go undetected because of the lack of an identifiable phenotype (cf. petite mutants in yeast). There is however one cytoplasmically inherited trait, which in most cases is almost certainly due to mutations in the mtDNA and can be recognised phenotypically by the failure of the mature plant to produce functional pollen (7, 16). This trait is termed cytoplasmic male sterility (CMS) and is found in a wide range of plant species including maize, sorghum, sugar beet and sunflower. It is used commercially to produce F_1-hybrid seed varieties by preventing self-fertilisation of the seed parent plant. The male sterile cytoplasms can be restored to fertility (and thus produce functional pollen) by specific nuclear restorer genes (Rf genes).

The cytoplasmic male sterile phenotype has provided us with one means of screening for mitochondrial mutations in a range of higher plants. The use of CMS in maize and sorghum breeding programmes has led to the availability of a range of male sterile cytoplasms which can be distinguished from each other by the specific nuclear,Rf genes which restore fertility (see 7 for review). This classification is confirmed by restriction endonuclease digestion of mtDNAs from the various male sterile cytoplasms found within a plant species. While most of the restriction fragments are common to both fertile and male sterile cytoplasms, each type of cytoplasm is characterised by a few specific DNA fragments. These various mitochondrial genotypes probably arose by recombinational and/or transpositional events which have become stabilised and selectively amplified. Similar rearrangements of the mitochondrial genome also occur at high frequency in plant tissue culture or during regeneration of plants from culture and are then stably inherited during the normal sexual cycle (20, 21).

While many of the changes in mtDNA organisation are probably silent, at least some of the changes may be related to the CMS phenotype and are correlated with synthesis of characteristic variant polypeptides by mitochondria isolated from the CMS lines in maize and sorghum (7, 16, 17, 18, 22).

Table 1. VARIANT MITOCHONDRIAL TRANSLATION PRODUCTS IN MAIZE AND SORGHUM

Cytoplasm Type	Phenotype	Molecular Weight of Variant Polypeptides Synthesised by Isolated Mitochondria
Maize		
N	MF	-
T	MS	13,000
C	MS	17,500
S	MS	42-85,000
Sorghum		
Kafir	MF	-
Milo	MS	65,000
IS1112C	MS	82,000 + 12,000
M35-1	MS	82,000
9E	MS	42,000

MF = male fertile, MS = male sterile. This polypeptide is a variant form of subunit I of cytochrome c oxidase - see Figure 4.

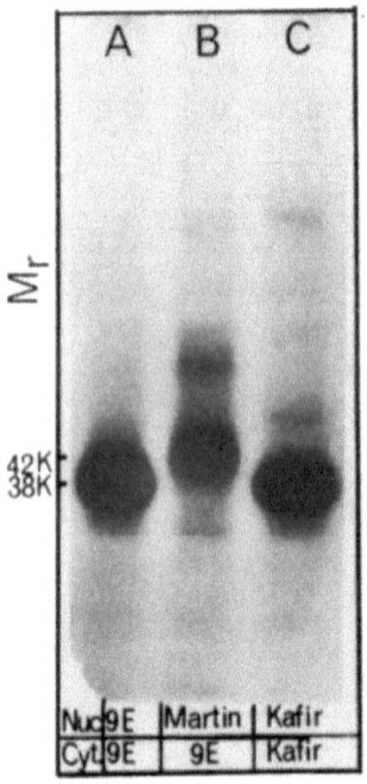

Figure 4. Immunoprecipitation of subunit I of cytochrome c oxidase from the mitochondrial translation products of fertile and cytoplasmic male sterile sorghum lines:(A) 9E nucleus: 9E cytoplasm-male fertile, (B) Martin (equivalent to Kafir) nucleus: 9E cytoplasm-male sterile, (C) Kafir nucleus : Kafir cytoplasm-male fertile. Mitochondrial translation products were labelled with [^{35}S]-methionine, immunoprecipitates were obtained using antibody raised against yeast cytochrome c oxidase subunit I, fractionated on 15% (w/v) SDS-polyacrylamide gels and autoradiographed (18).

In only one case has the variant polypeptide been identified and that is in the male sterile, 9E cytoplasm, of sorghum (18). Mitochondria from the cytoplasm of the fertile combination 9E nucleus - 9E cytoplasm, synthesise a normal spectrum of polypeptides, including subunit I of cytochrome c oxidase (M_r = 38,000) (Figure 4A). However when the 9E cytoplasm is transferred to any one of fourteen foreign nuclear backgrounds by back-crossing (e.g. a Martin nucleus) a variant form of cytochrome c oxidase, subunit I with a molecular weight of 42,000 is synthesised (Figure 4B). We are currently investigating how this variant form of cytochrome c oxidase subunit I actually causes male sterility, and together with Drs. D. Hanson and T. Fox, whether its synthesis is due to a mutation in the gene or associated control regions, or an effect on: RNA processing, initiation or termination of translation, or post-translational processing.

There is thus good circumstantial evidence linking mutations in mtDNA to differential or altered gene expression, as exemplified by the synthesis of variant polypeptides and the phenotypic expression of CMS in several plants. It will however require identification of the functional role of the variant polypeptides involved and isolation and sequencing of the mitochondrial genes which encode them, before we can fully understand how a particular mutation leads to an abnormal phenotype at a specific stage of plant development (e.g. pollen formation).

The phenomenon of CMS emphasises the importance of the interactions between nucleus and cytoplasm which are involved in regulating mitochondrial genome expression. In all cases examined when cytoplasms are transferred from their original nuclear background (in which their phenotype is fertile), to a foreign nuclear background (in which the phenotype is cytoplasmic male sterile), some change in mitochondrial gene expression results. This suggests that during the course of evolution nuclear and mitochondrial genes controlling mitochondrial biogenesis and function may have diverged so that a foreign nucleus cannot correctly regulate mitochondrial gene expression at all stages of plant development. It is conceivable that the failure in mitochondrial function and/or biogenesis during pollen formation may be in response to the 'developmental stresses' imposed during a time of very active mitochondrial biogenesis. The demonstrated effect of nuclear genes, which restore fertility, on mitochondrial genome expression (7, 16, 17), provides an exciting opportunity to turn our attention towards an analysis of the nuclear genes which encode mitochondrial proteins and their role in regulating the expression of the mitochondrial genome.

Acknowledgements: This work was supported by research grants to C.J.L. from the Science and Engineering Research Council and Agricultural Research Council. The seed used in all our studies was kindly provided by Pioneer Hi-Bred International.

References.

1. Lebacq, P., Vedel, F.: Plant Science Letts. 23, 1-9 (1981).
2. Ward, B.L., Anderson, R.S., Bendich, A.J.: Cell 25, 793-803 (1981).
3. Lonsdale, D.M., Hodge, T.P., Fauron, C, M.-R., Flavell, R.B.: in UCLA Symp. on Molecular and Cellular Biology. New Series, 12, ed. R.B. Goldberg. Alan R. Liss Inc. N.Y. 1983.
4. Stern, D.B., Lonsdale, D.M.: Nature 299, 698-702 (1982).
5. Stern, D.B., Palmer, J.-D., Thompson, W.F., Lonsdale, D.M.: in UCLA Symp. on Molecular and Cellular Biology. New Series, 12, ed. R.B. Goldberg. Alan R. Liss Inc. N.Y. 1983.
6. Lonsdale, D.M., Hodge, T.P., Howe, C.J., Stern, D.B.: Cell, in press (1983).
7. Leaver, C.J., Gray, M.W. Ann. Rev. Plant Physiol.: 33, 373-402 (1982).
8. Stern, D.B., Dyer, T.A., Lonsdale, D.M.: Nucleic Acid Res. 11, 3333-3337 (1982).
9. Spencer, D.F., Schnare, M.N., Gray, M.W.: submitted for publication.
10. Fox, T.D., Leaver, C.J.: Cell 26, 315-323 (1981).
11. Brennicke, A.: personal communication.
12. Anderson, S., De Bruijn, M.H.L., Coulson, A.R., Eperon, I.C., Sanger, F., Young, I.G.: J. Mol. Biol. 156, 683-717 (1982).
13. Nobrega, F.G., Tzagoloff, A.: J. Biol. Chem. 255, 9828-9837 (1980).
14. Hensgens, L.A.M., Arnberg, A.C., Roosendaal, E., Van Der Horst, G., Van Der Veen, R., Van Ommen, G-J.B., Grivell, L.A.: J. Mol. Biol. 164, 35-58 (1983).
15. Bonitz, S.G., Coruzzi, G., Thalenfeld, B.E., Tzagoloff, A., Macino, G.: J. Biol. Chem. 255, 11927-11941 (1980).
16. Leaver, C.J., Forde, B.G., Dixon, L.K., Fox, T.D.: in Mitochondrial Genes. (Slonimski, P., Borst, P., Attardi, G., eds.), pp. 457-470. Cold Spring Harbor Laboratory, 1982.

17. Leaver, C.J., Forde, B.G.: in Genome Organisation and Expression in Plants. (C.J. Leaver ed.), pp. 407-425. Plenum Press, New York, 1980.

18. Dixon, L.K., Leaver, C.J.: Plant Mol. Biol. 1, 89-102 (1982).

19. Hack, E., Leaver, C.J.: EMBO J. in press (1983).

20. Gengenbach, B.G., Connelly, J.A., Pring, D.R., Conde, M.F.: Theor. Appl. Genet. 59, 161-167 (1981).

21. Kemble, R.J., Flavell, R.B., Brettell, R.I.S.: Theor. Appl. Genet. 62, 213-217 (1982).

22. Dixon, L.K., Leaver, C.J., Brettell, R.I.S., Gengenbach, B.G.: Theor. Appl. Genet. 63, 75-80 (1982).

GENE EXPRESSION AND ORGANIZATION IN TRYPANOSOME MITOCHONDRIA

Rob Benne, Manuela Agostinelli, Berend F.De Vries, Janny Van den Burg, Bep Klaver and Piet Borst*

Section of Medical Enzymology, Laboratory of Biochemistry, University of Amsterdam, P.O.Box 60.000, 1005 GA Amsterdam, The Netherlands.

Introduction

We are studying gene organization and expression of the highly unusual mitochondrial DNA of trypanosomes. This DNA consists of a complex network of catenated circles, in which two different size classes are found: small circles (mini-circles) of 1-3 kb and larger ones (maxi-circles) of 20-40 kb, depending on the species (for reviews see refs. 1-3).

In Trypanosoma brucei, 10^4 mini-circles and 50-100 maxi-circles occur in a specialized portion of its single mitochondrion which is called the kinetoplast. Mini-circles are heterogeneous in sequence (4,5), they evolve rapidly (6), they are not transcribed (7-10) and have an as yet unidentified role in trypanosome mitochondria. The maxi-circles are homogeneous, conserved in evolution and they are transcribed (7,10,11). It is generally accepted, therefore, that the maxi-circle represents the trypanosome equivalent of mitochondrial DNA and contains the mitochondrial genes. We have set out to identify these genes and the way they are organized. Our aim is:

1) To demonstrate the existence of a separate mitochondrial

---

* Present address: The Netherlands Cancer Institute, Antoni van Leeuwenhoekhuis, Plesmanlaan 121, 1066 CX Amsterdam, The Netherlands.

Abbreviations: CAP, D-chloramphenicol; cDNA, complementary DNA; rRNA, ribosomal RNA; RF, reading frame; kb, kilobase pair(s).

Mitochondria 1983

protein synthesizing machinery and to characterize its products.

2) To precisely map mitochondrial transcripts (both poly(A)$^+$ and poly(A)$^-$) and study their mode of processing.

3) To determine the nucleotide sequence of maxi-circle segments that may contain genes.

This paper briefly summarizes the state of the art in each area.

Protein synthesis in trypanosome mitochondria

The classical approach to identify mitochondrially encoded proteins makes use of the different sensitivity of the cytoplasmic and mitochondrial ribosomes to certain antibiotics such as cycloheximide and D-chloramphenicol (CAP). Cycloheximide-resistant protein synthesis in T. brucei mitochondrial preparations has been reported (12), but the level of incorporation in these experiments was about two orders of magnitude lower than that observed in mitochondria from other organisms. Moreover, a detailed analysis of the products by gel electrophoresis was lacking in this study. We have, therefore, performed, such an analysis in the insect trypanosome Crithidia fasciculata. Cells were labelled, in vivo, with [^{35}S] methionine in the presence or absence of cycloheximide and anisomycin. Fig.1 shows a gel analysis of the labelled proteins. It is clear that all cellular protein synthesis is blocked by these inhibitors, since no cycloheximide- or anisomycin- resistant protein bands occur, not even after long exposure of the autoradiograms.

In earlier work it was shown that 99% of protein synthesis in Crithidia is resistant to inhibitors of mitochondrial protein synthesis such as CAP and ethidium bromide (13). We have repeated these experiments and checked the effect of a large number of inhibitors such as CAP, ethidium bromide, carbomycin, erythromycin and tetracycline. The results were essentially the same: no inhibition was observed (experiments not shown). Moreover, the rate of growth of Crithidia is unaffected by the addition of high concentrations (1-2 mg/ml) of CAP, tetracycline or erythromycin to the growth medium. A possible explanation of these observations could

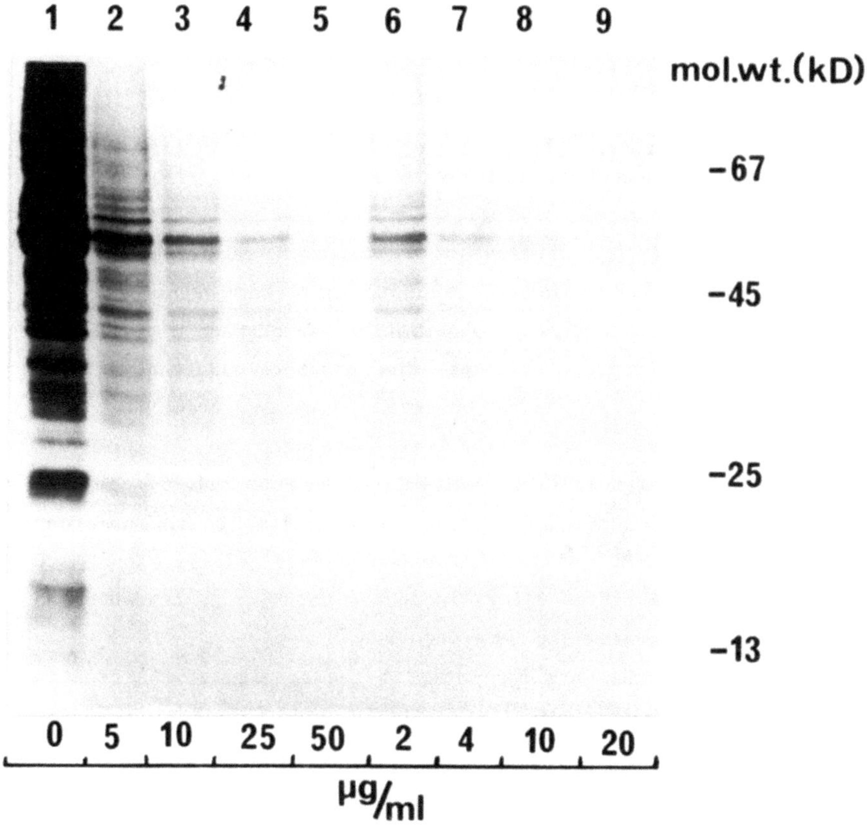

Fig.1: Protein synthesis in <u>Crithidia fasciculata</u>. The procedure employed was essentially as described in ref.13. <u>C. fasciculata</u> cells were grown in 50-ml batches in Boné medium. After reaching a density of 10^7 cells/ml, the cells were collected by centrifugation (10 000 rpm, for 5 min, at 4°C), washed twice and resuspended at 4 x 10^7 cells/ml in Locke's solution (see ref.12). Amino acid incorporation was measured for 30 min at 25°C after addition of 2 µC [^{35}S] the growth medium. A possible explanation of these observations could be that the cellular methionine (all lanes) and amounts of cycloheximide (lanes 2-5) and anisomycin (lanes 6-9) as indicated. The reaction was stopped by the addition of 50 µl Sodium dodecyl sulphate sample buffer (14) and the radioactivity in protein bands was analyzed on a 10% polyacrylamide gel according to Laemlli (14). The figure shows an overnight autoradiogram.

be that the cellular and/or mitochondrial membrane are impermeable for these substances. We favour the hypothesis, however, that we are dealing with a very unusual mitoribosome and are supported in this view by the outcome of sequence analysis of the ribosomal genes (15, see below). As discussed extensively by Eperon et al. (15), regions on the large ribosomal RNA (rRNA), thought to be involved in binding CAP to the ribosome, are altered in trypanosomes. Mutations in this area result in a CAP-resistant ribosome in other organisms (see ref.15, Fig.5). The same type of observation can be made with respect to erythromycin resistance. The suggestion has been made that an A to G transition in a highly conserved area upstream of the CAP-resistance site on the large rRNA results in erythromycin resistance of the ribosome (16). Table I (taken from ref.16) compares the sequence of this area in a number of organisms to an analogous sequence at position 883 of the 12S sequence (see ref.15), about 100 nucleotides upstream of the CAP resistance site. Also in trypanosomes an A to G transition seems to have occurred.

These observations, are in line with the results of the protein synthesis experiments, predicting a highly unusual mitoribosome in trypanosomes, resistant to CAP and erythromycin and - possibly - sensitive to cycloheximide. Unfortunately these predictions are hard to verify, since numerous attempts by ourselves and others to isolate these ribosomes have failed. The identification of protein genes, therefore, must rely on other lines of experimentation.

Transcript mapping

Detailed transcription maps of the maxi-circle have recently been reported for Leishmania tarentolae (10) and T. brucei (11). Our own transcription map is an extension of the one discussed in ref.7 and is shown in Fig.2. The map gives the approximate positions of the rRNAs of 12S and 9S, already mentioned in the previous section and extensively discussed by Eperon et al. (15). No evidence for the existence of tRNA genes has been obtained, but both the size and

TABLE I

Comparison of the erythromycin sensitivity region of the large rRNA gene from various sources

Organism	Sequence	Position	Erythromycin
E.coli	GACGG A AAGAC	2058	sensitive
Tobacco Chloroplast	GACAG A AAGAC	2067	sensitive
S.cerevisiae Mitochondria (E^S)	GACGG A AAGAC	1951	sensitive
idem (E^R)	GACGG G AAGAC	1951	resistant
S.cerevisiae Cytoplasm	GGAAA G AAGAC	2398	resistant
Mouse Mitochondria	GACGA G AAGAC	1065	?
A.nidulans Mitochondria	AACGA G AAGAC	ca. 1790	resistant?
T.brucei Mitochondria	AAAGA G AACAT	883	resistant?

number of transcripts would be sufficient to account for a normal set of other mitochondrial genes. It is remarkable that many of the transcripts overlap. The relative amounts of members of some of the overlapping pairs (and trios) varies somewhat per preparation, most likely a reflection of subtle differences in growth conditions, harvesting time, etc. Their occurrence is presumably related to intricate processing pathways that may exist in synthesis and/or breakdown of these RNAs. The variable region, which contains a repetitive sequence that evolves rapidly (6), is poorly transcribed (see also ref.10).

For five of the major RNAs - 12S, 9S, RNAs 1, 2 and 3, (see Fig.2) - we have determined the exact map positions at the nucleotide level with a combination of Northern blot analysis with strand- and area-specific maxi-circle DNA segments cloned in phage M13, S_1 nuclease analysis with the same DNAs as probe and RNA sequencing with oligonucleotide primers (provided by Dr.J.H.Van Boom, Leiden). A

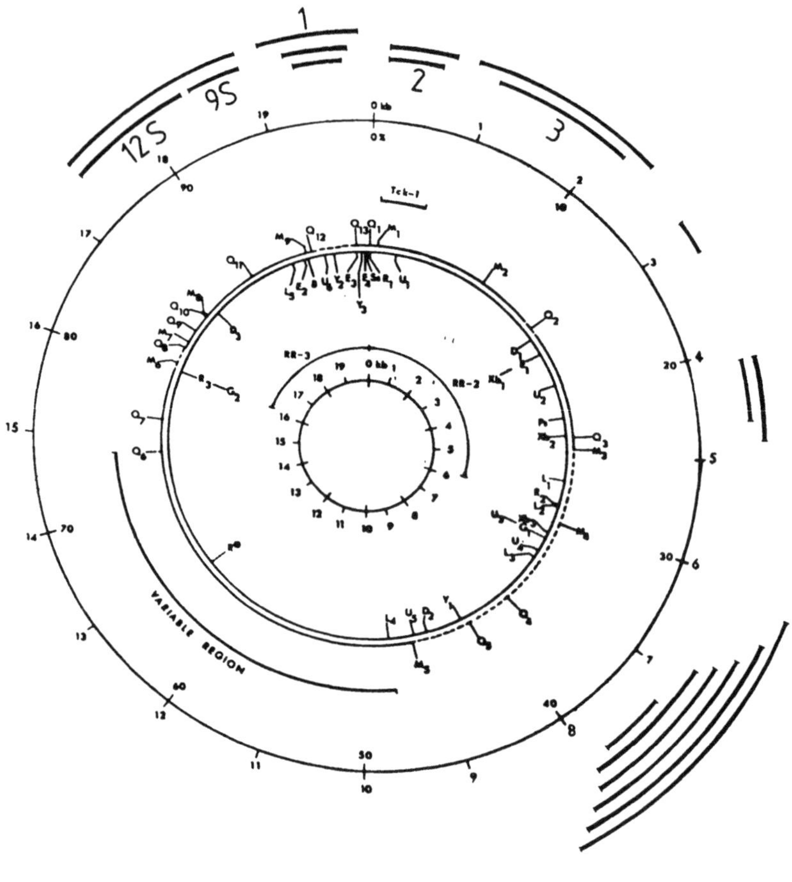

Fig.2: Transcription map of the maxi-circle of <u>T. brucei</u> 427. The location of the recognition sites for most of the restriction enzymes is from ref.17. Interrupted regions indicate maxi-circle segments with many sites of the enzyme bordering the region. Approximate RNA positions were determined by hybridizing RNA filters to various maxi-circle fragments as described (7): R3-R1; R1-R2; D3-D1; D1-D2; Q5-Q7. The position of most of the RNA species has been further defined with smaller sub-probes of the fragments mentioned, except for those RNAs localized in the R2-D2 region. RNA positions are indicated by the heavy bars. RNAs which have been precisely mapped (see Table II) are indicated: the 9S and 12S RNA species and three poly(A)$^+$ RNAs: 1, 2 and 3. The exact position of the other RNAs has not been determined and are given by approximation only. The position of a cDNA (Tck-1, ref.1) and the variable region (ref.18) have also been indicated. Abbreviations: B, BamHI; D, HindIII; E, HaeIII; G, BglII; L, HhaI; M, MboI; Ps, PstI; Q, TaqI; R, EcoRI; Ss, Sst; U, Sau96I; Xb, XbaI; Y, HapII.

TABLE II

Precise mapping of the rRNAs and three major poly(A)$^+$ species on the maxi-circle of T. brucei.

RNA	Size in Fig.2 (nucleotides)	Map Positions on D3-D1 fragment[1)]	Techniques used
12S	1230	32-1183	a,b,c
9S	640	1209-1807	a,b,c
RNA 1	860	2600-3400[2)]	a,b
RNA 2	510	3450-3950	a,b,c
RNA 3	1200	3950-5070	a,b

The data were obtained with the following lines of experimentation:

a) Northern blot analysis of glyoxylated RNA from T. brucei culture form was performed as described (ref.7). The blots were probed with single-stranded M13 clone DNA, containing strand- and area- specific maxi-circle inserts (not shown in detail).

b) S_1 nuclease protection experiments were performed as described by Berk and Sharp (21) as outlined in detail (7). 0.5 µg of M13 DNA, containing strand and area specific maxi-circle inserts (see above), and 5 µg of RNA was used. A typical experiment is shown in Fig.3.

c) Sequence analysis.

Primed cDNA synthesis using a synthetic dodecadeoxynucleotide as primer was carried out in the presence of chain terminators essentially as described (22). The primers were situated at a 75-100 nucleotide distance from the putative 5'-end of the RNAs, as determined in S_1 nuclease protection experiments. 12S primer, 5' TTTACTACACGG 3'; 9S primer, 5' ATTAACTATTCC 3'; RNA 2 primer, 5' CTGGCGGTGTCC 3'.

1) Map positions are given in nucleotides from the HindIII site D3 (7), taking the second A of the AAGCTT cutting sequence as nucleotide 1.
2) The 3'-end of RNA 1 has not yet been precisely determined.

compilation of these data is given in Table II.

With the exact positions of the rRNAs we hope to construct more accurate secondary and tertiary structure models than the ones in ref.15, which were devised when the exact 5'- and 3'-end positions of these RNAs were not yet available. This is particularly relevant for the smaller rRNA of 9S, since its new location is shifted approximately 100 nucleotides towards the HindIII site D1, relative to the map position inferred by Eperon et al. (15). The map coordinates of the poly(A)$^+$ species are of relevance for the localization of putative protein genes, which will be discussed in the next section.

All RNAs hybridized to the same maxi-circle strand, inserted into single-stranded M13 DNA, indicating that the direction of transcription is the same for all the RNAs discussed here (clockwise in Fig.2). Moreover, no major transcripts were detected with M13 clones containing inserts of opposite polarity, suggesting that at least this section of the maxi-circle is asymmetrically transcribed. The genes on this part of the maxi-circle seem to be tightly packed: only a few nucleotides separate the 12S from the 9S RNA, RNA 1 from RNA 2 and RNA 2 from RNA 3. A similar situation is observed for the gene organization in mammalian mitochondrial genomes, which are of comparable size (18-20).

No evidence for RNA splicing was obtained for the rRNAs from S_1 nuclease protection analysis (15). We have done similar experiments for RNA 2 and RNA 3, which are shown in Fig.3. Total poly(A)$^+$ from T. brucei was hybridized to M13 DNAs containing the area of the maxi-circle on which these RNAs are mapped, i.e. the area between the EcoRI site R1 and the HindIII site D1 (see Fig.2; Table II and ref.7).

The entire R1-D1 segment protects two major fragments of 500 and 1110 nucleotides against nuclease action (Fig.3, lane 1). In the experiment of lane 2, a smaller insert is used, as indicated, In this case only the larger fragment is protected, which localizes RNA 2 on the missing segment (see Table II). Comparison of lanes 2 and 3 reveals the approximate localization of RNA 3. A protected fragment of about 400 nucleotides is visible in lanes 2 and 3. This fragment is derived from a minor RNA species situated downstream of RNA 3; it is not further discussed here.

The sizes of the protected fragments are only slightly smaller than the values obtained for the corresponding RNAs by Northern blot analysis (see Table II) and this indicates that no major rearrangements, such as splicing, occur in the maturation process of these RNAs. This conclusion has been directly confirmed for RNA 2 by

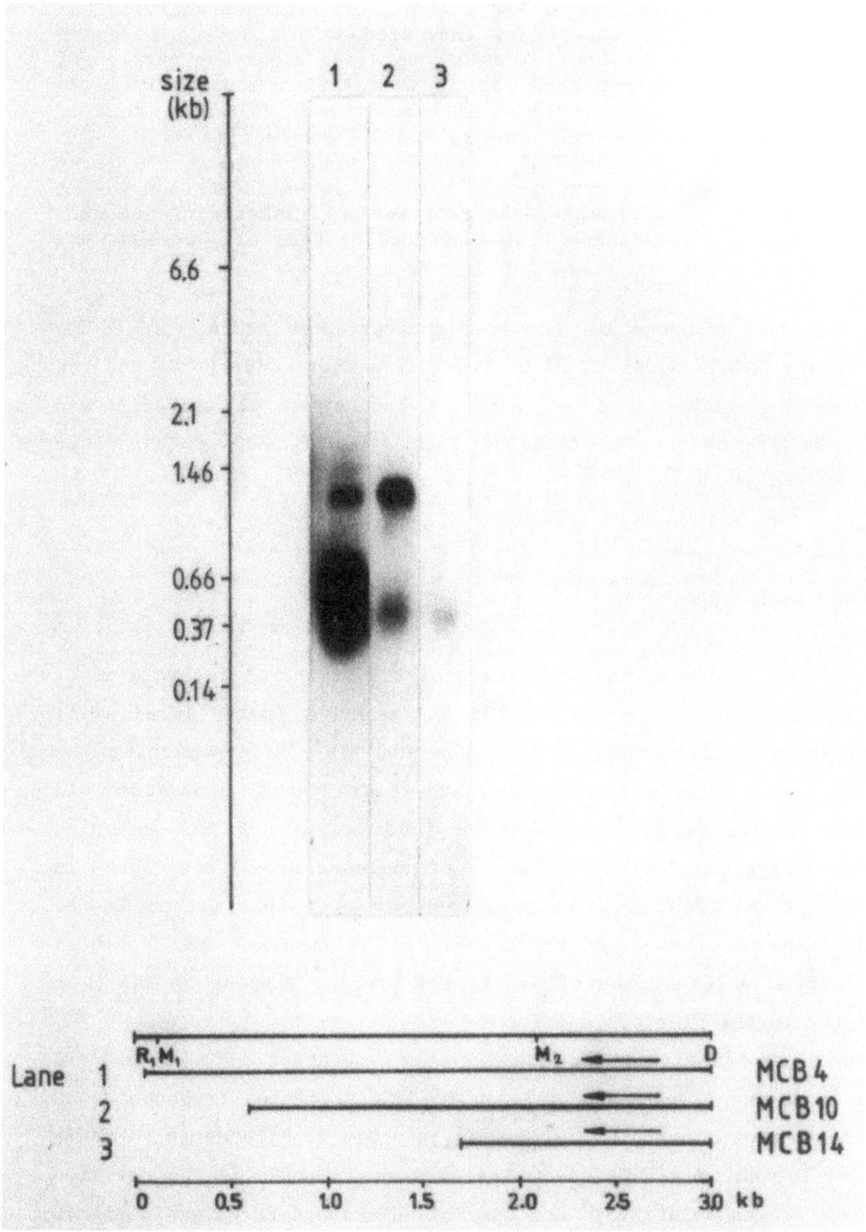

size (kb)
1 2 3
6.6
2.1
1.46
0.66
0.37
0.14
R₁ M₁
M₂
D
Lane 1
2
3
MCB 4
MCB 10
MCB 14
0
0.5
1.0
1.5
2.0
2.5
3.0 kb

Fig.3: S_1 nuclease analysis of RNA 2 and 3. S_1 nuclease analysis was performed according to ref.21, as indicated in the legend to Table II. M13 DNAs used in the hybridization contained a varying portion of the EcoRI/HindIII segment R1-D1 in an orientation complementary to the RNA (as deduced from Northern blot analysis, Table II). Lane 1, MCB 4, nucleotide 3306 - D1; lane 2, MCB 10, nucleotide 3911 - D1; lane 3, MCB 14, nucleotide 5012 - D1. The probe used was the large TaqI fragment Q1-Q2 (see Fig.2; ref.7). S_1 nuclease incubation was at 30°C for 30 min. Size markers were restriction fragments of pBR 322. Control lanes (S_1 incubation in the absence of RNA) did not show any hybridization.

comparing the sequence of its complementary DNA (cDNA; cloned as recombinant DNA in E. coli, clone Tck-1, see Fig.4 and ref.7) and the corresponding maxi-circle segment. No indication of splicing was found. We are currently extending this type of analysis to other areas of the maxi-circle.

Sequence analysis

We have continued the sequence analysis of the maxi-circle of T. brucei, initiated at the 9S and 12S RNA genes by Eperon et al. (15) in a clockwise direction (see Fig.2) and are now approaching the HindIII site D1. As a general approach we use rapid, non-random M13 cloning procedures as described by Poncz et al. (23), in which fragments progressively shortened by exonuclease Bal-31 are cloned in M13 mp8 or mp9 (24). An additional advantage of this method is the acquisition of area- and strand-specific S_1 nuclease and Northern blot probes, which can be of use in the precise mapping of RNAs, as discussed in the previous section.

We have translated the DNA sequences obtained into amino acid sequences using a genetic code in which UGA specifies tryptophan, as it does in most mitochondria (25). In this way we have detected three long (> 100 amino acids) open reading frames (RFs) and the position of these frames relative to and the RNAs discussed is given in Fig.4. The amino acid sequences are given in Fig.5 and salient points of the amino acid composition of these putative proteins are highlighted in Table III.

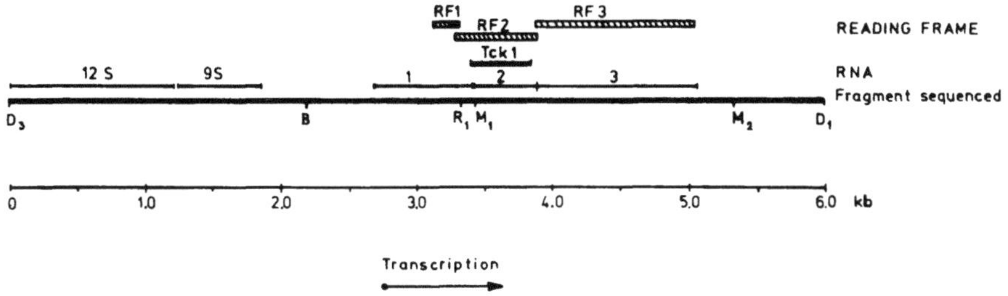

Fig.4: Position of open reading frames (RFs) on the transcription map. The fragment D3-D1 (see Fig.2) of which sequence analysis is nearly completed, is depicted together with the position of those RNAs (12S, 9S, RNA 1, 2 and 3, see Table II) which have been accurately mapped and that of long (> 100 residues) open reading frames: RF1, nucleotide 3088-3474; RF2, 3284-3973; and RF3, 3930-5018 on the nucleotide map of the fragment.

The amino acid composition of RF3 is similar to that of a bona fide mitochondrial protein (Table III) and in fact, as shown by computer analysis, the sequence (Fig.5) displays 25% homology to the genes for apocytochrome b from mammalian and fungal sources (results not shown). If the T. brucei sequence is aligned with the other apocytochrome b genes in a manner resulting in maximal homology, only a few deletions and insertions have to be assumed. The protein has a size of 350 amino acid residues and starts most likely at the methionine of position 14.

In this alignment, 6 out of the 6 proline residues and 9 out of the 14 glycine residues of the T. brucei sequence are lined up with invariant proline and glycine residues, respectively, in the other apocytochrome b genes examined. These amino acids are thought to disrupt an α-helix structure. Moreover, 50% of the substitutions in the T. brucei sequence are conservative (e.g. arginine for lysine, leucine for valine for isoleucine and vice versa, etc.). The

```
               10         20         30         40         50         60
RF 1   TGRRERGIDS GFGAERIDYL EGGREEGRVE GFLIFVKEKE GSRFERDSER EKDGGFLGVK

               70         80         90        100        110        120
       RKPVLDQRKG GAGNSALWNT PKDLRNFWGS STAGGALFEE GGSRRLSEER GDRDRWRLFF

              130
       GYWGLFKIE*
```

```
               10         20         30         40         50         60
RF 2   TKERGGPGIQ LCGTPQRIWG IFGGARRRAE HYLRRAGAEG FLRKEGTEID EGYFLVIEDC

               70         80         90        100        110        120
       LKLNKKAFWK GIFGGHRQRR RVLEEFVLRG GFEGRGERER ERNGPERRVE EGGFEGEGRL

              130        140        150        160        170        180
       SDQGKEGRLR KGKTICEGEG FLEGFWEERG FGETRWDCLQ KQRGFWAKEY NLQRGESGRR

              190        200        210
       NTGGKTGFRK RERGEGKGFS WNEEVVCRKL RIMVINFI*
```

```
               10         20         30         40         50         60
RF 3   ILYKSGEKRK GLLMSGCLYR IYGVGFSLGF FIALQIICGV CLAWLFFSCF ICSNWYFVLF

               70         80         90        100        110        120
       LWDFDLGFVI RSVHICFTSL LYLLLYIHIF KSITLIILFD THILVWFIGF ILFVFIIIIA

              130        140        150        160        170        180
       FIGYVLPCTM MSYWGLTVFS NIIATVPILG IWLCYWIWGS EFINDFTLLK LHVLHVLLPF

              190        200        210        220        230        240
       ILLIILILHL FCLHYFMSSD AFCDRFAFYC ERLSFCMWFY LRDMFLAFSI LLCMMYVIFI

              250        260        270        280        290        300
       NWYFVFHEES WVIVDTLKTS DKILPEWFFL YLFGFLKAIP DKFMGLFLMV ILLFSLFLFI

              310        320        330        340        350        360
       LNCILWFVYC RSSLLWLTYS LILFYSIWMS GFLALYVVLA YPIWMELQYW VLLLFLLIVC

       RLD*
```

Fig.5: Amino acid sequence of RF1, RF2 and RF3

TABLE III

Amino acid composition of RF1, 2 and 3.

Residues	Occurrence (as % of all amino acids) in RF1	RF2	RF3	apocytochrome _b_
glycine	17.7	18.3	4.4	6.3
basic	20.0	22.8	6.9	7.1
acidic	17.7	16.0	4.9	4.0
polar	12.3	12.8	15.3	22.3
hydrophobic	32.3	30.1	68.1	60.3

The values for apocytochrome _b_ from human mitochondria have been taken from ref.19. Basic amino acid residues are: His, Arg and Lys; acidic residues, Asp and Glu; polar residues, Ser, Thr, Asn, Gln and Cys; the others are considered hydrophobic (see ref.26).

T. brucei protein, therefore, shows a very apocytochrome _b_-like hydrophobicity profile (see ref.26).

We also infer from the rather unambiquous alignment, that no deviations from the universal genetic code occurs in trypanosome mitochondria, except for UGA specifying tryptophan. AUA, for example, lines up much more frequently with isoleucine than with methionine. No definitive assignment could be made for AUU, however, since it appears to line up with 13 different residues, without clear preference (see below).

The direct amino acid and nucleotide homology with other apocytochrome _b_ genes is rather low. A comparison of the yeast (27) and _T. brucei_ nucleotide sequence, for example, reveals a homology of 48%, whereas the longest stretch of completely homologous nucleotides is only 12 residues. This would explain our failure to detect any cross-hybridization between this area of the maxi-circle and yeast apocytochrome _b_ probes (results not shown). Such a cross-hybridization has been reported for the maxi-circle of _Leishmania tarentolae_ (28).

The small extent of direct homology is surprising though, since the apocytochrome _b_ protein is one of the better conserved proteins in mitochondria (a comparison of e.g. yeast (27) and human apocytochrome b (19) gives 50% homology). It shows that the low degree of similarity observed between trypanosomal rRNA genes and those of other organisms, also extends to protein-coding genes (see also below).

The amino acid compositions predicted by RF1 and RF2 are highly unusual for mitochondrial proteins. These proteins would predominantly consist of glycine, arginine, lysine, and glutamic acid and therefore would be highly-charged, hydrophilic proteins without long stretches of hydrophobic residues. They do not resemble known mitochondrial proteins and URFs from other organisms. Not surprisingly, their sequences do not display homology with any of the mitochondrial protein genes and URFs as shown by computer analysis. Also, these proteins do not seem to start with methionine, but it cannot completely be ruled out that in trypanosome mitochondria AUU specifies methionine (see above), in which case the isoleucine at position 17 of RF1 and the isoleucine at position 57 or 72 of RF2 (occurring immediately after the 5'-end of RNA 2, see Table II), would represent the start of the proteins.

It could be argued that these areas of the maxi-circle do not contain protein genes and that RNA 1 and 2 have another (structural?) function. We are dealing, however, with abundant poly(A)$^+$ RNA species. In analogy with the yeast mitochondrial system, it has been proposed that RNA 2, which is the most abundant poly(A)$^+$ RNA in *T. brucei* mitochondria, codes for subunit II of cytochrome *c* oxidase (10). Although homology with the *Cox2* gene is not found, it is not unreasonable to expect that this area codes for another, albeit unusual, mitochondrial protein, since long open reading frames of this size are not likely to be present by accident in DNA, which is presumably evolving rapidly. For RNA 1 the situation is less clear, since the reading frame is much smaller and appears to cover only a small (3') part of the RNA (see Table II and Fig.4).

Conclusions

Protein synthesis experiments, transcript mapping and sequence analysis have been used to initiate the identification of the T. brucei mitochondrial genes, their organization and the mode of transcript processing. In addition to its highly unusual DNA structure this mitochondrial genetic system appears to have some very distinct features:

- The rRNAs are extremely small. The large rRNA is only one third of the size of the 23S Escherichia coli rRNA and a ribosome containing such a RNA is likely to be unusual. Although the ribosome has not been isolated, sequence analysis and protein synthesis experiments are in line with this production and show an abnormal sensitivity/resistance pattern of the protein-synthetic machinery towards inhibitors such as CAP, cycloheximide, etc.

- Hybridization studies and sequence analysis give no evidence for tRNA genes. Of course, this could imply that all tRNAs are imported from the cytoplasm, as has been postulated for some of the mitochondrial tRNAs in Tetrahymena (29), but a more likely explanation is that an unusual structure of these tRNAs in T. brucei mitochondria allows them to go undetected in both types of analysis (compare mammalian mitochondrial $tRNA_{ser}$, missing the complete DHU arm; 30). Therefore, like the rRNAs the tRNAs may be very different from what is usually observed.

- The protein genes may also be abnormal, as deduced from sequence analysis of 30% of the maxi-circle. Although the genetic code, the mode of gene organization and expression do not appear to be very much different from what is found in other organisms, the proteins themselves may be less well conserved (apocytochrome b) or even be totally different (see RF1 and 2) from those so far seen in other mitochondria.

The outcome of further sequence analysis may, therefore, be of considerable interest. It should not only give us a complete picture of this highly unusual mitochondrial genetic system, but it may also show the minimal requirements in terms of amino acid sequence and composition of the functional domains of the mitochondrially-encoded proteins of the respiratory chain.

Acknowledgements

We thank Drs. L.A.Grivell and H.F.Tabak for critical comments on the manuscript and Dr.J.H.Van Boom (Department of Organic Chemistry, Gorlaeus Laboratories, State University Leiden, Leiden, The Netherlands) for preparing the oligonucleotide primers used in the RNA mapping studies. This work was supported in part by a grant to P.B. from the Foundation for Fundamental Biological Research (BION), which is subsidized by the Netherlands Organization for the Advancement of Pure Research (ZWO).

References

1. Simpson, L. (1972) Int.Rev.Cyt. 32, 139-207.
2. Borst, P. and Hoeijmakers, J.H.J. (1979) Plasmid 2, 20-40.
3. Englund, P.T. (1981) In: Biochemistry and Physiology of Protozoa (Levandowsky, M. and Hutner, S.H., Eds), 2nd edn. Vol. 4, Academic Press, New York, pp. 333-383.
4. Steinert, M. and Van Assel, S. (1980) Plasmid 3, 7-17.
5. Chen, K.K. and Donelson, J.E. (1980) Proc.Natl.Acad.Sci.U.S. 77, 2445-2449.
6. Borst, P., Fase-Fowler, F., Hoeijmakers, J.H.J. and Frasch, A.C.C. (1980) Biochim.Biophys.Acta 610, 197-210.
7. Hoeijmakers, J.H.J., Snijders, A., Janssen, J.W.G. and Borst, P. (1981) Plasmid 5, 329-350.
8. Hoeijmakers, J.H.J. and Borst, P. (1978) Biochim.Biophys.Acta 521, 407-411.
9. Borst, P. and Hoeijmakers, J.H.J. (1979) In: Extrachromosomal DNA: ICN/UCLA Symposia on Molecular and Cellular Biology (Cummings, D.J., Borst, P., Dawid, I.B., Weissmann, S.M. and Fox, C.F., Eds), Vol. 15, Academic Press, New York, pp. 515-531.
10. Stuart, K.D. and Gelvin, S.B. (1982) Mol.Cell.Biol. 2, 845-852.
11. Simpson, A.M., Simpson, L. and Livingston, L. (1982) Mol.Biochem.Parasitol. 6, 237-252.
12. Spithill, T.W., Shimer, S.P. and Hill, G.C. (1981)

Mol.Biochem.Parasitol. 2, 235-255.

13. Kleisen, C.M. and Borst, P. (1975) Biochim.Biophys.Acta 390, 78-81.
14. Laemmli, U.K. (1970) Nature 227, 680-685.
15. Eperon, I.C., Janssen, J.W.G., Hoeijmakers, J.H.J. and Borst, P. (1983) Nucl.Acids Res. 11, 105-125.
16. Sor, F. and Fukuhara, H. (1982) Nucl.Acids Res. 10, 6571-6577.
17. Borst, P. and Fase-Fowler, F. (1979) Biochim.Biophys.Acta 565, 1-12.
18. Bibb, M.J., Van Etten, R.A., Wright, C.T., Walberg, M.W. and Clayton, D.A. (1981) Cell 26, 167-180.
19. Anderson, S., Bankier, A.T., Barrell, B.G., De Bruijn, M.H.L., Coulson, A.R., Drouin, J., Eperon, I.C., Nierlich, D.P., Roe, B.A., Sanger, F., Schreier, P.H., Smith, A.J.H., Staden, R. and Young, I.G. (1981) Nature 290, 457-465.
20. Anderson, S., De Bruijn, M.H.L., Coulson, A.R., Eperon, I.C., Sanger, F. and Young, I.G. (1982) J.Mol.Biol. 156, 683-717.
21. Berk, A.J. and Sharp, P.A. (1977) Cell 12, 721-732.
22. Van der Ploeg, L.H.T., Liu, A.Y.C., Michels, P.A.M., De Lange, T., Borst, P., Majumder, H.K., Weber, H., Veeneman, G.H. and Van Boom, J. (1982) Nucl.Acids Res. 10, 3591-3604.
23. Poncz, M., Solowiejczyk, D., Ballantine, M., Schwartz, E. and Sunney, S. (1982) Proc.Natl.Acad.Sci.U.S. 79, 4298-4302.
24. Messing, J. (1981) In: Third Cleveland Symposium on Macro-molecules: Recombinant DNA (Walton, A. Ed), Elsevier, Amsterdam, pp.143-153.
25. Grivell, L.A. (1983) Sci.Amer. 248, 78-89.
26. Hopp, T.P. and Woods, K.R. (1981) Proc.Natl.Acad.Sci.U.S. 78, 3824-3828.
27. Nobrega, F.G. and Tzagoloff, A. (1980) J.Biol.Chem. 255, 9828-9837.
28. Simpson, L., Spithill, T.W. and Simpson, A.M. (1982) Mol.Biochem.Parasitol. 6, 253-264.
29. Suyama, Y. and Hamada, J. (1976) In: Genetics and Biogenesis of Chloroplasts and Mitochondria (Bücher, Th., Neupert, W., Sebald, W. and Warner, S., Eds), North-Holland, Amsterdam, pp. 763-770.

30. De Bruijn, M.H.L., Schreier, P.H., Eperon, I.C., Barrell, B.G., Chen, E.Y., Armstrong, P.W., Wong, J.F.H. and Roe, B.A. (1980) Nucl.Acids Res. 8, 5213-5222.

ORGANISATION OF THE ASPERGILLUS NIDULANS MITOCHONDRIAL GENOME

C. Scazzocchio

Department of Biology, University of Essex, Wivenhoe Park, Colchester CO4 3SQ, U.K.

T.A. Brown, R.B. Waring, J.A. Ray and R. Wayne Davies

Department of Biochemistry and Applied Molecular Biology, UMIST, Manchester M60 1QD, U.K.

Introduction

Virtually the entire 33 kb mitochondrial genome of the ascomycete *Aspergillus nidulans* has now been sequenced (1,2,3,4,5), revealing that the genetic capacity of this molecule is greater than predicted either by direct genetic analysis or by analogy with *Saccharomyces cerevisiae* (Figure 1). In addition to genes coding for known mitochondrial translation products (i.e. cytochrome oxidase subunits I, II and III, apocytochrome b, ATPase subunits 6 and 9), the genome contains eleven unidentified reading frames (URFs), including four homologues of mammalian mt URFs and a reading frame equivalent to the *S.cerevisiae* URF *aapl*. As well as these protein-coding genes, the *A.nidulans* mitochondrial genome contains genes for the 16S and 23S rRNA molecules, and for 28 tRNAs. Overall, the genome is tightly packed, non-coding intergenic regions accounting for less than 15% of the total DNA sequence.

Genetic Markers

A number of mitochondrially inherited mutations have been described in *A.nidulans* by our group and Dr. G. Turner and co-workers (6,7,8,9,10 and references in reference 6). Mutations to oligomycin resistance map at two sites, *oli*A and *oli*B, which

Mitochondria 1983

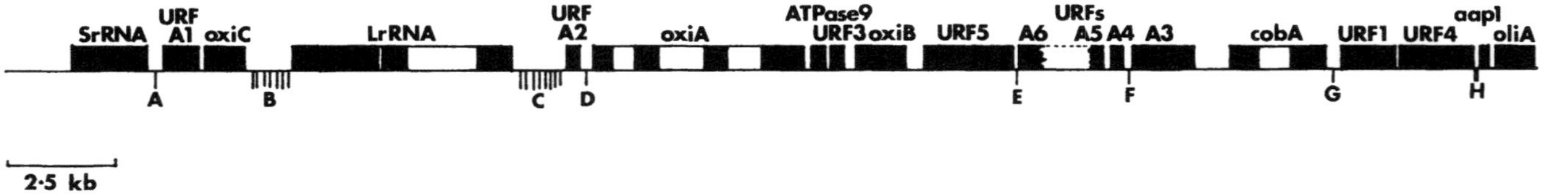

SrRNA = 16S rRNA.
LrRNA = 23S rRNA.
oxiA,B,C = cytochrome c oxidase subunits I, II, III.
cobA = apocytochrome b.
oliA = ATPase subunit 6.

tRNA genes:
A: tyr.
B: lys, gly1, gly2, asp, ser1, ser2, trp, ile, pro.
C: thr, glu, val, met1-3, leu1, leu2, ala, phe, gln.
D: his.
E: cys.
F: asn.
G: cys.
H: arg, asn.

Figure 1. Genetic organisation of the *A. nidulans* mitochondrial genome.

are very closely linked. These probably represent mutations in the gene coding for subunit 6 of the ATPase, although recent DNA sequence data open the unlikely formal possibility that they map in the gene coding for subunit 9 of the ATPase. The DNA sequence of the A.nidulans gene coding for subunit 6 is identical to the homologous S.cerevisiae wildtype gene at both the sites known to mutate to oligomycin resistance in this gene of S.cerevisiae.

Two closely linked sites can mutate to chloramphenicol resistance, and those presumably map in the gene specifying the large rRNA. A cold-sensitive mutation and a mitochondrially inherited suppressor of this mutation have been described. These map eight units apart, and thus could well represent mutations within the same gene. We have presented indirect evidence that the cold-sensitive mutation, cs-67, affects the mitoribosome. This fact, and the detectable linkage to cam A 112 (chloramphenical resistance) suggest that cs-67 might also map in the 21S rRNA or in the open reading frame of the intron in the 21S rRNA.

Recently, G. Turner has isolated mucidin-resistant mutants. The mutations presumably, by analogy with S.cerevisiae, map in the cobA gene. In our laboratory, T. Langdon has isolated a number of as yet unmapped, mitochondrially inherited cold-sensitive mutations.

Formal Genetics

The ordering of markers in mitochondrial crosses presents some difficulty in an obligate aerobe, where deletions affecting essential genes are a priori impossible. Waring and Scazzocchio (6) have discussed the methodology of the formal genetics of the mitochondrion of A.nidulans. The most reliable method is to perform 4 (or more) point crosses in which two markers are used to select recombinant mitochondrial genomes, and the other two (or more) markers can be scored among the recombinants in both wildtype and mutant forms. It should be pointed out that in *A.nidulans* (6) as in *S.cerevisiae*, the existence of multiple rounds of replication and recombination results in mutations that are relatively close in molecular terms appearing not to be linked. Thus detection of any genetic linkage means that mutations are close together in the DNA sequence, which is invaluable for the establishment of the approximate positions of mutations prior to DNA sequence determination.

The DNA Sequence

1) Genes coding for known translation products.

The physical organisation of the *A.nidulans* mt genome was studied initially by probing *Hind*III and *Hae*III restriction digests with defined pieces of mt genes prepared from *S.cerevisiae* petite strains (11). In this way the approximate locations of the genes for apocytochrome b (*cob*A), cytochrome oxidase subunits I, II and III (*oxi*A,B,C) and ATPase subunit 6 (*oli*A) were determined, providing a starting point for restriction fragment subcloning and eventual characterisation of the DNA sequence. The identities of reading frames have been confirmed by comparisons of nucleotide and derived amino acid sequences with equivalent genes from *S.cerevisiae*, *Neurospora crassa* and *H.sapiens*.

The *oxi*A gene, coding for cytochrome oxidase subunit 1, is very homologous with the *S.cerevisiae* equivalent from the fourth codon onwards, but does not have an AUG or any other candidate for a start codon at or near the position of the AUG in the yeast gene (12). Either there is a small, structurally atypical, intron linking the clearly homologous sequence to an upstream AUG, or a very unusual start codon is used, such as AUA or AUU isoleucine codons, which may be used in mammalian mitochondrial genomes (13). The equivalent *cox*I gene of *N. crassa* may also have a non-standard start codon (14). All the other *A.nidulans* genes start with AUG except for *oxi*C, which starts with GUG (4).

Two mosaic protein-coding genes are present in the A.nidulans mtDNA. The gene for cytochrome oxidase subunit I contains three introns, the second and third of which have long open reading frames that could code for maturase-like proteins (15), as could an open reading frame in the single intron in the apocytochrome b gene (16).

Towards the 3' end of the A.nidulans oxiA gene there is a stretch of DNA coding for 16 amino acids that are not present in the homologous genes. This corresponds roughly in position to the two extra stretches of 7 and 15 amino acids, compared with the human gene sequence, in the oxi-3 gene of S.cerevisiae . As a similar homologous stretch of 16 amino acids is found in the mature cox1 protein of N.crassa (14) it is likely that these sequences do not correspond to introns as proposed previously (17) but represent divergence between species towards the carboxy-terminal end of the gene. Although the cobA intron is equivalent in position to intron B3 of the S.cerevisiae cob gene, none of the three oxiA introns appear to have positional homologues in yeast. All four introns have the 'core' RNA secondary structure found in a class of introns fitting a splicing model proposed by Davies et al. (18). This core RNA structure may play an active as well as structural role in intron splicing since it is also found in the self-splicing RNA of the Tetrahymena rRNA intron (19). In addition a sequence in each intron could act as a guide to align the splice sites (19,20).

An intact copy of the ATPase subunit 9 gene with correct mitochondrial cocon usage is present in the A.nidulans mt genome even though convincing genetic evidence exists for the nuclear location of the functional version of this gene.(21). The A. nidulans mitochondrial subunit 9 gene shows 86% amino acid homology with the N.crassa subunit 9 sequence as determined by direct amino acid sequencing of the proteolipid. The presence of such similar mitochondrial genes for subunit 9 in A.nidulans and N.crassa argues against these genes being silent.

The A.nidulans gene for cytochrome oxidase subunit II (5) is very homologous in amino acid sequence to the corresponding gene of Neurospora crassa (22) except at the amino-terminal coding end where only two of the first twelve amino acids are conserved between the two species. The S.cerevisiae gene is different from both filamentous fungal genes in this region (22). It has been argued that this region does not appear in the final cytochrome oxidase subunit 2 protein.

2) Unidentified reading frames

a) Homologues of mammalian URFs. Genes corresponding to four of the mammalian URFs have been located in the A.nidulans mt genome. These genes correspond to URFs 1 and 4 (1), 3 and 5 (5). In each case there is convincing amino acid homology between the derived fungal and mammalian translation products (Table 1), and with URFs 1 and 4 the comparison has been extended to show conservation of predicted secondary structures in the fungal and human polypeptides (1). These analyses have demonstrated beyond doubt that the fungal URF 1 and URF 4 genes are very closely related to the mammalian URFs, and have indicated that the genes code for functionally important polypeptides that occur in the mitochondria of widely divergent species.

AMINO ACID HOMOLOGY BETWEEN URFs OF A.NIDULANS AND OTHER ORGANISMS

URF 1 (352 codons)	
Homo sapiens	39%
Drosophila	35% (176 codons sequenced)
Neurospora crassa	80% (81 codons sequenced)
Zea mays	27% (125 codons sequenced)
URF 4 (488 codons)	
Homo sapiens	26%
URF 3 (about 75% sequenced)	
Homo sapiens	34%
URF 5 (about 40% sequenced)	
Homo sapiens	38%

b) URFs unique to A.nidulans. Six possible URFs not as yet identified in other organisms have been located in the A. nidulans mtDNA (5) (Table 2). URFs A 5 and A 6 may in fact be parts of the same gene, since the sequence between them has not been completed. Three genes (URF A1, URF A3 and URF A6) are of substantial size and it seems very unlikely that they are not expressed in the mitochondrion. However, the remaining three URFs are well less than 90 codons in length, raising a doubt as to whether or not they are functional genes. In practice they are considered to be URFs on the basis of two criteria: firstly they possess standard AUG initiation codons, and secondly they display typical codon usage for tenes in this mitochondrial DNA, showing a very marked bias, where a choice is available, for codons encosing in A or U. Several additional reading frames of up to about 100 codons in length are not considered likely to be URFs because of a failure to satisfy one or other of these rules.

URFs UNIQUE TO A.NIDULANS

	Codons	M.W.
URF A1	228	25100
URF A2	27	2970
URF A3	380	42000
URF A4	87	9600
URF A5	72+	8000+
URF A6	175+	19250+

c) aapl. A reading frame originally called URFx (2) has 48% amino acid homology with the yeast gene aapl. As in S.cerevisiae, this gene lies immediately upstream from the ATPase subunit 6 reading frame, emphasising the possibility of a functional relationship between these two genes (23).

3) rRNA and tRNA genes

All the relevant data concerning the 16S and 23S rRNA genes, as well as the tRNA gene clusters surrounding the 23S rRNA genes, have been published by H. Küntzel and his coworkers (4). Our sequencing work has shown that additional tRNA genes occur outside of the main tRNA clusters (Figure 1). The fairly even distribution of these tRNA genes around the mtDNA leads to the suggestion that, as proposed for mammalian mitochondrial systems (13), the tRNAs act as recognition signals for enzymes involved in processing of primary RNA transcripts.

4) DNA duplications in the genome

At least two major DNA duplications are present in the A.nidulans mt genome (Figure 2). The first, D1, is 300 bp in length and contains the tRNA gene for asparagine, a 116 bp intergenic region and 108 bp of open reading frame. The reading frame forms the first 36 codons of both URFA3 and aapl, which are consequently 83% homologous in this region. We have argued (23) that the duplicated peptide carries a functional domain that is important in the activity of both the URFA3 and aapl gene products. The second duplication, D2, involves two copies of the $tRNA^{cys}$ gene plus the surrounding intergenic regions. The duplicated tRNA genes in both D1 and D2 are very highly conserved (1 mismatch in $tRNA^{asn}$, 2 in $tRNA^{cys}$) even though the adjacent regions show relatively high loss of nucleotide homology. None of the mismatches occur in base-paired regions of the tRNA molecules. The implication is that both copies of the duplicated tRNA genes are functional, either as tRNAs or as transcript processing signals.

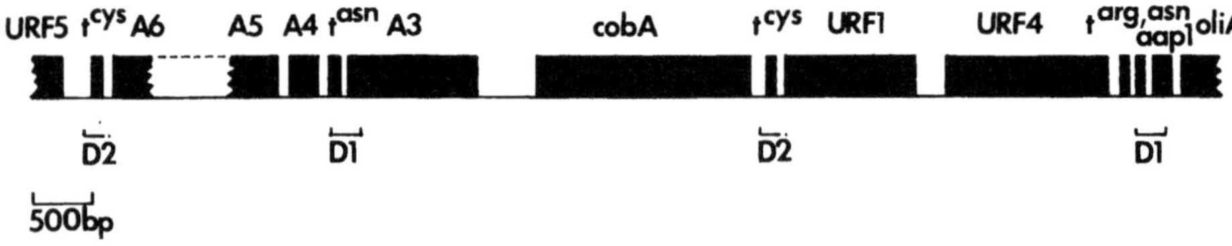

Figure 2. Major duplications within the genome.

Conclusion

The almost complete definition of the mitochondrial genome of A.nidulans by DNA sequencing, combined with further mutational analysis, provides a basis for studying problems such as the function of URFs, the nature and control of transcription and the mechanism of RNA splicing.

References

1. Brown, T.A., Davies, R.W., Ray, J.A., Waring, R.B. and Scazzocchio, C.: The EMBO Journal 2, 427-435 (1983).
2. Grisi, E., Brown, T.A., Waring, R.B., Scazzocchio, C. and Davies, R.W.: Nucleic Acids Res. 10, 3531-3539 (1982).
3. Waring, R.B., Davies, R.W., Lee, S., Grisi, E., Berks, M. M. and Scazzocchio, C.: Cell 27, 4-11 (1981).
4. Netzker, R., Köchel, H.G., Basak, N. and Küntzel, H.: Nucleic Acids Res. 10, 4783-4794 (1982).
5. Unpublished results of the UMIST laboratory.
6. Waring, R.B. and Scazzocchio, C.: Genetics 103, 409-428 (1983).
7. Rowlands, R.T., and Turner, G.: Mol. Gen. Genet. 132, 73-88 (1973).
8. Gunatilleke, I.A.U.N., Arst, H.N., and Scazzocchio, C.: Mol. Gen. Genet. 137, 269-276 (1975).
9. Rowlands, R.T. and Turner, G.: Mol. Gen. Genet. 141, 69-79 (1975).
10. Waring, R.B. and Scazzocchio, C.: J. Gen. Microbiol 119, 297-311 (1980).
11. Macino, G., Scazzocchio, C., Waring, R.B., McPhail Berks, M. and Davies, R.W.: Nature 288, 404-406 (1980).
12. Waring, R.B., Scazzocchio, C., Brown, T.A., and Davies, R.W.: submitted (1983).
13. Anderson, S., Bankier, A.T., Barrell, B.G., de Bruijn, M.H.L., Coulson, A.R., Drouin, J., Eperon, I.C., Nierlich, D.P., Roe, B.A., Sanger, F., Schreier, P.H., Smith, A.J.H., Staden, R. and Young, I.G.: Nature 290, 457-465 (1981).
14. Burger, G., Soriven, C., Machleidt, W. and Werner, S.: The EMBO Journal 1, 1385-1391 (1982).

15. Lazowska, J., Jacq, C., and Slonimski, P.P.: Cell 22, 333-348 (1980).

16. Waring, R.B., Davies, R.W., Scazzocchio, C. and Brown, T.A.: Proc. Natl. Acad. Sci. U.S.A. 79, 6332-6336 (1982).

17. Bonitz, S.G., Coruzzi, G., Thalenfield, B.E., Tzagoloff A. and Macino, G.: J. Biol. Chem. 255 , 11,927-11,941 (1980).

18. Davies, R.W., Waring, R.B., Ray, J.A., Brown, T.A., and Scazzocchio, C.: Nature 300, 719-724 (1982).

19. Waring, R.B., Scazzocchio, C., Brown, T.A. and Davies, R.W.: J. Molec. Biol. 167, 595-606 (1983).

20. Davies, R.W., Waring, R.B., Brown, T.A. and Scazzocchio, C.: this volume.

21. Turner, G., Imam, G. and Küntzel, H.: Eur. J. Biochem. 97, 565-571 (1979).

22. van den Boogaart, P., van Dijk, S. and Agsteribbe, E.: FEBS Lett. 147, 97-100 (1982).

23. Brown, T.A., Davies, R.W., Waring, R.B., Ray, J.A. and Scazzocchio, C: Nature 302, 721-723 (1983).

SEQUENCE OF THE MITOCHONDRIAL DNA, ARRANGEMENT OF GENES AND PROCESSING OF THEIR TRANSCRIPTS IN SCHIZOSACCHAROMYCES POMBE

B. Franz Lang, Fred Ahne, Sonja Distler, Helga Trinkl, Fritz Kaudewitz, Klaus Wolf

Institut für Genetik und Mikrobiologie der Universität München, Maria-Ward-Str. 1a, D-8000 München 19, Germany (FRG)

Introduction

In this paper we will present a short review on the mitochondrial genome of S. pombe and make a comparison to other known mitochondrial systems. Details will be published elsewhere.

Our reasons for analyzing mitochondrial genes and processing of their transcripts in S. pombe were initially based on the availability of various mitochondrial point- and deletion mutations and on the advantage of dealing with one of the smallest mitochondrial genomes among fungi (1-3). We suppose that we might more easily detect principles of fungal gene organization and expression by analyzing such a small genome.

The 19 kbp mitochondrial DNA of S. pombe is now almost completely sequenced and most of the transcripts, mainly with the exception of tRNAs, have been precisely mapped by the S1-nuclease technique. In addition to genes coding for known ubiquitous mitochondrial translation products (apocytochrome b, ATPase subunits 6 and 9, and the three cytochrome oxidase subunits), we have found the two rRNA and so far 22 tRNA genes. Although in the small mitochondrial genomes of mammalia eight conserved, unidentified reading frames (URFs) have been found (4,5), only two URFs are present in S. pombe. One of the URFs is homologous with the aap 1 gene in S. cerevisiae (6) and

Mitochondria 1983

possibly with the human URF A6L, whereas the other reading frame has no homology to any of the known mitochondrial URFs. A gene equivalent to var 1 in S. cerevisiae has not been detected. Despite the small size of the mitochondrial genome the cob and cox1 genes contain protein coding introns. They show clear homologies with respect to consensus sequences, secondary structures of the intron transcripts, and to conserved protein regions in other fungi (7-9). One of the introns will be discussed in more detail, because the putative intron protein is more than 60% homologous with the Aspergillus nidulans cox1 intron protein. This is an exceptional situation because protein homologies between fungal introns of this class are, in general, much lower. In many cases only two conserved decapeptides can be recognized (7-9). The highly homologous S. pombe and A. nidulans introns are moreover inserted at the identical position in the cox1 genes. This is currently the most convincing indication for exchange of mitochondrial sequences between different fungal species.

We will also report on the possible mechanism of RNA processing which combines the principle of tRNA punctuation of genes typical for mammalian mitochondria and the use of signals specific for S. pombe.

Results and Discussion

1) Organization and structure of genes.

The hybridization with mitochondrial gene probes from S.cerevisiae to mitochondrial restriction fragments from S.pombe has been used for mapping genes. This approach has already fixed the direction of transcription for cob, cox1, large rRNA and cox3, and has also predicted the existence of introns in the cob and cox1 gene. It was also obvious from the weakness of the hybridization signals that the

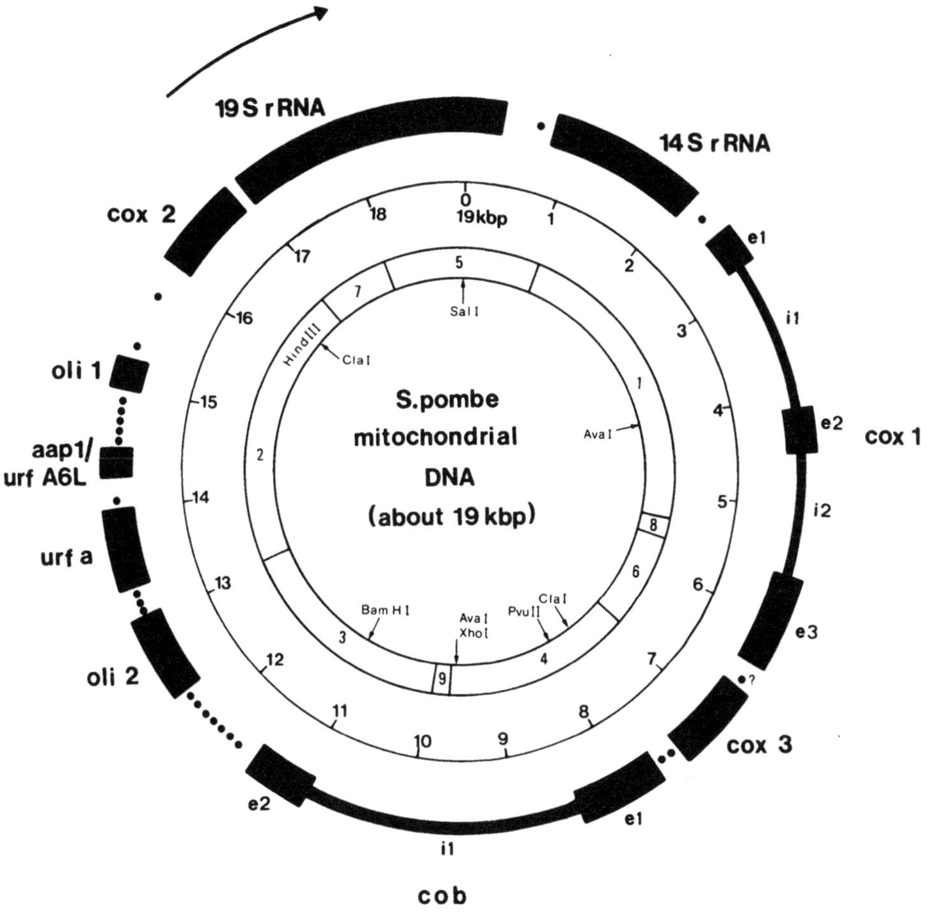

Fig.1: Organization of genes on the mitochondrial DNA of S. pombe

All genes are transcribed from the same DNA strand. The direction of transcription is indicated by an arrow. Dots represent tRNA genes. e1, e2 and e3 are the different exon regions of the split genes; i1 and i1 the introns. cox1, 2 and 3 are structural genes coding for the three mitochondrial subunits of cytochrome oxidase, cob codes for apocytochrome b, and oli1 and oli2 for subunit 6 and 9 of the ATPase complex, respectively.

homology would be very low between S. cerevisiae and S. pombe. The current progress in sequencing the mitochondrial DNA of S. pombe (about 95% is already known) has confirmed the hybridization data and added information about further genes (Fig.1).

So far, we have localized the genes coding for apocytochrome b, the three mitochondrial cytochrome oxidase subunits, subunits 6 and 9 of ATPase, the two ribosomal RNAs and 22 of the 24 expected tRNAs. Open reading frames are found in the single cob intron and in both cox1 introns. One URF has also been detected, coding for a putative protein of 48 amino acid residues, which is about 50% homologous with the aap1 gene. However, in contrast to the aap1 gene (6) and its homologous counterparts in A. nidulans (11,12) and N. crassa (G. Macino, pers. communication), the gene is not situated upstream but about 1000 bp downstream from the oli2 gene. Since the S. pombe gene is flanked by tRNA genes, we assume that it is expressed independent of the oli2 gene. The homologous gene in mammalia is most probably URF A6L. It is also located in front of the oli2 gene. Although overall homologies of the deduced amino acid sequence between mammalia and fungi are very low, it has been pointed out that there is a striking identity of seven N-terminal amino acids among fungi (11-13). This was also found in the S. pombe gene. The first four amino acids are as well identical between the mammalian proteins. Comparing all available data on the mammalian (4,5) and fungal genes (S.pombe, S.cerevisiae and A.nidulans) revealed similarities of the rest of the proteins only with respect to the apolarity. The amino acid sequence in this region is highly variable. Therefore, it is rational to assume that the human URF A6L is the corresponding gene in mammals.

The second URF (URF a) has a coding capacity of about 220 amino acid residues and has no significant homology to any of the as yet known mitochondrial URFs. Analysis of the polarity of the putative polypeptide reveals that it does not look like a membrane protein. A possible function of this URF could be related to a yet unmapped extrachromosomal, most likely mitochondrial, marker: the mutator mutation. This mutation gives rise to a high number of spontaneous, exclusively mitochondrial, point-and deletion mutations and seems to be a unique feature of S. pombe (2). The molecular basis for the mutator phenotype may be an alteration of a protein, involved in either replication or repair of the mitochondrial DNA. In order to show whether or not URF a is identical with the mutator gene, we started to sequence this DNA region in different mutator strains. Sequencing gaps which might contain unexpected structures remain between the cox1 and cox3 gene (about 200 bp), and in the 3' region of the sequence coding for the large rRNA (about 450 bp). We believe that the large rRNA, which is much shorter in S. pombe than in S. cerevisiae, does not contain a "4.5 S like" sequence at the 3' end, as found in S. cerevisiae (10). There are indications that a distinct 5S RNA is encoded in the gap at the end of the large rRNA gene (see the specific chapter on this topic).

A point which has often been stressed in comparing different fungal mitochondrial genomes, is the divergency in number and arrangement of genes. The S. pombe sequences revealed a further type of different gene organization among mitochondrial genomes. In fungi similar principles of gene order have been only observed concerning the rRNAs and the aap1 homologous sequences: (i) the small rRNA gene is normally followed by the large rRNA gene and is separated by a tRNA or additional genes. In S. pombe, however, the orientation of rRNA genes is inverted; (ii) the aap1/URF A6L homologous genes are always found in front of the oli2 sequence, not separated by a further gene. As mentioned above, both genes are located separately in S. pombe. Insofar, S. pombe represents the first example of a completely different gene order.

2) Cob and cox1 are split genes

In fungi three genes have been found to be preferred candidates for intron insertions: the cob, cox1 and large rRNA genes. Although S. pombe contains a short genome, there exists one intron of class II in the cob gene (intron classification according to ref. 7), which is 2529 bp long. In the cox1 gene there are two introns of class I (each about 1100 bp long). Therefore, about 25% of this small genome is already occupied by intron sequences, leaving only about 14 kbp for coding of the structural genes. The open reading frames of all three introns are in phase with their upstream exons and contain the consensus sequences, secondary structures and conserved protein coding regions described for introns of other fungi. In Fig.2 the conserved R and S consensus sequences (as defined in ref.7,8) of the two S. pombe cox1 introns are shown. Only four instead of five bases may pair between the R and S sequences of the first cox1 intron. Therefore, this intron is the first exception among more than 20 known introns of class I.

In Fig.3 the conserved decapeptide sequences of both cox1 introns from S. pombe are compared with typical intron sequences from S. cerevisiae (13,14) and A. nidulans (15).

Since class II introns are rare, a more detailed insight in the conservation of protein sequences and of RNA secondary structure was possible by including the data of the S. pombe cob intron. The three introns of class II, which contain open reading frames are the first two introns in the cox1 gene of S. cerevisiae and the S. pombe cob intron. Their comparison reveals: (i) three stretches, comprising

	"R" sequence	"S" sequence
S. cerevisiae, cob, I4:	TCAGAGACTACA	AAGATATAGTCC
S. cerevisiae, cox1, I4:	TCAGAGACTACA	AAGATATAGTCC
S. pombe, cox1, I1:	TCAGAGACTCGT	AAGGTATAGTCC
S. pombe, cox1, I2:	TCAGAGACTTAC	AAAATAAAGTCC
A. nidulans, cox1, I2:	TCAGAGACTATA	AAAATATAGTCC

Fig.2: Comparison of R and S consensus sequences in S.pombe, S.cerevisiae and A.nidulans introns of class I

Only R and S sequences are compared, not P and Q (definition acc. to ref.7,8), because only these show considerable homologies with the S.pombe cox1, intron 1 sequence. The possible base pairing between R and S sequences is indicated by arrows.

	"LAGLI" region	"DADG" region
S. cerevisiae, cox1, I3:	YLAGLIEGDG	WLAILTDADG
S. cerevisiae, cox1, I4:	WLAGLIDGDG	WVFGFFDADG
S. cerevisiae, cob, I4:	WLAGLIDGDG	WFIGFFDADG
S. pombe, cox1, I1:	WLVGFTDGDG	n.d.
S. pombe, cox1, I2:	YLAGLIDGDG	WLAGFSDADA

Fig.3: Conservation of decapeptide sequences in the S.pombe cox1 introns

The regions of protein homology (as defined in ref.7,9) are aligned, comparing the two cox1 intron sequences from S.pombe with three examples from S.cerevisiae (13, 14). It should be stressed that in S.pombe the second intron of cox1 is much better conserved than the first one. This first cox1 intron is also exceptional in not allowing to construct an internal guide RNA secondary structure (according to ref.8) and in preferring UGG codons for coding tryptophane.

half of the reading frame, show an amino acid homology of about 45% (Linda Bonen, pers.communic.); (ii) in addition to the described conserved core sequence in the secondary structure (7), two further consensus sequences were defined.

The cob intron and the first cox1 intron are inserted at positions in the exon, different from that used in other fungi. However, the second cox1 intron is inserted at the same position as the third cox1 intron in A. nidulans (16). A comparison of the two putative intron proteins revealed more than 60% homology. This is an unexpected finding, since usually the homology of class I intron proteins is restricted to only two decapeptide stretches (out of about 300 amino acid residues). Homologies of structural genes between S. pombe and A. nidulans are, in most cases, even much lower than 60% (e.g.comparison of oli2 genes; Table I). Only a small, GC-rich insert in the A.nidulans intron sequence, which interrupts the open reading frame, is not found in the S. pombe intron. By aligning the two highly homologous conserved sequences the exact extension of the GC-rich A. nidulans mini insert was obvious (16,17).

3) Ribosomal RNA genes

The genes for the two ribosomal RNA subunits are located in close proximity, separated by $tRNA_{asn}$, and a short sequence of yet unknown function. As mentioned above, the order of the rRNA genes is inverted: the large rRNA sequence is transcribed before the small rRNA. Such an inversion has only been described for the highly unorthodox mitochondrial DNA equivalent in Trypanosoma species (18). Im comparison with other fungi there exist also differences in length (only about 19S and 14S in S.pombe), and deviations from the generally accepted universal RNA secondary structure models (19,20).

We have determined the sequence of the large rRNA gene almost completely, and we have found that the highly conserved region, where introns are inserted in various other fungi, is uninterrupted in S. pombe. A comparison of the S. cerevisiae DNA sequence (10) with that of S. pombe reveals a perfect alignment of the two genes, including the "5.8 S like" region in the 5' region of the genes. However, the mature rRNA in S. pombe is considerably shorter (19S) than the rRNA of S. cerevisiae (21S). One possible explanation is that after transcription of the gene, the 5' end of the RNA is removed and exists as a separate 5.8 S RNA species. The other possibility is that the "4.5 S like" sequence at the 3' end, which is part of the 21S RNA in S. cerevisiae, exists as a distinct RNA species in S. pombe. Both explanations are in accordance with the occurrence of a mitochondrial RNA of about 5S length of yet unknown origin in S. pombe (R. Cedergren, pers. communication). Preliminary S1 nuclease mapping data favour the idea that the "5.8 S like" sequence is part of the large rRNA in S. pombe, but not the "4.5 S like" region at the 3' end.

We have used the DNA sequence of the small rRNA gene to analyze conservation of "invariant" sequences, and we attempted to build up a secondary structure according to the general model of Stiegler et al. (20). A domain in the 5' region, containing a secondary structure highly conserved in pro- and eukaryotes and believed to be essential for the ribosomal function, was not at all detectable in S.pombe.

4) Codon usage and tRNA genes

From the DNA sequences we can conclude that S. pombe follows the universal code in translation of structural genes but makes, in addition, use of UGA as a tryptophane codon in introns and in URF a. 22 of the expected 24 tRNA genes, which are necessary for translation, have been deduced from the DNA sequence by the possibility of the respective transcript to form the typical clover leaf structure. In some cases the expression of the tRNA genes has been proven by mapping the respective 4S RNA molecules with the Northern technique (21). By comparing the degree of conservation of secondary structures and the occurrence of invariant

nucleotides between S. cerevisiae, N. crassa and S. pombe tRNAs we have found some extraordinary aspects of the S. pombe genes: (i) in S. pombe, a $tRNA_{arg}$ is found which can read all CGN codons because there is a U in the wobble position of its anticodon (22). The use of all CGN codons in open reading frames is in agreement with the occurrence of the respective tRNA. However, in structural genes only CGA is used (at present 6 times), and in introns CGA is preferred (13 CGA out of 21 CGN codons); (ii) $tRNA_{trp}$ has -in contrast to S. cerevisiae or N. crassa (23,24)- a C in the wobble position of the anticodon, suggesting that only UGG is efficiently recognized by this tRNA, but not UGA. Indeed, in all structural genes (cob, cox1, 2 and 3, oli1, oli2 and aap1) exclusively UGG codons have so far been found (at present 29 times).
Only in the open reading frames of the cob intron, the second cox1 intron and in URF a, there are some UGA codons. Whether or not they act as stop codons is not yet clear. Three alternative explanations can be brought forward: (i) another tRNA recognizing UGA exists in a yet unsequenced region; (ii) the tRNA is imported from the cytoplasm; or, (iii) the C in the wobble position is able to pair strong enough also with an A, so that translation may work (maybe less efficient).

The tRNA genes coding for $tRNA_{ser}$ (AGPy) and $tRNA_{lys}$ (AAPu) have not yet been identified. We assume that they are located in the two sequencing gaps between cox1 and 3, and between the large and the small rRNA genes.

5) The mechanisms of transcription and RNA processing

It has already been mentioned that all mitochondrial genes identified so far in S. pombe, are coded on the same DNA strand. Such a situation suggests an even less complicated mechanism of transcription than in mammalia: starting from one or more promoters -most probably upstream the large rRNA- the complete genome could be transcribed, fol-

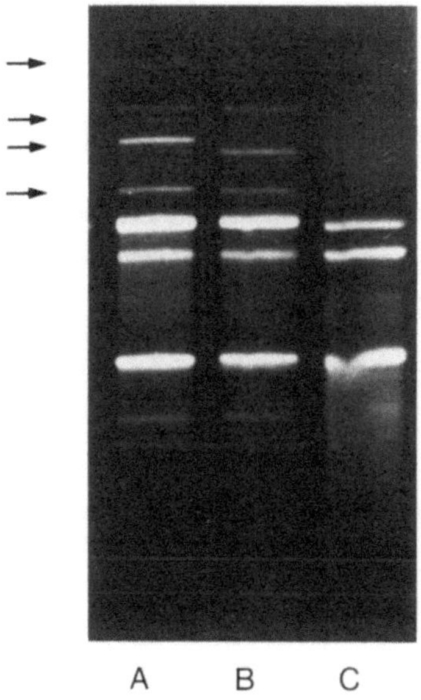

Fig.4: Accumulation of mitochondrial RNA precursors in mit^- mutants

RNA extracts were prepared from mitochondria of the wild type and two mutants (deleted in two different regions of the cob intron) and the RNA separated on an agarose/urea gel. In the mutant RNAs a ladder of faint high molecular weight precursors up to 7 kb long (i.e.spanning more than 1/3 of the genome) are visible in the ethidium bromide staining.

A: mutant diu 145; B: mutant diu 123; C: wild type, 50 h^-

lowed by the excision of tRNAs and mRNAs. This idea is based on the following observations:

a) All genes, including the rRNA genes, are separated by tRNA sequences which were proved to act as processing signals. Only upstream of the large rRNA no tRNA is found. This would be expected when transcription would, in fact, start in front of the rRNA genes as also observed in mammalia.

b) In case of several protein coding genes (oli1, oli2, cox1 and cox2) S1 nuclease experiments have shown transcripts obviously generated by processing precisely at the 3' end of the preceding tRNA. A separate initiation of transcription between tRNA and these transcripts is therefore very unlikely.

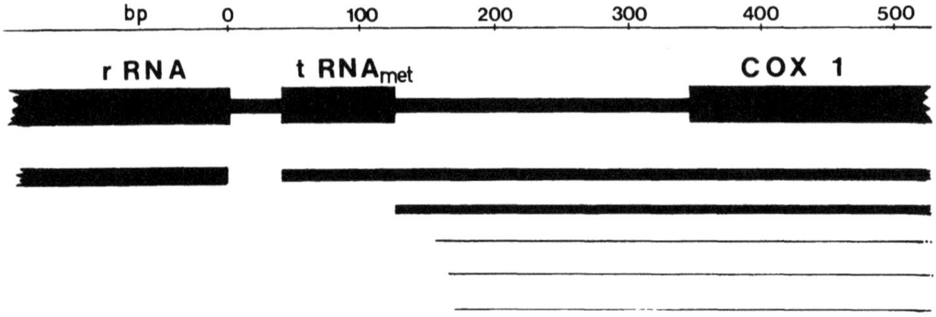

Fig.5: Transcript mapping in the 5' region of the cox1 gene

The figure is a diagrammatic presentation of various transcripts and their concentrations, originating from a region of the genome between the small rRNA and the cox1 gene. The upper bar represents the DNA, with genes indicated by black boxes. The lower bars represent the transcripts. Their relative concentrations, determined in an S1 nuclease experiment, are symbolized by the thickness of the bars. Interestingly, the strongest transcript of the cox1 gene includes the preceding tRNA met. A further processing step removes the tRNA from this precursor by precise processing at its 3' end.

c) In most mitochondrial deletion mutants of S. pombe, the kinetic of processing is generally slowed down, resulting in the accumulation of large RNA precursors, covering various portions of the genome (Fig. 4).

d) Even in the wild type, some RNA precursors have been identified by S1 nuclease experiments, including the upstream tRNA which is processed correctly at its 5' end (Fig. 5). We have also detected a minor oli2 transcript containing the complete tRNA at its 3' end (Fig. 6)

As also shown in Fig. 6, an oligo T/C stretch follows immediately the stop codon of the oli2 gene. Similar sequences are found at the 3' ends of all protein coding genes and of the small rRNA gene. They may be used as general recognition sites for 3' end processing in S.pombe.

A further characteristic of the S. pombe mitochondria is the occurrence of a slow processing of the transcripts of oli2 ⟶

5'.. ATT CAA GCT TAT GTA TTC GCT ATT TTA ACT TGT GGA TTC ATT AAT

tRNA$_{his}$

GAT AGT TTA AAC CTT CAC TAA (stop) TTTTCTCCCCCTATTTTTATGGTGATTGTAATTCAA

TGGTAGAATGCTTATTTGTGGCATAAGAAGTTCTTGGTTCGATTCCAAGTAATCACCCAATCAA

Fig.6: Processing signals in the 3' region of the oli2 gene

The DNA sequence of this region is shown to demonstrate the maturation of an oli2 precursor RNA. The precursor includes the complete tRNA, following the gene and is processed after an oligo T/C run (boxed sequence). The arrows above the DNA sequence indicate the processing points as seen in an S1 nuclease experiment, using a 3' labelled single stranded DNA beginning in the oli2 sequence. Note that the tRNA is not excised at the points expected from the general model of tRNA structure, as indicated by the brackets. The extensions by a G in the 5' end and CCA at the 3' end are posttranscriptional modifications of the cytoplasmic tRNA$_{his}$ in eukaryotes (28). Here, these extensions seem to be encoded by the gene.

protein coding genes, directly in front of the AUG initiation codon. Although there is no experimental proof at the moment, it is rational to assume that the RNA starting with the initiation codon is the mature, translated RNA, analogous to the mammalian transcripts. A similar example of a mitochondrial transcript in fungi, starting directly in front of the AUG codon, has been recently found for URF 1 in N. crassa (Burger and Werner, these Proceedings).

As in all other fungal systems, regulation of the mRNA concentration can be achieved by the kinetics of various maturation processes, like trimming of the 5' and 3' ends or excision of the introns. As in mammalian mitochondria, the abundance of the rRNAs, in relation to all other transcripts, has to be regulated. An extra promoter site upstream of the two rRNA genes, which allows a more efficient transcription of the rRNA cluster could also exist in S. pombe.

tRNA asp   tRNA gly

5'..GTGATGGATGCCCACTCGTACTGGGTCAGATTCG]TTTAA[GCAGATATAAGTCAATTGGCA

GACCTTCTGATTTCCACTTAGATTATGTTCGTTCGAATCGGACTATCTGC]ATTAAATAA[GCCT

tRNA leu1

AAATGCTGGAATTTGGTAGACAGAACAAACTTAAGATTTGTCGCTTTCGAGCGTGTTGGTTCG

tRNA trp

AGTCCAACTTTAGGTA]TAACTAAAAAAAAAAA[AAGAGTATAGTTTAATT...3'

Fig.7: Short AT-rich sequences separate the tRNA genes

Part of the DNA sequence of the tRNA cluster is shown which follows the cob gene (see Fig.1). tRNA genes are separated by short "AT cluster" of variable length. Comparable AT-rich sequences in S. cerevisiae are from three to some hundred bp long (26).

6) Evolutionary aspects

Several results presented in this article indicate clearly that the mitochondrial genome of S. pombe is derived from a fungal-type ancestor. The main similarities with other fungi lie in (i) a relatively high AT content of structural genes (about 65-70%); (ii) the occurrence and conservation of intron sequences; and (iii) separation of genes by AT-rich sequences (Fig.7). Although nothing is known about the function of 5' and 3' extensions of transcripts, such leader- or trailer-like sequences occur in the mitochondrial mRNA of all fungi studied so far. In S. pombe, these sequences are also present, however, sometimes reduced to only some bases.

The low conservation rate of the protein sequence of structural genes -as deduced from the DNA sequence (example shown in Table I)- and the differences in gene organization and transcript processing puts S. pombe at far distance from other fungi. The most uncommon feature of S. pombe is

	bovine	man	S. pombe	S. cer.	E.coli
A. nidulans	23	25	37	45	19
E. coli	12	16	14	19	
S. cerevisiae	27	28	39		
S. pombe	25	24			
man	62				

Table I: Conservation of the protein sequence of subunit 6 of the ATPase complex

The protein sequences, as deduced from the DNA sequence of the genes in S.cerevisiae (27), A.nidulans (11,12), HeLa, beef (4), E.coli (29), and S.pombe, were aligned and the percentage of identical amino acid residues compiled. Similar homologies are obtained, when comparing the respective cob genes.

the apparent use of the universal code in the main structural genes. This, taken together with the exceptionally high conservation of the second intron of the cox1 gene and the respective third one in A. nidulans, support the following hypothesis. The original set of mitochondrial genes in S. pombe, perhaps also in an ancestor of fungal mitochondrial genomes, make use of the universal code. Intron sequences using additionally UGA for coding tryptophane may then be recognized as a modern invasion, tolerated by the mitochondrial DNA, possibly because of benefits in regulation of expression. The highly conserved cox1 introns in S. pombe and A. nidulans could represent an example for an interspecific intron transposition.

Acknowledgements

The authors thank G. Burger, J. Kreike and T. Pillar for helpful comments on this manuscript, G. Macino, R. Waring and F. Michel for providing manuscripts and results prior to publication. The generous gift of rho$^-$ strains and plasmids, obtained from T. Fox, H. Fukuhara and R.J. Schweyen, has enabled us to isolate the various specific DNA probes used for gene localization. We are also indebted to I. Nerkorn for typing of the manuscript and to E. Praetzel for excellent technical assistance. This work was supported by the Deutsche Forschungsgemeinschaft.

References

1. Wolf, K., Lang, B.F., Del Giudice, L., Anziano, P.Q., Perlman, P.S.: In: Mitochondrial Genes. Cold Spring Harbor Symposium, Cold Spring Harbor Laboratory, New York, pp. 335-360 (1982)
2. Seitz-Mayr, G., Wolf, K.: Proc. Natl.Acad.Sci.USA 79, 2618-2622 (1982)
3. Anziano, P.Q., Perlman, P.S., Lang, B.F., Wolf, K.: Curr. Genet. 7, 273-284 (1983)
4. Anderson, S., Bankier, A.T., Barrell, B.G., deBruijn, H.L., Coulson, A.R., Drouin, J., Eperon, I.C., Nierlich, D.P., Roe, B.A., Sanger, F., Schreier, P.H., Smith, J.H., Staden, R., Young, I.G.: In: Mitochondrial Genes. Cold Spring Harbor Symposium, Cold Spring Harbor Laboratory, New York, pp. 5-43 (1982)
5. Bibb, M.J., Van Etten, R.A., Wright, C.T., Walberg, M.W., Clayton, D.A.: Cell 26, 167-180 (1981)
6. Macreadie, I.G., Novitski, C.E., Maxwell, R.J., John, U., Ooi, G.B., McMullen, G.L., Lukins, H.B., Linnane, A.W., Nagley, P.: Nucleic Acids Res. 11, 4435-4451 (1983)
7. Michel, F., Jacquier, A., Dujon, B.: Biochimie 64, 867-881 (1982)
8. Waring, R.B., Scazzocchio, C., Brown, T.A., Davies, R.W.: J.Mol.Biol. 167, 595-605 (1983)
9. Hensgens, L.A.M., Bonen, L., de Haan, M., van der Horst,G., Grivell, L.A.: Cell 32, 379-389 (1983)
10. Sor, F., Fukuhara, H.: Nucleic Acids Res. 11, 339-348 (1983)
11. Netzker, R., Köchel, H.G., Basak, N., Küntzel, H.: Nucleic Acids Res. 10, 4783-4794 (1982)
12. Grisi, E., Brown, T.A., Waring, R.B., Scazzocchio, C., Davies, R.W.: Nucleic Acids Res. 10, 3531-3539 (1982)

13. Nobrega, F.G., Tzagoloff, A.: J.Biol.Chem. 255, 9828-9837 (1980)

14. Bonitz, S.G., Coruzzi, G., Thalenfeld, B.E., Tzagoloff, A., Macino, G.: J.Biol.Chem. 255, 11927-11941 (1980)

15. Waring, R.B., Davies, R.W., Lee, S., Grisi, E., McPhail Berks, M., Scazzocchio, C.: Cell 27, 4-11 (1981)

16. Waring et al., submitted for publication in J.Mol.Biol.

17. Lang, B.F., in preparation (for simultaneous publication in J.Mol.Biol., as ref. 16)

18. Borst, P., Hoeijmakers, J.H.J., Frasch, A.C.C., Snijders, A., Janssen, J.G.W., Fase-Fowler, F.: In: The Organization and Expression of the Mitochondrial Genome (Kroon, A.M. and Saccone, C., eds.) Elsevier, North-Holland Biomedical Press, Amsterdam

19. Noller, H.F., Kop, J., Wheaton, V., Brosius, J., Gutell, R.R., Kopylov, A.M., Dohme, F., Herr, W., Stahl, D.A., Gupta, R., Woese, C.: Nucleic Acids Res. 9, 6167-6189 (1981)

20. Stiegler, P., Carbon, P., Ebel, I.P., Ehresmann, C.: Eur.J.Biochem. 120, 487-495 (1981)

21. Alwine, J.C., Kemp, D.J., Parker, B.A., Reiser, J., Renart, J., Stark, G.R., Wahl, G.M.: In: Methods in Enzymology (Wu, R., ed.) Vol. 68, pp. 220-242, Academic Press New York (1979)

22. Grosjean, H.J., de Henau, S., Crothers, D.M.: Proc.Natl. Acad.Sci.USA 75, 610-614 (1978)

23. Bonitz, S.G., Berlani, R., Coruzzi, G., Li, M., Macino, G., Nobrega, F.G., Nobrega, M.P., Thalenfeld, B.E., Tzagoloff, A.: Proc.Natl.Acad.Sci.USA 77, 3167-3170 (1980)

24. Heckman, J.E., Sarnoff, J., Alzner-DeWeerd, B., Yin, S., RajBhandary, U.L.: Proc.Natl.Acad.Sci.USA 77, 3159-3163 (1980)

25. Grivell, L.A.: Scientific American 248, 60-73 (1983)

26. Miller, D.L., Martin, N.C.: Curr.Genet. 4, 135-143 (1981)

27. Macino, G., Tzagoloff, A.: Cell 20, 507-511 (1980)

28. Cooley. L., Appel, B., Söll, D.: Proc.Natl.Acad.Sci.USA 79, 6475-6479 (1982)

29. Gay, N.J., Walker, J.E.: Nucleic Acids Res. 9, 3919-3926 (1981)

NUCLEOTIDE SEQUENCE AND TRANSCRIPT MAPPING OF A mtDNA SEGMENT COMPRISING CO 1, $tRNA_{arg}$, AND SEVERAL UNIDENTIFIED READING FRAMES IN NEUROSPORA CRASSA:

Gertraud Burger and Sigurd Werner

Institut für Physiologische Chemie, Physikalische Biochemie
und Zellbiologie der Universität München,
Goethestrasse 33, D-8000 München 2, FRG

Introduction

The filamentous fungus *Neurospora crassa* (strain 74A) has a mtDNA of about 60 kbp length (1). Fig.1 shows a simplified map for the localisation of the mitochondrial genes coding for rRNA and tRNA species and for several structural proteins.

We have focussed our interest on a mtDNA segment of about 8 kbp including the gene for subunit 1 of cytochrome oxidase (CO 1).

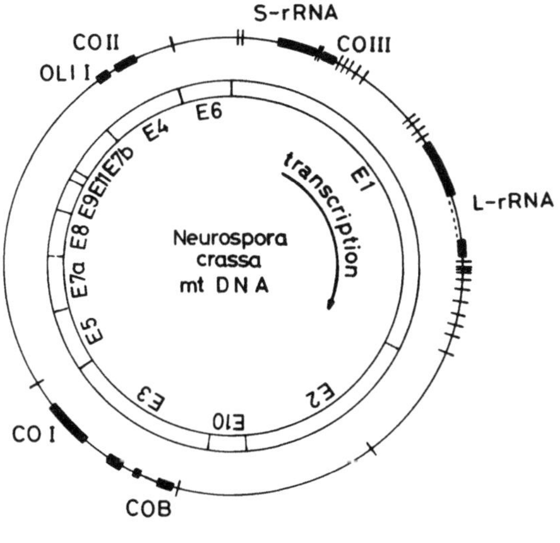

Fig.1. Genetic and physical map of the mtDNA of N. crassa (strain 74A). The inner ring represents the Eco RI restriction fragments. On the outer ring the loci for mitochondrial genes are depicted, and are indicated either as black bars or in the case of tRNA genes as thin lines. Data assembled are taken from references 1 - 7.

This particular polypeptide is believed to play a key role in the biogenesis of the terminal respiratory complex (8). It is synthesized in the form of a larger precursor polypeptide, displaying an extended polypeptide chain at the N-terminus (9,10). Studies of mutants indicate that the removal of the prepiece is necessary to initiate the assembly process of complex IV (9).

Results and discussion

THE GENE SEGMENT CODING FOR THE MATURE SUBUNIT 1

The mature subunit 1 is coded by a sequence of 1665 bp specifying 555 amino acids (5,11). The N-terminal residue is serine (10). There are several lines of evidence (e.g. direct comparison with partial amino acid sequences of the protein) that this portion of the gene is not interrupted by intervening sequences. This is not only in contrast to findings in other lower eukaryotes like *Saccharomyces cerivisiae* (12), *Aspergillus nidulans* (13), or *Schizosaccharomyces pombe* (14), but is also in contrast to the situation in other wild type strains of *Neurospora*. As reported very recently, some of these strains contain up to four optional introns within the region coding for the mature subunit 1 (15).

THE GENE SEGMENT CODING FOR THE PREPIECE

The coding sequence for the N-terminally extending prepiece of the precursor is obviously separated from the DNA region which codes for the mature polypeptide. This conclusion has been drawn from the following facts:
Only one ATG triplet has been detected upstream and within the same reading frame of the sequence coding for the mature subunit. The ATG is found at position -2 (Fig.2). It is unlikely, that this particular triplet codes for the required

```
                                                                                5'.........GGGGGACGGGGGGAT
                                                                                            G  G  R  G  D

-450 ATGTGGATAAGGGTAGCGAGAAAAGTTAATTTACTACATGAGACTAATGGAATATTGGATAGCTGACTTAATATAGCAAGGCACTTCGTTGGTTCTAATA
     [M] W  I  R  V  A  R  K  V  N  L  L  H  E  T  N  G  I  L  D  S  W  L  N  I  A  R  H  F  V  G  S  N

-350 CTGATTTAGCCGTGATTCCGCCTTTTTTTTTTTTTTTTTTTTTTTAAAAAAAAAAAAAAAAAAGAAAGTCCACAAAACATTGAAAGGCTAGTGTGTTTTT
      T  D  L  A  V  I  P  P  F  F  F  F  F  F  F  L  K  K  K  K  K  R  K  S  T  K  H  W  K  A  S  V  F  F

-250 TAGTAAGTCAGTCAATCACCTACGTAGGTGATCCAGAATTACGTTACCGGGGATATTCTGTTGGTCGGTTTACAGAATATCCTGGTTAGCATTGATTGTA
       S  K  S  V  N  H  L  R  R  W  S  R  I  R  L  P  G  I  F  C  W  S  V  Y  R  I  S  W  L  A  L  I  V

     -50                                     -40                                     -30
-150 TAA TCA ATA AGA TGT TGT AAT CTT ATT TTA TTT TTA TTT ATA AGA TAC GTA ACC ATT ATA ATA AAA ACC ATT ATA
      *  ser ile arg cys cys asn leu ile leu phe leu phe ile arg tyr val thr ile ile ile lys thr ile ile

                         -20                                     -10                                 -1
 -75 TTA TTT AAT CAA TTA AAT AGT GAG GAG TTT GGG CTC TCT TTA AAT TCA AGT AAA AGA AGT GTC GGG TTA ATG TCG
     leu phe asn gln leu asn ser glu glu phe gly leu ser leu asn ser ser lys arg ser val gly leu [met] ser

      1                                      10                                      20
  +1 TCA ATT TCG ATT TGA ACT GAA AGA TGA TTT TTA TCA ACG AAT GCT AAA GAT ATA GGA GTT CTT TAT CTA ATC TTC
     SER ILE SER ILE TRP THR GLU ARG TRP PHE LEU SER THR ASN ALA LYS ASP ILE GLY VAL LEU TYR LEU ILE PHE
```

Fig.2. Nucleotide sequence of the 5′end region of CO 1 in N. crassa. The nucleotide sequence has been translated into protein sequences. The one letter abbreviation is used for Urf U, the three letter abbreviation for the CO 1 reading frame. Those amino acid positions also determined by protein sequencing are marked by the capitalized three letter abbreviation. Nucleotide positions are given in the left margin. Numbering starts at the N-terminus of the mature subunit 1 - amino acid positions are shown above the lines. Tandem sequences are indicated by arrows. ATG triplets are boxed.

formyl-methionine of the precursor polypeptide, because

i) amino acid sequencing of the precursor provided evidence that the prepiece includes more than 12 residues (Machleidt et al., unpublished results)

ii) the electrophoretic behavior of the precursor in the presence of SDS suggests an elongation of about 40 residues compared with the mature product (9).

Since the region coding for the prepiece is obviously located at some distance from the gene segment coding for the mature subunit, a linking of these separated pieces by RNA splicing has been considered.

DOES THE CO 1 TRANSCRIPT UNDERGO SPLICING?

Northern hybridization and S1 nuclease experiments were performed in order to test whether DNA fragments adjacent to the region coding for subunit 1, are present in the mature tran-

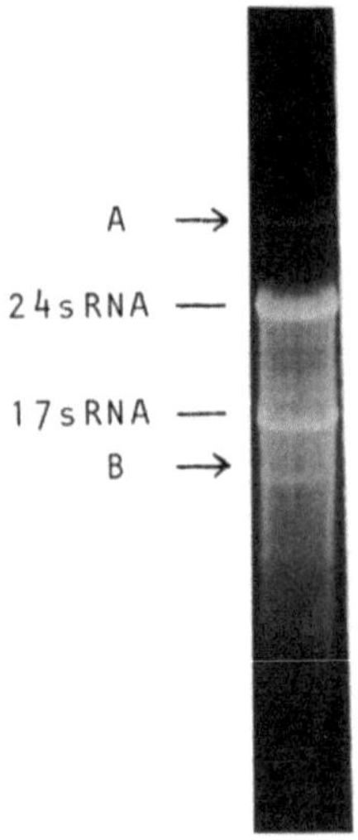

Fig.3. Pattern of total mtRNA of N. crassa. The RNA was denaturated by methyl mercuric hydroxide, separated on an agarose gel and stained with ethidium bromide. Positions of the large and of the small rRNA are indicated. Arrow A points to the CO 1 transcript. Single stranded DNA probes derived from CO 1, Urf U and Urf N hybridized to this RNA species. Arrow B points to the smallest Urf 1 transcript observed.

script. It turned out that single stranded DNA probes overlapping the 5' and 3' boundaries of the CO 1 reading frame were completely protected by mtRNA against S1 digestion. DNA probes originating from adjacent regions outside of the CO 1 reading frame all hybridized to the same single RNA species of about 6 kb, as do fragments derived from the coding region of the mature subunit (Fig.3). Both results are inconsistent with the idea that the regions coding for the prepiece and the mature polypeptide of subunit 1 are linked together by splicing the transcript prior to translation.

MAPPING OF THE CO 1 TRANSCRIPT

The identified CO 1 transcript could code for about four genes of the size of subunit 1. Although no RNA splicing signal could be detected, the region coding for the prepiece of the precursor polypeptide should be located within the DNA segment which is covered by the CO 1 transcript. Therefore, the extension of the transcript was determined. The 3' end has been located 2.3 kb downstream of the terminator triplet of CO 1, more precisely between a GC-rich palindrome and a tRNA sequence (Fig.4).

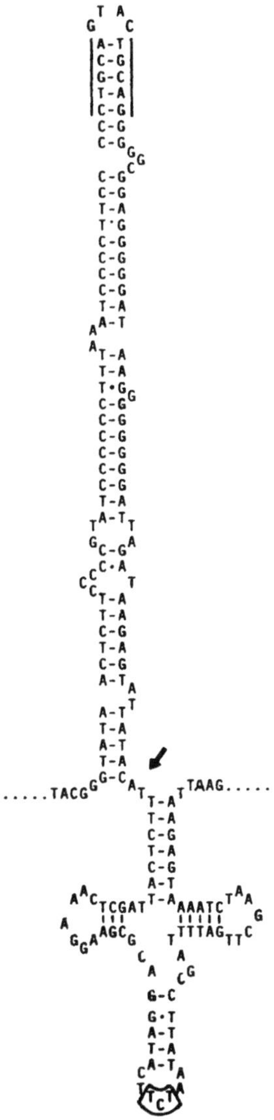

Considering the length of about 6 kb, the transcript should start near the 3' end of the cob gene. In order to determine the prepiece coding region, the DNA segment, extending from a position 1.5 kbp upstream of the 5'end of CO 1 to a position 2.3 kbp downstream of its 3'end has been sequenced. The results are briefly reviewed.

THE CO 1 TRANSCRIPT INCLUDES THREE MAJOR READING FRAMES

Preceeding the CO 1 reading frame, a further reading frame was found that ends with a UAA codon at nucleotide position -150 (see Figs.2 and 5). This DNA region, designated here as Urf U, contains striking dyad symmetries which could build up hairpin structures of the RNA by pairing of AU-rich as well as GC-rich sequences. In addition to the AUG, found at nucleotide position -450, we have recently detected a second potential initiator codon some 100 bp upstream in the same reading frame. However, it is not yet clear, whether part of this reading frame or one of

Fig.4. Putative secondary structures of the mitochondrial tRNA specifying arginine and the preceeding GC-rich palindromic sequence encompassing a double Pst I site (marked by vertical lines). The arrow points to the 3'end processing site of the CO 1 transcript, determined by S1 nuclease mapping. For this purpose, single stranded DNA fragments extending from positions 4.48 to 5.60 and from positions 5.60 to 4.70 (see nucleotide scale in Fig.5) were used.

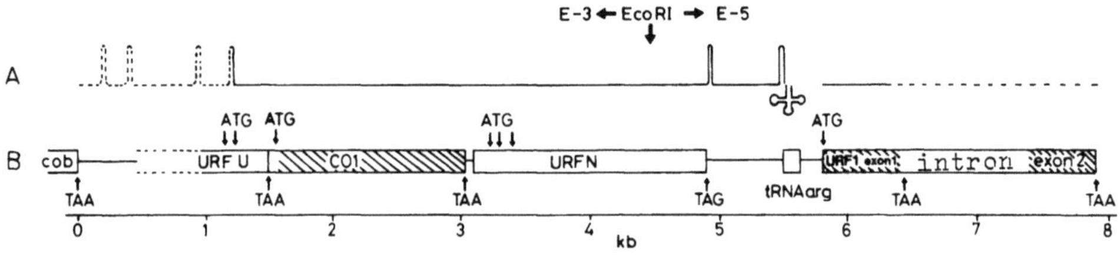

Fig.5. Map of genes, unidentified reading frames and transcripts of the analysed mtDNA of N. crassa. (A) Transcripts: The locations of GC-rich palindromic sequences are marked by loops. Dashed parts of the sequence have not yet been mapped. (B) DNA: Bars represent reading frames; possible initiator and terminator codons are indicated by arrows. The exon regions of Urf 1 and the portion of the CO 1 reading frame that codes for mature subunit 1 are hatched.

various other minor reading frames, found more upstream in the interval between CO 1 and cob do code for the prepiece of subunit 1. This question must be answered by further results concerning the protein sequence of the N-terminal portion of the precursor polypeptide.

Another reading frame was identified in the CO 1 transcript: Urf N starts 17 bp downstream of CO 1 and has a length of about 1.7 kbp. No homology to genes, Urfs or introns of other organisms has been detected so far. The first AUG codon that may serve as initiator of a polypeptide occurs at nucleotide position 138. Codons in the first third position of the reading frame specify predominantly apolar amino acids. The other part of Urf N shows an extremely high percentage of codons specifying charged residues.

This finding is illustrated in Fig.6, were a hydrophilicity profile of the deduced amino acid sequence is plotted. Concatemeric repetitions of oligonucleotides and the corresponding amino acid sequences are the most striking feature of Urf N. For example, the sequence pro-pro-lys-pro-his, present in the C-terminal part of CO 1 is repeated in Urf N 40 bp downstream of its 5'end. A unit of 15 triplets is repeated three times in the middle region of Urf N. In addition, palindromic sequences

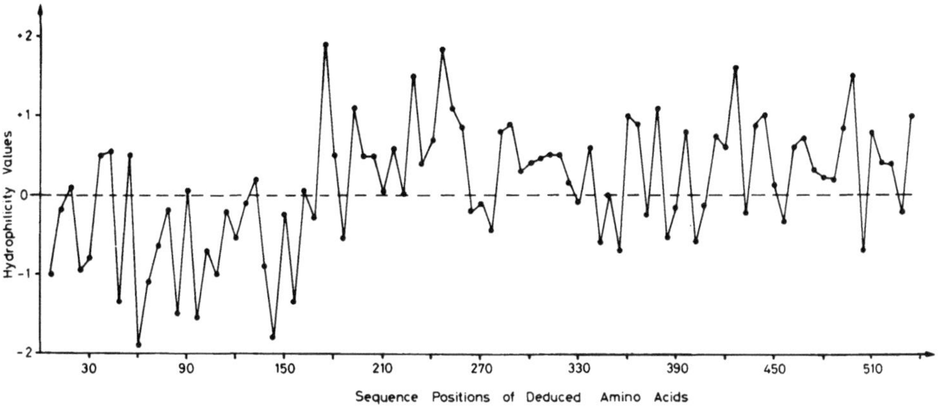

Fig.6. Polarity profile of the deduced amino acid sequence of Urf N. The hydrophilicity values have been calculated according to Hopp and Woods (16). Values greater than zero indicate "polar regions" of the polypeptide.

Table :

		URF N	CO 1			URF N	CO 1			URF N	CO 1			URF N	CO 1
Phe	UUU	3.7	4.3	Ser	UCU	2.3	2.5	Tyr	UAU	1.8	3.1	Cys	UGU	0.5	o.4
	UUC	0.5	4.3		UCC	1.o	o.2		UAC	o.3	1.3		UGC	-	-
Leu	UUA	4.0	9.5		UCA	2.3	3.1	Ter	UAA	-	0.2	Trp	UGA	1.4	2.7
	UUG	1.0	0.7		UCG	-	0.2		UAG	0.2	-		UGG	0.2	-
	CUU	1.6	1.3	Pro	CCU	2.3	2.9	His	CAU	1.0	1.8	Arg	CGU	1.3	0.2
	CUC	4.8	0.2		CCC	-	0.2		CAG	0.3	0.7		CGC	0.2	-
	CUA	1.3	1.3		CCA	1.8	2.0	Gln	CAA	2.7	2.2		CGA	0.3	-
	CUG	0.2	0.2		CCG	0.2	-		CAG	1.4	0.2		CCG	-	-
Ile	AUU	2.1	4.0	Thr	ACU	2.7	1.3	Asn	AAU	3.7	3.2	Ser	AGU	1.3	3.6
	AUC	0.6	1.1		ACC	0.3	-		AAC	1.3	-		AGC	0.3	0.5
	AUA	1.6	3.8		ACA	1.9	3.4	Lys	AAA	6.3	1.4	Arg	AGA	1.4	2.3
Met	AUG	1.9	3.1		ACG	0.6	0.2		AAG	2.1	-		AGG	0.3	0.2
Val	GUU	0.6	2.5	Ala	GCU	2.9	2.9	Asp	GAU	3.2	1.8	Gly	GGU	2.7	4.1
	GUC	-	0.2		GCC	0.3	0.5		GAC	0.8	0.5		GGC	0.2	-
	GUA	1.4	2.9		GCA	1.9	2.7	Glu	GAA	4.0	1.3		GGA	1.9	3.2
	GUG	0.5	0.9		GCG	0.6	0.9		GAG	1.6	0.4		GGG	-	1.8

Frequencies of codons in Urf N compared with those used in CO 1, which codes for a typical membrane protein. Values are given in percent of the total number of codons found in the respective reading frame.

allowing various clover leaf structures in the transcript were found.
A large number of codons specifying basic amino acids were observed in Urf N. Unusual codons, such as the CGN family as well as the UAG terminator are present (Table). In these features, Urf N resembles introns of other fungi.

DOES THE CO 1 TRANSCRIPT REPRESENT A POLYCISTRONIC MESSENGER?

Three major reading frames are comprised in the CO 1 transcript, Urf U, CO 1, and Urf N. This raises the question, of whether the large CO 1 transcript indeed represents the mature messenger. If it is the mature transcript, distinction between coding and non-coding regions or between different genes, may occur at the level of translation by a mechanism similar to that proposed for the reading of polycistronic messages in bacteria (17). On the other hand, the large transcript could represent a stable precursor. In this case, the mature messenger may not have been detected if the concentration was too low for Northern hybridization and S1 nuclease experiments. However, experimental evidence, favouring either hypothesis has not been obtained yet.*

DO GC-RICH PALINDROMIC SEQUENCES ACT AS PROCESSING SIGNALS?

Six GC-rich palindromic sequences, encompassing Pst I sites, are present in the CO 1 transcript: three upstream of Urf U, one within Urf U, and two downstream of Urf N (Fig.5). Originally, GC-rich inverted repeat sequences were described to separate clustered tRNA genes and to be present near the intron/exon boundaries of the large rRNA gene in *N. crassa* (4). Palindromes found in the DNA segment covered by the CO 1 transcript are very similar to those observed in the rRNA region (see Fig.4). It has been discussed that these hairpin

---

* *A less serious hypothesis is illustrated in a cartoon at the end of the paper.*

structures may serve (among other possibilities) as signals in the processing of transcript precursors of rRNA and tRNA genes (4), and also of cox 2 and cob (Macino et al., these proceedings).

We have indeed found that 3'end processing of the CO 1 transcript takes place directly behind a Pst palindrome (see arrow in Fig.4). Two facts, however, do not support this model. First, four of the six GC-rich sequences are not located near to the 5' or 3'ends of the transcripts. Second, Urf U, CO 1, and Urf N are sequentially arranged without being separated by Pst palindromes or similar sequences. If processing of the CO 1 transcript might actually occur, a different type of signal should then be used. In this context, it is more likely that the tRNA gene, encoded immediately downstream of the 3'end of the CO 1 transcript, acts as the processing signal, rather than the adjacent GC-rich palindrome. In the mtDNA of mammalia (e.g. 18) and of *S. pombe* (14), for example, structural genes are usually flanked by tRNA genes, regulating processing of the primary transcription product into the mature messages.

THE CO 1 REGION IS FOLLOWED BY A tRNA GENE AND URF 1

As mentioned above, a tRNA is encoded directly adjacent to the 3'end of the DNA segment, corresponding to the CO 1 transcript. This tRNA is the only one known so far to specify arginine. The sequence is included in Fig.4.

200 bp downstream of the tRNA, a sequence homologous to Urf 1 found in mammalia (18) and *A. nidulans* (19) was detected. The *N. crassa* sequence is four triplets longer at the 5'end than that of *A. nidulans*, otherwise the homology on the basis of the deduced amino acids is very high. Nevertheless, Urf 1 of *N. crassa* contains an intron of about 1.2 kbp, inserted in a highly conserved domaine at amino acid position 212. The reading frame of the intron is not in phase with the preceeding exon. A base pairing of the exon/intron boundaries according

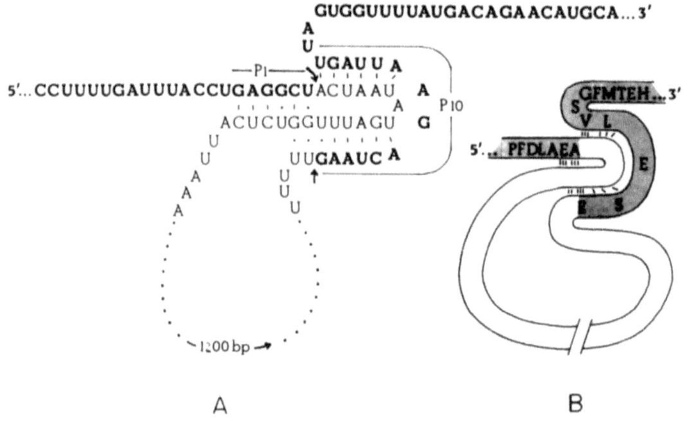

Fig.7. Alignment of exon/intron boundaries of Urf 1 for RNA splicing according to the model of Davies et al. (20). (A): Nucleotide pairing. The bold letters indicate exon sequences, light letters the intron sequence. Arrows point to the proposed splice sites in the base-paired regions, labeled P1 and P10. (B): Part of the exon RNA sequence shown in (A) is displayed as an amino acid sequence in order to draw attention to the junction of the two exon regions (shaded).

to the model proposed by Davies et al.(20) is shown in Fig.7 featuring the P1 and P10 sequences, characteristic for introns of type I. Following this theory, an internal guide RNA sequence serves for the precise alignment of the intron ends and thus permits exact intron excision and exon ligation. Introns of type I are reported to occur also in the cob and oli 2 gene of *N. crassa* (Macino et al., these proceedings).

Northern experiments with Urf 1 probes revealed weak hybridization to several RNA species. The smallest transcript observed has a size of about 1.8 kb (see arrow in Fig.3). It cannot be excluded that even smaller transcripts may be detected after prolonged exposure of the autoradiograms. In contrast to the transcript of CO 1, that of Urf 1 starts within a very short region (max. three bases) in front of the first AUG codon (Fig.8). In this respect the Urf 1 transcript resembles those found in mammalia (21) and *S. pombe* (14).

```
                    Urf 1
               ↓↓↓↓ |——————————————————————————————————>
5'...TCTAATTTTTTTTTTTT ATG TTT TAT TCC CTA ACA ATA ATC TCG ...3'
                       met phe tyr ser leu thr ile ile ser
```

Fig.8. 5'end region of Urf 1 in N. crassa. the arrows indicate the location of the observed S1 signals. Single stranded DNA extending from positions 5.7 to 6.3 (see map in Fig.5) was used in this experiment.

It is noteworthy, that no GC-rich palindromic sequence separates the tRNA gene from Urf 1. (As mentioned above, the analog situation exists in the CO 1 transcript concerning the arrangement of Urf U, CO 1 and Urf N). Furthermore, the 5'end processing site of Urf 1 is located within a sequence which does not allow the formation of " conspicuous " secondary structures in the transcript.

Acknowledgements: We thank Drs. G.Macino and H.Bertrand for kindly providing clones, Drs. R.Waring and G.Macino for making available sequence data prior to publication, Dr. B.F.Lang for stimulating discussions and C.Scriven for reading the manuscript. We are grateful to Ms. Rothe and Ms. Zauner for excellent technical assistance. This work was supported by the Deutsche Forschungsgemeinschaft (We 706/5-1).

References

1. Hahn,U., Lazarus,C.M., Lünsdorf,H. and Küntzel,H.: Cell 17, 191-200 (1979).
2. Heckman,J.E. and RajBhandary,U.L.: Cell 17, 583-595 (1979).
3. Browning,K.S. and RajBhandary,U.L.: J.Biol.Chem. 257,5253-5256 (1982).
4. Yin,S.,Heckman,J. and RajBhandary,U.L.: Cell 26, 325-332 (1981).
5. Burger,G., Scriven,C., Machleidt,W. and Werner,S.: The EMBO J. 1, 1385-1391 (1982).
6. Macino,G. and Morelli,G.: J.Biol.Chem., in press, (1983).
7. Citterich,M.H., Morelli,G. and Macino,G.: J.Biol.Chem., in press, (1983).
8. Werner,S. in: Biological Chemistry of Organelle Formation (T.Bücher, W.Sebald, and H.Weiss, eds.), Springer Verlag, Berlin, pp. 43-58 (1980).

9. Werner,S. and Bertrand,H.: Eur.J.Biochem. 99, 463-470 (1979).

10. Werner,S., Machleidt,W., Bertrand,H.,and Wild,G. in: The Organisation and Expression of the Mitochondrial Genome (A.M.Kroon and C.Saccone, eds.) North-Holland Biomedical Press, Amsterdam, pp. 399-411 (1980).

11. De Jonge,J.C. and De Vries,H.: Curr.Gen. 7, 21-28 (1983).

12. Bonitz,S.G., Coruzzi,G., Thalenfeld,B.E., Tzagoloff,A. and Macino,G.: J.Biol.Chem. 255, 11927-11941 (1980).

13. Waring,R.B. et al., in preparation.

14. Lang,B.F., Ahne,F., Distler,S., Trinkl,H., Kaudewitz,F. and Wolf,K.: These proceedings

15. Collins,R.A. and Lambowitz,A.M.: Plasmid 9, 53-70 (1983).

16. Hopp,T.P. and Woods,K.R.: Proc.Natl.Acad.Sci.USA 78, 3824-3828 (1981).

17. Yanofsky,C. and Ito,J.: J.Mol.Biol. 21, 313-334 (1965).

18. Anderson,S. et al.: Nature 290, 457-465 (1981).

19. Brown,T.A., Davies,R.W., Ray,J.A., Waring,R.B. and Scazzocchio,C.: The EMBO J. 2, 427-435 (1982).

20. Davies,R.W., Waring,R.B., Ray,J.A., Brown,T.A. and Scazzocchio,C.: Nature 300, 719-724 (1982).

21. Montoya,J., Ojala,D. and Attardi,G.: Nature 290, 465-470 (1981).

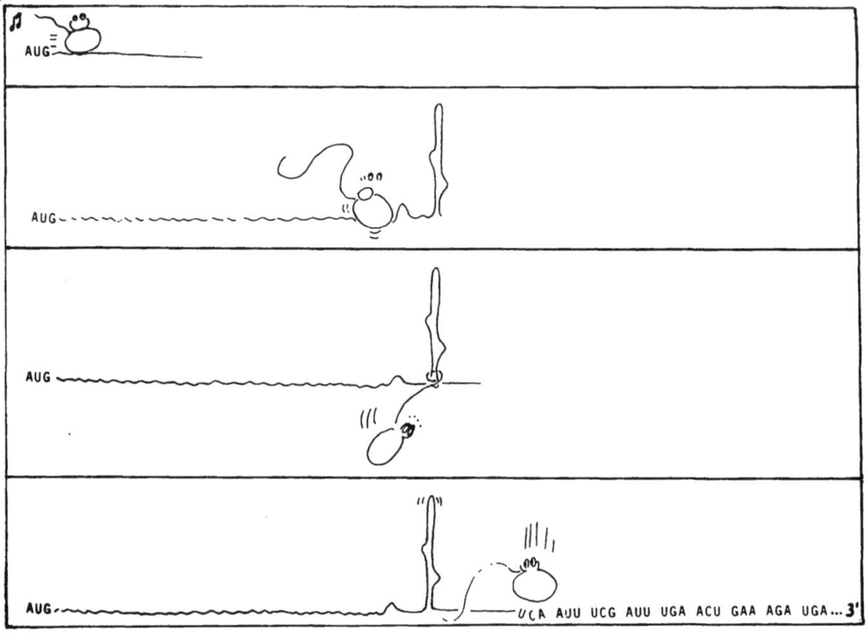

THE EXPRESSION OF THE MITOCHONDRIAL GENES FOR SUBUNIT 1 OF CYTOCHROME C OXIDASE AND FOR AN ATPase PROTEOLIPID IN NEUROSPORA CRASSA: NUCLEOTIDE SEQUENCES AND TRANSCRIPT ANALYSIS

Hans de Vries, Jenny C. de Jonge, Annika Arnberg[a], Ad A.C.M. Peijnenburg and Etienne Agsteribbe
Laboratory of Physiological Chemistry and (a) Biochemical Laboratory, State University, Groningen, The Netherlands.

Introduction

Our group has recently concentrated on the characterization of protein genes located on Neurospora mtDNA. One is the Mitochondrial ATPase-proteolipid-Like (MAL) gene, a complete open reading frame which potentially codes for a subunit 9-like protein (1). Such a protein is, however, not detectable among the mitochondrial translation products. The indetectability of the translation product, and the fact that the actual Neurospora ATPase proteolipid is encoded in the nucleus (2) and is synthesized on cytoplasmic ribosomes (3), have offered a formidable puzzle regarding the function of this gene. Possibilities are: the gene is silent, non-functional; the gene is expressed only under certain conditions, either in certain stages of the cell cycle or when the cell, or the nuclear gene product, is damaged; the gene product has escaped detection so far. The MAL gene is preceded by the ATPase 6 gene and followed by the CO II gene (4,5). Results obtained by hybridizing probes from this region to Northern blots of total mtRNA showed that the MAL and CO II genes are cotranscribed to a very long transcript (1). For CO II a smaller unique transcript of 1200 nucl probably is the mRNA, for MAL no such small unique transcript could be found. We want to report here on our further investigations of the expression of the MAL gene

Mitochondria 1983

and the surrounding genes.
The other gene under investigation is the gene for subunit 1 of cytochrome oxidase (CO I) (6). This gene is located on the Hind III fragment H7c, about opposite on the circular map of Neurospora mtDNA. The gene on the type II (7) mtDNA and part of the upstream region has recently been sequenced independently by Burger et al. (8) and by our group (6).
In the laboratory strains used, the CO I gene is uninterrupted by introns and is at least 1665 bp long. The C-terminal part of the gene is completely unlike that of any other known CO I gene in that it has an extension of about 70 basepairs. It is not known whether or not the additional 23-24 amino acid residues are part of the mature protein. At the N-terminal part of the gene a prepiece sequence has to be present, since the mature protein starts with a serine (9). Upstream from the TCA coding for this serine, an ATG was found at -6. This could mean that the N-terminal prepiece is only a dipeptide (met-ser). However, the precursor to CO I has an apparent molecular weight of 45000, about 4000 higher than that of the mature protein. Hence, it may be that the ATG at -6 is not the initiator codon for the gene, or that additional C-terminal processing of the precursor occurs, or that differences in apparent molecular weight do not reflect real molecular weight differences. To further characterize the upstream region we have sequenced about 1100 nucl in front of the CO I gene. Of these, 474 were already reported in ref. 6. Further upstream no other in-frame ATG codons could be found between a TAA at -150 and the ATG at -6. The TAA at -150 terminates a potential protein gene of at least 100 amino acids (6,8). Results obtained by Northern blotting revealed an RNA molecule of about 6000 nucl as the major transcript. It appears likely that this long molecule is the messenger for CO I. This messenger may be polycistronic, since both upstream (6,8) and downstream (mentioned in ref. 8 open reading frames are present.
In this paper we shall describe the further characterization of the nucleotide sequences upstream of the CO I gene, espe-

cially of the region where transcription initiates. Moreover, the long transcript has been further characterized by Northern blotting and S_1 nuclease mapping.

Results and Discussion

Nucleotide sequence of the region upstream from the CO I gene

Fig. 1 shows the nucleotide sequence of 1100 bp upstream from the CO I gene. Several striking features are indicated in the

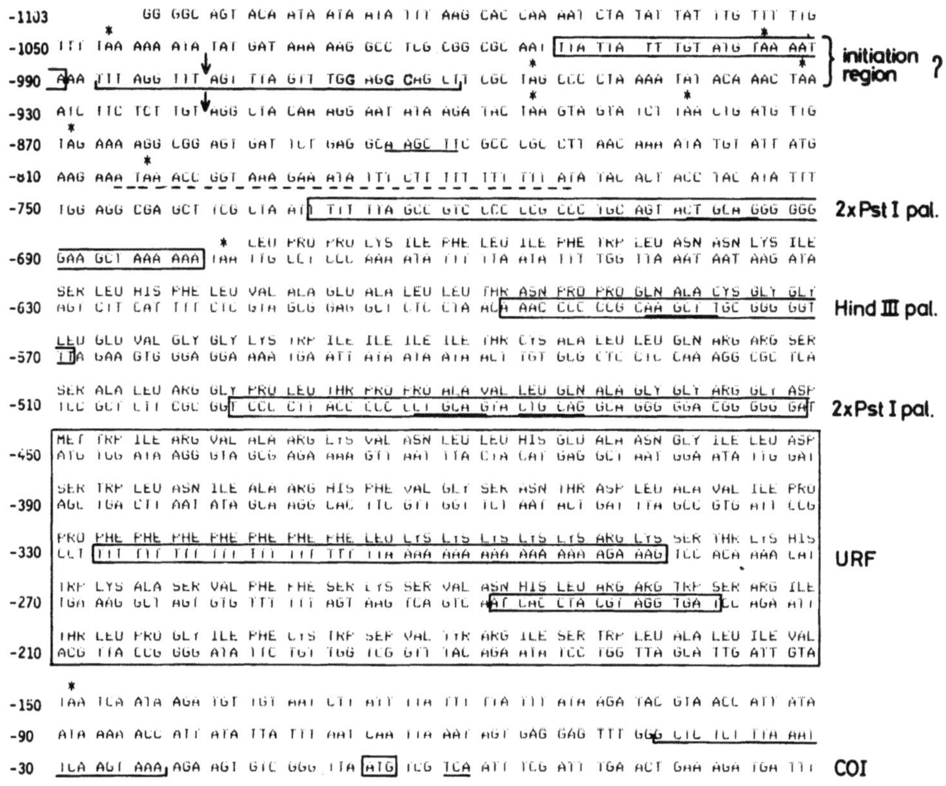

Fig. 1. Nucleotide sequence of the region upstream from the CO I gene. Palindromic structures are boxed. Two regions showing inverted complementarity are underlined. Arrows indicate the estimated start point(s) of transcription. Note: The sequence upstream from the initiation region is preliminary.

figure and will be discussed here. (i) The URF: the ATG codon already found earlier at -450 (6,8) is not preceded by any other in-frame ATG codon until a TAA at -712 is encountered. Hence, if this reading frame is really a protein gene, it maximally codes for a 100-aminoacid protein, particularly because for *Neurospora* mitochondria there is no evidence at all that other codons than AUG can be used for initiation. Moreover, immediately preceding the ATG at -450 a double-Pst I palindrome is present. These palindromic structures, first reported by RajBhandary's group (10), are particular to and abundant in *Neurospora* mtDNA. Since these structures have always been found surrounding, and never within coding sequences of genes, a processing function has been assigned to these palindromes. Within the URF two unusual palindromic structures are present, one being a long poly T stretch, followed by a poly A stretch. If 2 G's are also included, this structures provide a hairpin of 44 nucl. One could argue that such a structure makes it improbable that this URF codes for anything sensible. In the past year, however, it has been shown by Hudspeth et al. (11) that sequences as AT-rich as 90% may still code for an important protein, in their case yeast *var*-1. Therefore, our working hypothesis is that this URF is indeed coding for an extremely basic protein (18 basic, against 3 acidic residues). (ii) Palindromic structures: a further peculiarity of this region is the presence of several palindromic regions, two of which are double Pst-I palindromes. The third is a Hind III-palindrome as described by Burke and RajBhandary (12). Furthermore, two AT-rich palindromic structures of doubtful significance are present. Finally, at -981 a 22 bp region is present that is palindromic to the region, immediately preceding the CO I gene. The homology is 80%, or 90% if G-T pairing is included. It is not known what the significance of this complementarity is. One might speculate that in the messenger RNA both sequences can pair, thus leading the ribosomes directly towards the CO I gene. No such structural requirements have been found earlier, however.

Characterization of the CO I transcript

The long transcript of the CO I gene has been reported to be 6800 nucl (8) or about 6000 nucl (6). Furthermore, preliminary characterization in our laboratory and in that of Burger and colleagues by Northern hybridization suggested that the transcript is not spliced. As the CO I gene is only some 1700 nucl, the messenger is 3 to 4 times longer than seems necessary.

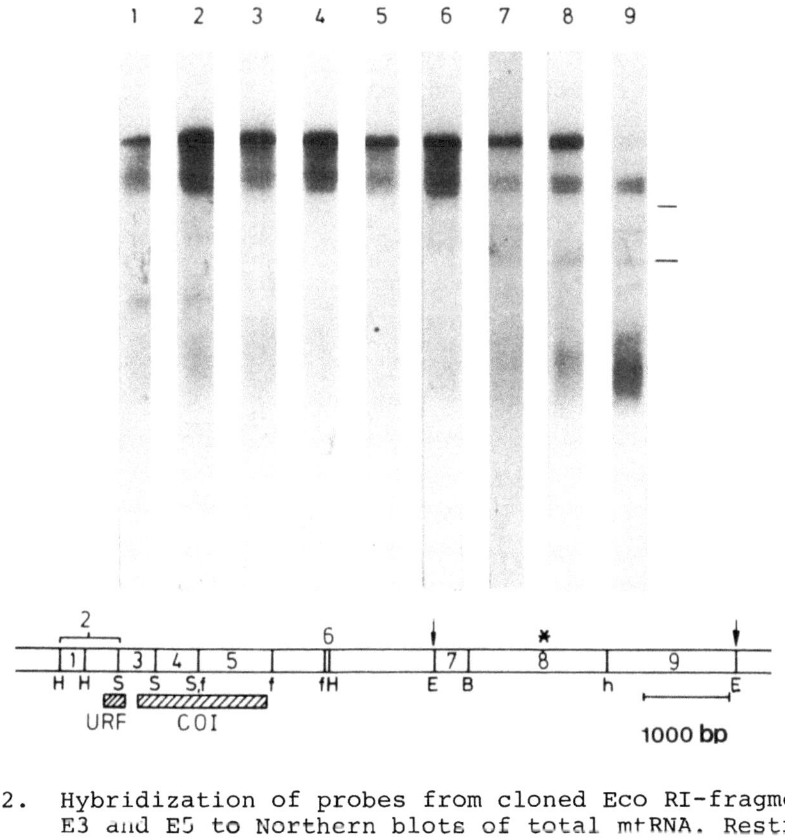

Fig. 2. Hybridization of probes from cloned Eco RI-fragments E3 and E5 to Northern blots of total mtRNA. Restriction sites are: E = Eco RI, H = Hind III, f = Hinf I, S = San 3AI, B = Bam HI, h = Hinc II. The asterisk indicates the 3' end of the CO I transcript as deduced from Fig. 3.

Therefore, we have characterized this molecule in more detail. Fig. 2 shows the hybridization of selected probes from Eco RI fragments E3 and E5 to Northern blots of total mitochondrial DNA. It is clear that the fragments 1 through 8, covering over 6000 bp, all show strong hybridization to the long transcript. Because fragment 9 (obtained from cloned E5 by cleavage with Hinc II) does not hybridize to the CO I transcript, the 3'-end of the transcript must lie on fragment 8, i.e. between the Bam HI and the Hinc II site of E5. The 5'-end then is either within probe 1 (the small Hind III-fragment preceding H7c, termed "H20", or even further upstream, in H2. To localize the ends more precisely, we have studied R-loops of the linearized plasmid pBE3 (consisting of E3 cloned into the Eco RI site of pBR322) with total mRNA. Fig. 3 is an electronmicrograph of such a hybrid. The length of the hybridized stretch is 4400 nucl, whereas the unpaired single-stranded RNA hanging out is about 1300 nucl. The R-loop itself measures 4650 nucl. Thus, the total length of the transcript is 5700 to 6000 nucl. This result indicates that the 3'-end of the transcript is

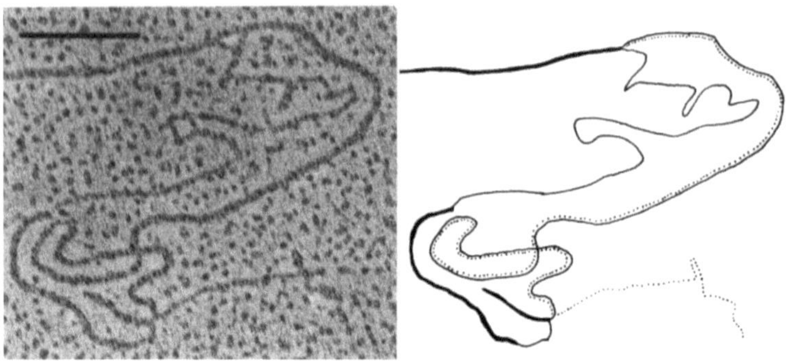

Fig. 3. Electronmicrograph and schematic drawing of R-loop of plasmid pBE3, linearized with Bam HI, and total mtRNA. The bar indicates a length of 0.2 μm. (The R-loop contains a double-stranded part, probably due to hybridization to fragmented DNA). The asymmetric location of the Bam HI site in the pBR322-part allowed the unambiguous assignment of the transcript to the CO I region of E3.

indeed halfway the Bam HI and Hinc II sites of E5. This is indicated in Fig. 2 by an asterisk. The 5' terminus of the transcript can be localized within 400 bp from the H2-H20 Hind III site (Fig. 4). Additional characterization involved S_1 nuclease mapping of hybrids of total mtRNA to end-labeled H20 and H2 (not shown). Using the coding strand of fragment H20 we found no other band than full-length, which showed that it was entirely covered by RNA. Mapping of fragment H2 (5' labeled and cut with Xho I to yield a 800 bp fragment) gives a rather complicated pattern of bands. By far the most intense bands, though, are found at -135 and -75 nucl from the labeled end of H2. We therefore tentatively conclude that transcription starts around -970 in Fig. 1 with a possible second initiation point around -910. It must be emphasized that, due to the use of total mtRNA in the hybridization and S_1 mapping experiments, there is no rigorous proof that the 5' terminus found is that of the primary transcript. The electronmicroscopical data, and the fact that no larger transcript

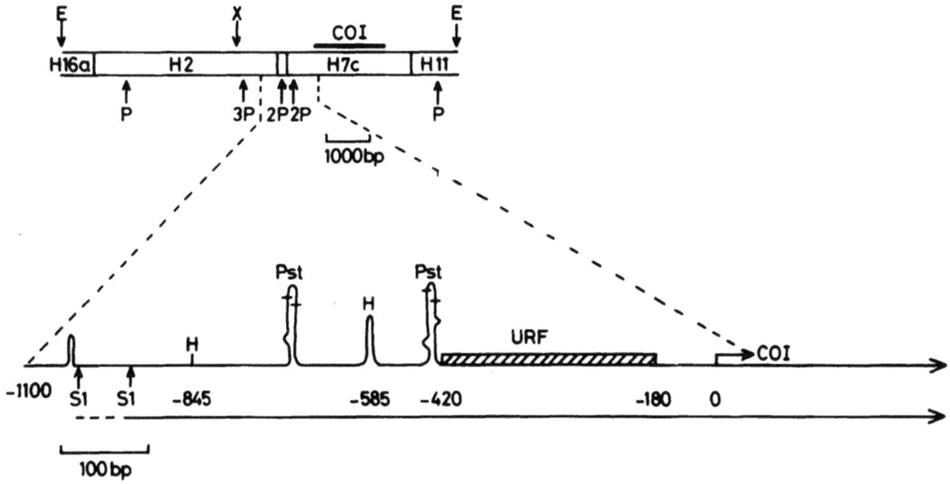

Fig. 4. Schematic map of the CO I region, and of the upstream part (enlarged). Symbols: E = Eco RI, P = Pst I (2P, 3P = double, resp. triple Pst I sites), H = Hind III, X = Xho I; S_1 = start of transcript as determined by S_1 mapping).

was picked up in Northern hybridizations, give us confidence that we are dealing with the primary transcript indeed. The sequence around -970 is TTTAGTTTA, whereas that around -910 is TGTAGGCTA. These sequences are rather homologous to each other and may both serve as a transcription origin. Immediately preceding the initiation site at -970 one more sequence of interest is present (Fig. 1): at -1008 an 18-bp sequence (boxed) starts that has 70% homology with the yeast mitochondrial promotor sequence TTATTAT(A)TATATAAGTA reported by Osinga and Tabak (13). At present, no possible mitochondrial promoters have been found for other lower eukaryotes than yeast. Therefore, calling the sequence at -1008 a promoter would be premature. Further sequencing and characterization of other mt transcripts will be necessary to find out whether or not similar sequences are present around initiation start points of Neurospora mt genes.

Expression of the genes encoding an ATPase proteolipid and subunit 2 of cytochrome aa_3

The genes encoding ATPase 6, a mitochondrial ATPase-proteolipid-like protein (MAL) and CO II are located in relatively close proximity on the mtDNA (Fig. 5). In our further search for the expression of this region, especially of the MAL gene, we have followed three lines.

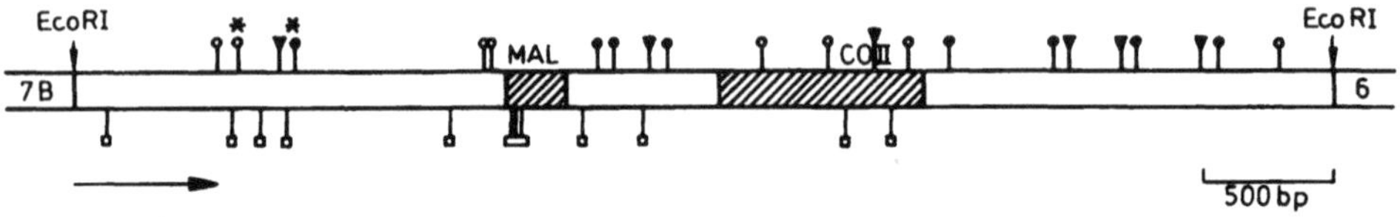

Fig. 5. Gene and restriction map of Eco RI fragment 4 of Neurospora mitochondrial DNA. MAL = mitochondrial ATPase proteolipid-like gene, CO II = gene encoding subunit 2 of cytochrome aa_3. The gene encoding subunit 6 of the mitochondrial ATPase is located around the Eco RI site between fragments 4 and 7B. Restriction sites are indicated as follows: O = Hinf I, ● = Hae III, ⊔ = Alu I and ∇ = Pst I. The arrow indicates the direction of transcription. The base sequence of the region between the two asterisks is shown in Fig. 7.

Table I. Neurospora strains tested for the presence of the MAL gene. Exotic strains were obtained from the Fungal Genetic Stock Center, Arcata, California.

Neurospora	crassa	ANT-1 (inl-89601)	
"	"	430	North Africa
"	"	2499	India
"	"	1776	interspec. hybr.
"	sitophila	2216	
"	tetrasperma	1270	
"	intermedia	2316	

(i) We have excluded the possibility that the presence of the MAL gene is a non-representative feature of our Neurospora crassa strain (inl-89601). Seven other Neurospora strains (see Table I) were tested for the presence of the MAL gene. Hybridization with an internal fragment of the MAL gene from Neurospora inl-89601 showed the presence of homologous sequences on the mitochondrial DNAs of all these strains.
(ii) Since it could be possible that the MAL gene is only active under certain conditions we have tried to induce its expression by adding antimycin to the growing mycelium and by heat-shock treatment of the organism. The rationale for the addition of antimycin is the effect it has in Neurospora on respiration and gene expression. Antimycin blocks the electron flow along the cytochromes, and soon after this inhibition a nuclear-coded alternate oxidase is produced (14). However, no concomitant induction of the MAL gene was observed by us. Heat-shock is known to induce both switching-off of genes and expression of hitherto silent genes (15) e.g. in yeast. Also this treatment, however, did not seduce Neurospora to start or to increase the expression of the MAL gene.
(iii) Transcript analysis was repeated under conditions optimized for the detection of minor transcripts. These conditions were: 1. The use of cloned DNA fragments in hybridization experiments to avoid cross-contamination of hybridization probes 2. Prolonged electrophoresis of total mitochondrial RNA to obtain better resolution of transcripts. The results of the

hybridization of ATPase 6, MAL and CO II probes against Northern blots of mitochondrial RNA are shown in Fig. 6. The tentative messengers for ATPase 6 and CO II have lengths of ca 2000 and 1200 nucleotides resp. The MAL gene has three unique transcripts distinct from the transcripts of the CO II gene. These are present in minor amounts. Their lengths are ca 1800, 1600 and 1400 nucleotides resp. We do not know yet whether or not

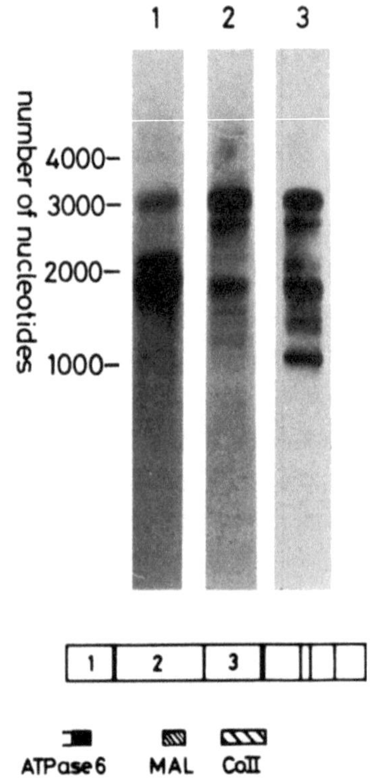

Fig. 6. Transcripts of the ATPase 6, MAL and CO II genes. Probes used for hybridization were Pst I subclones (1, 2 and 3) of Eco RI fragment 4 of _Neurospora_ mitochondrial DNA.

the smallest transcript acts as a messenger RNA. Nor can it be excluded that one of the larger more abundant transcripts, that also contain the information of the CO II gene, is a polycistronic messenger. In case the 1400 nucl transcript is the messenger RNA, then it seems likely that the translation product of the MAL gene will also occur in very small quantities inside the cell. At present we are pursuing the immunological detection of a translation product of the MAL gene. Antibodies will be obtained by using as an antigen a synthetic antigenic site determined as from the predicted amino acid sequence (16) and secondly the translation product itself, synthesized, however, in E. coli by means of an protein expression vector. As mentioned before a peculiar feature of the Neurospora mitochondrial genome is that genes are in general preceded and followed by tandemly repeated Pst I sites (10). These Pst I sites are often part of longer palindromes. It has been suggested that these palindromic sequences fulfil a role in RNA processing (10). S_1-nuclease mapping revealed that a transcript of the MAL gene region maps with

GANTCCGGAG TTCCCACCAC CATGGGTGGC GTGCCTCGAG AAGGGGGCGT GCCTCTGGAT
GCTTGCGCTT TTTTTTTTTT TTTGCCGTCC CCCCGCCCTG CAGTACTGCA GGGGGGGGGA
AGCTAAAAAA AATTAAAGAA AGGCC

A

B

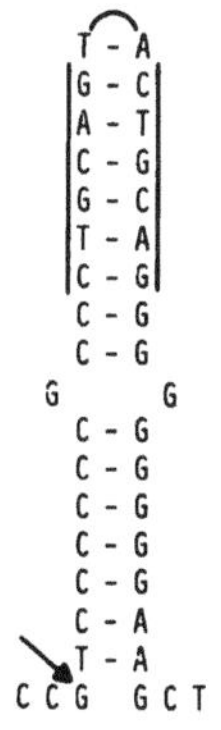

Fig. 7A. Base sequence of the region upstream of the MAL gene with tandemly repeated Pst I sites. The exact position of this sequence is indicated in Fig. 5. The Pst I restriction sites are indicated by solid lines above their recognition sites.

Fig. 7B. Palindromic sequence upstream of the MAL gene encompassing the tandemly repeated Pst I sites. The arrow indicates the 5' end of a transcript of the MAL region as obtained by S_1-nuclease mapping.

its 5' end at the base of the Pst I palindrome upstream of the MAL gene (Fig. 7). It may be inferred that the mapping is an S_1 nuclease artefact caused by instability of the DNA molecule at the base of the stem. This objection is ruled out, however, by our result, mentioned above that no S_1 bands were found at the double Pst I sites upstream from the CO I gene. Therefore, our conclusion is that indeed transcript processing occurs at the Pst I palindrome upstream from the MAL gene. Further characterization of sequence and transcription of this region is in progress.

Concluding remarks

The data presented above show that simple, intron-less, mitochondrial genes do not _a priori_ have predictable, simple transcripts. Two examples are given here: a) the CO I transcript, which potentially contains the message for at least two additional proteins, and therefore may be a polycistronic mRNA. A further peculiarity of this transcript is the presence of double Pst I palindromes which do not function as processing sites. b) the 6-MAL-CO II transcripts: the primary transcripts appear at first also, polycistronic. Subsequently, however, there is processing to separate transcripts. In this case a double Pst I-palindrome functions as a processing site. One may wonder what determines whether or not a particular Pst I-palindrome is used as a processing site. The structures themselves give no clue to the answer, hence other factors contributing to conformation of RNA molecules must play a role. Research in the near future in our laboratory will be directed towards further understanding of mitochondrial transcription and processing, e.g. by looking for common sequences in the transcription initiation regions of the protein genes on E4 and E3. Moreover, the pursuit of the MAL gene product is being continued, mainly by immunological methods.

Acknowledgements

We thank Prof. Dr. A.M. Kroon for his criticism and encouragement, Rinske Kuperus for typing the manuscript, Theo Deddens and Bert Tebbes for graphical and photographical help. This work was supported in part by a grant to A.M. Kroon from the Netherlands Foundation for Chemical Research (SON) with financial aid from the Netherlands Organization for the Advancement of Pure Research (ZWO).

References

1. Van den Boogaart, P., Samallo, J. and Agsteribbe, E.: Nature 298, 187-189 (1982).
2. Sebald, W., Hoppe, J. and Wachter, E. in: Function and Molecular Aspects of Biomembrane Transport (ed. Quagliariello, E.), pp. 63-74 (Elsevier, Amsterdam 1979).
3. Jackl, G. and Sebald, W.: Eur. J. Biochem. 54, 97-106 (1975).
4. Macino, G.: J. Biol. Chem. 255, 10563-10565 (1980).
5. Van den Boogaart, P., van Dijk, S. and Agsteribbe, E.: FEBS Lett. 147, 97-100 (1982).
6. De Jonge, J.C. and de Vries, H.: Curr. Genet. 7, 21-28 (1983).
7. Mannella, C.A., Pittenger, T.H. and Lambowitz, A.M.: J. Bacteriol. 137, 1449-1451 (1979).
8. Burger, G., Scriven, C., Machleidt, W. and Werner, S.: EMBO J. 1, 1385-1391 (1982).
9. Werner, S., Machleidt, W., Bertrand, H., Wild, G. in: The organization and expression of the mitochondrial genome (ed. Kroon, A.M. and Saccone, C.), pp. 399-411 (Elsevier/North-Holland, Amsterdam 1980).
10. Yin, S., Heckman, J., Sarnoff, J. and RajBhandary, U.L. in: The organization and expression of the mitochondrial genome (ed. Kroon, A.M. and Saccone, C.), pp. 307-310 (Elsevier/North-Holland, Amsterdam 1980).
11. Hudspeth, M.E.S., Ainley, W.M., Shumard, D.S., Butow, R.A. and Grossman, L.I.: Cell 30, 617-626 (1982).
12. Burke, J.M. and RajBhandary, U.L.: Cell 31, 509-520 (1982).
13. Osinga, K.A. and Tabak, H.F.: Nucl. Acids Res. 10, 3617-3627 (1982).

14. Lambowitz, A.M. and Zannoni, D. in: Plant Mitochondria (ed. Ducet, G. and Lance, C.), pp. 283-291 (Elsevier/ North-Holland, Amsterdam 1978).

15. Lindquist, S.: Nature 293, 311-314 (1981).

16. Hopp, T.P. and Woods, K.R.: Proc. Natl. Acad. Sci. U.S. 78, 3824-3828 (1981).

EXPRESSION OF SPLIT GENES OF THE NEUROSPORA CRASSA MITOCHONDRIAL GENOME

Manuela Helmer Citterich, Giorgio Morelli, Mary Anne Nelson, and Giuseppe Macino

Department of Genetics and Molecular Biology, University of Rome "La Sapienza" 00185 Rome, Italy

Introduction

Wide diversity exists in the organization of mitochondrial genomes. Mammalian mitochondrial genomes are very compact; spacer regions between genes are very short or nonexistent, and the genes do not contain intervening sequences (1,2). In contrast, many fungal mitochondrial genomes contain large spacer regions and split genes. Introns have been noted within the fungal genes for the large rRNA (3,4,5), cytochrome oxidase subunit 1 (6) and cytochrome *b* (7,8,9). Comparison of the structures of fungal mitochondrial introns have permitted the identification of short regions of shared sequence homology; these introns may fold into similar secondary structures (10,11). Many of these introns contain extended open reading frames that potentially encode proteins with short regions of homology; these proteins may play a role in the splicing of introns (12).
In *S. cerevisiae* mitochondria, large precursor transcripts have been detected for several genes. Mature mRNA species are derived from these transcripts after multiple processing events (13). Intron nucleotide sequences may serve as guides or signals for the splicing of introns (10,11). Also, GC-rich clusters that flank coding sequences have been proposed to play a role in the processing of precursor RNAs (14). The large rRNA (4), cytochrome *b* (15), and ATPase subunit 6 (Morelli and Macino, in preparation) genes of *N. crassa* contain intervening sequences. Expression of the large rRNA gene has been extensively studied (16,17). In this paper we will review the organization of the cytochrome *b* and ATPase subunit 6 genes, and we will present information relating to their expression. RNA sequences that may serve as signals for the splicing of introns and the processing

Mitochondria 1983

of precursor transcripts are described.

Structure of the cob and oli2 Genes of N. crassa

The nucleotide sequences of the apocytochrome b (cob) genes of S.cerevisiae (7,8), A. nidulans (9), and N. crassa (15) have been determined. The cob genes of these closely-related organisms all contain intervening sequences. The "long" form of the S. cerevisiae cob gene contains five introns, three of which contain long open reading frames (ORFs). The second and fourth introns probably encode maturase proteins that are necessary for the splicing of the cob gene (18). The cob gene of A. nidulans contains a unique intron; this intron lies in the same position as the third intron of the S. cerevisiae cob gene, and the two introns share extensive nucleotide homology. The unique intron of the A. nidulans cob gene contains an ORF that may encode a maturase protein (12). The cob gene of N. crassa is interrupted by two intervening sequences of about 1260 bp each (Figure 1). The two introns are located at sites different from those of introns in the

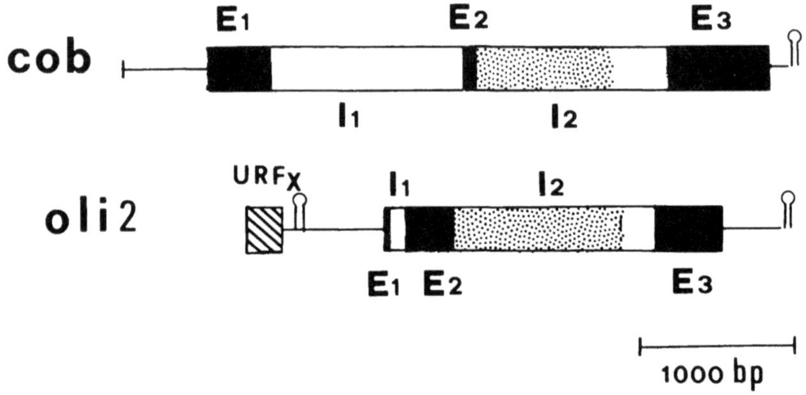

Figure 1. Structure of the cob and oli2 genes. Shaded regions indicate the exon sequences, open regions the intron sequences, and dotted regions the open reading frames within the second introns of both genes. The URFx gene is represented by a slashed box. The location of PstI and PstI-like palindromes is indicated by stem and loop structures.

genes of A. nidulans and S. cerevisiae. The N. crassa cob introns share no regions of extensive homology with the introns of the A. nidulans and S. cerevisiae cob genes. In contrast, the exon sequences of the N. crassa and A. nidulans cob genes are highly homologous. Thus, introns of these closely-related species have very different nucleotide sequence and positioning within the cob genes, while those of the less closely-related A. nidulans and S. cerevisiae show considerable conservation of intron structure and position (12).

The sequences of the ATPase subunit 6 (oli2) genes of S. cerevisiae (19), A. nidulans (5,20) and N. crassa (Morelli and Macino, in preparation) have been determined. The oli2 gene of N. crassa contains two introns (Figure 1); while the corresponding genes of A. nidulans and S. cerevisiae are not interrupted by intervening sequences. The two introns are 93 and 1370 bp long, respectively. The first intron is positioned after only ten bp of coding sequence. Very short exons have also been noted in other fungal genes, including the nuclear-encoded actin gene of S. cerevisiae, which is interrupted by an intron after the third codon (21).

An unknown reading frame (URFx) is located 650 bp upstream from the N. crassa oli2 gene. URFx is highly homologous to URF sequences present in similar positions within the mitochondrial genomes of A. nidulans (5,20) and S. cerevisiae (22,23); these URFs show limited homology to the similarly-located mammalian mitochondrial URFA6L (1,2). The conservation of this URF just upstream from the subunit 6 gene in such diverse organisms suggests a relationship between the URFx and subunit 6 gene products.

S. cerevisiae strains that harbor mutations in URFx contain reduced levels of the ATPase subunit 6, and are defective in the incorporation of this subunit into the ATPase complex (22,23).

Structure of N. crassa Mitochondrial Introns

Many fungal mitochondrial introns share short regions of homology (10 to 12 bp long); these regions (called P, Q, R and S) are always present in the same respective order (10). The P and Q sequences can base pair, as can the R and S sequences. As proposed by Davies et al. (10) these conserved sequences may direct the formation of a core secondary structure that brings the intron ends into close proximity. Other sequences, including ones called E and E' (10) are conserved as to position and ability to base pair, but differ as to sequence. Cis-dominant mutations in the R, S and E' sequences that reduce the strength of base-pairing between R and S or E and E' cause splicing deficiencies (10,24,25). Revertants that restore base-pairing also allow normal splicing (10,24,25). Davies et al. (10) have proposed that an internal guide RNA sequence (IGS), specific to each intron, might serve to precisely align intron ends and thus permit exact intron splicing.

No regions of extensive homology exist among the introns of the cob and oli2 genes of N. crassa. However, the introns of the cob gene and the second intron (I2) of the oli2 gene contain the conserved P, Q, R and S sequences (Table 1). Deviations from the consensus P, Q, R and S sequences in N. crassa introns are compensated by complementary alterations, so that the ability to base pair is retained. The P and Q sequences of the second intron of the cob gene provide a dramatic example of complementary alterations; three out of five bases are altered in P, and three compensatory changes are evident in Q. The proposed secondary structure of this intron is shown in Figure 2.

	P		Q		R		S
Consensus	AUGCUGGAAA		AAUCAGCAGG		UCAGAGACUACA		AAGAUAUAGUCC
cob I1	ACGGUUAAAA	-20-	UAAGACCGUU	-17-	CUGCAGACUGGG	-28-	AAUGUACAGUCG
	* * *		* * *		*		*
cob I2	UUACAAGGAC	-30-	UAUUUGUAGC	-39-	UCAACGACUAAU	-64-	AAGAUAUAGUCU
	* **		** *				
oli2 I2	GUGCUGGGAA	-60-	AAUCAGCAGG	-36-	UCAGAGACUACA	-46-	AAGAUAUAGUCC

Table 1. Conserved nucleotide sequences within the introns of the oli2 and cob genes compared to the P, Q, R and S consensus sequences defined by Davies et al. (10). Sequences are listed in the 5' to 3' direction. Asterisks mark the bases that differ from the consensus sequence and yet maintain the pairing between P and Q or R and S [complementary alterations (see text and Figure 2)]. Complementary sequences between P and Q, as well as between R and S, are underlined. Figures indicate the number of bases between the conserved sequences.

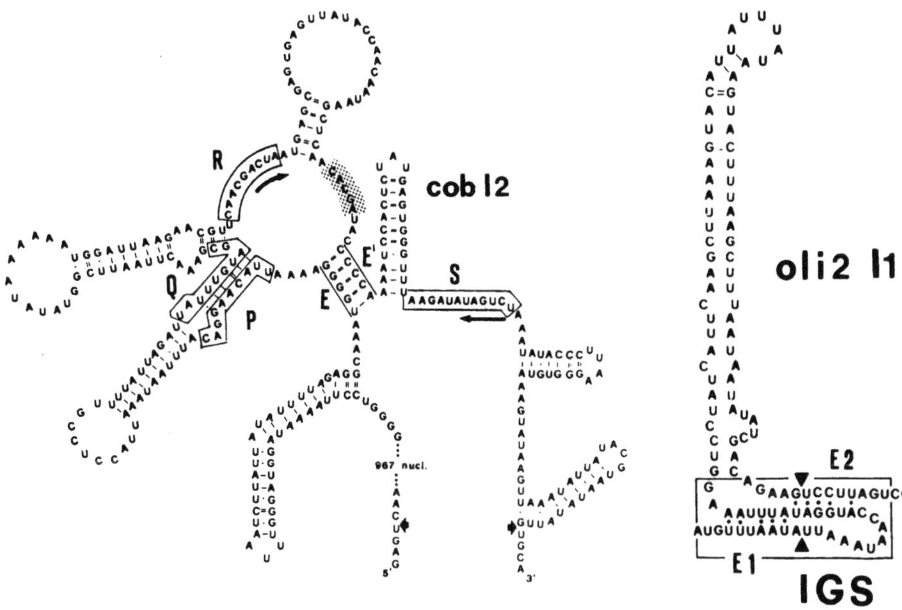

Figure 2. Secondary structure models of cob and oli2 introns. The conserved P, Q, R and S sequences of the second intron of the cob gene are boxed, and E and E' sequences are underlined. Complementarity between the R and S sequences is indicated by heavy arrows. The proposed IGS structure of the first intron of the oli2 gene is boxed. Short arrows point to the proposed splice junctions.

The second intron of the N. crassa cob gene contains a 50 nucleotide long region that is nearly identical to sequences found in all ribosomal introns of fungal mitochondria sequenced to date (3,4,5), as well as the fifth intron of the "long" form of the S. cerevisiae cob gene (8). The R sequences are identical in these introns. Also, palindromes that vary in sequence but are conserved as to position immediately follow the R sequences. The loop between the palindromes (of 18 to 20 bases) is highly maintained at the nucleotide level. The maintenance of this region of the introns of different species suggests that these introns are derived from a common ancestor. The first intron (93 bp) of the oli2 gene is unlike other fungal mitochondrial introns, both as to length and potential secondary structure (Figure 2). All other known mitochondrial introns are at least 700 bp long.

However, I1 maintains some of the conserved features of mitochondrial introns. I1 contains a 53 nucleotide long palindromic sequence that could direct the formation of a compact secondary structure, thus bringing intron ends into close proximity. An internal guide sequence (similar to that proposed for other introns) that might direct the precise alignment of intron ends (thought to be essential for proper splicing) is shown in Figure 2. As a consequence, I1 might be recognized and spliced by the same enzymes that remove the larger mitochondrial introns. The characteristics that I1 shares with the majority of fungal mitochondrial introns (compact secondary structure and IGS sequence) suggest that I1 might have arisen from the deletion of a large internal portion of a precursor intron.

Intron-encoded Proteins (Maturases?)

The second introns of the *N. crassa cob* and *oli2* genes contain extended open reading frames (of 290 and 365 codons, respectively), continuous and in frame with the upstream exon sequences (15; Morelli and Macino, in preparation). The first introns of the *cob* and *oli2* genes (1260 and 93 bp long, respectively) contain no ORFs of appreciable length. Many other introns of *S. cerevisiae* and *A. nidulans* contain long open reading frames (6,7,8,9). Mutations of the *cob* gene of *S. cerevisiae* that disrupt the intron reading frames and cause defects in the splicing of mRNA precursors have been described (26,27,28). It has been proposed that the ORFs of the second and fourth introns of the "long" form of the *S. cerevisiae cob* gene encode maturase proteins whose functions are required for the splicing of both *cob* and *oxi3* mRNA precursors (18,24,29,30).
Extensive homology does not exist among most of these intron reading frames, but two conserved decapeptide sequences (present in most fungal mitochondrial ORFs) have been noted (12). The two decapeptides are homologous to each other, and they are usually separated by about 100 to 150 amino acid residues. The intron-encoded proteins that contain this pair of decapeptides may act as maturases, and the conservation of the decapeptides may reflect the enzymatic activities common to maturase proteins (12). The ORFs of the second introns of the *N. crassa cob* and *oli2* genes contain the

conserved decapeptide sequences; these sequences show high homology to other decapeptide sequences (Table 2).

Transcript Analyses of N. crassa Split Genes

Six transcripts corresponding to the N. crassa cob gene have been detected (Figure 3). Two large transcripts (6.5 and 5.5 kb) contain the three exons (1155 nucleotides long), two introns (of about 1260 nucleotides each), and a substantial amount of flanking sequence. The 4.0 kb transcript contains all exon sequences and at least 550-600 bases 5' to the start codon; probes corresponding to both introns hybridize to this band. The 4.0 kb species is probably comprised of a mixture of two transcripts, each containing only one of the two introns (since the transcript is too small to contain 5' and exon sequences as well as both introns). The 2.6 and 1.8 kb species are potential mature mRNAs, since neither contains intervening sequences. The smaller of the potential mature mRNA species (1.8 kb) contains at least 550-600 bases 5' to the first exon, the three exon sequences (1155 bp), and perhaps a short 3' tail (of about 100 nucleotides). The flanking sequences

			I		II
S. cerevisiae	cob	I4	WLAGLIDGDG	-96 -	WFIGFFDADG
S. cerevisiae	oxi3	I4	WLAGLIDGDG	-106-	WFVGFFDADG
S. cerevisiae	oxi3	I3	YLAGLIEGDG	-134-	WLAILTDADG
A. nidulans	cobA	I1	YLVGLFEGDG	-132-	WLVGFIEAEG
N. crassa	oli2	I2	YITGFVDGEG	-160-	WIRGFIEGEG
N. crassa	cob	I2	FLAGLVDGDG	-121-	WIVGFTYSKG

Table 2. Conserved decapeptide sequences in the open reading frames of fungal mitochondrial introns. The first (I) and second (II) occurrences of the decapeptides are noted; figures indicate the number of amino acid residues between the decapeptide-encoding sequences. The single letter amino acid code is used. The intron sequences of the N. crassa cob and oli2 genes are compared with those of the S. cerevisiae cob (8) and oxi3 (6) genes, and the A. nidulans cob gene (12).

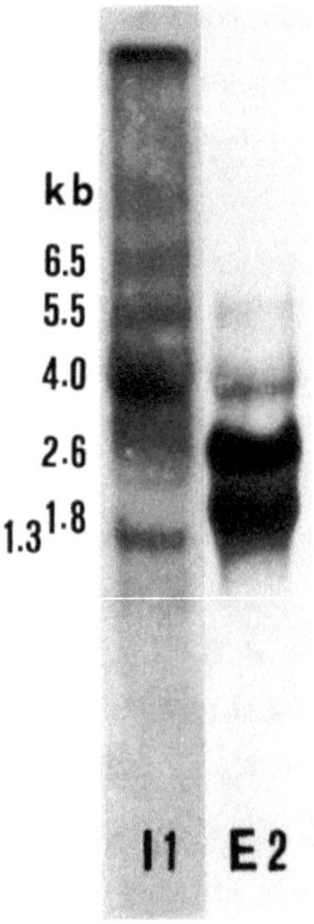

	Probe specific to					
kb	5'	E1	I1	E2	I2	E3
6.5	+	+	+	+	+	+
5.5	+	+	+	+	+	+
4.0	+	+	+[a]	+	+[a]	+
2.6	+	+	-	+	-	+
1.8	+	+	-	+	-	+
1.3	-	-	+[b]	-	+[b]	-

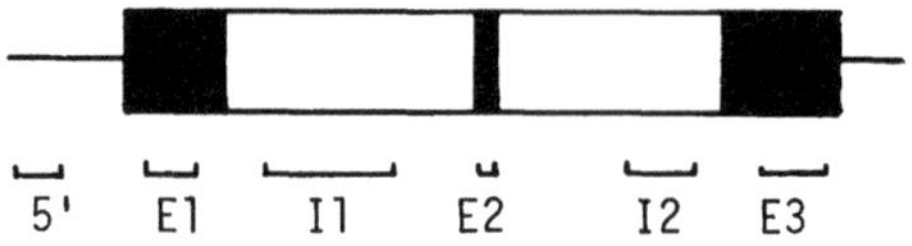

Figure 3. Analysis of the transcripts of the N. crassa cob gene. Mitochondrial RNA from the wild type strain SL74A of N. crassa was extracted as described by Bonitz et al. (6), denatured with glyoxal (39), and fractionated on agarose gels. The RNA was then transferred to diazobenzyloxymethyl paper (40), and analyzed by hybridization to nick-translated or 5'-end labelled probes. Results are shown for the first intron and the second exon of the cob gene (A). Hybridization (+) or lack of hybridization (-) to the six transcripts with other probes is indicated (B), as are the positions of the six probes (C).

a. The 4.0 kb species probably consists of a mixture of two transcripts, each containing one of the two introns (see text).
b. I1 and I2 are both about 1260 bp long, and so they hybridize to transcripts of the same size.

of this transcript are considered further below (in the section on PstI palindromes). The 2.6 kb transcript includes additional flanking sequences. The two introns of the N. crassa cob gene (of nearly identical size) are moderately stable, as shown by hybridization to a 1.3 kb band (Figure 3). The excised I1 accumulates to a greater extent than I2 (data not shown). Transcription of the oli2 gene of N. crassa has also been analyzed (manuscript in preparation), and a complex pattern of transcripts was detected. A large precursor molecule [3 kb (containing both intron and exon sequences)], as well as several smaller transcripts, were detected. Sequences corresponding to the URFx gene are included in the 3 kb transcript and in most of the smaller transcripts. The excised second intron is quite stable, as indicated by the accumulation of a 1.35 kb transcript containing the entire intron sequence and no exon sequences.

Two classes of fungal mitochondrial introns (stable and unstable after excision) have been described (11). Stable introns lack the conserved core secondary structure described by Davies et al. (10). These introns of S. cerevisiae exist in circular form after excision (31); moreover, they are stable during vegetative growth but are degraded in sporulating cells (32). No function has yet been assigned to the stable introns, but the control of their stability in different physiological conditions may reflect their importance. In contrast, introns of S. cerevisiae that contain the conserved P, Q, R and S sequences and maintain the potential to form the core secondary structure are degraded after excision, with the exception of the intron of the large rRNA gene (33).

The introns of the large rRNA gene (16) and the cob gene of N. crassa are moderately stable after excision, while the second intron of the oli2 gene is quite stable (manuscript in preparation). The stability of these introns might be controlled by a specific mechanism in N. crassa mitochondria. These introns might exist in circular form after excision; experiments that are in progress address this question.

PstI Palindromes as Signals for RNA Processing

All of the N. crassa mitochondrial genes whose structure has been determined (including those for ribosomal and transfer RNAs) are flanked by highly conserved GC-rich palindromic sequences (34,35,36). These sequences (from 50 to 150 bp long) have been called PstI palindromes by Yin et al. (34), because they contain two closely-spaced PstI restriction sites (Figure 4). Some highly homologous sequences in other parts of the genome lack the PstI restriction sites but maintain the postulated secondary structure [PstI-like palindromes; (36)]. A similar phenomenon has been noted in the mitochondrial genome of S. cerevisiae, which contains about 100 GC-rich clusters (37,38). These clusters may assume secondary structures like that

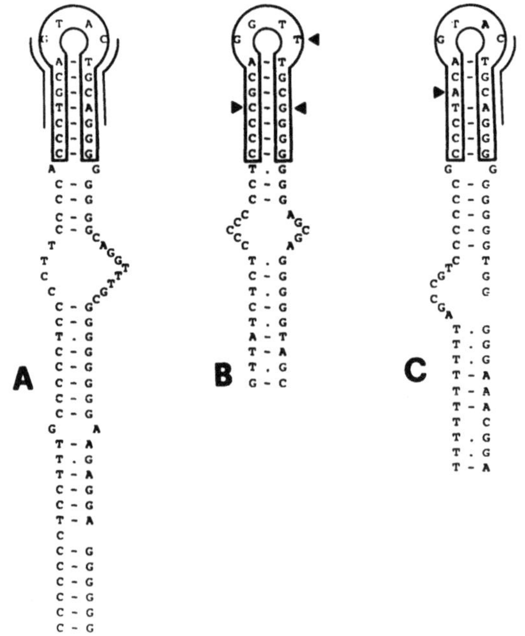

Figure 4. Structure of PstI and PstI-like palindromes. (A) This PstI palindrome has been described by Yin et al. (34). Postulated secondary structures of the PstI-like palindromes located 100 bp downstream of the cox2 (B) and cob (C) genes are shown. The eighteen nucleotides containing the PstI or PstI-like sites are boxed. The PstI sites are underlined. Arrows indicate the differences between the PstI and PstI-like sites.

of the PstI palindromes of N. crassa, and their positioning within the genome suggests that they might play a role in RNA processing (14). The PstI palindromes of N. crassa probably serve as signals for processing of precursor transcripts, including those of the cytochrome oxidase subunit 2 gene [cox2 (36)] . S1 nuclease analysis has shown that large precursor transcripts of the cox2 gene are processed immediately adjacent to a PstI palindrome located about 330 nucleotides upstream from the initiation codon. The precursor species are probably also processed adjacent to a PstI-like palindrome at a site about 100 nucleotides 3' to the terminator codon. A PstI palindrome is situated about 100 bp downstream from the stop codon of the N. crassa cob gene [Figure 1 and 4 (15)] . Since one of the potential mature mRNA species (1.8 kb) contains between 550 and 600 nucleotides of leader sequence and a 1155 base coding region, this transcript is probably generated by a processing event just 5' to the downstream PstI palindrome (as discussed above).

Transcript analysis has suggested that PstI-like palindromes also serve as signals for the processing of transcripts of the N. crassa oli2 gene (manuscript in preparation). This study has also suggested that other, as yet unknown, signals are recognized in the processing of precursor transcripts. Extensive analysis of the oli2 transcripts and of transcripts in other regions of the genome may allow us to identify these signals.

Acknowledgements

Part of this research was supported by a grant from the "Istituto Pasteur/ Fondazione Cenci Bolognetti" and from "Progetti Finalizzati, Ingegneria Genetica del C.N.R.".

References

1. Anderson, S., Bankier, A.T., Barrell, B.G., de Bruijn, M.H.L., Coulson, A.R., Drouin, J., Eperon, I.C., Nierlich, D.P., Roe, B.A., Sanger, F., Schreier, P.H., Smith, A.J.H., Staden, R., Young, I.G.: Nature 290, 457-465 (1981).
2. Anderson, S., de Bruijn, M.H.L., Coulson, A.R., Eperon, I.C., Sanger, F., Young, I.G.: J. Mol. Biol. 156, 683-717 (1982).
3. Dujon, B.: Cell 20, 185-197 (1980).
4. Burke, J.M., RajBhandary, U.L.: Cell 31, 509-520 (1982).
5. Netzker, R., Kochel, H.G., Basak, N., Kuntzel, H.: Nucl. Acids Res. 10, 4783-4794 (1982).
6. Bonitz, S.G., Coruzzi, G., Thalenfeld, B.E., Tzagoloff, A., Macino, G.: J. Biol. Chem. 255, 11927-11941 (1980).
7. Lazowska, J., Jacq, C., Slonimski, P.P.: Cell 22, 333-348 (1980).
8. Nobrega, F.G., Tzagoloff, A.: J. Biol. Chem. 255, 9838-9847 (1980).
9. Waring, R.B., Davies, R.W., Lee, S., Grisi, E., McPhail Berks, M., Scazzocchio, C.: Cell 27, 4-11 (1981).
10. Davies, R.W., Waring, R.B., Ray, J.A., Brown, T.A., Scazzocchio, C.: Nature 300, 719-724 (1982).

11. Michel, F., Jacquier, A., Dujon, B.: Biochimie 64, 867-881 (1982).

12. Waring, R.B., Davies, R.W., Scazzocchio, C., Brown, T.A.: Proc. Nat. Acad. Sci. USA 79, 6332-6336 (1982).

13. Grivell, L.A., Hensgens, L.A.M., Osinga, K.A., Tabak, H.F., Boer, P.H., Crusius, J.B.A., van der Laan, J.C., de Haan, M., van der Horst, G., Evers, R.F., Arnberg, A.C.: in Slonimski, P.P., Borst, P., Attardi, G. (eds.) Mitochondrial Genes, Cold Spring Harbor Laboratory, New York, 225-239 (1982).

14. Tzagoloff, A., Nobrega, M., Akai, A., Macino, G.: Current Genet. 2, 149-157 (1980).

15. Helmer Citterich, M., Morelli, G., Macino, G.: EMBO J., in press (1983).

16. Green, M.R., Grimm, M.F., Goewert, R.R., Collins, R.A., Cole, M.D., Lambowitz, A.M., Heckman, J.E., Yin, S., RajBhandary, U.L.: J. Biol. Chem. 256, 2027-2034 (1981).

17. Bertrand, H., Bridge, P., Collins, R.A., Garriga, G., Lambowitz, A.M.: Cell 29, 517-526 (1982).

18. Jacq, C., Lazowska, J., Slonimski, P.P.: C.R. Acad. Sci. Paris, Ser. D 290, 89-92 (1980).

19. Macino, G., Tzagoloff, A.: Cell 20, 507-517 (1980).

20. Grisi, E., Brown, T.A., Waring, R.B., Scazzocchio, C., Davies, R.W.: Nucl. Acids Res. 10, 3531-3539 (1982).

21. Gallwitz, D., Sures, I.: Proc. Nat. Acad. Sci. USA 77, 2546-2550 (1980).

22. Macreadie,I.G., Choo, W.M., Novitski, C.E., Marzuki, S., Nagley, P., Linnane, A.W., Lukins, H.B.: Biochem. Int. 5, 129-136 (1982).

23. Novitski, C.E., Macreadie,I.G., Maxwell, R.J., Lukins, H.B., Linnane, A.W., Nagley, P.: in Nagley, P., Linnane, A.W., Peacock, W.J., Pateman, J.A. (eds.) Manipulation and Expression of Genes in Eukaryotes, Academic Press, Sydney, 257-268 (1983).

24. De La Salle, H., Jacq, C., Slonimski, P.P.: Cell 28, 721-732 (1982).

25. Weiss-Brummer, B., Holl, J., Schweyen, R.J., Rodel, G., Kaudewitz, F.: Cell 33, 195-202 (1983).

26. Church, G.M., Slonimski, P.P., Gilbert, W.: Cell 18, 1209-1215 (1979).

27. Halbreich, A., Pajot, P., Foucher, M., Grandchamp, C., Slonimski, P.P.: Cell 19, 321-329 (1980).

28. van Ommen, G.J.B., Boer, P.H., Groot, G.S.P., de Haan, M., Roosendaal, E., Grivell, L.A.: Cell 20, 173-183 (1980).

29. Church, G.M., Gilbert, W.: in Joseph, D.R., Schultz, J., Scott, W.A., Werner, R. (eds.) Mobilization and Reassembly of Genetic Information vol. 17, Academic Press, New York, 379-395 (1980).

30. Netter, P., Jacq, C., Carignani, G., Slonimski, P.P.: Cell 28, 733-738 (1982).

31. Hensgens, L.A.M., Arnberg, A.C., Roosendaal, E., van der Horst, G., van der Veen, R., van Ommen, G.J.B., Grivell, L.A.: J. Mol. Biol. 164,

35-58 (1983).

32. Schroeder, R., Breitenbach, M., Schweyen, R.J.: Nucl. Acids Res. 11, 1735-1746 (1983).

33. Bos, J.L., Osinga, K.A., van der Horst, G., Hecht, N.B., Tabak, H.F., van Ommen, G.J.B., Borst, P.: Cell 20, 207-214 (1980).

34. Yin, S., Heckman, J.E., RajBhandary, U.L.: Cell 26, 325-332 (1981).

35. Browning, K.S., RajBhandary, U.L.: J. Biol. Chem. 257, 5253-5256 (1982).

36. Macino, G., Morelli, G.: J. Biol. Chem., in press (1983).

37. Prunell, A., Kopecka, H., Strauss, F., Bernardi, G.: J. Mol. Biol. 110, 17-52 (1977).

38. Cosson, J., Tzagoloff, A.: J. Biol. Chem. 254, 42-43 (1979).

39. Carmichael, G.G., McMaster, G.K.: Methods in Enzymology 65, 380-390 (1980).

40. Alwine, J.C., Keny, J.D., Stark, G.R.: Proc. Nat. Acad. Sci. USA 74, 5350-5354 (1977).

MUTATIONS OF THE RAT MITOCHONDRIAL GENOME

Katsuro Koike*, Masanori Taira*, Yoshiyuki Kuchino**, Katsuyuki Yaginuma*, Toyozo Sekiguchi**, Midori Kobayashi*
Cancer Institute*, Kami-Ikebukuro, Toshima-ku, Tokyo 170,
National Cancer Center Research Institute**, Tsukiji, Chuo-ku, Tokyo 104, Japan

Introduction

The mitochondrial genome is thought to be a sensitive target for certain mutagens and/or carcinogens and to be altered in tumor cells. Some of the metabolites of these carcinogens are covalently linked to more mtDNA than nuclear DNA (1). This uniqueness of the mammalian mitochondrial genome has prompted investigation of its primary structure. If mutations of mtDNA occur in tumor cells, the following questions arise: (i) What kind of mutations occur in mitochondrial genes ? (ii) Is a certain mutation characteristic of the tumor ? (iii) How are mutations in the mitochondrial and nuclear genomes cooperatively involved in carcinogenesis ? To answer these questions, we tried to find mutations in the mtDNA of existing rat mutant and tumor cell lines, that is, Morris Hepatoma(#5123D), Yoshida sarcoma(YS), ascites hepatomas(AH130 and AH7974), chloramphenicol-resistant myoblastic cell($L6TG^RCAP^R$). In a series of experiments, we found insertions, deletions or substitutions of nucleotides in the genes coding for rRNA, tRNA(2), polyA-containing mRNAs and in the D-loop region.

Experimental Procedures

Four tumor cells, Morris hepatoma(#5123D) Yoshida sarcoma(YS)

Mitochondria 1983

ascites hepatomas(AH-130 and AH-7974) were obtained as described previously (2). The CAP-resistant cell was the rat myoblastic cell($L6TG^R CAP^R$) derived from CAP-sensitive cell($L6TG^R$) and grown normally in the presence of 150 µg/ml CAP (3).

Mitochondrial DNAs were prepared from rat tumor and CAP-resistant cells and digested with restriction endonuclease BamHI (4). The resulting BamHI fragments, BamA (ca.11 kb) and BamB (ca.5.2 kb), were joind to the BamHI site of PBR322 by T4 ligase and subjected to *E. coli* HB101 transformation (2,5).

The clones used in the present experiments were two types of recombinant plasmids, carrying the fragments BamA and BamB of rat mitochondrial DNA (2). The recombinant plasmid containing the BamHI fragment of rat mtDNA was cultivated and each BamHI fragment was obtained by BamHI digestion followed by 0.7% agarose gel electrophoresis. DNA was extracted from the gel by homogenization and incubation with 0.5 M ammonium acetate/ 0.1% SDS/0.1 mM EDTA.

5'-End labeling of the DNA fragment, strand separation and sequence analysis by a chemical method were performed as described by Maxam and Gilbert (6). Chemical RNA sequencing was performed as described by Peattie (7).

Polyacrylamide gel electrophoresis of the restriction DNA fragments was carried out on vertical slab gel (1.0mm x 14cm x 40cm) in 45 mM Tris-borate (pH 8.2), 2 mM EDTA at 400-500 V. Agarose gel electrophoresis was carried out on horizontal slab gel in 40 mM Tris-HCl (pH 7.8), 20 mM sodium acetate, 2 mM EDTA 0.5 µg/ml ethidium bromide at 60-70 V.

Results

Changes in nucleotide sequences in tRNA genes from tumor cell mitochondria. To detect changes in nucleotide sequences in BamHI fragments of mtDNAs from normal liver and the four different tumor cells, we first compared the electrophoretic mobility of the restriction fragments on a 5 % polyacrylamide gel. The results clearly indicated that the mtDNAs of AH-130 and AH-7974 differed from those of normal liver with respect to the restriction enzyme cleavage site and the electrophoretic mobility of the restriction fragments. Only a few fragments of YS differed in electrophoretic mobility. Some of the changes were found in the region of the HindB/EcoC fragment of BamA, where five tRNA genes are clustered about the L-strand replication origin (O_L) (2).

In AH-130 and AH-7974, a new HaeIII site was created in the $tRNA^{Tyr}$ gene and a TaqI and HinfI site was lost in the URF2

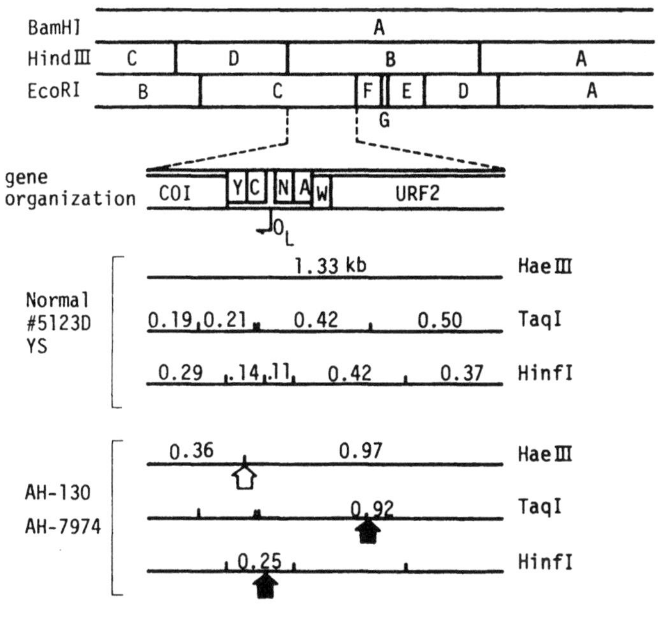

Fig. 1. Restriction maps of the HindB/EcoC fragment of mitochondrial DNA from rat liver and tumor cells.

Top: Restriction maps of the BamA fragments of rat liver mtDNA. Center: Gene organization of the HindB/EcoC fragment, where the tRNA genes are indicated by the one-letter amino acid code. Genes for $tRNA^{Tyr}$(Y), $tRNA^{Cys}$(C), $tRNA^{Asn}$(N) and $tRNA^{Ala}$ (A) are located in the H strand, while genes for $tRNA^{Trp}$(W), cytochrome c oxidase subunit I(COI) and unidentified reading frame 2(URF2) are in the L strand. Bottom: Restriction maps of the HindB/EcoC fragments of mtDNAs from rat liver and tumor cells. In AH-130 and AH-7974, a new HaeIII site appeared (⇧) and all of the TaqI and HinfI sites disappeared (⬆).

gene and tRNA^{Cys} gene, respectively (Fig. 1). On HinfI digestion of the HindB/EcoC fragment, the 0.14 kb fragment from YS had slower mobility than that of the fragment from normal liver (2). The 0.14 kb fragment contained genes for the N-terminal region of cytochrome C oxidase subunit I(COI), tRNA^{Tyr} and the 5'-half of tRNA^{Cys} (Fig. 1). It was also observed that the mobility of the 0.42 kb HinfI fragment from AH-130 or AH-7974 was less than that of the fragment from normal liver (2), in which the genes for tRNA^{Ala}, tRNA^{Trp} and the C-terminal region of URF2 are located (Fig. 1). On the other hand, mtDNA from hepatoma #5123D was almost identical to normal liver in its restriction map and the electrophoretic mobility of the restriction fragments so far examined.

To determine what kinds of changes occur in the tumor cell mtDNA, we then compared the regions containing the five tRNA

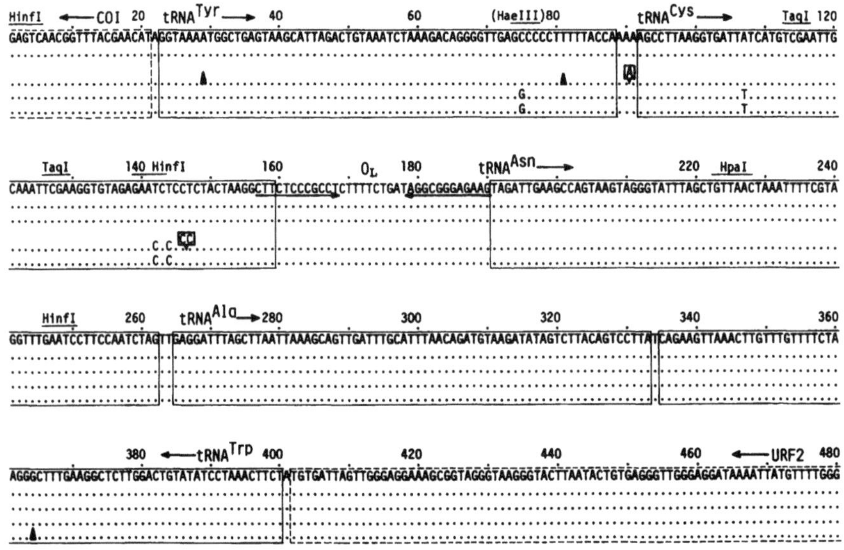

Fig. 2. Nucleotide sequences of the DNA region containing a cluster of five different tRNA genes.
The first sequence is the H-strand sequence of the rat liver DNA. The tRNA genes are shown in boxes with solid lines. The 5'-end and 3'-end of COI and URF2 genes, respectively, are indicated by broken lines. The polarities of genes are indicated by arrows. Identical nucleotides are shown by dots. Black triangles indicate deletions. [symbol] is an inserted nucleotide sequence.

genes from normal liver and tumor cells. When the nucleotide sequences of 480 bp from the HinfI site within the 5'-end region of the COI gene were compared, significant changes in nucleotide sequences were detected in the genes for $tRNA^{Tyr}$, $tRNA^{Cys}$ and $tRNA^{Trp}$ of three tumor mitochondria except #5123D (Fig. 2). In AH-130 and AH-7974, the C→G transversion at position 75 created a new HaeIII site in the $tRNA^{Tyr}$ gene and the T→C transition at position 142 eliminated one of the HinfI sites in the $tRNA^{Cys}$ gene. In AH-130, dinucleotide CC was inserted at a position next to 146th nucleotide in the $tRNA^{Cys}$ gene. The 364th nucleotide G of the $tRNA^{Trp}$ gene was deleted

(a) $tRNA^{Tyr}$

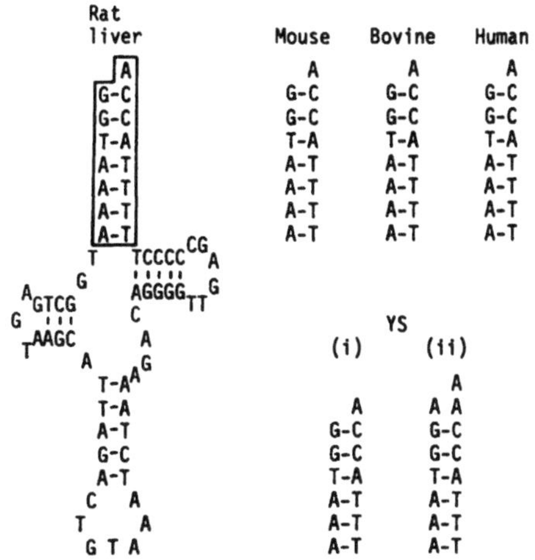

Fig. 3. Comparison of secondary structures of mitochondrial $tRNA^{Tyr}$ genes from rat liver, tumor cells and mammals.
The divergent region in the tRNA gene from the tumor is shown as a boxed-in region.

in AH-7974. In the YS sequence, two nucleotides A and T were deleted at positions 29 and 81, respectively, of the $tRNA^{Tyr}$ gene, while one nucleotide was inserted into the spacer region next to the 90th nucleotide, resulting in a change in mobility of the 0.14 kb-HinfI fragment on polyacrylamide gel (2). In addition, the trinucleotide TGG was inserted into the URF2 gene region of the 0.42 kb-fragment of AH-130 and AH-7974, resulting in reduced mobility of this fragment (2).

Structural feature of mitochondrial tRNA genes from rat tumor cells and mammals. To examine the differences in mitochondrial tRNA genes of tumor cells and those of normal liver, we compared the secondary structures of the genes for $tRNA^{Tyr}$, $tRNA^{Trp}$ and $tRNA^{Cys}$, as deduced from the nucleotide sequences of these four tumor cells, with those of rat liver, mouse, bovine and human cell. As shown in Fig. 3, the aminoacyl acceptor stem of the $tRNA^{Tyr}$ (boxed region) is completely conserved in the four mammalian species, but one of the seven base pairs in the stem structure is deleted in the YS $tRNA^{Tyr}$ gene, that is two bases A and T at positions 29 and 81, respectively (Fig.2). Consequently, the product of this gene would have a shorter aminoacyl acceptor stem consisting of six base pairs (Fig. 3 YS (ii)). If an A-A pair is formed, seven base pairs may be constructed using spacer nucleotides adjacent to the gene.

The anticodon stem of $tRNA^{Trp}$ in the mammalian species is well conserved, but one nucleotide C on the right hand side of the stem in the AH-7974 gene was deleted (2). This deletion might result in the formation of two alternative stems of either an abnormal, or different structure, where one nucleotide from the extra-loop region could compensate for this deletion in the anticodon stem. In the latter case, the extra loop would consist of three nucleotides, the size of which has been reported in only three tRNAs (8,9).

The size and sequence of the "TψC" loop of the mitochondrial tRNA gene were variable in the mammalian species. The "TψC" loop of the $tRNA^{Cys}$ gene consisted of eight nucleotides in rats and six nucleotides in other species (2). In AH-130, however, this loop was ten nucleotides long because of a dinucleotide CC insertion. In comparison with this tRNA gene in other mammals and even with other tRNA genes of mammalian mitochondria, the loop consisted of nine nucleotide at most (8,9,10-15).

In summary, several mutations of the mitochondrial tRNA genes were demonstrated in tumor cells, such as YS, AH-130 and AH-7974, but not in #5123D. Nucleotide deletions occurred in sequences of the YS $tRNA^{Tyr}$ and AH-7974 $tRNA^{Trp}$ genes which were,

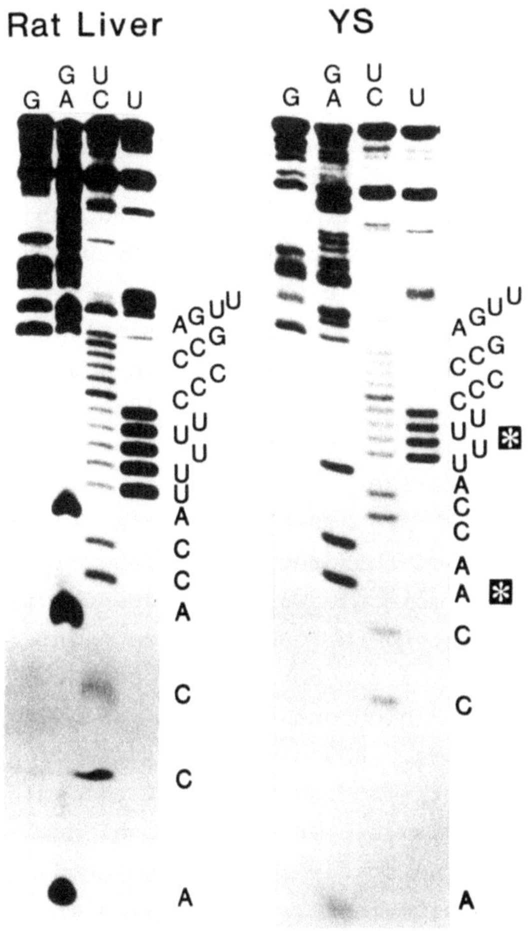

Fig. 4. Nucleotide sequence of the 3'-terminal region of the tRNA^{Tyr} molecule from Yoshida sarcoma mitochondria.

* indicates insertion of nucleotide A and deletion of nucleotide U in the YS tRNA^{Tyr} molecule.

in turn, well conserved in rat, mouse, bovine and human mitochondria. In addition, the insertion of two nucleotides resulted in the formation of a "TψC" loop of the AH-130 tRNA^{Cys} gene containing ten nucleotides which was longer than the size of "TψC" loops in either mitochondrial or cytoplasmic tRNAs (8-14). Therefore, these changes seem to be mutations characteristic of each respective tumor.

<u>Nucleotide sequence of the tRNA^{Tyr} molecule from Yoshida sarcoma mitochondria</u>. In order to obtain direct evidence for existence of mutated tRNA genes, we directed our attention on deletion occurring in the nucleotide sequence of the tRNA^{Tyr}

molecule from YS mitochondria. Extraction and purification of 4S RNA from YS and normal mitochondria were performed as described by Zubay (16). Purified 4S RNA was 3'-end-labeled by the method of Peattie using [5'-^{32}P]pCp and T4 RNA ligase (7) and then purified on three different polyacrylamide gels (10%→20%→15%+7M urea). RNA was recovered from the gels and subjected to base-specific cleavage followed by electrophoresis on 20% polyacrylamide gel containing 7 M urea.

As shown in Fig. 4, the sequence of the 22 3'-terminal nucleotides was revealed. In the $tRNA^{Tyr}$ molecules, a CCA sequence was found at the 3'-end in contrast to the $tRNA^{Tyr}$ gene; that is, there was no coding sequence for CCA in the DNA of this region. We also found deletion of nucleotide U in the region upstream from the CCA sequence of $tRNA^{Tyr}$ from YS mitochondria, as demonstrated in the DNA sequence of this region. Another interesting finding of this $tRNA^{Tyr}$ molecule was the presence of an additional nucleotide A adjacent to the CCA sequence. This is in contrast to the DNA sequence of the aminoacyl acceptor stem of the $tRNA^{Tyr}$ gene from YS, in which one of seven base pairs in the stem structure was deleted. RNA sequencing data clearly indicate that the length of the 3'-end region of the aminoacyl acceptor stem was 7 nucleotides. This additional nucleotide A likely originated from the spacer sequence adjacent to the 3'-end of this gene, resulting in seven base pairs instead of six using a possible A-A pair.

Transcription of the mutated $tRNA^{Tyr}$ gene. Previous experiments suggest that up to 100% of the $tRNA^{Tyr}$ molecule were synthesized in the YS cells under the direction of this mutant gene. Accordingly, the mitochondrial 4S RNA fraction was prepared and labeled at 3'-end. To permit simple identification of the $tRNA^{Tyr}$ molecule and rule out any possible contamination of other tRNA molecules, labeled 4S RNA molecules from YS and normal liver were separated by two dimentional polyacrylamide gel electrophoresis, first an 18% gel + 3.5 M urea and second, a 10% gel. The results obtained from parallel runs indicate that there were significant differences not only in the mobility but also location of the $tRNA^{Tyr}$ molecule. In the case of YS mitochondria, we observed only the YS $tRNA^{Tyr}$ band which was slightly behind the $tRNA^{Tyr}$ from normal liver mitochondria. This observation was confirmed by the fact that the $tRNA^{Tyr}$ molecule synthesized in the normal liver mitochondria was only coded by the normal $tRNA^{Tyr}$ gene.

A quantitative comparison of the trascription efficiency was then made between the normal liver and YS $tRNA^{Tyr}$ genes. The total mitochondrial 4S RNA fraction from YS was labeled and subjected to two-dimentional gel electrophoresis as before. After the run, two spots, $tRNA^{Tyr}$ and $tRNA^{Lys}$, resolved out of many spots, were taken out and their radioactivities compared. In a control experiment, the same comparison was made between $tRNA^{Tyr}$ and $tRNA^{Lys}$ from normal liver. By comparing the ratios of these two tRNAs, a significant decrease was found in the relative number of the $tRNA^{Tyr}$ molecule in the YS mitochondria. Our results indicate that YS mitochondrial $tRNA^{Tyr}$, once aminoacylated, is functional in protein synthesis, but this mutated $tRNA^{Tyr}$ gene is poorly expressed in mitochondria. This may pose an important problem in the regulatory mechanism of mitochondria.

<u>Structural features of the H-strand replication origin.</u> Mammalian mtDNA contains a unique DNA sequence (D-loop region) that has been defined by analysis of species specificity as a requirement for the initiation of unidirectional DNA replication. This region consists of about 900 bp of DNA between $tRNA^{Pro}$ and $tRNA^{Phe}$ genes, where the D-loop DNA (630-680 nucleotides) is synthesized. Unlike the rest of the genome, this region is highly variable in sequence and in size except several small blocks of homology between mammalian species (8,10,13, 17)(Fig. 5). From the functional point of view, the homologous portions coule be separated into two areas such as upstream and downstream regions from the 5'-end of the D-loop DNA (O_H). Unlike many of the reading frames, the DNA sequence of this

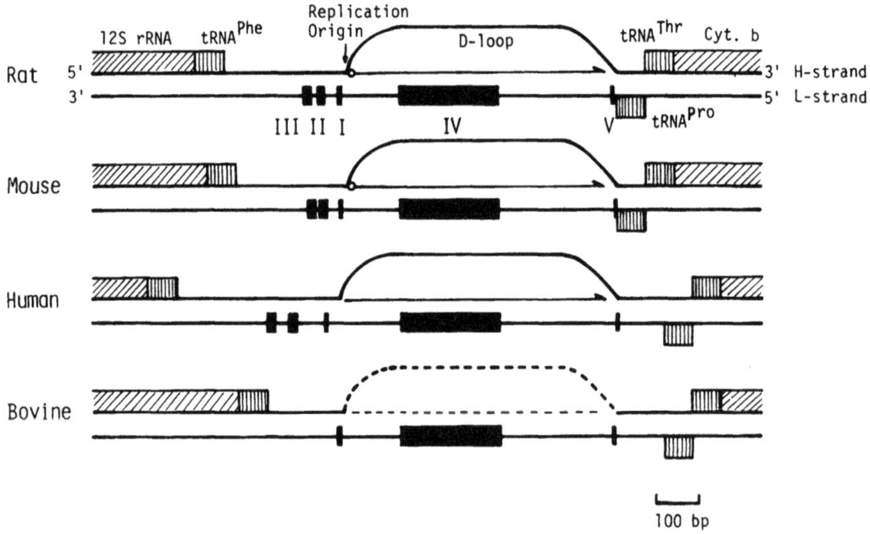

Fig. 5. Structural comparisons of the H-strand replication origin between mammalian mitochondrial DNA.
Five different homology regions (I, II, III, IV, V) are indicated by closed blocks.

region does not contain any significant reading frame but may be transcribed as a small RNA.

In the 290 nucleotide region between the 5'-end of the D-loop DNA and the tRNA^{Phe} gene, where sequence variability is the most extreme, three significant homology sequences between the

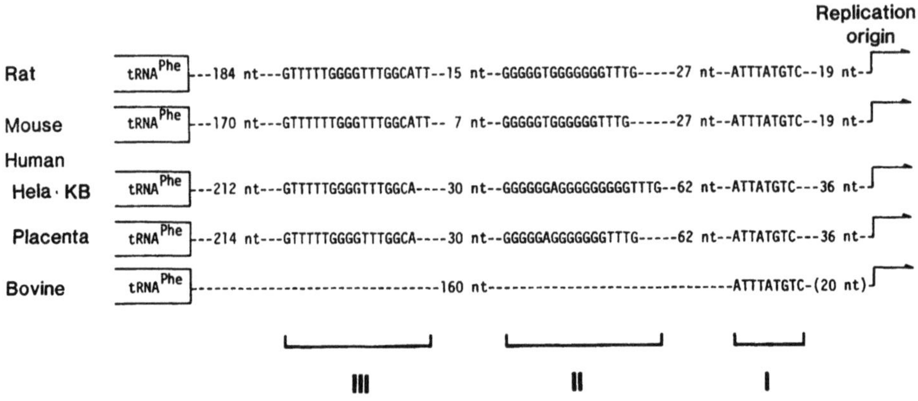

Fig. 6. Comparisons of the nucleotide sequences of three different homology blocks, I, II and III in mammalian mitochondiral DNA.

mammalian genomes occur in short conserved tracts I, II, III (Fig. 6). The conserved sequence I is 19 nucleotides upstream from the replication origin (O_H) in rat mtDNA. The conserved sequence II is a 17 nucleotide cluster consisting of G and T and located 27 nucleotides upstream from the conserved sequence I in rat. It is of interest to note that there is a significant difference in this conserved sequence II between human placenta and HeLa cells. In tumor cells, the size of the G cluster is shorter than that of placenta by three G nucleotides. The conserved sequence III 18 nucleotides in size and consists of a mosaic structure from nucleotides G and T. In the case of bovine mtDNA sequences similar to II and III have not been found. This is probably due to the sequence shorter than the corresponding region in other mammals. It is intriguing to speculate that these three conserved sequences I, II, III, may be involved in regulating the synthesis and processing of primer RNA for H-strand replication.

When aligning these conserved regions from normal liver and AH-130 mtDNA, all of the homology blocks are perfectly conserved not only in sequence but size as well. It seems probable that another tumor mtDNA has the same homology blocks, indicating that similar nuclear-coded proteins interact in common. However, it is not clear yet whether the base replacement found in the variable regions would be associated with the specific property of the tumor cells.

In the D-loop region between the 5'-end of the D-loop DNA and $tRNA^{Pro}$, the middle of the region was more conserved than its surroundings (Fig. 5). This conserved sequence IV represents a mosaic structure consisting of several homology blocks. These blocks are highly conserved in other mammalian mtDNA sequences and are about 120 nucleotides downstream from the replication origin. Hence, it is likely that nuclear coded protein regulating the L- or H-strand transcription interacts here. Another conserved sequence V recognized near the 5'-end of $tRNA^{Pro}$ gene may be a signal governing the termination of D-loop DNA (Fig. 5). When these conserved sequences IV and V of the rat liver mtDNAs were compared with corresponding regions of the tumor mtDNAs, at least a perfect homology between rat liver and #5123D or AH-130 was found, although there are no complete sequences available for the four different tumor mtDNAs. Therefore, it cannot be said whether the sequences between these homology blocks are well-conserved or not.

Structural features of L-strand replication origin. The sequence of the putative L-strand replication origin of rat mtDNA was compared with that of other mammals (2,8,10,11). The rat L-strand origin of replication, like that for other mammlian mtDNA, can be drawn in the form of a stem-loop structure (Fig. 7). The genome position, size and sequence of the stem show highly conserved features (2). Apart from the conserved stem structure, the sequences of the loop regions between mammals diverge markedly. Nevertheless, this stem-loop structure was perfectly conserved in AH-130. Therefore, it is likely that mitochondria-coded species-specific protein regulating L-strand replication recognizes this loop region. In contrast, the stem structure held in common with other mammals is necessary for the function of nuclear-coded protein involved in the ragged initiation of L-strand synthesis. However, there is no direct evidence available for nuclear-coded protein(s) and mitochondria-coded protein(s) as well.

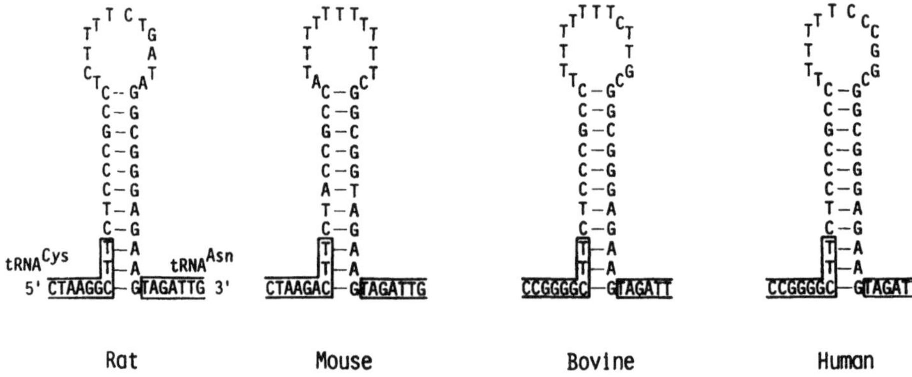

Fig. 7. Stem-loop structures of the L-strand replication origin of mitochondrial DNA.

A new locus in the large ribosomal RNA gene governing chloramphenicol resistance. Chloramphenicol (CAP) resistance in mammalian cells has been reported as correlated with a single nucleotide change in two adjacent highly-conserved regions near the 3'-end of the large ribosomal RNA (rRNA) gene of mtDNA (18-20). The recombinant plasmid containing the BamA fragment of mtDNA from L6TG^R or L6TG^{R}CAP^R was cultivated and each BamA fragment was obtained by BamHI digestion. To analyze the difference between the two BamA fragments from L6TG^{R}CAP^R and L6TG^R, the fragments were digested with restriction enzyme HinfI or TaqI. The digests were then analyzed by 5% polyacrylamide gel electrophoresis to compare the mobility of the digested fragments. Only 570 bp TaqI or 160 bp HinfI fragments were found to exhibit a change in mobility between L6TG^{R}CAP^R and L6TG^R. Furthermore, a 400 bp TaqI/HaeIII fragment obtained by the digestion of a 570 bp TaqI fragment with HaeIII also showed a difference in mobility. Based on the fine restriction map, these fragments are known to derive from the same region of the large 16S rRNA gene (Fig. 8).

To determine the DNA sequence in the relevant region, the

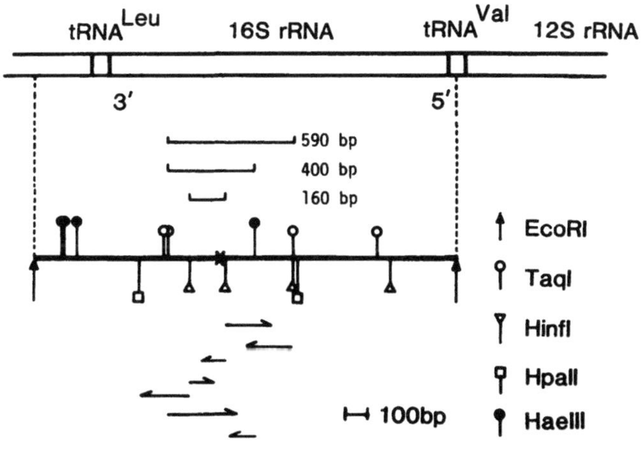

Fig. 8. Restriction maps of the EcoD fragment of mitochondrial DNA from CAP-resistant and CAP-sensitive cells.
The schematic map of genes is shown relative to the EcoD fragment of mtDNA. The site of nucleotide difference in the G to A transition is indicated by x.

EcoD fragment containing the entire 16S rRNA gene was purified. The 400 bp TaqI/HaeIII and 160 bp HinfI fragments were then fractionated and subjected to DNA sequencing. The nucleotide sequences indicated that the 400 bp TaqI/HaeIII fragments from $L6TG^{R}CAP^{R}$ and $L6TG^{R}$ differed by only a single nucleotide at residue 1063 from the 5'-end of the 16S rRNA gene. A single nucleotide change in G to A was confirmed by DNA sequencing of the 160 bp HinfI fragment. This sequence change reflected only the detected difference in mobility on polyacrylamide gel electrophoresis. We could find no other differences in the sequences in the region near the 3'-end of the 16S rRNA gene consisting of two highly-conserved blocks, where a single nucleotide substitution is the mutation responsible for the CAP-resistant phenotype.

Compared with other sequence data of large rRNA genes, the present single nucleotide change was found in the sequence block perfectly identical to human mtDNA (21), bovine mtDNA (8), mouse mtDNA (11), rat mtDNA or E. coli DNA (22) (Fig. 9). It seems unlikely that this nucleotide change appeared by chance in this highly-conserved region even though there were no differences whatsoever in other conserved or non-conserved sequences. This perfect conservation probably indicates functional constraints imposed on this region of the large rRNA gene. Thus, it is suggested that the sequence alteration in this block is responsible for the CAP-resistant phenotype. We tentatively named (termed) this newly-identified region as C locus and the other two changes previously identified were named A and B loci from the 3'-end of the 16S rRNA gene.

Chloramphenicol is known to bind the large ribosomal subunits and block the binding of the 3'-terminus of aminoacyl tRNA (23). It may be assumed that these three loci, which are responsible for the CAP-resistant phenotype and are well separated in the primary sequence of the large rRNA gene, probably approach each other in active ribosomes so as to participate in the binding of aminoacyl tRNA. Therefore, we investigated a possible secondary structure in this portion of the 16S rRNA

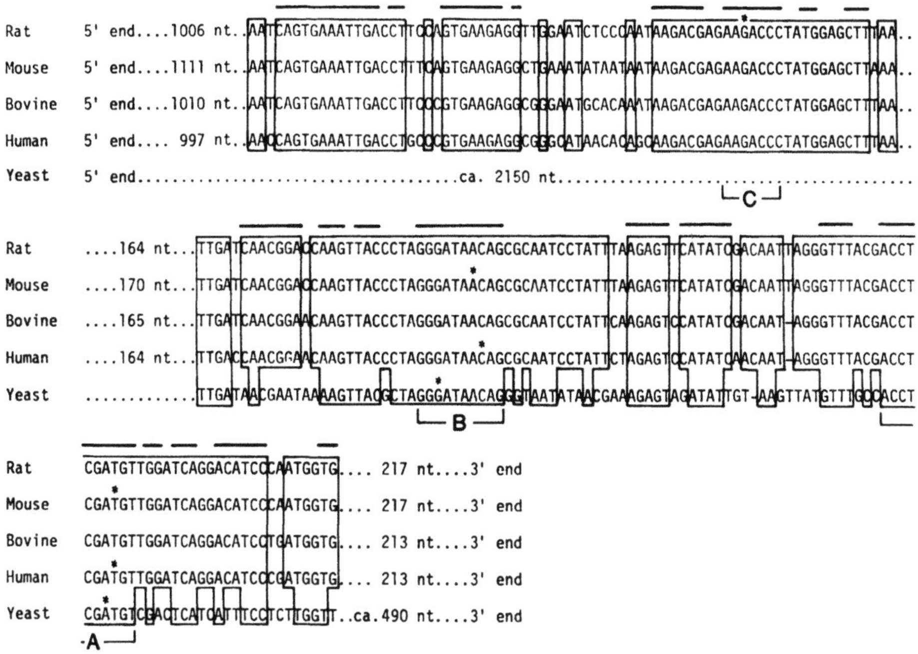

Fig. 9. Sequence homology of the large rRNA genes of chloramphenicol-resistant rat and other mitochondria.
The sequence data of yeast mitochondria is only partially available (25). The completely homologous sequences are boxed, and the identical sequences with E. coli 23S rRNA gene are indicated by thick bars. Asterisks above the nucleotides indicate the position of the mutation correlated with CAP-resistance. A, B and C loci, as tentatively named, are indicated beneath the sequences.

gene of rat mitochondria.

Recently, Glotz et al. constructured a secondary structure model of E. coli 23S rRNA by experimental analysis of short base-paired RNA fragments and fragments containing intra RNA cross-links (26). Following their model, we also constructed a potential secondary structure of the rat mitochondrial 16S rRNA gene. As shown in Fig. 10, the two loci B and C, separated by 200 nucleotides in the primary sequence, come close to each other in the secondary structure because of base pairing. This interaction further confirms that the newly-discovered C locus, besides A and B loci, is responsible for the CAP-resist-

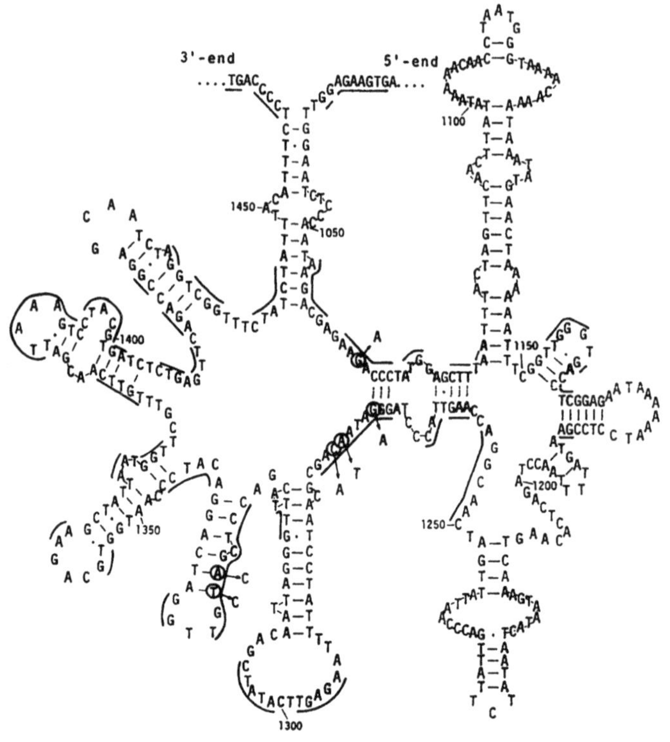

Fig. 10. The possible secondary structure of the region of rat mitochondrial 16S rRNA gene responsible for chloramphenicol resistance.
Circled nucleotides indicate the position of mutations found in CAP-resistant cells. Arrows indicate nucleotide changes correlated with CAP-resistance. Thick bars indicate identical sequences with E. coli 23S rRNA gene. Nucleotide numbers are from the 5'-end of the 16S rRNA gene of rat mitochondria.

ant phenotype. This secondary conformation of the large rRNA gene with stem and loop structures, in which the three conserved regions correlated with CAP resistance are positioned adjacent to one another, can be built up from the E.coli 23S rRNA gene to human mitochondrial 16S rRNA gene. Also in E.coli 23S rRNA, the potential puromycin-reactive site (24) and the protection sites from Kethoxal modification on association of 50S and 30S ribosome subunits (22) have been found to be present in the circle area of this secondary structure. Based on these observations, we assume that this circle conformation of the secondary structure of the large rRNA may be involved in the formation of the peptidyltransferase center and play an essential role in protein synthesis from prokaryote to human mitochondria.

References

1. Backer,J.M., Weinstein,I.B.: Science 209, 297-299 (1980)
2. Taira,M., Yoshida,E., Kobayashi,M., Yaginuma,K., Koike,K.: Nucl. Acids Res. 11, 1635-1643 (1983)
3. Sekiguchi,T., Tosu,M., Yoshida,M.C., Oikawa,A., Ishihara,K. Fujiki,H., Tsumuraya,M., Kameya,T.: Somatic Cell Genetics 8, 605-622 (1982)
4. Kobayashi,M., Koike,K.: Gene 6, 123-136 (1979)
5. Bolivar,F., Rodriguez,R.L., Greene,P.J., Betlach,M.C., Heyneker,H.L., Boyer,H.W., Crosa,J.H., Falkow,S.: Gene 2, 75-93 (1977)
6. Maxam,A.M., Gilbert,W.: Meth. Enzym. 65, 499-560 (1980)
7. Peattie,D.A.: Proc. Natl. Acad. Sci.USA 76, 1760-1764 (1797)
8. Anderson,S., deBruijn,M.H.L., Coulson,A.R., Eperon,I.C., Sanger,F., Young,I.G.: J. Mol. Biol. 156, 683-717 (1982)
9. Sprinzl,M., Gauss,D.H.: Nucl. Acids Res. 10, r1-r55 (1982)
10. Anderson,S., Bankier,A.T., Barrell,B.G., de Bruijn,M.H.L., Coulson,A.R., Drouin,J., Eperon,I.C., Nierlich,D.P., Roe, B.A., Sanger,F., Schreier,P.H., Smith,A.J.H., Staden,R., Young,I.G.: Nature 290, 457-465 (1981)
11. Van Etten,R.A., Walberg,M.W., Clayton,D.A.: Cell 22, 157-170 (1980)
12. Kobayashi,M., Seki,T., Yaginuma,K., Koike,K.: Gene 16,297-307 (1981)
13. Koike,K., Kobayashi,M., Yaginuma,K., Taira,M., Yoshida,E., Imai,M.: Gene 20, 175-183 (1982)
14. Sprinzl,M., Gauss,D.H.: Nucl.Acids Res. 10, r57-r81 (1982)
15. Celis,J.E., Piper,P.W.: Nucl.Acids Res. 10,r83-r91 (1982)
16. Zubay,G.: J. Mol. Biol. 4, 347-356 (1962)
17. Sekiya,T., Kobayashi,M., Seki,T., Koike,K.: Gene 11, 53-62 (1980)
18. Kearsey,S.E., Craig,I.W.: Nature 290, 607-608 (1981)
19. Blanc,H., Wright,C.T., Bibb,M.J., Wallace,D.C., Clayton,D. A.: Proc. Natl. Acad. Sci.,USA 78, 3789-3793 (1981)
20. Blanc,H., Adams,C.W., Wallace,D.C.: Nucl.Acids Res. 9,5785-5795 (1981)
21. Eperon,I.C., Anderson,S., Nierlich,D.P.: Nature 286, 460-467 (1980)
22. Brosius,J., Dull,T.J., Noller,H.F.: Proc.Natl.Acad.Sci.USA 77, 201-204 (1980)
23. Nierhauss,D., Nierhauss,K.H.: Proc.Natl.Acad.Sci. USA 70, 2224-2228 (1973)
24. Eckermann,D.J., Symons,R.H.: Eur.J.Biochem. 82, 225-234 (1978)
25. Dujon,B.: Cell 20, 185-197 (1980)
26. Glotz,D., Zweib,C., Brimacombe,R., Edwards,K., Kossel,H.: Nucl. Acids Res. 9, 3287-3306 (1981)

ORGANIZATION OF THE MITOCHONDRIAL 21S rRNA GENE IN *Saccharomyces cerevisiae* : MUTANTS OF THE PEPTIDYL TRANSFERASE CENTRE AND NATURE OF THE *omega* LOCUS

Bernard Dujon and Alain Jacquier

Centre de Génétique moléculaire du CNRS, Laboratoire associé à l'Université Pierre et Marie Curie 91190 GIF SUR YVETTE, FRANCE

Introduction

Chloramphenicol inhibits peptidyl transfer in bacterial and organelle ribosomes while cytoplasmic ribosomes remain unaffected (see 1 for review). DNA sequencing of two mitochondrial chloramphenicol resistant mutants (C^R) of yeast, has previously shown that single nucleotide changes in the distal section of the large rRNA gene determine the resistance to chloramphenicol (2). These nucleotide changes occur in two short sequence blocks (10 bp each) which are highly conserved between mitochondrial and bacterial rRNAs, suggesting their importance for the function of the ribosome. DNA sequencing of the corresponding regions of the large rRNA gene from mitochondria of mouse and human C^R mutants shows that, also in mammals, single base substitutions in these two blocks are responsible for the resistance to chloramphenicol (3-5). In yeast, the upstream block is split by the insertion of the intron of the 21S rRNA gene (intron rl) in *omega*$^+$ strains but not in the *omega*$^-$ strains (2, 6-8). Efficient introduction of the rl intron into its insertion site in the *omega*$^-$ form of the gene is observed after crosses with *omega*$^+$ strains (this results in a polarity of recombination between flanking genetic markers (9)). One of the two C^R mutations sequenced so far ($C^R{}_{323}$) is only three nucleotides upstream from the intron insertion site and, interestingly, it simultaneously confers the *omega*^{n} character (i.e. it abolishes the polarity of recombination), stressing the importance of the upstream block

Mitochondria 1983

for the integration of the rl intron (2). Because about half of the spontaneous C^R mutants isolated from omega$^-$ strains become omega^n as the results of single step mutations (10), it can be concluded that the C^R mutations are equally distributed between the two blocks and that only the upstream block gives rise to omega^n mutants (2). In a symmetrical manner, omega^n mutations have also been found in about half of the spontaneous C^S "fast" revertants isolated from some omega$^-$ strains carrying a specific C^R mutation (C^R_{336}) which confers a slow growth on glycerol medium (10). We report here, the determination of the nucleotide sequences of these C^S "fast" revertants and discuss the nature of the omega locus and the possible direct catalytical role of RNA sequences at the peptidyl transferase center. A possible mechanism for the insertion/deletion and transposition of introns, based on the analogy with the yeast mating type locus, will also be presented.

Results

We have determined the nucleotide sequence of six independent "fast" C^S revertants previously isolated from the "slow" growing C^R_{336} mutant (10). These revertants have been characterized genetically in details, two are

STRAINS	GENOTYPE		PHENOTYPE
IL324 -3A	a ural	$\omega^- C^R_{336}$	slow growth, chloramphenicol resistant
IL324-3A/2	- " -	$\omega^n_2\ C^R_{336}$	fast growth, chloramphenicol sensitive
IL324-3A/6/1	- " -	$\omega^n_{61}\ C^R_{336}$	fast growth, chloramphenicol sensitive
IL324-3A/8/2	- " -	$\omega^- C^S$	fast growth, chloramphenicol sensitive
IL324-3A/9/2	- " -	$\omega^n_{92}\ C^R_{336}$	fast growth, chloramphenicol sensitive
IL324-6B	a ural hisl	$\omega^- C^R_{336}$	slow growth, chloramphenicol resistant
IL324-6B/2/2	- " -	$\omega^- C^S$	fast growth, chloramphenicol sensitive
IL324-6B/2/3	- " -	$\omega^n_{23}\ C^R_{336}$	fast growth, chloramphenicol sensitive

Table 1. Strains used for DNA sequencing of the C^S "fast" revertants. The isolation and genetic characterization of these strains have been published previously (10).

omega$^-$ and four are omega^n (Table 1). Using the sequencing strategy outlined in figure 1, sequences were read directly from rho$^+$ mtDNA of these mutants along segments of ca. 400 bp centered around the unique SalI site and covering the chloramphenicol region. The results are summarized

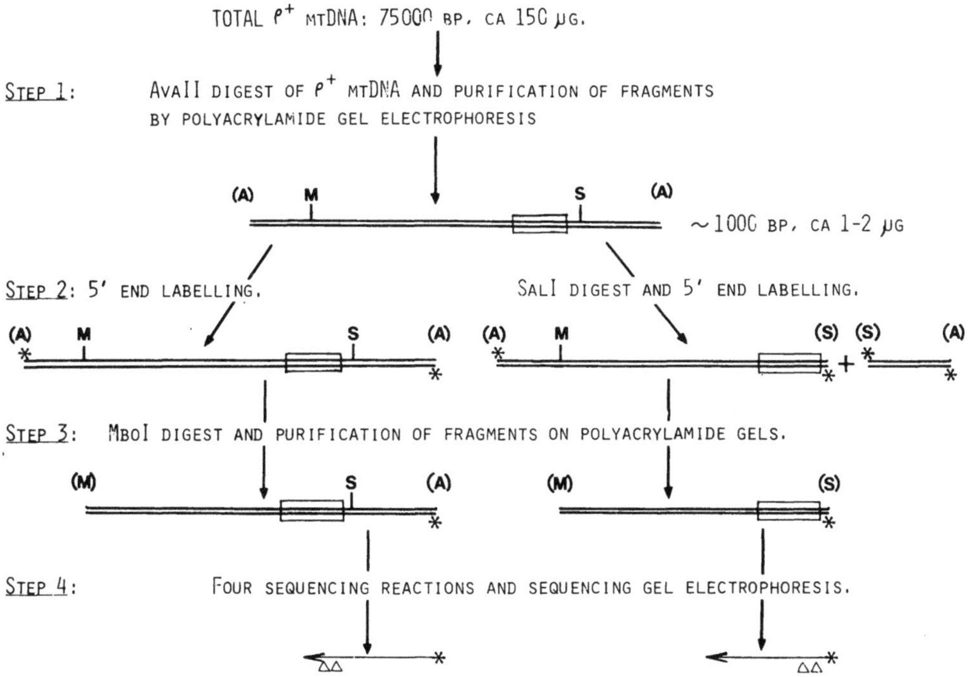

Fig. 1: Strategy used for direct rho$^+$ mt DNA sequencing of C^S "fast" revertants. Highly purified high molecular weight total rho$^+$ mt DNA has been prepared using a simple and efficient method described in (11). DNAs were prepared from 3 liters late stationnary phase glycerol cultures (usually 500 µg are obtained). Digestion of ca. 150 µg of total rho$^+$ mtDNA by AvaII produced 70-80 fragments which were separated on a 6%(1:20) polyacrylamide gel. After step 1, ca.1-2 µg of the fragment containing the region of interest were recovered from the gel. In some cases the fragment was contaminated by other fragments migrating at the same position in the gel but such contaminating material was easily eliminated at step 3 since the SalI site is unique in rho$^+$ mtDNA and the contaminating fragments happen not to contain MboI sites. Because of the proximity of the chloramphenicol region (box) to the SalI site, the samples were separated in two parts at step 2, one part being cut at SalI, the other not. This procedure ensures a continuous reading of sequences from the distal AvaII site to beyond the chloramphenicol region (ca. 400 bp in total). A=AvaII, S=SalI, M=MboI, *=^{32}P end label, △ △= mutational sites. Arrows indicate extent of sequences read.

in figure 2 (only the region of the two chloramphenicol blocks is shown). Sequences of all four <u>omega</u>^{n} C^{S} "fast" revertants were found identical and differ from the two <u>omega</u>$^{-}$ C^{S} "fast" revertant sequences (which are themselves identical) by only two base substitutions: a C to T transition in the upstream chloramphenicol block and a T to C transition in the downstream chloramphenicol block (positions 3 and 55 respectively in fig. 2). The first substitution occurs three bases downstream from the intron insertion site. No other differences could be found between the six strains examined in the ca. 400 bp long segment sequenced. Comparing these sequences with the wild type <u>omega</u>$^{-}$ sequence previously determined (2) reveals that the <u>omega</u>$^{-}$ C^{S} "fast" revertants are true back mutations of the original C^{R}_{336} mutation while the <u>omega</u>^{n} C^{S} "fast" revertants arise from a second site mutation (here called <u>sup</u>) which suppresses the resistance to chloramphenicol conferred by the original C^{R}_{336} mutation. The two sites are closely linked to each other in the 21S rRNA gene as was

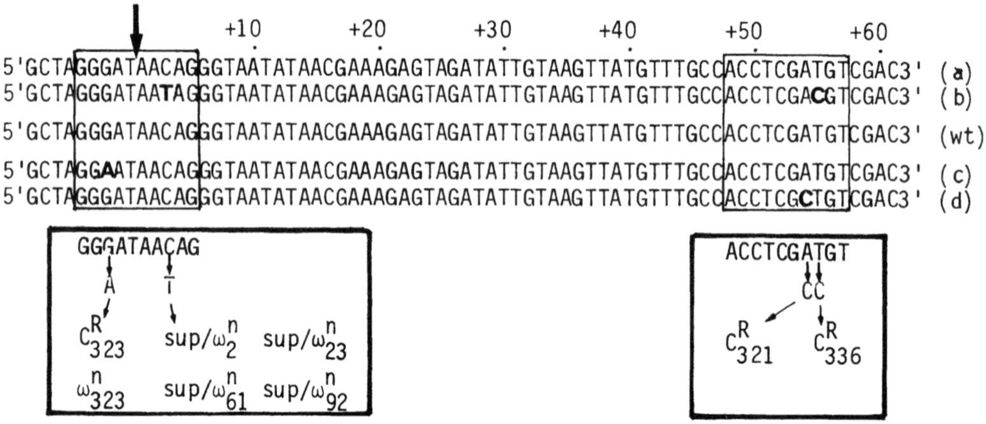

Fig. 2: Sequence of the non-transcribed strand of a fragment of the 21S rRNA gene from an <u>omega</u>$^{-}$ wild type (wt), from <u>omega</u>$^{-}$ and <u>omega</u>^{n} C^{R} mutants and from <u>omega</u>$^{-}$ and <u>omega</u>^{n} C^{S} revertants. Wild type sequence is from (2). (a) sequences of strains IL324-3A/8/2 and IL324-6B/2/2 (<u>omega</u>$^{-}$ C^{S} revertants). (b) sequences of strains IL324-3A/2, IL324-3A/6/1, IL324-3A/9/2 and IL324-6B/2/3 (<u>omega</u>^{n}C^{S} revertants). (c) sequence of strain IL9-8A (<u>omega</u>^{n}C^{R}_{323}). (d) sequence of strain IL16-11D (<u>omega</u>$^{-}$$C^{R}_{321}$. Sequences (c) and (d) are from (2). Positions are numbered from the intron insertion site in <u>omega</u>$^{+}$ strains (arrow). Boxed sequences are conserved in bacterial, chloroplast and mitochondrial RNAs and show only 1 or 2 substitutions in nuclear RNAs of various eucaryotes. Positions +15 and +37 show polymorphic variations in <u>S.cerevisiae</u> strains (2) but all strains used here are of the same form. Heavy boxes summarize all mutations sequenced so far that map to the conserved blocks.

predicted from recombination data (10). Now, which of the two nucleotide substitutions represents the first site C^R mutation, and which the suppressor, can be unambiguously determined because <u>rho</u>$^-$ deletion mapping previously reported (12) shows that the suppressor mutation must be either upstream or very closely linked to the intron insertion site. From this we conclude that the T to C transition at position 55 (fig. 2) represents the C^R_{336} mutation while the C to T transition at position 3 represents the <u>sup</u> mutation. Since the original C^R_{336} mutant is <u>omega</u>$^-$ (10) we further conclude that the C to T transition at position 3 is also responsible for the <u>omega</u>^{n} character of the "fast" C^S revertants. These results, when compared to previous result on the <u>omega</u>^{n} C^R_{323} mutant, define the <u>omega</u> locus as a short sequence (5' GATAAC 3') that lies symmetrically on both sides of the intron insertion site and in which single base substitutions affect the transfer of the rl intron. This sequence is split in two halves in the <u>omega</u>$^+$ strains and altered in the <u>omega</u>^{n} mutants.

Discussion

1) Single base substitutions simultaneously determine the resistance/sensitivity to chloramphenicol and the <u>omega</u>^{n} character.

Our results confirm and extend the previous finding that a single base substitution, within the upstream conserved block of the 21S rRNA gene, simultaneously determines the resistance/sensitivity to chloramphenicol and abolishes the polarity of recombination of the rl intron. We now show that, out of six C^S revertants independently isolated from the C^R_{336} mutation, the two <u>omega</u>$^-$ ones are true back mutations while the four <u>omega</u>^{n} ones are all due to the same C to T transition, three nucleotides downstream from the intron insertion site. That these isolates are truly independent is not, however, questionable because: 1) the original C^R strains were subcloned prior to isolation of revertants and a single revertant was picked from each subclone and 2) the four <u>omega</u>^{n} mutants arise from two different strains. The four times occurence of the same C to T transition indicates, therefore, that only this type of mutation is able to simultaneously suppress the resistance to chloramphenicol and

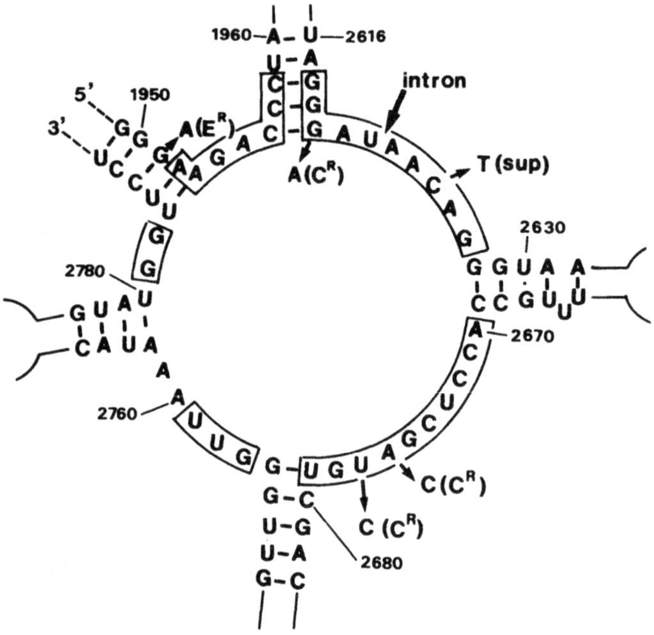

Fig. 3: Section of a secondary structure model of the mitochondrial 21S rRNA from S.cerevisiae showing the base substitutions that determine the resistance/sensitivity to chloramphenicol of the peptidyl transferase center. The structure has been arranged as in (13-16) with minor modifications, taking advantage of the conservation of primary sequences and double helical regions between bacterial, chloroplast and mitochondrial RNAs. Sequence from position 2616 to position 2787 is from (2) and this work, sequence from position 1949 to position 1960 is from (17). Mutations that cause resistance to chloramphenicol are a G to A transition at position 2620 (C^R_{323}), a A to C transversion at position 2676 (C^R_{321}) and a T to C transition at position 2677 (C^R_{336}). The C to T transition at position 2625 (sup) suppresses the resistance caused by mutations C^R_{336} and C^R_{321} and gives pseudo wild-type chloramphenicol sensitive phenotypes. The G to A transition at position 1951 which causes resistance to erythromycin is given for comparison (17). The arrow indicates the intron insertion site in the omega$^+$ gene. Boxed sequences are conserved in eubacterial, chloroplast and mitochondrial RNAs. Positions are numbered from the 5' end of the mature 21S rRNA as determined by (18, 19), assuming a gene sequence as in (20) but without the 65 bp mini-insert (2).

determine the <u>omega</u>^{n} character. This conclusion is fully consistent with the fact that the <u>omega</u>$^{-}$ to <u>omega</u>^{n} mutations (which were not selected for) always arised simultaneously to the C^R to C^S mutations (that were selected for) by single step mutations (10).

2) The interaction of two highly conserved non double helical RNA segments in the peptidyl transferase centre determines the resistance/sensitivity to chloramphenicol.

Elaborate secondary structure models have recently been proposed for the large rRNA molecule of bacteria, chloroplasts and mitochondria (13-15). The models are supported by the conservation of primary sequences and of double helical parts with compensatory changes. The mitochondrial 21S rRNA molecule of yeast can be folded into a secondary structure model which closely resembles that of the <u>E. coli</u> rRNA. Fig. 3 represents the portion of this model corresponding to the peptidyl transferase centre and summarizes all types of base substitutions that determine the resistance/sensitivity to chloramphenicol. Two main features emerge from our results which suggest that, in the peptidyl transferase center, the RNA is directly involved into the catalytical function and not through its binding to proteins. Firstly, the C^R mutations occur within the non double helical parts of the RNA. Therefore, they do not affect the stability of the secondary structure of the ribosome. To the contrary, the only two E^R mutations sequenced so far (the same G to A transition, see 17) occurs within a double helical part (fig 3, note, however, that this segment is not double helical in <u>E.coli</u>: see 13-15). Secondly, the different substitutions confer different phenotypes to the cell and to the isolated ribosome (21): they differ by the level of resistance to chloramphenicol, the cross resistance to erythromycin and spiromycin, the growth rate in glycerol medium lacking the antibiotic and the interaction with the E^R mutants. $C^R{}_{321}$ and $C^R{}_{323}$ mutants both grow normally on glycerol containing medium and are highly resistant to chloramphenicol (although $C^R{}_{323}$ is slightly less resistant than $C^R{}_{321}$). They are both weakly resistant to spiromycin and, in addition, $C^R{}_{323}$ shows a slight resistance to erythromycin (21). The $C^R{}_{336}$ mutation, on the contrary, grows poorly on glycerol (it is precisely this feature which allowed the isolation of the C^S "fast" revertants) and shows a strong interaction with many E^R

mutations such that the C^RE^R recombinants formed in crosses are rapidly and quantitatively converted into rho$^-$ cells (22 and Netter and Dujon, unpublished). The sup mutations acts equally well on the two C^R mutations of the downstream block and restores pseudo wild-type phenotypes (the strains with sup can be distinguished from true chloramphenicol sensitive wild types by a slight growth after long incubation in the chloramphenicol containing medium). This can be deduced, in the case of C^R_{321}, from previous crosses to specific rho$^-$ mutants (12). Unfortunately, we could not test the action of the sup mutation on the C^R_{323} mutation because of the difficulty to construct recombinants between the two sites (they are separated by only 5 bp). Finally, we do not know whether the sup mutation alone determines a specific phenotype.

Comparison of our results to the mammalian mutants (fig. 4) strengthens the idea of a direct involvment of the RNA in the function of the peptidyl transferase center because two different substitutions, at the same nucleotide, result in opposite phenotypes (resistance to chloramphenicol for the C to A transversion and suppression of resistance for the C to T transition). Figure 4 also reveals that greater constraints exists on the downstream block than on the upstream one. The four times occurence of the same T to C transition in organisms as different as yeast

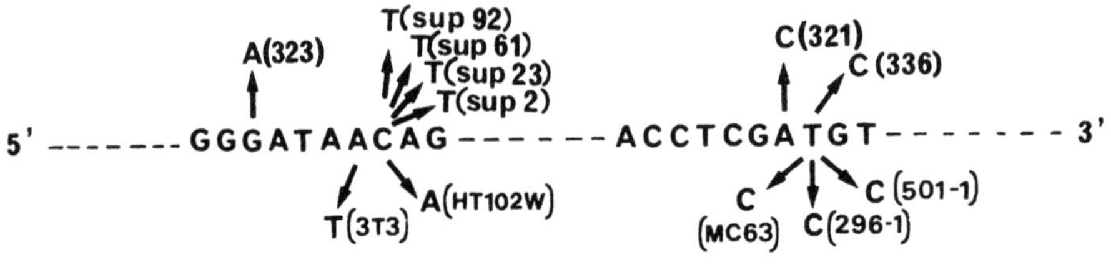

Fig. 4: A summary of all sequenced mutations that affect resistance to chloramphenicol in mitochondrial ribosomes. The figure represents the two short stretches of the rRNA gene which are highly conserved in various organisms and contain mutations affecting chloramphenicol resistance in yeast (top), human and mouse (bottom). Data taken from (2) for yeast mutants 321 and 323, from present work for other yeast mutants, from (4) for mouse mutant 3T3 and human mutant MC63, and from (3 and 5) for mouse mutant 501-1 and human mutants HT102W and 296-1

and mammals suggests that very few mutations in the downstream conserved block are capable of conferring resistance to chloramphenicol and/or are compatible with the function of the mitochondrial ribosome. A larger diversity of mutations is found in the upstream block where each C^R mutation is different.

3) The omega locus is a stretch of a few nucleotides lying on both sides of the intron insertion site.

Our results allow a better definition of the nature of the omega locus. The two types of $omega^n$ mutations limit a 6 bp long sequence (5' GATAAC 3') that lies on both sides of the intron insertion site. We propose that this sequence constitutes the omega locus i.e. represents a recognition site for the mechanism which triggers the insertion of the intron into the 21S rRNA gene during crosses. With the exception of the two nucleotides defined by the $omega^n$ mutants, neither the exact extent of this site nor the importance of each nucleotide it contains can, of course, be further precised here. But the important result is that the omega locus extends on both sides of the intron insertion site. The recognition sequence is, therefore, split by the presence of the intron in the $omega^+$ strains and this creates the dissymetry required for the efficient integration of the intron into the copies of the 21S rRNA gene originating from the $omega^-$ parent and not vice versa.

The mechanism by which such an integration takes place is not clear. Precise hypotheses, based on classical recombination models involving heteroduplex formation as the initiating step of a gene conversion process, have been put forward (23, 24). However, we favor hypotheses that do not involve long heteroduplex formation and allow conversion of both strands at once for the following reasons. Firstly, generalized recombination, as measured between other genetic markers (or even between the genetic markers flanking the rl intron), does not create enough recombinants to quantitatively explain the polarity of recombination at the omega locus. This has already been discussed in details (25-27) from calculations of the average number of mating rounds and maximal probabilities of gene conversion at each pairing. Secondly, branch migration, in the heteroduplex hypothesis, should be arrested when reaching the intron because no homology is found between the $omega^+$ and the omega^{-} strands at this point, hence preventing gene conversion. We, therefore, prefer the idea that the integration of the rl intron into the omega^{-} gene is not initiated by heteroduplex formation but is mediated by another pathway of recombination. The double strand break model that has recently been proposed (28) is one of them.

4) The omega locus in mitochondria shows striking similarities to the Y-Z1 border in the nuclear MAT locus.

An unexpected and intriguing observation seems to give credit to this last idea. The omega locus, as defined above, shows a striking resemblance to the Y-Z1 border in the MAT locus (fig 5). The stretch of 5 nucleotides (5' AACAG 3') which immediately follows the intron insertion site in the

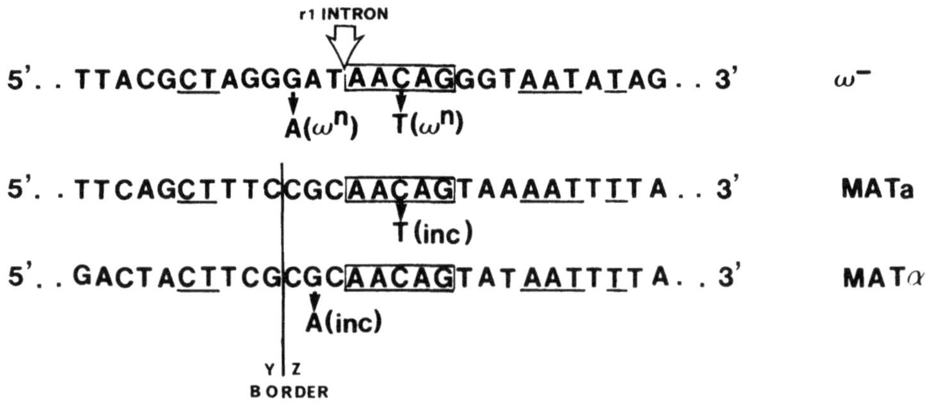

Fig. 5: Sequence similarities between the omega locus and the cleavage site at the MAT locus. The omega^{-} sequence is from (2). MATa and MATalpha sequences are from (32). In all cases the sequence of the non-transcribed strand is given. Sequences have been aligned from the common 5' AACAG 3' pentanucleotide (boxed). Other sequence homologies on either side of the pentanucleotide are underlined. Positions and natures of mutations are respectively inc4-28 for the C to T transition in MATa and inc3-7 for the G to A transition in MATalpha (29). Omega^{n} mutations are respectively omega$^{n}_{323}$ for the G to A transition and omega$^{n}_{2}$, omega$^{n}_{23}$, omega$^{n}_{61}$ and omega$^{n}_{92}$ for the C to T transitionS. Initiation of homothallic switching at MAT involves a four base pairs staggered cut within the pentanucleotide (29, 30) with one end corresponding to the intron insertion site.

mitochondrial 21S rRNA gene is present at the begining of the Z segment of the MAT locus. Such a 5 bp match is, obviously, not significant enough by itself (such a sequence occurs at random once in every ca.1800 bp for a DNA with 18% G+C, the average value of mtDNA, and once in every ca.1000 bp for 50% G+C). We regard, however, this homology as intriguing for the following reasons. Firstly, it is the very same mutation (a C to T transition) within this pentanucleotide which is responsible for the $omega^n$ character in mitochondria and for the incapability of MAT to switch in the nucleus (fig. 5). Secondly, the other mutant at MAT, the sequence of which has been determined, is a G to A transition occuring one nucleotide next to the G to A transition found in the other $omega^n$ mutant, when the two sequences are aligned from the pentanucleotide (fig. 5). Thirdly, the cassette switch at MAT is initiated by a double strand staggered cut within the 5' AACAG 3' pentanucleotide (29, 30), leaving a 4 bp 3' extension, and one of the cuts occurs exactly at the intron insertion site in mitochondria. Further sequence similarities may also be seen on both sides of the 5' AACAG 3' pentanucleotide (fig. 5) but their significance is obviously not certain.

5) A possible mechanism for intron insertion/deletion or transposition based on double stranded breaks at specific recognition sites.

The similarities between the omega locus and MAT led us to suspect that the mechanism responsible for the polarity of recombination of the rl intron in the $omega^+$ by $omega^-$ crosses might be similar to the mechanism of switching at the MAT locus. A double strand break at, or near, the $omega^-$ sequence (possibly within the sequence 5' GATAACAG 3') could be made by a specific endonuclease which should not recognize the $omega^+$ sequence because the site is split by the presence of the intron, hence the initial dissymetry. Then repair could take place by a variety of mechanisms, using the uncut $omega^+$ sequence as template. Such a mechanism is prone to explain that the genetic markers flanking the intron are also introduced from the $omega^+$ molecules into the $omega^-$ ones if repair extends to some distance on both sides of the original cut.

The analogy drawn here between the polarity of recombination in mitochondria and the switching at MAT is of mechanistic nature rather than of enzymatic nature. If the enzymes involved might similar, we consider it doubtful that both processes use the exact same set of enzymes because, unlike MAT switching, the polarity of recombination is not under the control of HO (data not shown, see ref. 31) and is not affected in crosses homozygous for the rad 52 mutation (38).

The nature of the template sequence for repair of the double stranded cut needs not be precisely defined here. It could obviously be the $omega^{+}$ gene itself but it could also be a precursor transcript (or reverse transcript) of the 21S rRNA or even the spliced intron RNA itself (or reverse transcript of it). If it is true that, in mitochondria, partial ribosome assembly precedes intron splicing (39), then only the precursor transcripts and the excised intron RNA should be available as template sequences for the mechanism postulated here, hence favouring intron insertion. The upstream junction hairpin and the internal guide sequence which have been proposed to help align intron-exon junctions for splicing (33-35) could serve, by a reverse mechanism, for integration of introns as well, if a cut is available in the recipient DNA at or near one end of the guide. This could exactly be the situation for the rl intron. Similarly, a double stranded cut at a specific site could be the initiating step for precise loss of introns (see 36, 37 and Gargouri et al., this volume) if the cut occurs within a gene close to an intron-exon junction. In addition, for mRNAs, unlike rRNAs, it is the mature transcript which should be primarily available as template sequence for the mechanism we postulate, hence favouring intron deletion. Finally, double stranded cuts at specific points may also trigger intron transposition if an appropriate intron nesting site is available in the vicinity of the cut site. To this respect, two segments bearing striking resemblance to the omega locus (13 bp and 16 bp long respectively with almost complete matching) exist in the mitochondrial DNA of S.cerevisiae. However, they do not seem appropriate for an intron insertion because one is in opposite orientation with respect to transcription (at the end of the cob-box gene) and the other is in the direct orientation but inside the "core" sequence of the second intron of the oxi 3 gene. But it is perhaps from another yeast species (Kluyveromyces sp.) that the first example of such a transposition will be available (A.Jacquier, in prep.).

Further experiments are critically needed to determine if there exists specific double stranded breaks in mitochondrial DNA (and what their frequency distribution would be), the exact nature of the intron nesting sites, if reverse transcription could be active in mitochondria and if, in addition to the *cis* acting site discussed here, mitochondrially encoded *trans* acting function(s) are also involved in the integration or transposition of introns. In particular, the role of the rl encoded ORF, which is evolutionarily conserved (11), remains to be clarified.

Acknowledgments. We thank H.Blanc, P.Dehoux, W.Fangman, M.Hours and F.Michel for their stimulating discussions. This work was supported by grants from CNRS "ATP Biologie Moléculaire du gène" and from INSERM "PRC 134004".

References

1. O'Brien T.W. and D.E.Matthews (1976) in Handbook of Genetics vol.5 pp535-580 R.C.King Ed. Plenum Press, New York

2. Dujon B. (1980) Cell 20, 185-197

3. Blanc H., C.T.Wright, M.J.Bibb, D.C.Wallace and D.A.Clayton (1981) Proc.Natl.Acad.Sc. USA 78, 3789-3793

4. Kearsey S.E. and I.W.Craig (1981) Nature 290, 607-608

5. Blanc H., C.W.Adams and D.C. Wallace (1981) Nuc.Acids Res. 9, 5785-5795

6. Bos J.L., K.A.Osinga, G.Van der Horst, N.B.Hecht, H.F.Tabak, G-J.B.Van Ommen and P.Borst (1980) Cell 20, 207-214

7. Bos J.L., C.Heyting, P.Borst, A.C.Arnberg and E.F.J.Van Bruggen (1978) Nature 275, 336-338

8. Faye G., N.Dennebouy, C.Kujawa and C.Jacq (1979) Molec.Gen.Genet. 168, 101-109

9. Bolotin M., D.Coen, J.Deutsch, B.Dujon, P.Netter, E.Petrochilo and P.P.Slonimski (1971) Bull.Inst.Pasteur 69, 215-239

10. Dujon B., M.Bolotin-Fukuhara, D.Coen, J.Deutsch, P.Netter, P.P.Slonimski and L.Weill (1976) Molec.Gen.Genet. 143, 131-165

11. Jacquier A. and B.Dujon (1983) Molec.Gen.Genet. (in press)

12. Michel F., C.Grandchamp and B.Dujon (1979) Biochimie 61, 985-1010

13. Noller H.F., J.A.Kop, V.Wheaton, J.Brosius, R.R.Gutell, A.M.Kopylov, F.Dohme and W.Herr (1981) Nucl.Acids Res. 9, 6167-6189

14. Glotz C., C.Zwieb, R.Brimacombe, K.Edwards and H.Kössel (1981)

Nuc.Acids Res. 9, 3287-3306

15. Branlant C., A.Krol, M.A.Machatt, J.Pouyet, J.P. Ebel, K.Edwards and H.Kössel (1981) Nuc.Acids Res. 9, 4303-4324

16. Garrett R. (1983) Trends in Biochem. Sc. 8, 189-190

17 Sor F. and H.Fukuhara (1982) Nucl.Acids Res 10,6571-6577

18 Osinga K.A. and H.Tabak (1982) Nucl.Acids res. 10, 3617-3626

19. Edwards J.C., D.Levens and M.Rabinowitz (1982) Cell 31, 337-346

20. Sor F. and H.Fukuhara (1983) Nucl.Acids Res. 11, 339-348

21. Grivell L.A., P.Netter, P.Borst and P.P.Slonimski (1973) Bioch.Biophys.Acta 312, 358-367

22. Netter P., E.Petrochilo, P.P.Slonimski, M.Bolotin-Fukuhara, D.Coen, J.Deutsch and B.Dujon (1974) Genetics 78, 1063-1100

23. Butow R.A., P.Terpstra and R.L.Strausberg (1979) in "Extrachromosomal DNA" D.J.Cummings et al. Eds Academic Press, New York pp269-286

24. Heyting C. and H.N.Menke (1979) Molec.Gen.Genet. 168, 279-281

25. Dujon B., Slonimski P.P. and L.Weill (1974) Genetics 78, 415-437

26. Dujon B., A.Kruszewska, P.P.Slonimski, M.Bolotin-Fukuhara, D.Coen, J.Deutsch, P.Netter and L.Weill (1975) Molec.Genet.Gen. 137, 29-72

27. Dujon B. (1976) PhD Thesis Univ. P. and M. Curie, Paris

28. Szostak J.W., T.L. Orr-Weaver and R.J.Rothstein (1983) Cell 33, 25-35

29. Weiffenbach B., D.T.Rogers, J.E.Haber, M.Zoller, D.W.Russell and M.Smith (1983) Proc.Natl.Acad.Sc. USA 80, 3401-3405

30. Strathern J.N., A.J.S.Klar, J.B.Hicks, J.A.Abraham, J.M.Ivy, K.A.Nasmyth and C.McGill (1982) Cell 31, 183-192

31. Jensen R., G.F.Sprague,Jr. and I.Herskowitz (1983) Proc.Natl.Acad.Sc. USA 80, 3035-3039

32. Astell C.R., L.Ahlstrom-Jonasson, M.Smith, K.Tatchell, K.A.Nasmyth and B.D.Hall (1981) Cell 27, 15-23

33. Michel F., A.Jacquier and B.Dujon (1982) Biochimie 64, 867-881

34. Michel F. and B.Dujon (1983) The EMBO J. 2, 33-38

35. Davies R.W., R.B.Waring, J.A.Ray, T.A.Brown and C.Scazzocchio (1982) Nature 300, 719-724

36. Jacq C., P.Pajot, J.Lazowska, G.Dujardin, M.L.Claisse, O.Groudinsky, H.De la Salle, C.Grandchamp, M.Labouesse, A.Gargouri, B.Guiard, A.Spyridakis, M.Dreyfus and P.P.Slonimski (1982) in "Mitochondrial genes" P.P.Slonimski et al. Eds. Cold Spring Harbor Laboratory, New York pp155-183

37. Labouesse M. and P.P.Slonimski (1983) The EMBO J. 2, 269-276

38. Dujon B., A.Baranowska and G.Dujardin (1977) in "Mitochondria 1977" W.Bandlow et al. Eds. W.De Gruyter and Co. Berlin and New York

39. LaPolla R.J. and A.M.Lambowitz (1979) J.Biol.Chem. 254, 11746-11750

Note added in proof: The sequence of a mitochondrial chloramphenicol resistant mutation from rat has been reported at this meeting (Koike et al. this volume). It is a single base substitution at a new position of the non double helical part of the peptidyl transferase center, upstream from the two blocks defined here.

THE CORRECTED SEQUENCE OF THE FIRST INTRON OF THE COB-BOX GENE IN THE YEAST STRAIN 777-3A

Jaga Lazowska, Ali Gargouri and Piotr P. Slonimski
Centre de Génétique Moléculaire, Laboratoire Propre du C.N.R.S., associé à l'Université P. et M. Curie, 91190 Gif-sur-Yvette, France

Introduction

In a preceding article (Lazowska *et al.*, 1980) we have reported the sequence of the box3 open reading frame located in the second intron, bI2, of the gene specifing cytochrome b in yeast mitochondria, and have shown that it encodes a trans-acting protein involved in splicing, the bI2 mRNA-maturase. At the same time we have determined the sequence of the first intron (bI1) and deduced from it that it cannot code for a protein. However, we have mentioned in this article that the sequence of this region was less reliable than that of the box3 region. Especially, uncertainties remained in the AT rich segment located between the XbaI and BglII sites (Lazowska *et al.*, 1980).

In this article we report the corrected sequence of the intron bI1. Several dispersed single base errors are corrected and the main error concerning the orientation of the Hinf-Hinf fragment 177 bp long and including the XbaI site is straightened out. The correct length of the intron bI1 is 767 bp (and not 765 bp, as previously reported) but the main conclusion remains the same : all the six frames are completely blocked so that the intron cannot code for any protein.

In addition, we confirm by comparaison with the isogenic strain our previous result that the mutation box4-2 (G171) is located close to the exon B1-intron bI1 junction and transforms a CAG sense codon into a TAG nonsense codon. We also pin point some polymorphic variations in the exon B1.

Mitochondria 1983

Results and Discussion

Figure 1 shows the restriction map and the strategy for sequencing the intron bI1 and its flanking regions. The restriction sites denoted in the Fig. 1 were found using a computer program. The presence of all these sites, except the AccI and NdeI, was verified on the mtDNA from the rho⁻ clone L29 which was used for sequencing. The unique difference between the previous restriction map and the present one is the relative position of the XbaI site within the small 177 bp Hinf-Hinf fragment. The position of the XbaI site on the right side of this fragment was determined as follows: on the one hand, rho⁻L29 mtDNA was digested with SfaNI + DdeI restriction endonucleases, the SfgaNI-DdeI fragment (on the left of Fig. 1) was eluted, labeled atits 5'-ends with γ-^{32}p-ATP (from Amersham) and polynucleotide kinase (from Boehringer) and digested with XbaI. The two labeled fragments resulting from this digestion were sequenced according to the method of Maxam and Gilbert (1980). On the

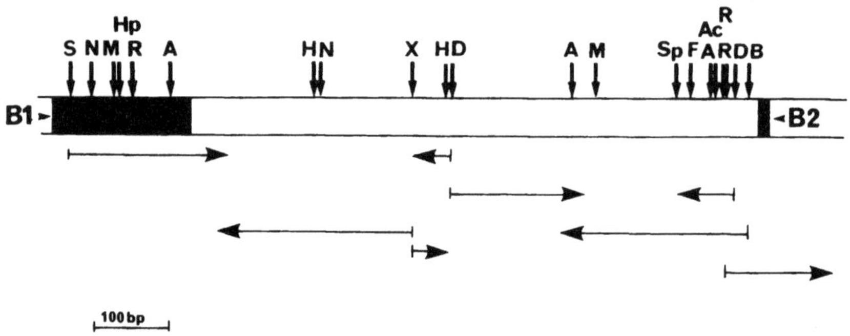

Figure 1: Strategy for sequencing the intron bI1

The upper part shows the restriction map of the sequenced region:A-AluI, Ac-AccI, B-BglI, D-DdeI, F-FokI, H-HinfI, Hp-HphI, M-MboI, N-NdeI, R-RsaI, S-SfaNI, Sp-SphI, X-XbaI. The arrows below indicate the direction and approximative length of the fragments sequenced. Black bars: a part of exon B1 and exon B2.

other hand, the rho$^-$L29 mtDNA was digested by XbaI, labeled and redigested with RsaI + DdeI endonucleases and the two labeled fragments were sequenced. From these two independant determinations it appeared clearly that the fragments XbaI-DdeI and DdeI-XbaI have the same length (49 bp), complementary sequences and permit to sequence across the right HinfI site. The sequencing gels are shown in figure 2.

Taking together the results reported previously (Lazowska et al. 1980) and those reported here, we conclude, that in the first determination of the sequence of this Hinf fragment, which was done after strands separation, we have inverted the fragment's orientation and therefore mistaken the coding for the noncoding strand.

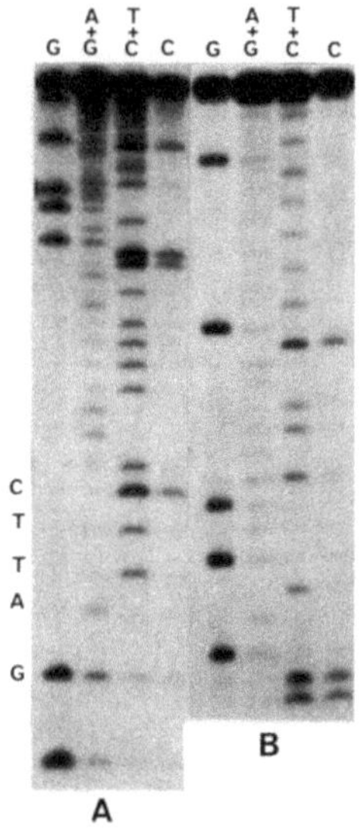

Figure 2: Sequencing gels of the small DdeI-XbaI fragment 5'-labeled at the DdeI site.

In A a 20% sequencing gel shows the HinfI restriction site. In B a 8% sequencing gel. The gels represent the sequence of the nonsense strand.

```
1 ATG GCA TTT AGA AAA TCA AAT GTG TAT TTA AGT TTA GTG AAT AGT TAT ATT ATT GAT TCA   60
  CCA CAA CCA TCA TCA ATT AAT TAT TGA TGA AAT ATG GGT TCA TTA TTA GGT TTA TGT TTA
  GTT ATT CAA ATT GTA ACA GGT ATT TTT ATG GCT ATG CAT TAT TCA TCT AAT ATT GAA TTA
  GCT TTT TCA TCT GTT GAA CAT ATT ATA AGA GAT GTG CAT AAT GGT TAT ATT TTA AGA TAT
  TTA CAT GCA AAT GGT GCA TCA TTC TTT TTT ATG GTA ATG TTT ATG CAT ATG GCT AAA GGT  300
  TTA TAT TAT GGT TCA TAT AGA TCA CCA AGA GTA CTA TTA TGA AAT GTA GGT GTT ATT ATT
       *                                                          *   ▼
  TTC CTT TTA ACT ATT GCT ACA GCT TTT TTA GGT TAT TGT TGT GTT TAT GGA CAG AGT GAG
  ACA AGT ATA AGT ATA TTA TTA TAA TAT CAT ACC ATT GAA TAA ATT ATT TTA ATG AAA TGA
  TTA TGT TTA TAT ATA ACA TAT ACC TAA TTA GAC ATG CAT TAT TAG TAA TAA TTT TAT ATG
  AAA CTC TAA TAA TAA TAA TTA TTA TTA ATT ATT AAG GTA AGA TTC ATA TGG ATA GCG TAA  600
  GTC AAT CTA ATA TTA TAA AAT ATC GTA ACA TAA ACA ATA TTT TTT TCT ATT ATT AAT TAA
  TAA ATA ATA ATA AAT AAA AAT AAT TAT ATG AGA AGT AAG ATA TTC AAT TCT GTC TAG AAT
  ACA TAT ATA TAC GTT AAT ACT CAT CGG TAT AAA ATT AGA ATC CTA AGT GAA TTA TTG AAA
  GTA TAA TAA TAT AAA CTT GGT AAG CCC AAT TAT TTC CAT ATA ATA TTA ATA TAA ATA TTA
  TAT GGT AGT TAT ATA TAA TAT TAT TAA ATA AAT AAT AAT AGA AAT TAT AAT ATA GAT AAG  900
  TGG GTA AAA GAC TAT TGA AAA AGC TAA AGA TTA TAT GTA ATG TAT AAT ATA GAT CAA ATT
  ATT ATA TAT TTT AAT AAA AAT ATA TTA ATA ATG GTT AAT ATT ATT ATT AAT TAA TTA ATT
  AAT TAA TTA ATA ATA ATA ACG AAT AAA TGA TTA ATG TGA AAG CAT GCT AAC TTC AAT ATA
  GGA TGA TTT ATA TAG TAT ATA AAT TGT TTG AGC TGT ATA CTA TGA AAG TAG TAC GTA CAG 1140
  TTC TGA GTG GGG GAA AAT TTG TAA AGA TCT ACC TAT CAC AAT TG TCA CAT TGA GGT
```

Figure 3. Corrected nucleotide sequence of the first intron and its flanking exons of cytochrome b long gene in the wild type strain 777-3A.

The sequence of the sense strand is displayed from the 5' to the 3' direction. The sequence of intron bI1 is underlined. The polymorphic changes are denoted by asterisk and black triangle indicates the position of the <u>box</u> <u>4-G171</u> mutation.

The <u>rho</u>⁻L29 mtDNA was also used for determination of the sequence downstream from the XbaI site. As seen in the Fig. 1, it was performed with the labeled DdeI-DdeI fragment (from the double digestion mentioned above) cut with SphI; with labeled BglII fragment cut by HinfI; and with labeled RsaI-RsaI fragment cut by BamHI (only 5' end of this fragment is shown). This has permitted, thanks to the better quality of the sequencing gels to correct several scattered errors. The corrected sequence of the first intron is shown in figure 3. It is 767 bp long instead of 765 bp as previously reported. In spite of these changes, the main conclusion remains the same: all the six reading frames are completely blocked. They contain 27, 30 and 24 stop codons on the sense strand and 20, 18 and 18 stop codons on the nonsense strand. The longest ORF has 52 codons, which is compatible with a Poisson distribution for an average value of 9 codons between two stops.

Several additional informations have been obtained in the present study. We have repeated sequence determinations of exon B2 and of the maturase coding part of intron bI2. No discrepancies were found with the previously reported sequence (Lazowska et al. 1980).

We have completed the entire sequence of exon B1 from the long gene present in the strain 777-3A. For this purpose we have sequenced the HphI-HphI fragment (sequencing strategy not shown) from two isogenic $\underline{rho}^-$ clones L29 and L12 and the SfaNI-DdeI fragment from $\underline{rho}^-$L29. This sequence is shown at the begginning of the Fig. 3 and is identical to the one reported previously by us (Lazowska et al. 1980) and by Nobrega and Tzagoloff (1980) for the strain D273-10B. We confirm our previous finding that the respiratory deficient mutation box 4-G171 carried by the $\underline{rho}^-$L12 and absent from the $\underline{rho}^-$L29 is located close to the junction B1-bI1 and transforms a CAG codon at the position 412 into a TAG stop codon.

Two other changes, which we believe represent functionally silent polymorphic variations, were found. The first one concerns position 364. In two instances we found a C (codon CTT = threonine) and in one instance an A (codon ATT = leucine). The first two cases concern the $\underline{rho}^-$L12, which carries the box 4-G171 mutation, and a rho^+ revertant derived from the box 8-P366 mutant (see Gargouri et al., this volume). The second instance concerns $\underline{rho}^-$L29 which carries the box 3-G1909 mutation and also applies to the strain D273-10B (Nobrega and Tzagoloff, 1980). Since the same A $\rightleftharpoons$ C substitution at position 364 is found in various respiratory sufficient wild type strains (or their $\underline{rho}^-$ derivatives) we do not think that it could lead to a functional impairement. It is rather due to sequence polymorphism. Another variation concerns position 410. In three instances (L29, L12 and a box 8-P366 revertant) we found an A instead of a G present in D273-10B (Nobrega and Tzagoloff, 1980). However, this observation has to be taken with caution, until the opposite strand sequence is established.

Acknowledgements

This work was supported by grants from the CNRS, DGRST, INSERM and ligue Nationale Française contre le Cancer. A.Gargouri acknowledges the receipt of a Bourse de Cooperation du Gouvernement Français.

References

1. Lazowska, J., Jacq, C., Slonimski, P.P.: Cell 22, 333-348 (1980).

2. Maxam, A., Gilbert, W. Meth. Enzymol. 65, 499-560 (1980).

3. Nobrega, F., G., Tzagoloff, A.: J. Biol. Chem. 255, 9828-9837 (1980).

NUCLEAR MUTANTS DEFECTIVE IN PROCESSING OF cob AND oxi3 TRANSCRIPTS IN SACCHAROMYCES CEREVISIAE

Tim Pillar, Jan Kreike, B. Franz Lang, Fritz Kaudewitz
Institut für Genetik und Mikrobiologie der Universität München, Maria-Ward-Str. 1a, D-8000 München 19, Germany (FRG)

Introduction

Both mitochondrial and nuclear mutants of S. cerevisiae have been isolated, which are defective in the correct excision of individual introns from cob and oxi3 transcripts (1-5). This suggests that processing of the mitochondrial cob and oxi3 transcripts depends on the participation of mitochondrial and nuclear encoded products. The simultaneous defect in cob and oxi3 processing (box-effect) as observed in some mitochondrial mit^- mutants, has been characterized by several groups (6-8). These mutants are defective in the production of a functional cob I4 maturase, which is necessary for the excision of cob I4 (bI4) itself, and apparently also for the excision of oxi3 I4 (aI4) (9,10). We are interested in the potential control mechanisms of cob and oxi3 processing and the interaction of these processes. Therefore we have screened our collection of pet^- strains for mutants defective either simultaneously in cob and oxi3 processing (box type), or in the excision of aI4 alone.

We have found three classes of such mutants:
(i) nuclear box type mutants, in which the bI4 maturase cannot be synthesized because upstream introns are not excised from the cob transcript. As a consequence also oxi3 transcripts cannot be fully processed;
(ii) nuclear box type mutants, in which bI4 is excised from the cob transcript, but a nuclear encoded product is defective,

Mitochondria 1983

involved both in the excision of bI5 and the oxi3 RNA processing;
iii) nuclear mutants, which are only deficient in the synthesis of cytochrome c oxidase subunit I. They appear to be defective in the excision of aI4.
In order to study the function of these nuclear genes and the cooperation of their gene products with the mitochondrially encoded maturases in the splicing process, we have transformed two of our mutants, using the yeast gene bank obtained from F. Lacroute. In this paper we describe the effects of the three classes of mutations on the processing of cob and oxi3 transcripts, and we will present a preliminary characterization of the plasmids complementing the nuclear mutations in two nuclear box type mutants.

Results and Discussion

Nuclear box type mutants defective in the production of bI4 maturase

We have isolated and characterized a number of nuclear pet^- mutants in which the correct splicing of cob and oxi3 transcripts is defective. The nuclear mutant MK2, containing the long form mitochondrial genome (MK2 long form), in which the cob gene contains five introns (Fig.1), has the box phenotype. This mutant is unable to excise introns bI2, bI3 and bI4 from the cob RNA, enriching an intermediate of 32S. oxi3 RNA processing is also blocked. This failure to excise bI2, bI3 and bI4 from the 32S cob intermediate may be interpreted in two ways:
(i) the nuclear encoded factor defective in this mutant is required for the correct excision of all three introns;
(ii) a polar effect exists, in which the non-excision of bI2 and/or bI3 is the primary effect of the mutation and this prevents the removal of the downstream intron bI4.

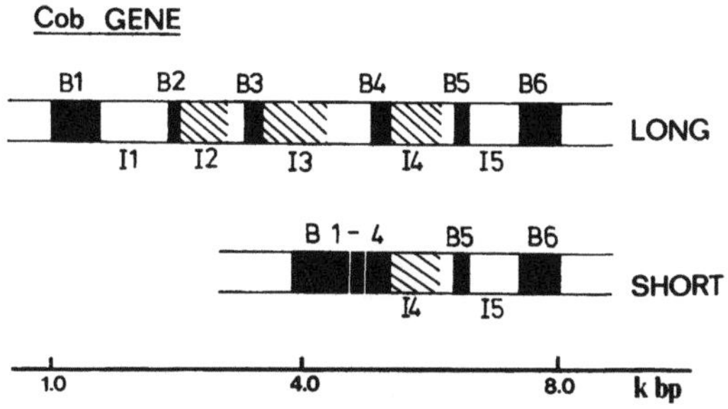

Fig.1: A diagrammatic representation of the cob gene from S. cerevisiae showing the long (KL14-4A) and short (D273-10B) forms.
Black boxes denote exon sequences (e.g.B1), open boxes denote intron sequences (e.g.bI1). Diagonal lined areas denote the approximate extent of known open reading frames, in phase with the preceding exon (11,12). At the 3' ends of bI2, bI3 and bI4 are closed reading frames.
Exons B1,2,3 and 4 of the long form genes are equivalent to the first exon of the short form (B1-4).

To determine which of these interpretations is correct, we introduced by cytoduction the short mitochondrial genome, in which the cob gene lacks bI1, bI2 and bI3 (Fig.1), into a nuclear background of MK2. If the failure to excise bI4 is due to a polar effect, it would be expected that the mutant containing the short cob gene would excise this intron from the cob RNA. This is found not to be the case. The mutant remains cytochrome b deficient due to the inability to excise bI4, enriching a cob intermediate of 24S (Fig.2), but interestingly is now able to process oxi3 RNA to the mature mRNA, which is translated (Fig.3,4). We conclude from these results that the mutation in MK2 prevents the correct excision of bI2, bI3 and bI4 from cob RNA, in other words, the nuclear factor defective in this strain might be required for the excision of three cob introns.

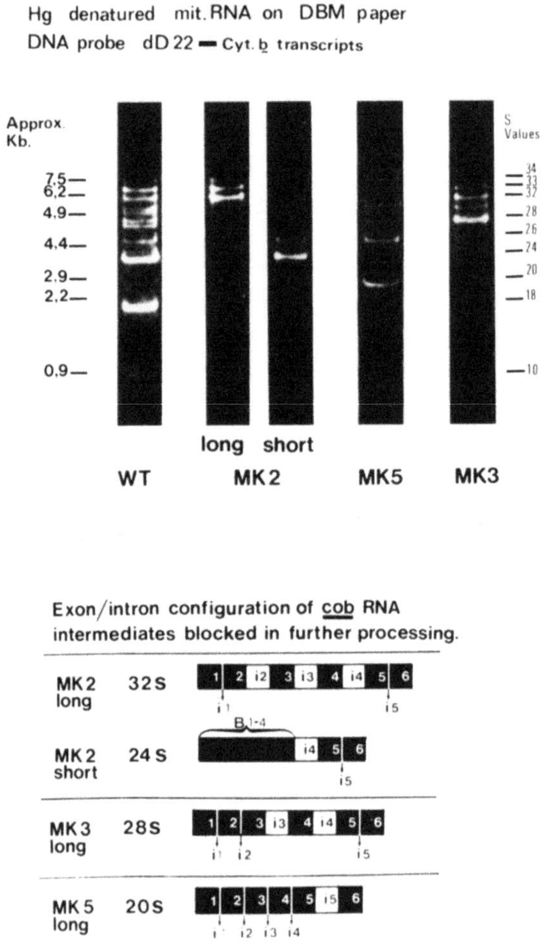

Fig.2: Nuclear mutants defective in cob RNA processing

Whole mitochondrial RNA was denatured on dimethyl mercury gels and blotted onto DBM paper as described by Alwine et al. (16). Cob RNA transcripts were visualized with cob DNA probes. "Long" and "short" denote, which mitochondrial genome is contained in the MK2 nuclear background (see Fig.1). WT = wild type. MK2, MK5 and MK3 are nuclear mutants defective in cob as well as in oxi3 RNA splicing (box type) (see Fig.3). The putative mature cob messenger RNA is approximately 2.2 kb long and bands at 18S.
The results of cob RNA transcript mapping are shown diagrammatically (5). The black boxes are exon sequences, white boxes introns. Excised introns are given below the intermediate, indicated by arrows. The respective S values are shown left of the intermediates.

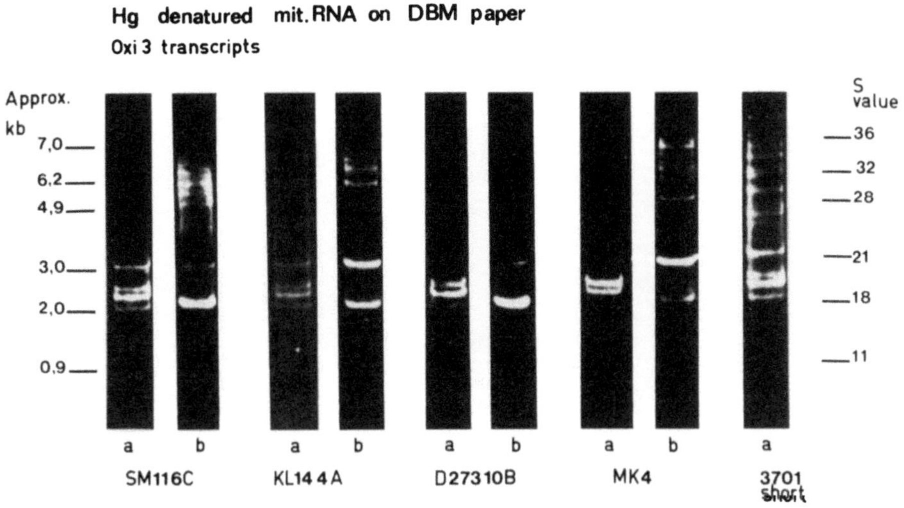

Fig.3: Identification of the mature oxi3 messenger RNA

DBM blots of mitochondrial RNA from three wild type strains and two nuclear mutants were hybridized with the following DNA probes: lane a) containing the whole oxi3 gene, b) spanning exon 4 and intron 4 of the oxi3 gene (see Fig.6). SM 11-6C (wild type, long form), KL 14-4A (wild type, long form), D 273-10B (wild type, short form), MK4 leaky box type nuclear mutant. 3701 (short form) nuclear mutant deficient in cytochrome b. MK4 is leaky, producing reduced amounts of the subunit I polypeptide. This is accompanied by reduced amounts of the 18S (2.1 kb) oxi3 RNA species on this gel system. It is likely that this species represents the translatable subunit I mRNA.

How can the box phenotype, seen in MK2 (long form) change to cob^-, when the mutant harbours the short gene? It has been proposed (9,10) that the bI4 maturase is involved in the excision of aI4 from the oxi3 transcript. In MK2 (long form) the bI4 maturase cannot be synthesized, since upstream introns bI2 and bI3 are not spliced out. They prevent translation from exon B1 downstream through the bI4 maturase sequence due to closed reading frames at the 3' ends of these introns (Fig.1). As a consequence oxi3 RNA processing is also blocked. The smallest oxi3 RNA species in MK2 (long form) is 3.0 kb long and hybridizes with a specific oxi3 aI4 DNA probe (Fig.5). This is

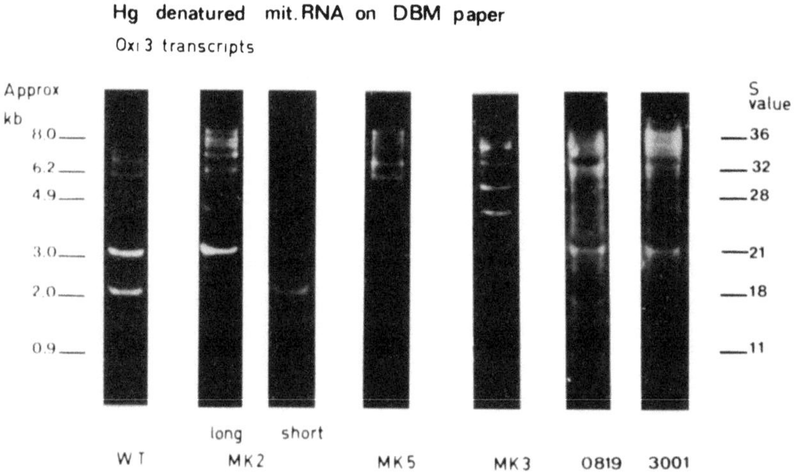

Fig.4: Nuclear mutants defective in oxi3 RNA splicing

DBM blots of RNA isolated from nuclear mutants were hybridized with a DNA probe, containing exon 4 and intron 4 of the oxi3 gene. WT = wild type. MK = nuclear mutants of box phenotype. "long" and "short" denote which mitochondrial genome is contained in the nuclear background of MK2. 0819 and 3001 are nuclear mutants deficient in subunit I of cytochrome oxidase (see Fig.7). The oxi3 mRNA bands at 18S (2.1 kb).

in line with the proposed dual function of the bI4 maturase (9,10). The size of this RNA would coincide with the size of an intermediate containing all exons (2.0 kb) plus intron aI4 (1.0 kb) (Fig.6). Whether this 3.0 kb RNA species does indeed correspond to this intermediate awaits further experimental proof.

The inability to excise bI4 from the cob RNA in MK2 (short form) leads to the accumulation of an intermediate of 24S, from which bI4 maturase can be translated. This is accompanied by the complete processing of oxi3 RNA (Fig.4). On basis of the results described above, we conclude that the box phenotype seen in MK2 (long form) is caused by a primary inhibition of cob RNA processing. The absence of a functional bI4 maturase prevents oxi3 RNA maturation as a secondary effect.

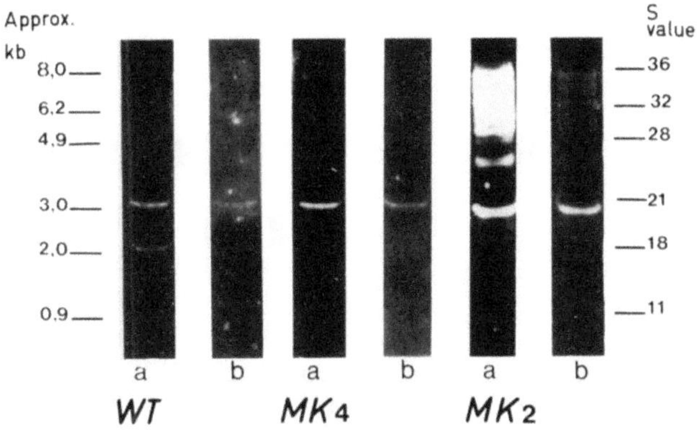

Fig.5: Hybridization with oxi3 aI4 DNA probe

Whole mitochondrial RNA from wild type (long form) and nuclear box type mutants was blotted onto DBM paper. The blots were hybridized with DNA probes: a) containing the oxi3 exon 4 and intron 4 (BamH I/Hind III); b) specific for aI4 (Hae III/Hind III). The oxi3 RNA intermediate of 3.0 kb (21S) hybridizes with probe b), showing aI4 homologous sequences are contained in this RNA species.

Nuclear box mutants having a functional bI4 maturase

Mutant MK5 belongs to a second type of nuclear box mutants. In this mutant the cob RNA intermediate, blocked in further processing contains intron bI5 (Fig.2). Since bI4 is excised, the bI4 maturase is apparently functional in this mutant. Processing of the oxi3 transcripts is blocked in this mutant at an intermediate of 30S, still containing at least four introns. Introns aI1 and aI2 are spliced out, since the two stable RNA species of 19.2S and 19.0S, corresponding to these excised introns, can be detected by hybridization (13) (results not shown). Therefore we consider that the nuclear factor defective in MK5 is not only required for the excision of bI5, but also for the oxi3 RNA processing. It is clear then, that the box effect in this mutant is distinct from that discussed for MK2, in that a single nuclear factor, defective in MK5, is necessary for both cob and oxi3 RNA processing.

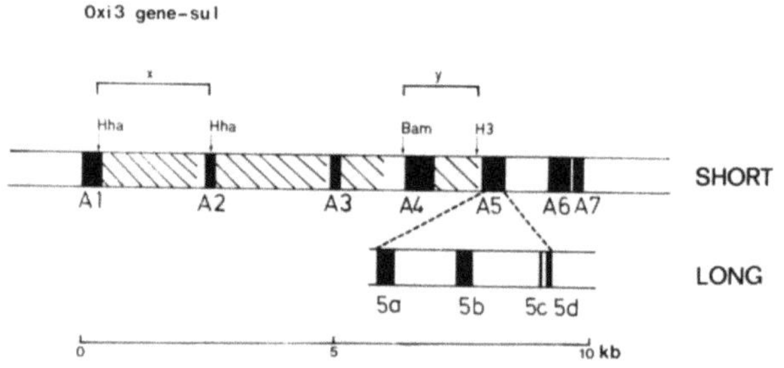

Fig.6: A diagrammatic representation of the oxi3 gene, showing the long (KL 14-4A) and short (D 273-10B) forms. Black boxes are exon sequences. White and dashed boxes represent introns, the dashed area denoting the extent of open reading frames in phase with the preceding exons.

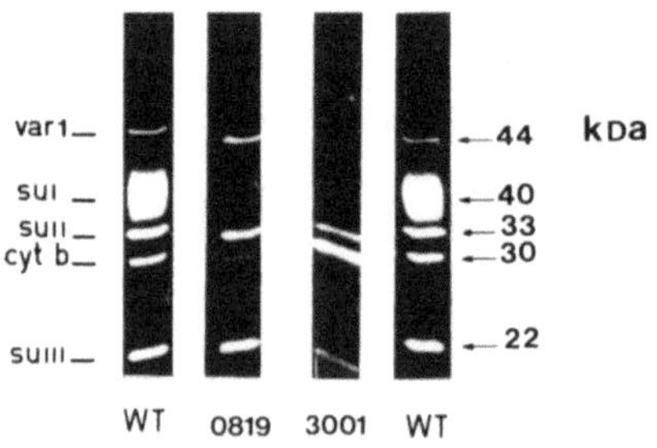

Fig.7: Mitochondrially coded polypeptide products of nuclear mutants deficient in SU I of cytochrome oxidase

^{35}S labelled mitochondrial products were separated on SDS acrylamide gels. Var = var I protein. SU I, II and III = subunits of cytochrome oxidase. Cyt b = apoprotein cytochrome b. kDa = kilodalton. WT = wild type, 0819 and 3001 = nuclear mutants deficient in cytochrome oxidase.

Nuclear mutants deficient in subunit I of cytochrome oxidase

We have also characterized mutants defective exclusively in the oxi3 RNA processing (Fig.7). A number of these mutants (e.g. 3001 and 0819) are particularly interesting, because they also accumulate the 3.0 kb oxi3 RNA intermediate, observed in

MK2 (long form) (Fig.4). This intermediate is the smallest RNA species hybridizing with specific aI4 probes. Probably in 3001 and 0819 a nuclear encoded factor is defective, that is necessary for the proper excision of oxi3 intron aI4.

The above data indicate that nuclear encoded products influence the processing of cob and oxi3 transcripts either indirectly via the production of bI4 maturase (MK2), or directly by a nuclear factor required for both cob and oxi3 processing. In the case of oxi3 intron aI4 we have data that suggest the participation of a nuclear encoded element in the excision of this intron. Since the bI4 maturase is probably also involved in the excision of oxi3 aI4, there might be an interaction between the nuclear encoded product, defective in our mutants, and the mitochondrially coded maturase. To investigate this cooperation we are planning several approaches, including the selection of mitochondrial mutations, which suppress the nuclear pet^- mutation (defective in aI4 excision), and the isolation of the nuclear gene, which will be used to set up an in vitro splicing system.

Isolation of nuclear genes involved in the processing of cob and oxi3 transcripts

Strain MK3 and strain 15-3, which has the same RNA pattern as MK2, but belongs to a different complementation group (Fig.2 and 4), have been crossed to introduce the $ura3^-$ and $leu2^-$ auxotrophic markers. We selected haploid strains (MK3* and 15-3*), which were indistinguishable from the original parental strains with respect to growth on glycerol, cytochrome spectra and mitochondrial RNA pattern. We transformed them with the yeast gene bank, kindly provided by F. Lacroute (14). From independent pet^+ transformants we have isolated three different plasmids complementing strain 15-3*, and two plasmids complementing MK3*.

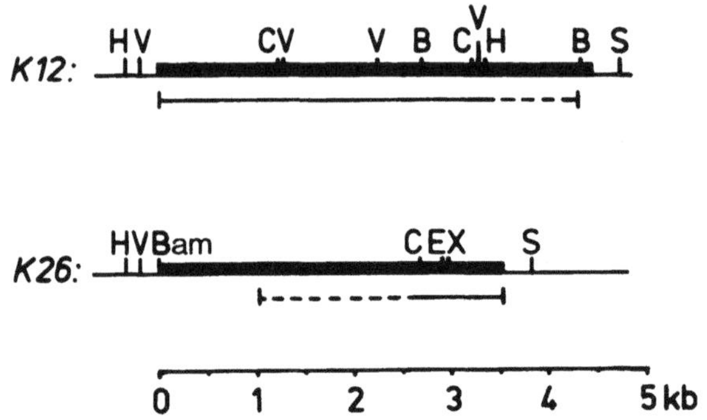

Fig.8: Restriction maps of plasmids K12 and K26

The restriction sites of the DNA inserts (thick lines) and the surrounding pFl 1 vector sequences (thin lines) are shown. The horizontal bars indicate the sequences homologous to other plasmids complementing the same mutant. The regions of homology, not yet certain, are indicated by dashed lines. Enzymes: B = Bgl II, Bam = BamH I, C = Cla I, E = EcoRI, H = Hind III, S = Sal I, V = EcoRV, X = Xho I.

The smallest plasmid complementing strain 15-3* (K12), has an insert of 4.5 kbp (Fig.8), of which the 2.1 kbp Cla I fragment and part of the 1.75 kbp Bgl II fragment are also present in the other two plasmids. To localize the gene in plasmid K12 we have removed the Bgl II fragment by digestion and religation. The resulting plasmid is no longer capable to complement the pet^- strain: the transformed cells do not grow on glycerol containing media. However, after prolonged incubation (3-5 days) pet^+ colonies appear with high frequency. We assume that the shortened plasmid does contain at least part of the structural information, which restores the pet^+ phenotype by integration or gene rescue. The observation that on non-selective media such transformants readily lose the $ura3^+$ phenotype, but remain pet^+, is in line with this idea.

The two plasmids complementing strain MK3* (K23 and K26) have inserts of 5.5 and 3.5 kbp, respectively. They only have in common the region around the EcoRI, the Xho I and the Cla I sites. First DNA sequencing data have revealed that the Cla I site lies within an open reading frame of at least 240 bp. These data suggest that a protein coding gene is located around the Cla I site of this plasmid, which is presumably involved in the processing of cob RNA. Surprisingly, after transformation with these plasmids, not the expected 100% of the $ura3^+$ transformants, but only 50-90% of them grow on glycerol. It is very unlikely that this phenomenon is caused simply by impurity of the DNA preparations, since it has been observed using several independent E. coli transformants as source of the DNA preparations. Moreover, restriction analysis showed that all preparations were essentially free from contaminating DNA. We are currently investigating this phenomenon in more detail.

Concluding Remarks

The data presented in this paper give information about the function of nuclear encoded elements in the processing of cob and oxi3 transcripts. We have shown that three nuclear encoded elements exist:

a) one involved in the excision of several introns from cob RNA;
b) another involved in the removal of aI4 from the oxi3 transcript; and
c) a third involved in the excision of bI5 from cob RNA and at least one intron from the oxi3 RNA.

Interestingly, the mutation in strain MK2 prohibiting the excision of three class I introns from the cob RNA (intron classification according to ref. 15), has no effect on the excision of other class I introns, e.g. in oxi3. When there are in this complementation group other mutants having exactly the same phenotype as MK2, this would imply that the class I introns can be subdivided into two different groups. We are currently investigating such mutants to test this possibility.

Acknowledgements

We thank R.J. Schweyen and G. Rödel for helpful comments, and I. Nerkorn for the typing of this manuscript. This work was supported by the Deutsche Forschungsgemeinschaft.

References

1. Dujon, B.: in "Molecular Biology of the Yeast Saccharomyces Life Cycle and Inheritance" (Strathern, J.N., Jones, E. and Breach, J.R., eds.) pp. 505-635, Cold Spring Harbor Laboratory (1981).

2. Borst, P., Grivell, L.A.: Nature 290, 443-444 (1981).

3. Dieckman, C.L., Bonitz, S.G., Hill, J., Homison, G., McGraw, P., Pape, L., Thalenfeld, B.E., Tzagoloff, A.: in "Mitochondrial Genes" (Slonimski, P.P., Borst, P. and Attardi, G., eds.) pp. 213-223, Cold Spring Harbor Laboratory (1982).

4. Faye, G., Simon, M.: Cell 32, 77-87 (1983).

5. Pillar, T., Lang, B.F., Steinberger, I., Vogt, B., Kaudewitz, F.: J.Biol.Chem. 258, 7954-7959 (1983).

6. De la Salle, H., Jacq, C., Slonimski, P.P.: Cell 28, 721-732 (1982).

7. Mahler, H.R., Hanson, D.K., Lamb, M.R., Perlman, P.S., Anziano, P.Q., Glaus, K.R., Haldi, M.L.: in "Mitochondrial Genes" (Slonimski, P.P., Borst, P. and Attardi, G., eds.) pp. 185-199, Cold Spring Harbor Laboratory (1982).

8. Haid, A., Grosch, G., Schmelzer, C., Schweyen, R.J., Kaudewitz, F.: Curr.Genet. 1, 155-161 (1980).

9. Netter, P., Jacq, C., Carignani, G., Slonimski, P.P.: Cell 28, 733-738 (1982).

10. Groudinsky, O., Dujardin, G., Slonimski, P.P.: Mol.gen. Genet. 184, 493-503 (1981).

11. Nobrega, F.G., Tzagoloff, A.: J.Biol.Chem. 255, 9828-9837, (1980).

12. Lazowska, J., Jacq, C., Slonimski, P.P.: Cell 22, 333-348, (1980).

13. Hensgens, L.A.M., Arnberg, A.C.,Roosendaal, E., Van der Horst, G., Van der Veen, R., Van Ommen, G.B., Grivell,L.A.: J.Mol.Biol. 164, 35-58 (1983).

14. Fasiolo, F., Bonnet, J., Lacroute, F.: J.Biol.Chem. 56, 2324-2328 (1981).

15. Michel, F., Jacquier, A., Dujon, B.: Biochimie 64, 867-881 (1982).

16. Alwine, J.C., Kemp, D.J., Parker, B.A., Reiser, J., Renart, J., Stark, G.R., Wahl, G.M.: in "Methods in Enzymology" (Wu, R., ed.) Vol. 68, pp. 220-242, Academic Press, New York (1979).

INTERFERENCE OF S.CEREVISIAE NUCLEAR MUTATIONS WITH MITOCHONDRIAL rRNA BIOSYNTHESIS AND OXI 3-TRANSCRIPT MATURATION.

Jürgen Naggert and Eckhart Schweizer
Lehrstuhl für Biochemie der Universität Erlangen-Nürnberg
D-8520 Erlangen

Introduction

In *Saccharomyces cerevisiae*, nuclear genes are involved in mitochondrial biogenesis at different levels: firstly, by encoding the majority of structural and functional components of the organelle and secondly, by providing part of the machinery for the biosynthesis and maturation of mitochondrial gene products (1, 2). The mitochondrial genes OXI 3 (3, 4, 5), COB (6, 7) and in some strains also the 21 S rRNA gene (8) exhibit intron-containing mosaic structures. Thus, the primary transcripts of these genes must undergo posttranscriptional processing to become functionally mature RNA's. In particular for the expression of the COB and OXI 3 genes, genetic data suggest that for the correct maturation of yeast mitochondrial RNA both mitochondrial and nuclear genes are necessary (6, 7, 2, 10). In addition, transcription of yeast mitochondrial DNA may proceed from only a few promoter sites. Therefore, the maturation of mitochondrial RNA's may consist of both, intron splicing and cleavage of polycistronic into monocistronics RNA's (9).

In order to obtain further insight into the complex nucleo-cytoplasmic interactions it may be helpful to study appropriate nuclear *S. cerevisiae* mutants defective in the biosynthesis or maturation of mitochondrial RNA. In the past, a collection of temperature-sensitive nuclear *pet*-mutants has been isolated in this laboratory. By *in vivo* pulse-labeling studies, Michaelis et al. (11) demonstrated that a sizeable portion of these nuclear mutants was defective in either one or several mitochondrially encoded proteins. Studies to be reported in this paper indicate that in several

Mitochondria 1983

of these mutants either mitochondrial RNA synthesis in general, or the levels of specific mtDNA transcripts are affected at the non-permissive temperature.

Results

1) mtDNA transcription patterns in selected pet-mutants defective in mitochondrial protein synthesis.

In order to characterize gross defects of mitochondrial RNA synthesis in a series of selected pet ts-mutants, total RNA was extracted from the mitochondria of these cells after growth at the non-permissive temperature. Under similar experimental conditions these mutants were known to be defective either in overall mitochondrial protein synthesis or specifically in subunit I or in subunits I and II of cytochrome oxidase (11). When subjected to agarose-gel electrophoresis in the presence of urea, most mutant mtRNA preparations exhibited no detectable difference from wild type mtRNA. Nevertheless, in a certain number of mutants, a representative sample of which is shown in Fig. 1 the mtRNA banding patterns were distinctly different from the wild type. All these mutants were non-allelic as inferred from their complementation characteristics (E. Schweizer, unpublished results). Some of the mutants (e.g. ts-575 and ts-3418) lacked all mtRNA. One would expect mitochondrial RNA polymerase mutations to exhibit this phenotype. In other mutants, like ts-2523, ts-1242, ts-3767 and ts-2168 the 21 S mitochondrial rRNA component was either missing or reduced. Similarly, the 15 S rRNA was absent or diminished in several other mutants (cf. ts-1224, ts-1369, ts-611, ts-642, ts-2615 and ts-1186). In contrast, mutants like ts-4304 or ts-852, contained both rRNA's in normal amounts, but the usual background of non-ribosomal RNA's was greatly reduced. Finally, in two mutants, ts-297 and ts-1628 (not shown), the 15 S and 21 S rRNA components were barely detectable but, instead, two abnormal RNA components of lower molecular weight appeared as prominent bands. It remains to be studied whether these bands actually represent lower molecular weight variants of the 15 S and 21 S rRNA's. Except for the splicing of 21 S rRNA in ω^+-strains, no mechanisms are known, so far, to specifically interfere with either 15 S or 21 S rRNA biosynthesis. Therefore, one is tempted to interpret the results shown in Fig. 1 on terms of a preferential instability and degradation of either one of the two rRNA's. A variety of mitochondrial ribosomal protein alte-

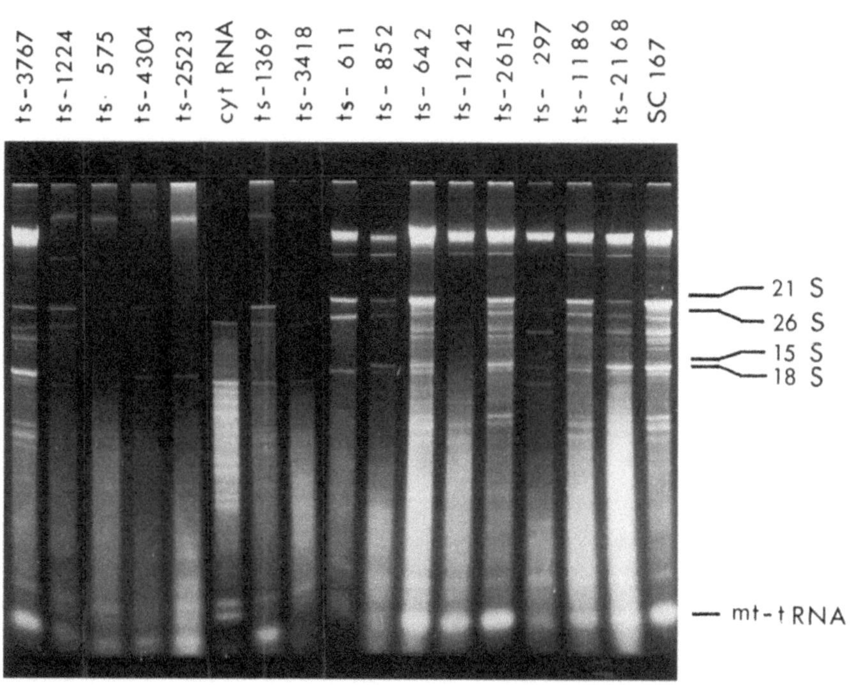

Fig. 1. Electrophoretic banding patterns of mtRNA's from wild type (SC 167) and selected pet ts-mutants.

Electrophoresis was performed in urea-containing agarose gels according to Locker (13). ts-Mutants were grown for 12 hr at 36°C in 1 % galactose, 1 % peptone, 1 % yeast extract and 0,01 % of each, adenine and guanine. Isolation procedures for mitochondria and mtRNA were adapted from Lang et al. (14), Hudspeth et al. (15) and Van Ommen et al. (16). cytRNA = cytoplasmic RNA.

rations may be responsible for this. In the same way, the putative simultaneous alteration of both rRNA components in ts-297 could be explained. Similar ts-mutants containing reduced amounts of mitochondrial 21 S rRNA at the non-permissive temperature have been described by Sor and Faye (12). These authors also invoke a mutationally altered ribosomal structure.

Drastic changes of mitochondrial ribosomal structure are expected to affect mitochondrial protein synthesis in general. Therefore, it was surprising to note that only two of the mutants shown in Fig. 1 (ts-2168 and ts-575) were lacking all mitochondrially made polypeptides, whereas most other mutants were defective only in the two cytochrome oxidase subunits I and II, some even only in subunit I (G. Michaelis, unpublished results). Therefore, additional and appropriate experiments will be necessary to reconcile these two sets of data. At the moment, it appears too early to speculate on ribosomal mutations exclusively affecting the synthesis of cytochrome oxidase subunit I. RNA patterns indicating a selective reduction of non-ribosomal RNA's are likely to reflect the longer half lifes of rRNA rather than a specific lesion of mRNA biosynthesis.

2) Nuclear mutations affecting oxi 3-transcript maturation.

Oxi 3, the structural gene of cytochrome oxidase subunit I, contains 7 - 8 exons and 6 - 7 introns (4, 5). Therefore, many specific processing reactions will be necessary to produce mature oxi 3-mRNA. In order to elucidate which of them are controlled by nuclear genes the accumulation of specific oxi 3-transcript processing intermediates was studied in cytochrome oxidase subunit I-deficient pet-mutants. These studies were performed by Northern-blot hybridization of total mtRNA using the three oxi 3-DNA probes (a), (b) and (c), shown in Fig. 2 A.

When hybridized to SC 167 mtRNA probe (c) almost exclusively reacted with the two excised introns aI1 and aI2 (Fig. 2B). Probes (a) and (b) preferentially hybridized with mature oxi 3-mRNA while higher molecular weight precursors responded only weakly. Obviously, in mutant ts-

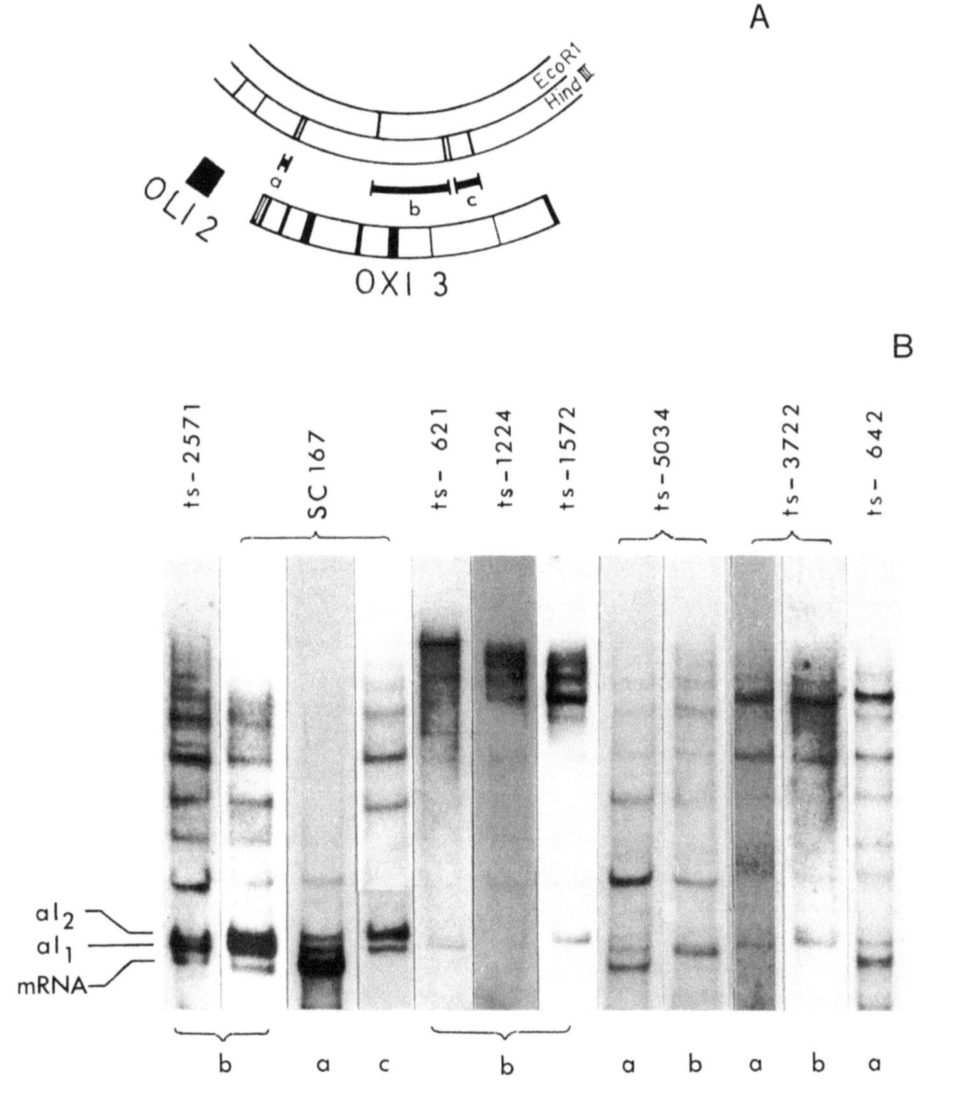

Fig . 2. Northern-blot hybridization of oxi 3-transcripts in cytochrome oxidase subunit I-deficient pet ts-mutants.

A. To be used as hybridization probes the EcoRI/HindIII fragment (b) and the EcoRI fragment (c) were cloned in pBR322 whereas fragment (a) was isolated by preparative gelelectrophoresis. Dark areas represent exons, open areas intron regions of the oxi 3 mosaic gene (3, 5).

B. mtRNA was isolated as indicated in Fig. 1. Electrophoresis was performed in formaldehyde-containing agarose gels according to Seeds (17). The DNA probes used are indicated at the bottom of the lanes. The presumptive positions of mature oxi 3 mRNA and the excised introns aI_1 and aI_2 are indicated.

2571 the steady state concentrations of these intermediates were higher than in wild type and more than a dozen distinct RNA-species may thus be distinguished. Among the cytochrome oxidase subunit I-deficient pet ts-mutants studied the eight mutants depicted in Fig. 2B were identified as containing no mature oxi 3-mRNA. Instead, oxi 3-mRNA maturation seems to be interrupted at four different stages. The highest molecular weight precursor accumulates in ts-621. It has an approximate length of 11 kb. Mutants ts-1224, ts-1572, ts-3722 and ts-642, though being nonallelic, synthesize two phenotypically identical patterns of intermediates. Finally, mutant ts-5034 contains a medium-size intermediate (approx. 3.3 kb) as its main oxi 3-transcript. From these data it is obvious that at least 4 different steps in oxi 3-transcript processing must be controlled by nuclear genes. Three of these steps represent rather early maturation events and the accumulation of high molecular weight precursors indicates that oxi 3-mRNA maturation has to proceed in a ordered sequence of steps. More refined hybridization studies with additional defined oxi 3-DNA probes will be necessary to elucidate the kinetic details of this process further.

Acknowledgements

This work was supported by the Deutsche Forschungsgemeinschaft and the Fonds der Chemischen Industrie.

References

1. Borst, P., Grivell, L.A.: Cell 15, 705-723 (1978)
2. Grivell, L.A., Borst, P.: Nature 298, 703-704 (1982)
3. Netter, P., Carignani, G., Jacq, C., Groudinsky, O., Clavilier, L., Slonimski, P.P.: Molec.gen.Genet. 188, 51-59 (1982).
4. Hensgens, L.A.M., Bonen, L., de Haan, M., van der Horst, G., Grivell, L.A.: Cell 32, 379-389 (1983).
5. Bonitz, S.G., Coruzzi, G., Thalenfeld, B.E., Tzagoloff, A., Macino, G.,: J.biol.Chem. 255, 11927-11941 (1980)
6. Weiss-Brummer, B., Rödel, G., Schweyen, R.J., Kaudewitz, F.: Cell 29, 527-536 (1982).

7. Claisse, M.L., Slonimsky, P.P., Johnson, J., Mahler, H.R.: Molec.gen. Genet. 177, 375-387 (1980).

8. Faye, G., Dennebouy, N., Kujawa, C., Jacq, C.,: Molec.gen.Genet. 168, 101-109 (1979).

9. Thalenfeld, B.E., Hill, J., Tzagoloff, A.: J.biol.Chem. 258, 610-615 (1983).

10. Dieckmann, C.L., Pape, L.K., Tzagoloff, A.: Proc.Natl.Acad.Sci.USA 79, 1805-1809 (1982).

11. Michaelis, G., Mannhaupt, G., Pratje, E., Fischer, E., Naggert, J., Schweizer, E.: in "Mitochondrial Genes", Slonimski, P., Borst, P., Attardi, G. (eds.), Cold Spring Harbor Laboratory Press, NY, pp. 311-321 (1982).

12. Sor, F., Faye, G.: Molec.gen.Genet. 177, 47-56 (1979).

13. Locker, J.: Anal.Biochem. 98, 358-367 (1979).

14. Lang, B., Burger, G., Doxiadis, I., Thomas, D.Y., Bandlow, W., Kaudewitz, F.: Anal.Biochem. 77, 110-121 (1977).

15. Hudspeth, M.E.S., Shumard, D.S., Tatti, K.M., Grossmann, L.I.: Biochim.biophys.Acta 610, 221-228 (1980).

16. van Ommen, G.-J.B., Groot, G.S.P., Grivell, L.A.: Cell 18, 511-523 (1979).

17. Seeds, B.: Genetic Engineering 4, 91-102 (1982).

PROCESSING OF THE OXI-3 PRE-MESSENGER RNA IN YEAST

Gérard Faye and Michel Simon
Institut Curie,Section de Biologie, Bât. 110,
91405 ORSAY, France

Introduction

Currently, more than forty three genes are known to be encoded by the yeast mitochondrial genome. At least three of them, namely the 21S ribosomal RNA, the cytochrome b and the cytochrome c oxidase subunit 1 genes (the oxi-3 locus) are split by intervening sequences (introns). Their primary transcripts are processed by a cut-and-ligation mechanism into a mature RNA. Though the excision of some of their introns requires mitochondrially encoded components such as the maturases (1,2), several nuclear genes which are necessary for intron removal or RNA splicing have now been identified (3,4,5). The transcription of the forty three mitochondrial genes appears to be initiated from a smaller number of promoter sites (6); this would imply that some mitochondrial transcription units contain several genes.

In this paper we describe five nuclear *pet* mutants which impede the maturation of the cytochrome c oxidase subunit 1 mRNA. Further, we present some arguments suggesting that the oxi-3 mRNA arises from a multigenic primary transcript.

Nuclear *pet* mutants involved in the RNA processing of the oxi-3 primary transcript

Among nuclear *pet* mutants unable to grow when glycerol is used as a carbon source, only those having lost the cytochrome c oxidase activity were studied. Mitochondrial RNA of eight mutants of Tzagoloff's collection (7)

Mitochondria 1983

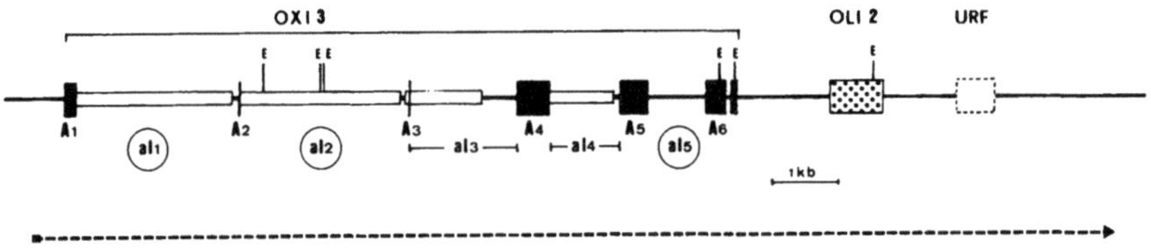

Fig. 1. Structure of the cytochrome c oxidase subunit 1 gene of strain D273-10B. A1-A5 and aI1-aI5 indicate exonic and intronic sequences, respectively. Dark bars are exons. Intron open reading frames are represented by open bars. Introns aI1, aI2 and aI5 are exciced as circular RNA. The oli-2 coding sequence is represented by a dotted bar. The long arrow denotes the oxi-3—oli-2 transcription unit.

and thirteen of ours were analysed by Northern blotting. As a probe, a labeled 350 bp DNA fragment from the exon A4 of the oxi-3 gene was used (Figure 1).

The nuclear *pet* mutants of Tzagoloff's collection were isolated from the wild type strain D273-10B (a short-gene strain), in which the oxi-3 gene is interrupted by five large introns and probably two very short ones located at its 3'-end (8) (Figure 1). Two mutants, E4-218 (4) and E2-116, disclosed a drastically altered Northern blot hybridization pattern when compared with blots of D273-10B/A RNA (Figures 2 and 3). In both mutants the mature mRNA and the shorter partially processed pre-mRNAs are missing whereas the larger processing intermediates are detected. The mutation carried by these two clones are referred to as mss51 and mss116, respectively (MSS, for mitochondrial splicing system). Despite the similarities of their autoradiograms, they belong to two different complementation groups.

We previously isolated numerous *pet* mutants from the wild type strain FF1210-6C (9). Three of the mutants studied displayed an abnormal Northern blot pattern. Mutant mss18 (which is slighly leaky) accumulates a 4,200 base long RNA intermediate. Mutants mss8 and mss9 have a very low level of oxi-3 transcripts (Figure 2), though rRNA and oli-2 RNA are transcribed at a normal rate (data not shown).

The mutation mss51 has been studied in more details. A yeast library

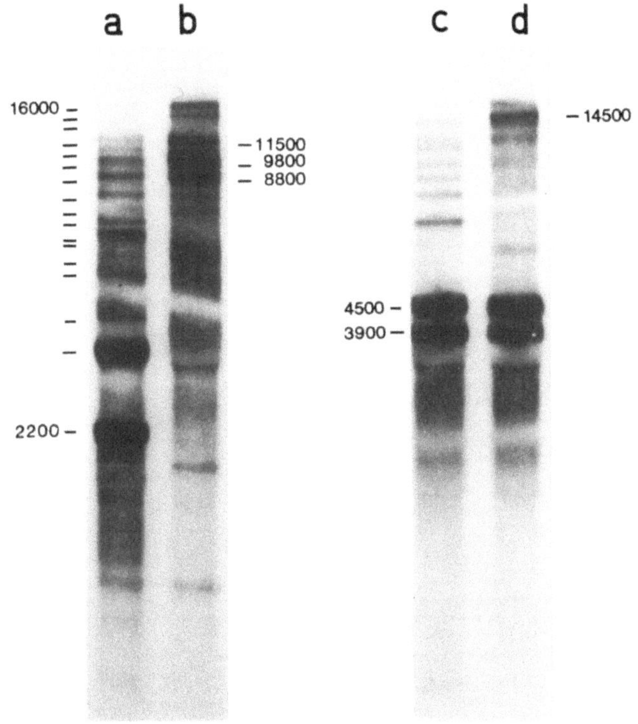

Fig.2. Northern blot analysis of mitochondrial RNA of nuclear *pet* mutants. RNA extracted from purified mitochondria, was denatured with glyoxal, electrophoresed on agarose gel and transferred to a nitrocellulose paper according to Thomas (17). The probe was a 350 bp HinfI-HinfI DNA fragment isolated from exon A4. Lane a: mitRNA of N6-78 (mutant isolated from D273-10B), lane b: E2-116 (mss116), lane c: N266 (mutant isolated from FF1210-6C), lane d: mss18, lane e: FF1210-6C, lane f: mss8, lane g: mss9.

constructed with the yeast-E. coli vector YEp13 and the whole fragmented genome of a wild type strain was used to transform a strain carrying the mutation mss51. A recombinant plasmid complementing this mutation has been isolated (11) and the shortest DNA fragment allowing complementation has been localized and sequenced. The MSS51 gene is transcribed and encodes a basic protein. Further we have attempted to know which steps in oxi-3 splicing are blocked. DNA fragments covering each oxi-3 exon-intron boundary were individually hybridized to wild type and mss51 mitochondrial RNAs.

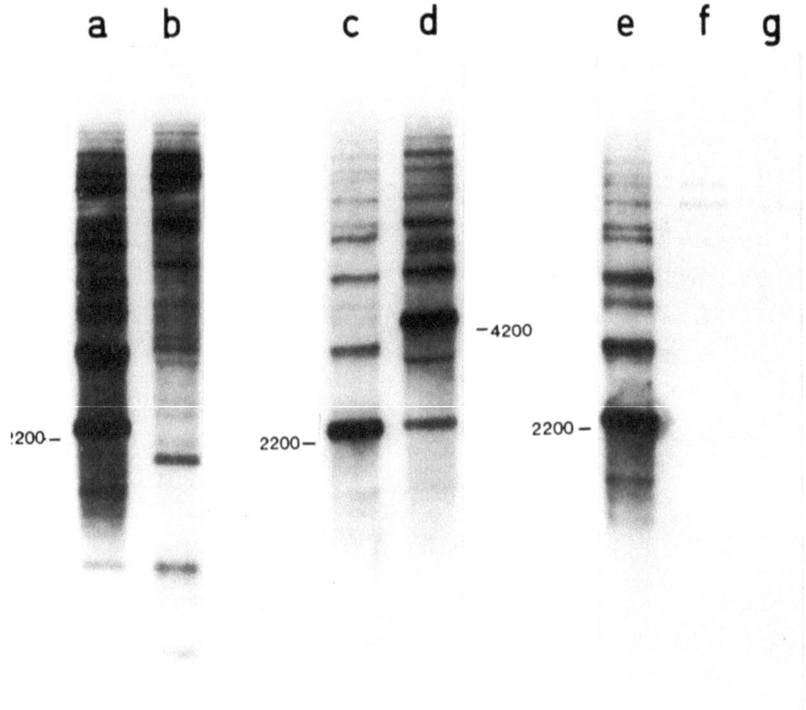

Fig. 3. Comparative analysis of mitochondrial transcripts detected with an exon A4 probe and an oli-2 probe. The four mitRNA samples were run alongside on an agarose gel. Lanes a and c: mitochondrial RNA purified from D273-10B/A; lanes b and d: mitochondrial RNA prepared from E4-218 (mss51). Lanes a and b: hybridization to exon A4 probe, lanes c and d: hybridization to oli-2 probe.

By analyzing the effects of S1 nuclease on the RNA-DNA hybrids thereof, we were able to determine those splice-junctions needing the MSS51 product in order to be cleaved. It appeared that the third intron was excised efficiently but the first two introns remained bracketed by their flanking exons. Further, the fourth and fifth introns were only partially split from their common exon and remained fused to their up- and downstream flanking exon, respectively (12). One revertant, non allelic to the mss51 mutation, has been isolated and is currently under study.

The oxi-3—oli-2 transcription unit

The mitochondrial gene encoding the subunit 1 of the cytochrome c oxidase is followed, some 1,400 to 1,500 bp downstream its 3'-end, by the gene of an ATPase subunit (the oli-2 locus) (13). Further downstream is lying an unassigned reading frame (URF). We suggest these sequences belong to the same transcription unit for the following reasons.

First, we estimated the size of the largest transcript hybridizing to various oxi-3 probes to be about 16,000-17,000 base long in strain D273-10B/A (11,12). This transcript is clearly larger than the oxi-3 gene the length of which is about 10,500 bases (14);the oxi-3 promoter is thought to be 400 to 500 bp upstream its ATG triplet, thus there is enough room left on this transcript for the oxi-3, oli-2 and URF sequences.

Second, we have run alongside on an agarose gel two samples of each D273-10B/A and E4-218 mitRNA. After transfer to a nitrocellulose paper, these RNA were hybridized with either a 1,100 bp DNA fragment carrying the oli-2 sequence or the 350 bp exon A4 probe, as a size controle (Figure 3). For both the wild type and the mutant RNA, the oli-2 probe revealed two major RNA species 3,900 and 4,500 base long. Nine larger transcripts were also detected with the D273-10B/A mitRNA; their estimated sizes are about 7,300, 8,500, 9,400, 10,000, 11,000, 12,000, 13,500, 14,500 and 16,500 bp (Figure 3, lane c). Only the four largest species are present in E4-218 (Figure 3, lane d), the 14,500 base long RNA giving a stronger signal than the corresponding transcript of the wild type strain. We believe this result cannot be accounted for by some sequence homology between oli-2 and oxi-3 genes for the strong 11,500, 9,800 and 8,800 base long RNA species hybridized only to the exon A4 probe (compare Figure 3, lanes b and d) but not to the oli-2 probe. Thus, these results are consistent with the possibility that oxi-3 and oli-2 are co-transcribed into a long multi-gene RNA. However, it is difficult to assign each intermediary processed RNA to a particular processing step for it is likely that some introns of the oxi-3 RNA portion are remove even before the oli-2 RNA portion is cut out, as indicated by the existence of the nine partially processed D273-10B/A transcripts that hybridized to the oli-2 probe. Further, since mature oli-2 RNA is detected in <u>mss51</u>, oxi-3 splicing is probably not a prerequisite to the oxi-3—oli-2 cleavage.

Discussion

The processing of a multi-genic transcript carrying a mosaic gene should require two systems: one that cleaves the primary transcript into single-gene RNA and another that excises the intervening sequences and splices the exons together. Essentially nothing is yet known about the former system but, a priori, some components could be shared by the two machineries.

Among twenty-one nuclear *pet* mutants tested, five affecting the splicing of the oxi-3 pre-mRNA were found. Since Northern blotting does not allow the detection of small alterations in RNA patterns, some splicing mutants may have been missed. Nethertheless, it seems likely that numerous genes are involved in the processing of this pre-mRNA.

At least two extreme situations may be considered to account for the removal of mitochondrial introns. Either each intron needs a particular set of components to be processed or a common cut and splice machinery excises the intronic sequences. On the other hand, despite the fact that yeast mitochondrial introns share common secondary (and tertiary) structures as well as with some chloroplast and nuclear(ribosomal) introns (15), it is unlikely that self-splicing, as was described for Tetrahymena (16) plays a preeminent part in mitochondrial processing for too numerous mitochondrial maturases and nuclear genes are involved in these reactions. Even though mitochondrial intronic RNA sequences possess some autocatalytic activity, proteic elements appear indispensable to ensure some crucial functions.

Aknowledgements:

We thank H. Fukuhara for the oli-2 probe and M. Bolotin-Fukuhara for her collaboration in the preliminary characterization of nuclear pet mutants. The investigation was supported by CNRS (ATP 3644), INSERM (ATP 72-79104 n° 10) and MRT (DMB-P53).

References

1. Jacq, C., Lazowska, J. and Slonimski, P.: C. R. Acad. Sci. Paris 290, 89-92 (1980).
2. De La Salle, H., Jacq, C. and Slonimski, P.: Cell 28, 721-732 (1982)
3. Groudinsky, O., Dujardin, G. and Slonimski, P.: Mol. Gen. Genet. 184, 493-503 (1981)
4. Faye, G. and Simon, M.: C. R. Acad. Sci. Paris 294, 245-248 (1982)
5. Dieckmann, C., Bonitz, S., Hill, J., Homison, G., McGraw, P., Pape, L., Thalenfeld, B. and Tzagoloff, A.: in Mitochondrial Genes (Slonimski, P., Borst, P. and Attardi, G. eds). Monograph 12, Cold Spring Harbor Laboratory Press, New York, pp 213-223 (1982)
6. Christianson, T., Edwards, J., Levens, D., Locker, J. and Rabinowitz, M. J. Biol. Chem. 257, 6494-6500 (1982)
7. Tzagoloff, A., Akai, A. and Needleman, R.: J. Biol. Chem. 250, 8228-8235 (1975)
8. Bonitz, S., Coruzzi, G., Thalenfeld, B., Tzagoloff, A. and Macino, G.: J. Biol. Chem. 255, 11927-11941 (1980)
9. Bolotin-Fukuhara, M., Faye, G. and Fukuhara, H.: Molec. Gen. Genet. 152, 295-305 (1977)
11. Faye, G. and Simon, M.: Cell 32, 77-87 (1983)
12. Simon, M. and Faye, G.: submitted (1983)
13. Macino, G. and Tzagoloff, A.: Cell 20, 507-517 (1980)
14. Hensgens, L., Arnberg, A., Roosendaal, E., van der Horst, G., van der Veen, R., van Ommen, G-J; and Grivell, L.: J. Mol. Biol. 164, 35-58 (1983)
15. Michel, F. and Dujon, B.: The EMBO Journal 2, 33-38 (1983)
16. Kruger, K., Grabowski, P., Zaug, A., Sands, J., Gottschling, D. and Cech, T.: Cell 31, 147-157 (1982)
17. Thomas, P.: Proc. Nat. Acad. Sci. USA 77, 5201-5205 (1980)

MITOCHONDRIAL INHERITANCE OF A VARIANT FORM OF THE oxi2 TRANSCRIPT IN SACCHAROMYCES CEREVISIAE

Thomas D. Fox and Michelle K. Reif

Section of Genetics and Development, Cornell University, Ithaca, NY 14853, USA

Introduction

To a first approximation at least, mitochondrial gene expression can be thought of as involving signals encoded in mtDNA that are acted upon by nuclearly encoded components of the mitochondrial genetic system. To analyse genetically the steps in mitochondrial gene expression it is necessary to have mutants that interrupt or alter the process. Several nuclear petite mutations of yeast are very interesting in this regard since they appear to specifically block the expression of particular mitochondrial genes (1). One of these mutations, pet494-1, blocks accumulation of the oxi2 gene product (cytochrome oxidase subunit III) at a post-transcriptional step (2).

During the course of our studies on the pet494-1 mutation, we noticed that some strains of yeast produce an oxi2 mRNA that is approximately 2kb shorter than the "standard" 3.6kb transcript. This shorter variant of the oxi-2 transcript has also been reported by others (3). We have investigated the inheritance of the genetic determinant(s) for this shorter oxi2 mRNA and report here that it is encoded in mtDNA. Furthermore the 3.6kb and 1.6kb forms of the oxi2 transcript appear to differ in both their 5' and 3' untranslated regions. These findings suggest that detailed anlaysis of the oxi2 genes and transcripts from strains producing the variants may yield information on signals in

Mitochondria 1983

mtDNA for transcription starts and stops and/or processing sites.

Results and Discussion

The oxi2 transcript (mRNA for cytochrome oxidase subunit III) can be easily detected in total yeast RNA by gel-blot hybridization using a labeled piece of oxi2 mtDNA as a probe. Such experiments reveal the presence of a single oxi2 transcript in cells of strain D273-10B, approximately 3.6kb in size (Fig. 1a, ref. 3). However, some yeast strains contain a substantially shorter variant oxi2 mRNA: strain LL20 has been reported to contain a transcript of 1.6kb (3), and the strain GRF18 (4) also contains a "short" oxi2 transcript similar, if not identical in size to that of LL20 (Fig. 1b). In at least the case of GRF18, this shorter transcript is fully functional since the strain grows normally on medium containing non-fermentable carbon sources. Moreover, GRF18 contains normal levels of cytochrome oxidase subunit III as judged both by labeling in the presence of cycloheximide and by detection of unlabeled subunit III in immune replica experiments (T.D.F. and T.L. Mason, unpublished results).

The genetic determinant(s) specifying the different length oxi2 transcripts could be located either in the nucleus or the mitochondrion. As a first step towards examining the inheritance of the determinant of oxi2 mRNA size we constructed diploid strains by crossing GRF18, or a rho^{o} derivative of it, to strain TF2 (mitochondrially isogenic to D273-10B (5) and containing the "long" oxi2 mRNA) or a rho^{o} derivative of TF2, in a reciprocal manner. The diploid strain DTG1 was formed by crossing TF2 (rho^{+}). Both DTG1 and DTG2 grew normally on medium containing non-fermentable carbon sources.

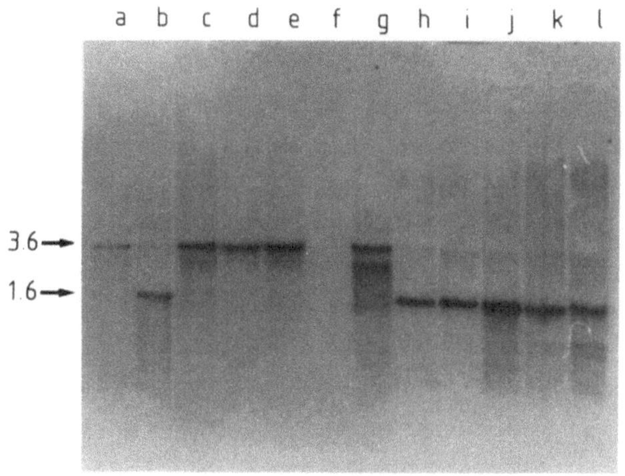

Figure 1. The size difference between oxi2 transcripts of D273-10B and GRF18 is inherited cytoplasmically. Total yeast nucleic acids were isolated, separated by electrophoresis in agarose gels containing formaldehyde and blotted to nitrocellulose as described (6). The filter was hybridized to a radioactively labeled 384bp HindIII fragment containing oxi2 sequences (7) that was isolated from plasmid pZHS3 (2). The RNAs were prepared from: a) D273-10B; b) GRF18; c) DTG1; d) DTG1a; e) DTG1b; f) DTG1c; g) DTG1d; h) DTG2; i) DTG2a; j) DTG2b; k) DTG2c; l) DTG2d. DTG1 was a diploid formed by crossing TF2 X GRF18(rho^{o}). DTG1a-d were the meiotic progeny of one tetrad from DTG1. DTG2 was a diploid formed by crossing TF2(rho^{o}) X GRF18. DTG2a-d were meiotic progeny of one tetrad from DTG2.

To examine the size of oxi2 transcripts present in DTG1 and DTG2, total RNA was prepared from the strains and analyzed by gel-blot hybridization using an oxi2 mtDNA probe. The diploid DTG1, carrying the mtDNA of TF2 (D273-10B), contained a 3.6kb oxi2 transcript indistinguishable from that of D273-10B (Fig. 1c). On the other hand, DTG2 contained a 1.6kb oxi2 transcript (Fig. 1h) indistinguishable from that of GRF18, the strain from which its mtDNA was derived. Since DTG1 and DTG2 are nuclearly isogenic to each other, but differ in their mtDNAs (and possibly other cytoplasmic elements curable by ethidium bromide) these results indicate that a mitochondrially encoded determinant plays an important role in specifying the length of oxi2 mRNA.

Although the experiment described above clearly demonstrated a role for mtDNA in determining the size of the oxi2 transcript, it still appeared possible that nuclear gene differences between the two strains might also be important. Such nuclear gene differences might be masked in the diploids DTG1 and DTG2 if, for example, they represented co-dominant nuclear alleles. To test whether any differences in oxi2 mRNA size might segregate in a Mendelian manner, DTG1 and DTG2 were sporulated and the oxi2 transcripts of the haploid progeny of one tetrad from each diploid were examined. (The nuclear marker leu2 segregated 2:2 in each tetrad.) The four meiotic progeny of DTG1 (Fig. 1d, e, f, & g) contained oxi2 mRNAs indistinguishable in size from DTG1. (Although the RNA run in lane f was severely degraded, a 3.6kb oxi2 transcript was nevertheless clearly detectable.) Similarly, the four haploid progeny of DTG2 contained 1.6kb oxi2 transcripts similar or identical to that of DTG2. The observation of non-Mendelian segregation of the determinant for oxi2 transcript length confirms that it is most likely located on mtDNA and suggests that no nuclear gene differences important for this phenotype, exist among these strains. Interestingly, small amounts of the 3.6kb transcript are often detectable in strains that contain predominantly the 1.6kb RNA.

The oxi2 transcript of D273-10B has been reported to have a 5' untranslated leader of 490bp and an untranslated 3' sequence 2450bβp in length (3). The fact that the oxi2 mRNA of strain GRF18 is approximately 2kb shorter than that of D273-10B clearly must mean that the former lacks sequences present in the 3' untranslated region of the latter. Whether the shorter 3' untranslated region of the GRF18 mRNA reflects a transcription termination or processing site that is closer to the structural gene than the corresponding site in D273-10B is at present unknown. The mtDNA of GRF18 may simply not contain transcribed sequences present in this region in D273-10B.

The fact that the 3' untranslated regions of the oxi2 transcripts of D273-10B and GRF18 differ, does not preclude the possibility that they might also differ at their 5' ends. To test this possibility we carried out S_1 protection experiments. A probe was prepared from cloned D273-10B mtDNA that was labeled at a single 5'-end within the oxi2 coding sequence. The probe was hybridized with total RNA prepared from either D273-10B or GRF18, digested with S_1-nuclease and electrophoresed on an alkaline agarose gel (Fig. 2). Hybridization with D273-10B RNA yielded a protected fragment terminating 610bp upstream from the start of translation while GRF18 RNA protected a fragment that was approximately 70 nucleotides shorter. Thus, either the 5'-end of the GRF18 oxi2 mRNA is slightly closer to the coding sequence than that of D273-10 , or the sequence of the GRF18 transcript diverges from that of the D273-10B-derived probe at that point. In any event, we conclude that neither the 5' nor the 3' untranslated sequences of the oxi2 mRNAs from these strains are identical. This finding raises the possibility, yet to be tested, that differences at opposite ends of the transcripts might be separable by recombination, producing intermediate forms of the oxi2 mRNA.

We determined the position of the 5' terminus of the D273-10B oxi2 transcript to be 610bp upstream from the start of the coding sequence, while Thalenberg, et al. (3) report that it is 120bp closer to the structural gene. We do not know the reason for this discrepancy, although we note that the probe used in their S_1 protection experiment did not extend far enough upstream to detect the terminus we report here.

We have been interested in studying the mechanism of action of the nuclear gene PET494, which is required to allow accumulation of the oxi2 gene product cytochrome oxidase subunit III (1). Recent work in this laboratory has indicated that the PET494 protein acts at a post-transcriptional step in gene expression, since strains carrying the

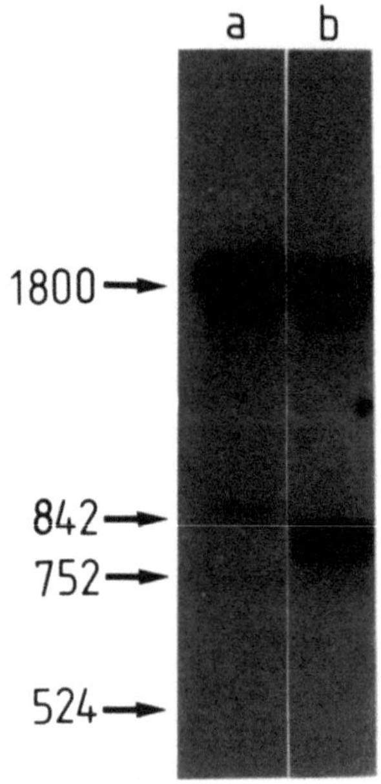

Figure 2. Protection of a D273-10B-derived 5'-end labeled oxi2 probe from S_1 digestion by RNA from D273-10B and GRF18. A probe was prepared from cloned D273-10B mtDNA that was radioactively labeled at a single 5'-end: the Sau3A terminus at position +232 of the oxi2 coding sequence (7). The probe extended 1.8kb upstream to an unlabeled HaeIII end. The probe was hybridized to total RNA from either a) D273-10B or b) GRF18. After S_1-nuclease digestion the products were analyzed by alkaline agarose gel electrophoresis. Size markers (positions indicated) were generated by cleavage of the probe with either XbaI (842), HinfI (752) or MspI (524).

pet494-1 mutation contain normal amounts of oxi2 trancript (2). This finding suggests the possibility that the 494 protein might directly interact with the oxi2 transcript to somehow promote translation of the message. (Other models for PET494 action, for example involving decreased stability of cytochrome oxidase subunit III in the mutant, have not been ruled out.)

The pet494-1 mutation was isolated and characterized in strains carrying mtDNA of D273-10B (8). The fact that the oxi2 transcript of strain GRF18 differs from that of D273-10B on both the 5' and 3' sides of the coding sequence raised the question of whether the pet494-1 mutation would lead to a pet⁻ phenotype in strains carrying GRF18 mtDNA. To answer this question we crossed GRF18(rho^+) with a pet494-1(rho^o) strain. The resulting diploid grew normally on non-fermentable carbon sources and sporulated. All of the 6 complete tetrads analysed from this cross showed 2:2 segregation of the pet^- phenotype. Since, by analogy to the crosses described above, all of these haploid segregants must have GRF18 mtDNA, the Mendelian segregation of the pet^- phenotype indicates that pet494-1 is not suppressed by the variant form of the oxi2 mRNA encoded by GRF18 mtDNA. Thus, if the PET494 protein interacts with the oxi2 mRNA, it most likely does so at a site present on both variant forms of this transcript. A suggestion for the possible location of such a site has come from the isolation of a mitochondrial mutation that partially suppresses pet494-1: in this mutation the 5' untranslated portion of the oxi2 gene has been replaced by fusion to sequences within the second intron of oxi3 (2). It is therefore of interest to point out that, at least within the limits of resolution of the S_1 protection experiment of Fig. 1, most of the 5' untranslated portions of the D273-10B and GRF18 oxi2 mRNAs are the same.

Acknowledgments

This work was supported by a Research Grant (GM29362) and a Research Career Development Award (HD00515) from the U.S. National Institutes of Health.

References

1. Cabral, F., Schatz, G.: J. Biol. Chem. 253, 4396 (1978).

2. Mueller, P., Reif, M.K., Shen Zong-hou, Sengstag, C., Mason, T.L., Fox, T.D.: in preparation

3. Thalenfeld, B.E., Hill, J., Tzagoloff, A.: J. Biol. Chem. 258, 610 (1983).

4. Donahue, T.F., Daves, .S., Luchinni, G., Fink, G.R.: Cell 32, 89 (1983).

5. Fox, T.D.: J. Mol. Biol. 130, 63 (1979).

6. Silverman, S.J., Rose, M., Botstein, D., Fink, G.R.: Mol. Cell. Biol. 2, 1212 (1982).

7. Thalenfeld, B.E., Tzagoloff, . J. Biol. Chem. 255, 6173 (1980).

8. Ebner, E., Mennucci, L., Schatz, G: J. Biol. Chem. 248, 5360 (1973).

A PRECURSOR TO SUBUNIT II OF CYTOCHROME OXIDASE IN SACCHAROMYCES CEREVISIAE

Gertrud Mannhaupt, Georg Michaelis, Elke Pratje
Universität Bielefeld, Fakultät für Biologie, D-48oo Bielefeld,
Federal Republic of Germany

Eckhart Schweizer
Lehrstuhl für Biochemie der Universität Erlangen-Nürnberg,
D-852o Erlangen, Federal Republic of Germany

Donald C. Hawthorne
Department of Genetics, University of Washington,
Seattle, Washington 98195, USA

Introduction

Cytochrome oxidase, the terminal complex of the respiratory chain, is an integral part of the inner mitochondrial membrane. The enzyme complex is composed of seven different subunits. The three largest subunits are of mitochondrial origin, whereas the four smaller subunits are encoded by nuclear genes. The interactions between the nuclear and mitochondrial genes regulating the biosynthesis and assembly of the cytochrome oxidase complex are mainly unknown. Studies on the nuclear encoded four subunits of cytochrome oxidase have revealed that at least three of them are synthesized as larger precursors in the cytosol and that energy is required for the import of these precursors into the mitochondria (for reviews see 1, 2).

We have shown that also the mitochondrially synthesized subunit II of cytochrome oxidase has to be processed in the yeast, Saccharomyces cerevisiae. The nuclear pet mutant ts2858 displaying temperature-sensitive growth on lactate medium has been described by us (3, 4). This nuclear mutation leads to the accumulation of a larger precursor to subunit II of cytochrome oxidase. A comparison of the N-terminal protein sequence of subunit II in wild-type and mutant was performed and the first complete presequence of a mitochondrially synthesized polypeptide described. The presequence consists of the following fifteen amino acids (5):

Met-Leu-Asp-Leu-Leu-Arg-Leu-Gln-Leu-Thr-Thr-Phe-Ile-Met-Asn-

The nuclear pet mutation ts2858 preventing post-translational processing of cytochrome oxidase subunit II seems to inactivate a nuclear gene coding either for a protease or an enzyme regulating a protease. Here we present evidence for a second nuclear gene involved in the proteolytic cleavage of the subunit II presequence.

Results

1) A second nuclear complementation group affecting the size of subunit II of cytochrome oxidase

Recently we have completed the analysis of the mitochondrial translation products in 313 nuclear temperature-sensitive pet mutants belonging to 239 different complementation groups (3, 4). In addition to the well characterized pet mutant ts2858 a second pet mutant with a larger subunit II of cytochrome oxidase could be identified. The two temperature-sensitive pet mutants ts2858 and ts1402 complement each other. In Figure 1 the mitochondrial translation products of ts1402, ts2858 and wild-type are compared. At the permissive temperature of 23°C mutant ts1402 synthesizes mainly the wild-type subunit II of cytochrome oxidase, and only trace amounts of the larger protein are visible. Ten minutes after the shift to 36°C, the restrictive temperature, the larger form of subunit II is the predominant product and only few wild-type subunit can be detected. When the cells are incubated for a longer period at 36°C the synthesis of subunit I of cytochrome oxidase is greatly reduced in addition and total mitochondrial protein synthesis declines rapidly. This situation is quite different from pet mutant ts2858. In this mutant the mitochondrial translation products can be labeled in vivo more easily at the non-permissive temperature and, strangely enough, the precursor protein replaces the mature subunit II both at the permissive and non-permissive temperature. The larger subunit II of mutant ts1402 cannot be distinguished from the

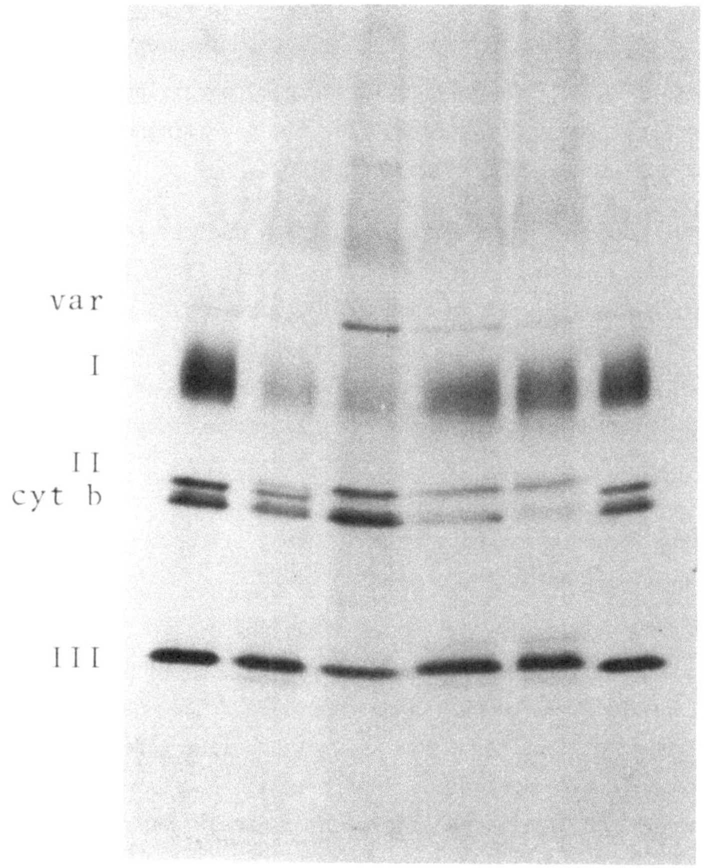

Figure 1. Mitochondrial translation products of wild-type, mutant ts1402, and mutant ts2858.
From left to right:
wild-type
ts1402 labeled at 23°C
ts1402 labeled at 36°C after a 1o minutes preincubation period at 36°C
ts2858 labeled at 23°C
ts2858 labeled at 36°C after a 14 hours preincubation period at 36°C
wild-type

Mitochondrial translation products were labeled with $^{35}SO_4^{--}$ at the indicated temperature in the presence of cycloheximide. Mitochondria were isolated, analyzed on a 1o-15% exponential gradient gel of LiDS-polyacrylamide, and visualized by fluorography as described previously (5). var = var1 protein; I, II, III = subunit I, II or III of cytochrome oxidase respectively; cyt b = cytochrome b.

precursor molecule of mutant ts2858 by LiDS-polyacrylamide gel electrophoresis. Furthermore, the two mutant subunits migrate faster than the wild-type protein in an LiDS-polyacrylamide gel containing 6M urea. Experiments are in progress to determine the first N-terminal amino acids of cytochrome oxidase subunit II in mutant ts1402.

To demonstrate that one single nuclear mutation confers temperature-sensitivity as well as the modification of subunit II to the cell, a tetrad analysis was performed. Mutant ts1402 was crossed with a wild-type strain and the resulting diploids were sporulated. The temperature-sensitive phenotype displayed a regular 2:2 segregation and only the temperature-sensitive spores transmitted the modified form of cytochrome oxidase subunit II.

2) The two nuclear pet mutations, which prevent processing of the precursor, map on different chromosomes

Pet mutants ts2858 and ts1402 are representatives of two different nuclear complementation groups. Nevertheless, the possibility of intragenic complementation cannot be excluded. As one approach to demonstrate that the two pet mutations are defects in two separate nuclear genes we decided to map them. The spo11 mapping procedure of Klapholz and Esposito (6) was used to determine the chromosomal location of the pet mutations. In this mapping method various diploids homozygous for spo11 and heterozygous for the unmapped mutation and a marker on each chromosome are sporulated. In such strains homozygous for the spo11 mutation chromosome segregation occurs without appreciable recombination. After sporulation the rare viable spores will survive an ether treatment whereas most of the diploid cells are killed. The spores can be recovered by plating them on complete medium. They are analyzed for their markers by replica plating to selective media and, in addition, by complementation testing with appropriate testers. Meiotic products expressing the unmapped recessive pet mutation will express markers in coupling on the same chromosome and will not express the genetic markers on the homolog contributed by the tester strain.

The spoll mapping procedure places pet-ts2858 on chromosome XIII and pet-ts1402 on chromosome V. We crossed a pet-ts2858 spoll strain to several spoll testers which all together carry at least one recessive marker for each chromosome. From among the pet-ts2858 spores examined, all the recessive markers of the tester strains were recovered except for lys7, the marker for chromosome XIII. When a pet-ts1402 ilvl spoll strain was crossed to the spoll testers, all the pet-ts1402 spores expressed the ilvl marker of chromosome V from the pet-ts1402 parent and not the ura3 or his1 markers for chromosome V from the tester strains.

The assignment of pet-ts2858 to chromosome XIII and pet-ts1402 to chromosome V was confirmed by a second mapping procedure in which interspecific hybrids between multiply marked strains of Saccharomyces cerevisiae and Saccharomyces douglasii (7) are constructed and sporulated. This new method (8) utilizes again the ability of the hybrid cells to undergo chromosome segregation with greatly reduced recombination during meiosis.

Tetrad analyses were undertaken to place the pet mutations on the linkage maps of Saccharomyces cerevisiae (9). We have placed pet-ts1402 on the right arm of chromosome V, distal to ilvl (18PD 3NPD 48T), and we found pet-ts2858 to lie between ilv2 (35PD 1NPD 18T) and rnal (14PD 4NPD 33T) on the right arm of chromosome XIII.

Acknowledgements

This work was supported by grants from the Deutsche Forschungsgemeinschaft to G.M. and E.S., and by an NIH grant, GM27624 to D.C.H.

References

1. Ades, I.Z.: Molecular and Cellular Biochemistry 43, 113-127 (1982).
2. Neupert, W., Schatz, G.: Trends Biochem.Sci. 6, 1-4 (1981).
3. Michaelis, G., Mannhaupt, G., Pratje, E., Fischer, E., Naggert, J., Schweizer, E.: In "Mitochondrial Genes", Slonimski, P., Borst, P., Attardi, G. (eds.), Cold Spring Harbor Laboratory Press, NY, pp311-321 (1982).
4. Mannhaupt, G., Pratje, E., Michaelis, G., Beyreuther, K., Schweizer, E.:

In "Endocytobiology II", Schenk, H.E.A., Schwemmler, W. (eds.), de Gruyter, Berlin - New York 1983, in press.

5. Pratje, E., Mannhaupt, G., Michaelis, G., Beyreuther, K.: The EMBO J. 2, 1049-1054 (1983).

6. Klapholz, S., Esposito, R.E.: Genetics 100, 387-412 (1982).

7. Riva, M., Buhler, J.-M., Sentenac, A., Fromageot, P., Hawthorne, D.C.: J.Biol.Chem. 257, 4570-4577 (1982).

8. Hawthorne, D.C.: manuscript in preparation.

9. Mortimer, R.K., Schild, D.: In "The Molecular Biology of the Yeast Saccharomyces. Metabolism and Gene Expression.", Strathern, J.N., Jones, E.W., Broach, J.R. (eds.), Cold Spring Harbor Laboratory Press, NY, pp639-650 (1982).

MITOCHONDRIAL RIBOSOMAL RNA MUTATIONS AND THEIR NUCLEAR SUPPRESSORS IN YEAST

Monique Bolotin-Fukuhara[§], Frédéric Sor[§§] and Hiroshi Fukuhara[§§]

§ Laboratoire de Biologie Générale, Bâtiment 400, Université Paris-Sud, Centre d'Orsay, 91405 Orsay, France

§§ Institut Curie, Section de Biologie, Bâtiment 110, Centre Universitaire Orsay 91405, France

Introduction

Mitochondrial ribosomes are a typical example of the necessary collaboration between nuclear and mitochondrial gene expression. Ribosomal RNAs (Faye et al., 1975, 1974) and the ribosomal protein Var 1 (Groot et al., 1979; Terpstra et al., 1979) are encoded by the mitochondrial genome, while other ribosomal components are specified by the nuclear genes. Since ribosomal proteins and ribosomal RNA (rRNA) interact in a very precise way to make functional ribosomes (see Nomura et al., 1974, and Chambliss et al., 1979 for general reviews), this implies a coordinated regulatory system for the biogenesis of mitochondrial ribosomes. We know very little at present about this regulation.

In order to study how the nuclear and mitochondrial genes interact in the biogenesis of ribosomes, we have chosen basically a genetic approach. We used the possibility offered by the mitochondrial system of yeast to select rRNA mutations and, in a second step, selected suppressors of these mutations. This approach attempts to identify nuclear encoded products that interact with rRNA, i.e., mostly ribosomal proteins and rRNA modifying enzymes.

In this paper, we will present some results illustrating this approach: mitochondrial rRNA mutations will be described in physiological and molecular terms. Revertants of these mutations will also be analyzed. In most cases these reversions are due to nuclear suppressors. The molecular mechanisms leading to the suppression will be discussed.

I. Genetic screening of mutations in the rRNA genes.

rRNA genes are redundant in all organisms from prokaryotes to eukaryotes.

Mitochondria 1983

This intramolecular redundancy makes difficult the selection of point mutations in these genes. Indeed mutations of rRNA genes have been extremely rare. Recently a few mutants have been obtained in E.coli by in vitro mutagenesis of a plasmid carrying a ribosomal RNA operon (Sigmund and Morgan, 1982; Gourse et al., 1982; Starck et al., 1982). However such mutants still contain wild type genes whose expression is competing with the mutant genes.

Yeast mitochondria appear to be a suitable system for the selection of rRNA mutations, because there is only one copy of each rRNA gene per genome. Point mutations should therefore be easy to obtain in separate clones and were indeed obtained as will be described here. The rRNA mutations belong to two different classes: non-defective and defective mutations.

I-1. Antibiotic-resistance mutations.

Some antibiotics inhibit selectively mitochondrial protein synthesis (chloramphenicol, erythromycin, and in some conditions paromomycin), and resistance mutations of mitochondrial origin have been isolated (Linnane et al., 1968; Thomas and Wilkie, 1968; Coen et al., 1970; Wolf et al., 1973). Genetic data were compatible with their localization in or close to the rRNA genes, but the final proof came only recently with sequencing data (Dujon, 1980; Sor and Fukuhara, 1982; Li and Tzagoloff, 1982; Martin et al., 1982). These mutations confer ribosomal resistance in vitro (Grivell et al., 1973; Spithill et al., 1979) but are not defective for ribosomal functions.

I-2. Conditional respiratory-deficient mutations.

Defective mutations of the rRNA genes should lead to a respiratory-deficient phenotype and be therefore detected as "petite colonie" mutants. However, no rRNA mutants have been obtained in this form. We reasoned that a deficiency in mitochondrial protein synthesis could lead to rho^- accumulation as observed by Williamson et al. (1974). For that reason we searched rRNA mutations among conditional "petite colonie" mutants able to grow on respiratory substrates at 28°C but not at 36°C (heat-sensitive) or 20°C (cold-sensitive), as described by Bolotin-Fukuhara et al. (1977). Mitochondrial conditional mutants did not appear with a high frequency (10^{-4} to 10^{-5} after $MnCl_2$ mutagenesis with our strain). We therefore devised a rapid genetic test which (a) descriminates between nuclear and mitochondrial mutants and (b) place the mitochondrial mutations on the genetic map. By this method we have obtained one mutation localized in the region of the 15S rRNA gene (Julou et al. in preparation) and many mutations localized in the region covering the C^R and E^R markers, i.e., the 21S rRNA gene and its flanking segments (Bolotin-Fukuhara, 1979; Zelikson and Bolotin-Fukuhara, in preparation).

II. Phenotypic and functional characteristics of the rRNA mutations.
We first examined our supposed rRNA mutants for their ability to make normal ribosomal subunits (Table 1). We can notice that the deficit of the large ribosomal subunit with respect to the small one, or vice versa, is correlated with the localization of the mutations in the region of the large or the small rRNA gene. It is also interesting to note that all these mutations, with three exceptions, ts87, ts44 and cs1309, present a characteristic deficit even at permissive temperature, and the deficit is enhanced at the restrictive temperature as expected.
The apparent deficit of ribosomal subunits may not necessarily be due to a mutation of the ribosomal genes, but due to a defective protein synthesis caused by other mutations. This question arises because some of our mutations are localized in a region carrying also tRNA genes. We therefore examined the ribosomes from a typical tRNA mutant (aspartyl tRNA gene mutation, ts832). We found that the ribosomal profile was normal, suggesting that the subunit deficit of the "ribosomal mutants" can be attributed to a change in the ribosomal components.
The anomaly of ribosomes was observed in two forms. In some mutants (for example ts18) the deficit in the large subunit seems to be due to an instability of the ribosome because one can find a large amount of 50S subunits under appropriate conditions of extraction. In some other mutants the deficit appears absolute, independent of methods of extraction, reflecting probably the in vivo situation.
We have examined in some cases the RNA transcription of the mutants. The mutation ts1297 showed a marked deficiency in the production of the small

Table 1. Phenotype of mitochondrial ribosomal mutants.

Mutants	Interaction with antibiotics	Ribosomal subunits 50S/37S ratio		Ribosomal RNA 21S/15S ratio	
		28°	36° or 20°	28°	36° or 20°
15S rRNA region:					
ts 1297	does not express C^R_{321}	3.8	n.d.	9.2	n.d.
Within 21S rRNA gene:					
ts 8	P^R(1 g/l); not compatible with P^R_{454}	0.18	§	1.34	§
ts R2b		0.18	n.d.	normal	n.d.
ts 902	not compatible with C^R_{321}; hyper-sensitive to erythromycin	0.22	trace	0.25	0.22
cs 909		1.23	1.05	normal	normal
cs 901		0.39	0.17	normal	normal
cs 1319	does not express C^R_{321}	1.50	0.64	n.d.	n.d.
21S rRNA gene flanking regions:					
ts 87	hyper-sensitive to paromomycin	normal	§	n.d.	n.d.
ts 44		normal	§	normal	n.d.
ts 812		0.45	n.d.	2.8	n.d.
ts 9		1.4	n.d.	1.6	n.d.
ts 18		§§	n.d.	normal	n.d.
ts 28c.25		0.50	n.d.	2.7	§
ts 29a.10		0.67	n.d.	0.89	n.d.
ts 2		0.43	n.d.	1.3	n.d.
cs-ts 978		1.1	n.d.	normal	n.d.
ts 22		§§	n.d.	normal	n.d.
ts 22b.16		0.80	n.d.	n.d.	n.d.
ts 932		0.91	n.d.	n.d.	n.d.
cs 1309		normal	n.d.	normal	n.d.
Wild type		1.7 ± 0.3		2.1 ± 0.2	

§ no subunits or no rRNA detectable; §§ normal in Ca^{++} but not in Mg^{++}.

ribosomal RNA (15S), while some mutations of the 21S rRNA gene region were deficient, to various degrees, in the production of the large rRNA. We did not find a mutation which alters simultaneously the transcription of both rRNA.

ts902 is a typical RNA deficient mutation. It is probably a transcription deficient mutation since we do not find any accumulation of precursor RNA. Interestingly the mutation is unambiguously localized within the structural gene for 21S rRNA, suggesting that some internal sequences may play a role in the correct or efficient initiation of transcription. The transcription termination is not aberrant because the small amount of RNA produced by the mutant appears to be of normal size.

Two other mutations (ts29a-10 and ts2), also deficient in 21S rRNA, have been localized in the region of the 3' extension sequence coding for the precursor RNA. These mutations may concern processing of rRNA.

We have also examined the behavior of the mutants in the presence of various concentrations of protein synthesis inhibitors (chloramphenicol, erythromycin and paromomycin). Results are presented in Table 1. Many of the mutants show an abnormal response to these inhibitors, conferring either resistance or hyper-sensitivity to the drugs. Thus the defective rRNA mutations often show other phenotypic anomalies for which they had not been selected. We can try to correlate these phenotypes with the position of the mutation on the rRNA sequence.

III. Molecular characterization of the mitochondrial mutations.

In order to study the relation between structure and function in the rRNA molecule, one wishes to identify the nucleotide change introduced by the mutation. To localize the mutations precisely enough for sequencing, we first constructed a collection of rho^{-} clones deleted for various parts of the 21S rRNA gene. These rho^{-} were analyzed by restriction mapping and, in parallel, characterized for their ability to restore a wild-type phenotype by recombination with each of the ribosomal mutants (Zelikson and Bolotin-Fukuhara, in preparation). This classical methodology allowed us to place the mutations as shown in Figure 1. Some of the mutations are clearly outside the structural gene sequence and some others within the gene or supposed to be so (ts8). We will describe here the latter class

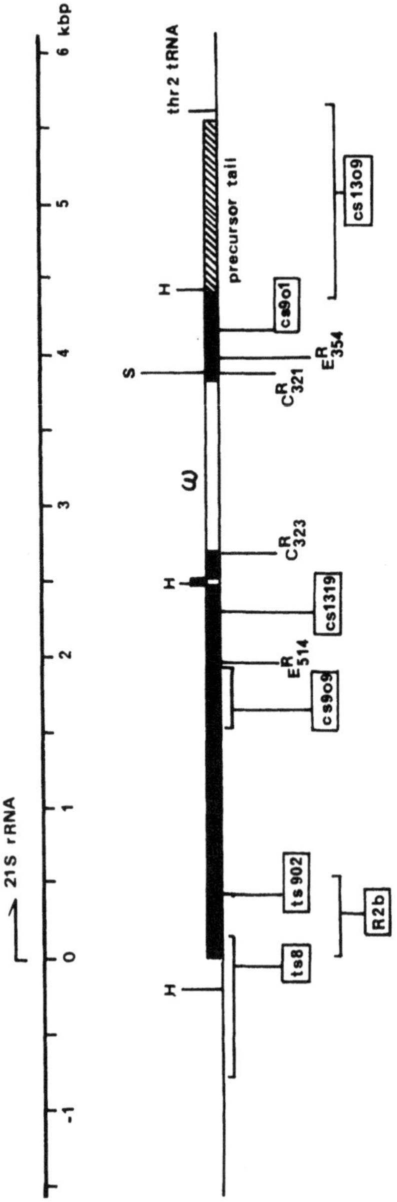

Figure 1. Large ribosomal RNA gene and localization of mutations.
H and S indicate the restriction sites for HpaII and SalI, respectively. Temperature sensitive mutations are boxed. Full bars: mature rRNA sequence; open bar: intron ω; hatched bar: 3' end sequence of precursor RNA. The central H site is present only in the optional mini-insert.

of mutations. These six mutations are very different in phenotype since some are cold-sensitive, some others heat-sensitive, and some also present a modified response to antibiotics (Table 1). They are scattered along the entire rRNA gene (Figure 1). All of them show anomalies in ribosomal subunits as described above.

Several of the mitochondrial mutations have been identified on the primary sequence of the 21S rRNA gene. While rho$^-$ derivatives of the wild-type DNA were useful in mapping the rRNA mutations, rho$^-$ DNA from the defective rRNA mutants were found to be unstable and could not be used for sequencing. The reason of this instability is not known. Therefore, the entire mitochondrial DNA carrying a defective mutation was first digested with HpaII endonuclease and cloned into the ClaI site of the plasmid pBR322. The cloned fragments were then sequenced and compared to the corresponding fragments of the wild-type DNA. Thus, cs 901 for example was localized in the domain VII of the 21S rRNA (cf. Branlant et al.'s nomenclature, 1981), cs909 in the domain V and ts902 in the domain I. Two erythromycin resistance mutations, E^R_{514} (locus rib 3) and E^R_{354} (locus rib 2) were found very close to each other within a well conserved loop structure of the secondary structure model of RNA, although the two mutations are about 900 basepairs apart on the DNA sequence (Sor and Fukuhara, 1982; see Figure 2)

IV. Interactions between mitochondrial mutations: bearing on the problem of the tertiary structure of rRNA.

Ribosomal RNAs have a highly conserved secondary structure through evolution (see Brimacombe et al., 1983, for recent review). A general model of secondary structure for each rRNA has been proposed (Woese et al., 1980; Stiegler et al., 1981; Glotz et al., 1981; Branlant et al., 1981; Noller et al., 1981). In contrast, informations on the tertiary structure are still scarce. The mitochondrial genetic system offers the possibility to reveal such spacial relations of sequences by the incompatibility or the mutual suppression of two rRNA mutations which belong to two different secondary structure domains.

For example, we have reported previously (Julou and Bolotin-Fukuhara, 1983) a genetic interaction between two mutations of the large rRNA gene, ts902 located in the domain I and C^R_{321} located in the domain VI. The

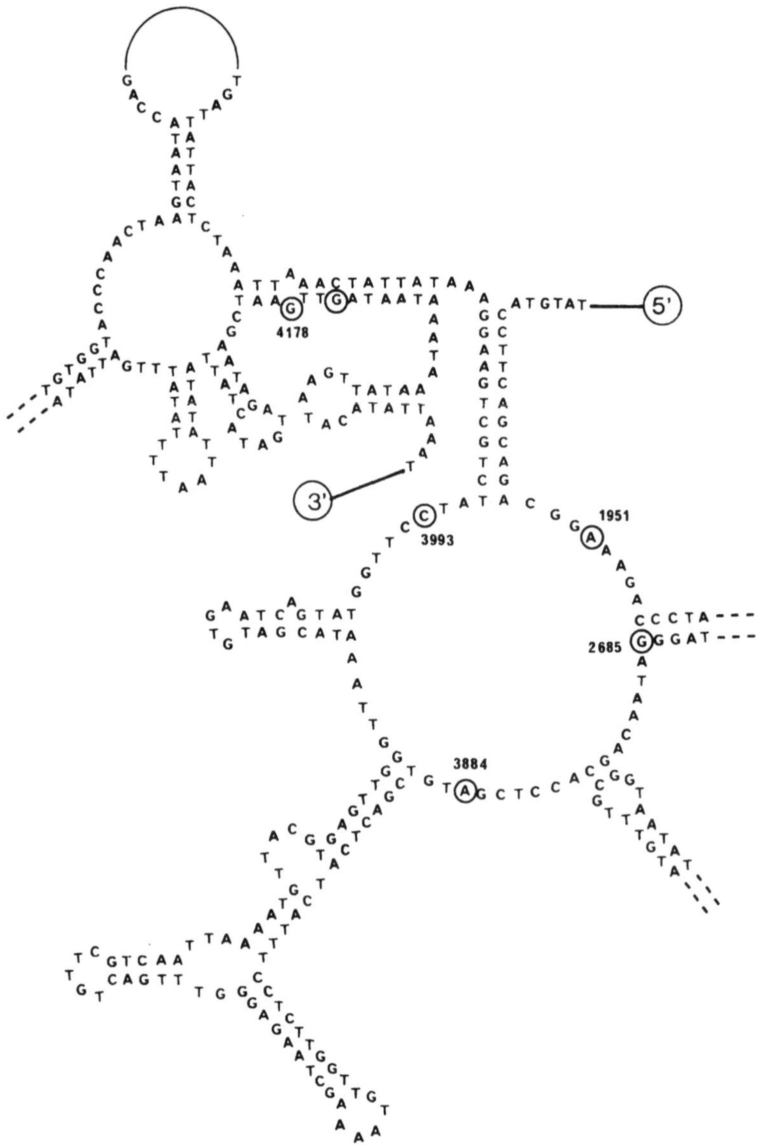

Figure 2. Part of the domain VI of the large ribosomal RNA.
Two erythromycin/spiramycin resistance loci are indicated at the position 1951 (locus rib 3, E^R_{514}) and at 3993 (locus rib 2, E^R_{354}). Chloramphenicol resistance mutations (Dujon, 1980) are at 2685 (C^R_{323}) and 3884 (C^R_{321}). The cold-sensitive mutation corresponds to a double nucleotide change at 4178 and 4180. These mutations occur within the stretches of sequences well conserved among different species. They are mostly in unpaired parts of the rRNA. The numbering includes the 1143 bp intron ω and 66 bp mini-insert (see Figure 1).

association of these two mutations in a cis configuration leads to the loss of mitochondrial functions of the cell, indicating that these sites on two different domains are functionally related. ts902 mutation also leads to a hyper-sensitivity to erythromycin, suggesting again the functional relation between the two domains. It is interesting to note that Stiege et al.(1983) have recently reported, using ultraviolet cross-linking methods, an interaction between the same two domains in E.coli rRNA molecule. By extending the study of interactions between rRNA mutations, it should be possible to reveal various sequences involved in the tertiary structure of rRNA.

V. Nuclear genes that interact with the rRNA mutations.

Selection of revertants from the rRNA mutations is an easy procedure because one can use a direct positive screening based on the search for temperature resistant colonies. We have looked for either spontaneous or $MnCl_2$-induced temperature-resistant revertants from various temperature-sensitive mutations. Since mitochondrial ribosomes are specified by nuclear genes and mitochondrial genes, we can expect to find, in addition to true revertants, nuclear and mitochondrial suppressor genes depending on the type of mutations one starts with and on the function it blocks or modifies.

V-1. Genetic determinism of the revertants.

The discrimination between the various possibilities leading to the suppression is made by crossing the revertant with rho^+ and rho^0 isonuclear tester strains and analyzing the meiotic and mitotic progeny of the cross. These analyses have been applied to five of the six mutations of the 21S rRNA gene, with the following results (Table 2). Most of the genes that suppressed the mitochondrial mutations were found to be nuclear genes. We obtained only one mitochondrial suppressor and no true revertants. The reason for this apparent bias is not known, but the preponderance of nuclear suppressors probably reflects the great number of nuclear genes that act on the mitochondrial rRNA as will be discussed further.

Table 2. Mitochondrial ribosomal RNA mutations and their suppressors.

Mitochondrial mutation	Type of suppression: Number of alleles examined	Number of genes involved	Genetic determinism
ts8	5	probably 1	Nuclear dominant
cs909	5	1	Nuclear dominant
	4	1	Nuclear recessive(cs)§
ts902	6	1	Nuclear dominant
	1		Mitochondrial
cs1319	many	not determined	Nuclear dominant or semi-dominant
tsR2b	3	not determined	Nuclear dominant

ts or cs symbolize the heat-sensitive or cold-sensitive phenotype due to the mitochondrial mutation. § A nuclear recessive suppressor of cs909 has by itself a cold-sensitive phenotype. Data from Guetari and Bolotin-Fukuhara (1982), Guetari (1983), Julou (1980), Contamine and Bolotin-Fukuhara (submitted) and unpublished results.

V-2. Possible mechanisms of suppression.

By selecting suppressors of rRNA mutations, one can detect and try to identify any gene the product of which interacts with rRNA. The reversion can arise by a variety of mechanisms.

The first class of possible interactions involves the classical interaction between ribosomal proteins and rRNA within the ribosome. There are some seventy proteins in the yeast mitochondrial ribosome (Faye and Sor, 1977), but we know, from the studies in E.coli, that all proteins do not interact with RNA but only part of them. Functionally equivalent proteins in yeast mitochondria should be identified by suppression effects. It is in fact very probable that some suppressors may prove to be structural genes for ribosomal proteins. Some suppressors we have studied seem to belong to this category. For example, among the revertants of cs909, we have identified one nuclear recessive suppressor which by itself confers a cold-sensitive phenotype for growth, associated with a cold-sensitive defect in ribosomal subunits (Contamine and Bolotin-Fukuhara, submitted). Among the suppressor genes that act upon ts8 mutation, there is one that confers a paromomycin-resistant phenotype and could well be due also to a ribosomal protein mutation.

The second class of suppressors concern RNA-RNA interactions. It is possible to imagine a suppression effect which is due to a second structural modification within the rRNA molecule. Such cases would appear as mitochondrial cis-acting suppressors, but have not been found so far.

Finally, a third class of suppression mechanism, which is not directly related with ribosomal architecture, involves the formation of rRNA itself. Effects of a mutation in the rRNA could be suppressed by post-transcriptional modifications such as changes in enzymes that methylate the RNA. These enzymes are probably encoded by the nucleus and their mutations may be rare since mitochondrial rRNA is not heavily methylated (Klootwijk et al., 1975). Some mutations could also be compensated by changes in transcription machinery if the mitochondrial mutations affect the transcription process. Transcription initiation defects or processing defects could be compensated by modification of enzymes involved in these activities. In these cases also the suppressor should be nuclear.

The third class of suppressors may be represented by the nuclear suppressor isolated from ts902, because it acts also on another rRNA deficient mutation ts1297 which is linked to the small rRNA gene (unpublished). Such a suppressor can well be a mutation of an enzyme involved in the rRNA transcription process.

Cloning of the suppressor genes is possible and will help to define their molecular nature and the mechanism of the suppression. The study of rRNA mutations and their suppressors should contribute to understand the nucleo mitochondrial relationship involved in the biogenesis of mitochondrial ribosomes.

Acknowledgements. We received financial supports from the CNRS (ATP Microbiologie) and the DGRST (81.E.1206).

References.

Bolotin-Fukuhara,M., Faye,G. and Fukuhara,H. (1977) Molec.Gen.Genet.152, 295-305.

Bolotin-Fukuhara,M. (1979) Molec.Gen.Genet.177, 39-46

Branlant,C., Krol,A., Machatt,M.A., Pouyet,J., Ebel,J.P., Edwards,K. and Kössel,H. (1981) Nucl.Acids Res. 9, 4303-4320

Brimacombe,R., Maly,P. and Zwieb,C. (1983) Progress in Nucleic Acid Research and Molecular Biology, 28, 2-43

Chambliss,G., Craven,G.R., Davies,J., Davis,K., Kahan,L. and Nomura,M. (1979) Ribosomes: Structure, Function and Genetics. University Park Press, Baltimore.

Coen,D., Deutsch,J., Netter,P., Petrochilo,E. and Slonimski,P.P. (1970) in Control of Organelle Development. Soc.Exp.Biol.Symp. vol.24 (Miller P.L. ed.) pp.449-496.

Contamine,V. and Bolotin-Fukuhara,M. (submitted)

Dujon, B. (1980) Cell 20, 185-197.

Faye,G., Kujawa,C. and Fukuhara,H. (1974) J.Molec.Biol. 88, 185-203.

Faye,G., Kujawa,C., Dujon,B., Bolotin-Fukuhara,M., Wolf,K., Fukuhara,H. and Slonimski,P.P. (1975) J.Molec.Biol. 99, 203-217.

Faye,G. and Sor,F. (1977) Molec.Gen.Genet. 155, 27-34.

Glotz,C., Zwieb,C., Brimacombe,R., Edwards,K. and Kössel,H. (1981) Nucl. Acids Res. 9, 3287-3306

Gourse,L.R., Starck,M.J. and Dahlberg,A.E. (1982) J.Molec.Biol. 159, 347-416

Grivell,L.A., Netter,P., Borst,P. and Slonimski,P.P. (1973)Biochim. Biophys.Acta 312, 358-367.

Groot,G.S.P., Mason,T.L. and Van Harten-Loosbroek,N. (1979) Molec.Gen. Genet. 174, 339-342

Guetari,M. and Bolotin-Fukuhara,M. (1982) Troisième Ecole de Biologie Moléculaire. Office des Publications Universitaires, Alger. pp.114-117.

Guetari,M. (1983) Thèse de Troisième Cycle, Université Paris-Sud.

Julou,C. and Bolotin-Fukuhara,M. (1983) Molec.Gen.Genet. 188, 256-260.

Julou,C. (1980) Thèse de Troisième Cycle, Université Paris-Sud.

Klootwijk,J., Klein,I. and Grivell,L.A. (1975) J.Molec.Biol. 97, 337-350.

Li,M. and Tzagoloff,A. (1982) J.Biol.Chem. 257, 5921-5928.

Linnane,A.W., Lamb,A.J., Christodoulou,C. and Lukins,H.B. (1968) Proc. Nat.Acad.Sci. USA. 59, 1288-1293.

Martin,R., Bordonné,R. and Dirheimer,G. (1982) Proceedings of the FEBS meeting on Cell Function and Differentiation, Athens. A.R.Liss, N.Y.

Noller,H.F., Kop,J.A., Wheaton,V., Brosius,J. Gutell,R.R., Kopylov,A.M., Dohme,F., Herr,W., Stahl,D.A., Gupta,R. and Woese,C.(1981) Nucl.Acids Res. 9, 6167-6189.

Nomura,M., Tissières,A. and Lengyel,P. eds. (1974) Ribosomes. Cold Spring Harbor Laboratory Monograph 4. Cold Spring Harbor.

Sigmund,C.D. and Morgan, E.A. (1982) Proc.Nat.Acad.Sci.USA. 79, 5602-5606.

Sor,F. and Fukuhara,H. (1982) Nucl.Acids Res.10, 6571-6577.

Spithill,T.W., Nagley,P. and Linnane,A.W. (1979) Molec.Gen.Genet. 173, 159-170

Starck, M.J., Gourse,R.L. and Dahlberg,A.E. (1982) J.Molec.Biol.159, 417-439.

Stiege,W., Glotz,C and Brimacombe,R. (1983) Nucl.Acids Res. 11,1687-1706.

Stiegler,P., Carbon,P., Ebel,J.P. and Ehresman,C. (1981) Eur.J.Biochem. 120, 487-495.

Terpstra,P., Zanders,E. and Butow,R.A. (1979) J.Biol.Chem. 254, 12653-12661.

Thomas,D. and Wilkie,D. (1968) Biochem.Biophys.Res.Commun. 30, 368-372.

Williamson,D.H., Maroudas,N.G. and Wilkie,D. (1971) Molec.Gen.Genet. 111, 209-223.

Woese,C., Magrum,L.J., Gupta,R., Siegel,R.B., Stahl,D.A., Kop,J., Crawford, N., Brosius,J., Gutell,R., Hogan,J.J. and Noller,H.F. (1980) Nucl.Acids Res. 8, 2275-2293.

Wolf,K., Dujon,B. and Slonimski,P.P. (1973) Molec.Gen.Genet. 125, 53-90.

CYTOPLASMIC SYNTHESIS OF PROTEINS OF THE MAMMALIAN MITOCHONDRIAL RIBOSOME

Gretchen L. Schieber and Thomas W. O'Brien

Department of Biochemistry and Molecular Biology
University of Florida, College of Medicine
Box J245 J. Hillis Miller Health Center
Gainesville, Florida 32610 USA

Introduction

Despite their similar function, the cytoplasmic and mitochondrial protein synthesis systems are completely distinct. When compared with cytoplasmic ribosomes, mitochondrial ribosomes (mitoribosomes) generally are smaller and appear in some ways to be more like prokaryotic ribosomes [1,2]. While there is variablility in size and physical-chemical properties among mitoribosomes of the other eukaryotes, the mitochondrial ribosomes of mammals all appear to be of the 55 S type [3,4]. The 55 S mammalian mitoribosomes are slightly larger than bacterial ribosomes [5] and their composition is much more protein rich (about 85 proteins, [6]), with their rRNA content correspondingly reduced.

Most of the proteins of all mitochondrial ribosomes must be synthesized on 80 S cytoplasmic ribosomes using mRNA transcribed from nuclear DNA (the nucleo-cytoplasmic protein synthesis system). However, it was shown in both yeast and Neurospora, using techniques which relied on differentiation of the two compartments by specific inhibitors [7-9] that there is one protein in the mitochondrial ribosome which is a mitochondrial product. This small subunit protein is implicated in assembly of the ribosomal subunit, since

Mitochondria 1983

impairing its synthesis results in incomplete assembly of small subunits [10-12]. An analogous protein has not been identified in any higher system.

In recent years, a general format for the mammalian mitochondrial genome was determined by complete sequencing of the genome of several mammals [13-16]. The genome codes for the mitochondrial rRNA's and tRNA's and for thirteen open reading frames. The identified products of mitochondrial protein synthesis have been assigned to five of the coding regions; the remainder are unidentified reading frames (URF's). While the sequences of putative proteins coded for by these regions are known, none of these "genes" has been assigned to a functional mitochondrial protein.

We have studied protein synthesis in cultured mammalian cells in order to determine whether or not mammalian mitochondria synthesize a protein analogous to the mitoribosomal protein already identified in the lower eukaryotes. That is, do any of the unidentified reading frames of the mammalian mitochondrial genome code for a protein of the mammalian mitochondrial ribosome?

Results and Discussion

Cells labeled in the presence of cycloheximide.

In order to inhibit labeling of the products of cytoplasmic protein synthesis, Madin-Darby bovine kidney (MDBK) cells [17] were labeled (2 h) with [^{35}S]methionine in the presence of cycloheximide (300 µg/ml). During a chase period, in the absence of both radioactivity and the inhibitor, any components required for assembly of ribosomes can be synthesized and the particles should assemble, but components made by the nucleo-cytoplasmic system will not be radiolabeled. The ribosomal large and small subunits were isolated from a mitochondrial lysate in sucrose gradients. For highest purity, the 55 S "monosomes" are ideally isolated first in a low salt, non-dissociating gradient (100 mM KCl, 20 mM $MgCl_2$, 10 mM triethanolamine, 5 mM

Figure 1. Fluorograms of Mitoribosomal Proteins Labeled in the Absence of Cytoplasmic Protein Synthesis. The proteins of mitochondrial ribosomal subunits, prepared from MDBK cells labeled in the presence of cycloheximide were analyzed by 2D-PAGE [6] and fluorography [18,19]. A, 28 S subunits, + CHM; B, 28 S subunits, no inhibitor; C, 39 S subunits, + CHM; D, 39 S subunits, no inhibitor.

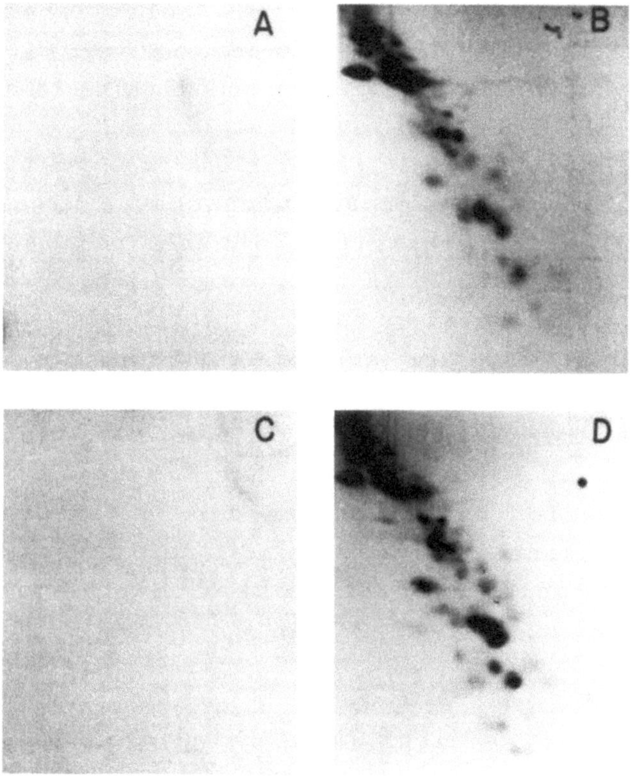

ß-mercaptoethanol, pH 7.5), and then the ribosomal subunits, 28 S and 39 S, are isolated in a high salt, dissociating gradient (300 mM KCl, 5 mM $MgCl_2$, 10 mM Tris, 5 mM ß-mercaptoethanol, pH 7.5). However, in order to maximize recovery, these subunits were prepared directly using a single dissociating sucrose density gradient. Additionally, unlabeled carrier mitoribosomes prepared from bovine liver were added to reduce adsorptive losses during the isolation procedure. The subunits were analyzed by two-dimensional electrophoresis [6] and fluorography [18,19].

The results are shown in fig. 1 together with the results from a parallel control (cells labeled in the absence of inhibitor). No spots are observed on fluorograms from either large or small ribosomal subunits labeled in the presence of cycloheximide (fig. 1, A and C). This is in contrast to the subunit preparations from parallel control cells (fig. 1, B and D) which show a great deal of incorporated radioactivity. The most prominent spots in these fluorgrams are, in fact, not ribosomal proteins, but are contaminating mitochondrial proteins of high specific radioactivity which are included in this preparation because the subunits were prepared by the single gradient method. The known proteins of the mitoribosome are visible as the background of paler spots. Thus, the conditions of the labeling period were adequate for the incorporation of [^{35}S]methionine into mitochondrial ribosomal proteins, but that incorporation is blocked by the presence of cycloheximide.

Cells labeled in the presence of chloramphenicol.

In a converse experiment to the one described above, MDBK cells were labeled with [^{35}S]methionine in the presence of chloramphenicol (100 µg/ml), allowing label to be incorporated only into those proteins produced by the nucleo-cytoplasmic system. As before, proteins made during the chase period (by the mitochondrial protein synthesis system) would not have any incorporated radioactivity. Since this inhibitor has a low toxicity to these cells, the

Figure 2. Fluorograms of Mitoribosomal Proteins Labeled in the Absence of Mitochondrial Protein Synthesis. The proteins of mitochondrial ribosomal subunits, prepared from MDBK cells labeled in the presence of chloramphenicol were analyzed by 2D-PAGE and fluorography. A, 28 S subunits, + CAP; B, 28 S subunits, no inhibitor; C, 39 S subunits, + CAP; D, 39 S subunits, no inhibitor.

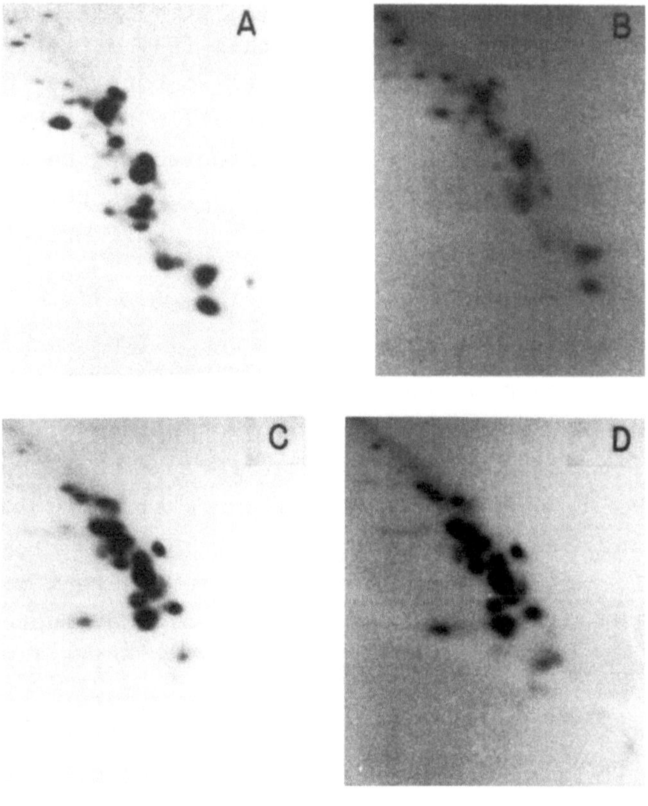

labeling period could be much longer (8-16 h), producing ribosomes of higher specific activity. This allowed the ribosomes to be prepared using both non-dissociating and dissociating sucrose density gradients. As before, the ribosomal subunit proteins were separated by two-dimensional PAGE and analyzed by fluorography.

The results are shown in figure 2. The fluorograms of 28 S and 39 S subunits labeled in the presence of chloramphenicol (fig. 2, A and C) show the expected pattern for mitoribosomal proteins. Indeed, the patterns show no significant differences from the fluorograms of ribosomal subunits from cells labeled in the absence of inhibitor (fig. 2, B and D). The patterns in these fluorograms are virtually the same as the stained pattern characteristically formed by the carrier bovine liver mitoribosomal proteins revealed by Coomassie blue stain [6], differing only in a few proteins which failed to label in both the inhibited and control ribosomes. This demonstration that the labeling of mitoribosomal proteins is unaffected by the inhibition of mitochondrial protein synthesis indicates that the proteins are all products of the nucleo-cytoplasmic protein synthesis system.

Effect of Inhibitors.

The interpretation of the ribosomal pattern is determined by the effect of the inhibitor under each condition. In order to demonstrate that the inhibitors produce the expected results, an aliquot of whole mitochondria from each labeling condition was analyzed directly by one-dimensional PAGE (equivalent to the second dimension of the system used above) and fluorography. In fig. 3, A is shown mitochondria from cells labeled in the presence of cycloheximide. The expected limited band pattern of mitochondrial products is produced from these cells. This confirms that cytoplasmic protein synthesis was eliminated by the cycloheximide but that mitochondrial protein synthesis continued. In a comparatively short exposure, the mitochondria from cells labeled in the presence of chloramphenicol (fig. 3, B) or from cells labeled in the absence of inhibitor (fig. 3, C) display the complex pattern of cytoplasmically produced proteins. Because of the overwhelming pattern produced by cytoplasmic protein synthesis, the absence of mitochondrial products is not observable.

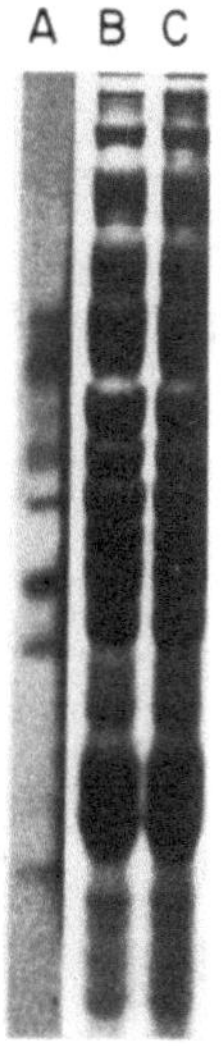

Figure 3. The effect of inhibitors on mitochondrial proteins. Mitochondria from cells labeled with [^{35}S]methionine in the presence of cycloheximide (A), chloramphenicol (B), and in the absence of inhibitors (C), were analyzed by one-dimensional SDS-PAGE and fluorography.

Several factors are included in this straightforward interpretation of the results, and control experiments were included to prove that these assumptions are, indeed, valid. The blank fluorograms which resulted with the inhibitor cycloheximide would also have been produced if the conditions for the labeling period (media, specific activity of radiotracer, etc.) or for recovery and analysis of ribosomal subunits (mitochondrial isolation, length of fluorography, etc.) were inadequate to visualize synthesis into mitoribosomal proteins. A parallel control experiment, however, where all such factors were the same except cycloheximide was omitted, demonstrated the the synthesis of mitoribosomal proteins was observable under these conditions. Additionally, it is known that mitochondrial protein synthesis is sensitive to the inhibition of cytoplasmic protein synthesis, that during prolonged incubations in the absence of cytoplasmic protein synthetic products, mitochondrial protein synthesis becomes inhibited and eventually mitochondrial products are degraded (20,21). In order to insure that the cycloheximide labeling period was not excessive, a sample of intact mitochondria from these cells was examined. It showed the expected limited band pattern of mitochondrially synthesized proteins, proving that conditions of the incubation supported mitochondrial protein synthesis. Finally, this work, of course, requires that any possible mitochondrial product contains methionine. Indeed, those proteins missing on fluorograms of mitoribosomes from cells labeled in both the presence and absence of chloramphenicol may simply lack methionine residues. Since the genome can be used to predict the sequence of any possible products, it has been determined that all mitochondrially synthesized proteins should be rich in methionine residues [14]. These unlabeled proteins, therefore, cannot be mitochondrial products.

The data in this paper are based on an analysis of the mitoribosomal proteins by a specific system of 2D-PAGE [6]. However, by analogy to lower eukaryotes, these data also indirectly address the possibility that a mitochondrially produced mitoribosomal protein might be missed in this analysis because it was not displayed in these 2D-PAGE separations. It is known that in yeast and Neurospora, the small ribosomal subunits formed in the absence of mitochondrial protein synthesis are defective. They migrate incorrectly in sedimentation analysis [11,12], and they specifically lack a few proteins which assemble after the mitochondrial product [10-12]. In our bovine system, the subunits examined were those which migrated to the position of assembled subunits in sucrose density gradients. If the particles made during the labeling in the presence of chloramphenicol were incompletely assembled, they would have been completed during the chase period. Proteins synthesized during the chase period would not be labeled, though. So the particles recovered from the gradients would be deficiently labeled in those proteins which were added after any mitochondrial product. Therefore, it is important to note that the subunits produced by labeling in the presence of chloramphenicol in the bovine system produce the same pattern of spot intensities as their parallel control.

Data presented in this paper indicate that the 55 S mitochondrial ribosomes of a mammalian (bovine) system contain only proteins which are products of the nucleo-cytoplasmic protein synthesis compartment. This was shown both by demonstrating that synthesis of mitoribosomal proteins was completely eliminated by cycloheximide, an inhibitor of cytoplasmic protein synthesis, and by demonstrating that synthesis of these proteins is not affected by chloramphenicol, an inhibitor of mitochondrial protein synthesis. This result is in contrast to the situation in lower eukaryotes, where one protein of the small mitoribosomal subunit is a mitochondrial product [7-9]. If indeed, the mitochondrially produced mitoribosomal protein in lower eukaryotes functions in ribosome assembly [10-12], that function might be performed in mammals by a protein or proteins imported from the cytoplasm. Because of the differences in the structure of the two classes of mitoribosomes, it might be that a protein of identical function is imported, or it is also possible that the ribosomal domain involved is sufficiently changed in mammals that a different assembly mechanism is used.

Thus, the mitochondrion is not obligated to make a set of specific types of proteins, but its protein synthetic products are evolving in response to, or in partnership with the proteins encoded on the nuclear genome.

Supported by USPHS grants GM-15438 and GM-25888.

Abbreviations

CAP: (D)threo-chloramphenicol
CHM: cycloheximide
Mitoribosome: mitochondrial ribosome
Monosome: monomeric ribosome, consisting of a small and a large subunit
MDBK cells: Madin-Darby bovine kidney cells
PAGE: polyacrylamide gel electrophoresis
SDS: sodium dodecyl sulfate
2D: two-dimensional
URF: unidentified reading frame

References

1. O'Brien, T.W., Matthews, D.E.: In; Handbook of Genetics. Vol. 5; Molecular Genetics (King, R. C., ed.) Plenum Press, New York, pp. 535-580 (1976).
2. Buetow, D.E., Wood, W.M.: Subcellular Bioch. 5, 1-85 (1978).
3. O'Brien, T.W.: J. Biol. Chem. 246, 3409-3417 (1971).
4. Sacchi, A., Ferrini, U., Londei, P., Cammarano, P., Miraldi, N.: Biochem. J. 168, 245-259 (1977).
5. Hamilton, M., O'Brien, T.W.: Biochemistry 13, 5400-5403 (1974).
6. Matthews, D.E., Hessler, R.A., Denslow, N.D., Edwards, J.E., O'Brien, T.W.: J. Biol. Chem. 257, 8788-8794 (1982).
7. Terpstra, P., Zanders, E., Butow, R.A.: J. Biol. Chem. 254, 12653-12661 (1979).

8. Groot, G.S.P., Grivell, L.A., VanHarten-Loosbroek, N., Kreike, J., Moorman, A.F.M., VanOmmen, G.J.B.: In; Structure and Function of Energy-Transducing Membranes (vanDam, K., vanGelder, B.F., eds.) Elsevier/North Holland, New York, pp. 177-186 (1977).

9. Lambowitz, A.M., Chua, N.-H., Luck, D.J.: J. Molec. Biol. 107, 223-253 (1976).

10. Terpstra, P., Butow, R.A.: J. Biol. Chem. 254, 12662-12669 (1979).

11. LaPolla, R.J., Lambowitz, A.M.: J. Molec. Biol. 116, 189-205 (1977).

12. Lambowitz, A.M., LaPolla, R.J., Collins, R.A.: J. Cell Biol. 82, 17-31 (1979).

13. Anderson, S., Bankier, A.T., Barrell, B.G., deBruijn, M.H.L., Coulson, A.R., Drouin, J., Eperon, I.C., Nierlich, D.P., Roe, B.A., Sanger, F., Schreier, P.H., Smith, A.J.H., Staden, R., Young, I.G.: Nature 290, 457-465 (1981).

14. Anderson, S., deBruijn, M.H.L., Coulson, A.R., Eperon, I.C., Sanger, F., Young, I.G.: J. Molec. Biol. 156, 683-717 (1982).

15. Bibb, M.J., VanEtten, R.A., Wright, C.T., Walberg, M.W., Clayton, D.A.: Cell 26: 167-180 (1981).

16. Attardi, G., Cantatore, P., Ching, E., Crews, S., Gelfand, R., Merkel, C., Montoya, J., Ojala, D.: In; International Cell Biology; 1980-1981 (Schweiger, H.G., ed.) Springer-Verlag, New York, pp. 225-238 (1981).

17. Madin, S.H., Darby, Jr., N.B.: Proc. Soc. Exp. Biol. Med. 98, 574-576 (1958).

18. Laskey, R.A., Mills, A.D.: Eur. J. Biochem. 56, 335-341 (1975).

19. Chamberlain, J.P.: Analyt. Biochem. 98, 132-135 (1979).

20. Costantino, P., Attardi, G.: J. Biol. Chem. 252, 1702-1711 (1977).

21. Yatscoff, R.W., Freeman, K.B.: Can. J. Biochem. 55, 1064-1074 (1977).

DIURON RESISTANCE IN SACCHAROMYCES CEREVISIAE

Anne-Marie Colson, Alan Fauconnier and Brigitte Meunier
Laboratoire de Génétique, Université de Louvain, Louvain-la-Neuve, Belgium

Introduction

In yeast, Diuron (3-(3,4-dichlorophenyl)-1,1-dimethylurea) inhibits the electron flow of the mitochondrial respiratory chain between the cytochromes b and c_1 (1). The first diuron-resistant mutants isolated after selection on glycerol plus diuron were found to be of mitochondrial heredity and showed an in vitro diuron-resistant respiration at the level of submitochondrial particles (2). These results indicated that a mitochondrial gene product being part of the mitochondrial inner membrane was involved in the expression of diuron-resistance. Extensive genetic analysis demonstrated that the mitochondrial diu^r mutations were carried by the cytochrome b gene (3). Therefore, cytochrome b became the best candidate as the diuron binding protein. So far, this has not been demonstrated. The mitochondrial diu^r mutations were found to belong to two loci diu1 and diu2 of the cytochrome b split gene (3). These loci were found to belong respectively to the first and the fourth exon of the gene (4). Antimycin-resistant mutations were also located in the cytochrome b gene (5, 6). They were found to belong to the same two loci as diu1 and diu2 (7). These results indicated that cytochrome b was also a good candidate as an antimycin binding protein. Cytochrome b, indeed, binds antimycin as it was demonstrated in (8, 10, 9). In the present paper, three approaches will be considered :

a) Analysis of the cross-resistance to antimycin of mitochondrial Diu^R mutants.
b) Isolation and properties of nuclear diuron-resistant mutants presenting a cross resistance to antimycin.
c) Description of some properties of diuron-resistant mutants selected on DL-lactate plus diuron.

Mitochondria 1983

Results and Discussion

a) Cross-resitance to antimycin of mitochondrial Diu^R mutants

In vivo cross-resistance to antimycin has been tested on the previously described mutants (2,3,7) as well as on a new set of mitochondrial mutants that we isolated and characterised. Out of ten mitochondrial mutants analysed 5 were sensitive to antimycin (D^SA^S) and 5 were resistant to antimycin (D^RA^R). The mutations which conferred the double resistance, D^RA^R, were located either at the diu2 locus or at least in the first exon of the cytochrome b gene. On the contrary, the diu^r mutations showing antimycin sensitivity belonged to the diu1 locus or at least to the fourth exon of the gene (fig. 1). The cosegregation of the double resistance was observed suggesting that the origin of the mutant phenotype was not due to double unlinked mutations. This observation might be interpreted as if the diu2 locus of the first exon was more involved in the binding of antimycin than the diu1 locus of the fourth exon. The mutant Diu2-728 characterised in Colson and Slonimski (3) was analysed for the binding of antimycin to the mitochondrial inner membrane by measuring the fluorescence of the drug (10). These authors found a reduced binding of antimycin to the mitochondrial particles of the mutant. This result indicated that the diu2-728 mutation located in the first exon of the cytochrome b gene affects the binding of antimycin to the mitochondrial membrane.
The mitochondrial mutants tested for in vitro resistance to diuron by measuring NADH-cytochrome c reductase activity were all found to be strongly diuron-resistant. They kept in the presence of diuron 70 to 100 % of the activity found without diuron. In contrast, different levels of antimycin-resistance were found in vitro in the mutants presenting the double resistance D^RA^R. In the presence of antimycin, the NADH-cytochrome c reductase activity was found to be either completely sensitive or presented up to 70 % of the activity observed without antimycin. The discrepancy between the in vivo and the in vitro antimycin-resistance phenotype of some D^RA^R mutants might suggest that this resistance is dependent on a particular configuration of the mitochondrial inner membrane i.e. a specific interaction between the cytochrome b and a nuclearly coded protein of the bc_1 complex which might be the antimycin binding protein of

Rieske (11). The isolation procedures could destroy this interaction leading to a sensitive phenotype in vitro.

b) Isolation and properties of nuclear diuron-resistant mutants

Most Diu^R mutants were isolated from strains such as KL14-4A which carries a long cytochrome b split gene (Fig. 1).

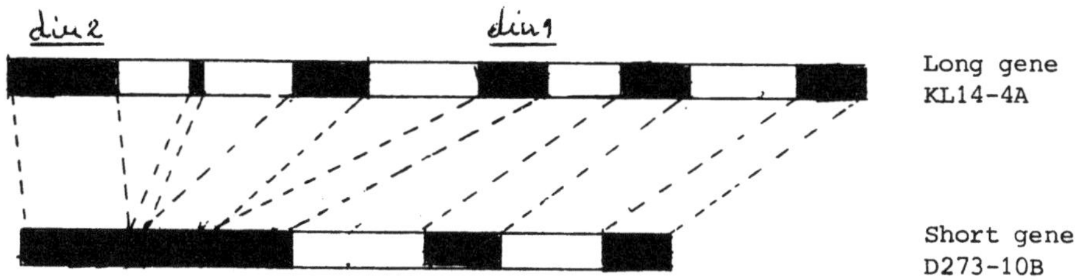

Figure 1 : Long and short cytochrome b gene
in black : exons
in white : introns

The diu^r mutations isolated from "long gene" strains were all found to belong to the mitochondrial cytochrome b gene. Diu^R mutants were also isolated from D273-10B, a strain carrying a "short" cytochrome b gene. Out of four mutants studied so far only one was of mitochondrial heredity (3). The three other mutants were found to be of nuclear heredity. Thus the only nuclear Diu^R mutants found so far were derived from a strain carrying a short cytochrome b gene. This nuclear diuron-resistance was found to be expressed at the level of the inner mitochondrial membrane. It is the first time that a nuclear drug-resistance to an inhibitor of a mitochondrial function can be detected in vitro at the level of the mitochondrion. A nuclear mutation which confers diuron-resistance to the mitochondrial inner membrane is most simply interpreted as if it belongs to a gene coding for a nuclear component of the bc_1 complex. If this was correct, then this protein would be a second candidate beside cytochrome b as a diuron binding protein. Thus diuron would bind to at least two proteins of the bc_1 complex. We found also that these nuclear Diu^R mutants presented a cross-resistance to antimycin both in vivo and in vitro (A.M. Colson, in preparation). Therefore, the nucleus coded protein involved might bind both diuron and antimycin.

Will the nuclear $\underline{diu}^r$ mutations affect the antimycin binding protein of Rieske (1976)?

c) Some properties of diuron-resistant mutants selected on DL-lactate plus diuron.

The selection for diuron-resistance using DL-lactate as carbon source produced mutants which exhibited resistance after 8 days of incubation on lactate plus diuron and sensitivity on glycerol plus diuron (12). In addition, resistant colonies appeared on glycerol at a frequency ranging between 10^{-3} to 10^{-4}. These resistant colonies were the result of secondary mutations which has accumulated in the mutant as if it had a "mutator" phenotype (Colson and Meunier, submitted). This apparent "mutator effect" was specific for diuron-resistance and did not affect mutations frequencies leading either to respiratory deficiency (petite or mit^- mutants) nor to drug-resistance phenotype (antimycin-, erythromycin- or cycloheximid- resistance). All secondary mutations so far tested (more than 10 mutations) belong to the mitochondrial cytochrome b gene. The primary mutation of one mutant of this class did not present either a nuclear or a mitochondrial heredity. (Colson and Meunier, submitted).

From our results, it appears that in addition to cytochrome b at least one nuclearly coded protein might also be involved in the binding of diuron and antimycin. A regulation at the level of the genetic expression seems to control the carbon source utilisation (glycerol-lactate) in the presence of diuron.

References

1. Convent, B. and Briquet, M. (1978) Eur. J. Biochem. 82, 473-481.
2. Colson, A.M., Luu The Van, Convent, B., Briquet, M. and Goffeau, A. (1977) Eur. J. Biochem. 74, 521-526.
3. Colson, A.M. and Slonimski, P.P. (1979) Molec. Gen. Genet. 167, 287-298.
4. Lazowska, J., Jacq, C. and Slonimski, P.P. (1980) Cell, 22, 333-348.
5. Michaelis, G. and Pratje, E. (1977) Molec. gen. Genet., 156, 79-85.

6. Lang, B., Burger , G., Bandlow, W., Kandewitz, F. and Schweyen, R.J. (1976) in : "Genetics and biogenesis of chloroplasts and mitochondria" (Th. Bücher, W. Neupert, W. Sebald and S. Werner, eds.) pp. 461-465 Amsterdam : Elsevier (North-Holland Biomedical Press 1976).

7. Colson, A.M., Michaelis, G., Pratje, E. and Slonimski, P.P. (1979) Molec. gen. Genet., 167, 299-300.

8. Roberto, H., Smith, S.C., Marzuki, S. and Linnane, A. (1980) Arch. Biochem. Biophys., 200, 387-395.

9. Meunier-Lemesle, D., Chevillotte-Brivet, P. and Pajot, P. (1980) Eur. J. Biochem., 111, 151-159.

10. Briquet, M. and Goffeau, A. (1981) Eur. J. Biochem., 117, 333-339.

11. Rieske, J.S. (1976) Biochem. Biophys. Res. Comm., 456, 195.

12. Colson, A.M. and Wouters, L. (1980) in : "The organisation and expression of the mitochondrial genome" (A.M. Kroon and C. Sacrone, eds.) Elsevier (North-Holland - Biomedical Press) pp. 71-74.

NUCLEAR MUTANTS AFFECTED IN THE RECOMBINATION OF MITOCHONDRIAL DNA IN SACCHAROMYCES CEREVISIAE.

Françoise Foury and Jan Kolodynski
Laboratoire d'Enzymologie, Louvain-la-Neuve, Belgium

Introduction

Recombination between ρ^+ and ρ^- genomes is the result of the organization of the ρ^- genome. Indeed the genome of a ρ^- mutant produced by the excision at specific repeated sequences (1, 2) of a segment of the wild-type DNA is identical in base composition to the original wild-type sequence but it is amplified by an unknown mechanism (3), up to a hundred times when the conserved sequence is as short as a few hundred base pairs. The orientation of the repeating-units may vary (4). They can be arranged in direct tandem abc abc (head-to-tail) or in palindrome (head-to-head, tail-to-tail), although the palindromic structure is always imperfect : abcde d'c'b'a' (5,6). More complex structures are met with "mixed-repeats" (5) in which tandem and palindromic arrangements are alternately present : abcdefghif'e'd' abcdefghif'e'd'
The amplification of the conserved sequence gives rise to a multimer. However under electron microscope, circular oligomeric series composed of the monomer, dimer, trimer, etc... of the conserved sequence have also been observed (7, 8). These oligomers resemble the recombination intermediates formed in plasmids by intermolecular recombination (9, 10).
The genome of the ρ^- mutants recombine with the ρ^+ genome by a process involving genetic exchanges. However the frequency of transmission of ρ^- markers into the ρ^+ diploid progeny varies highly from a ρ^- to another (11, 12). Only a few general rules have emerged so far from the recombination data analysis of ρ^+ x ρ^- crosses (13). Hypersuppressive* ρ^- which have a repli-

Mitochondria 1983

cation origin (ori or rep sequences) (14, 15) have low marker transmission efficiency, whereas ρ^- with low suppressiveness recombine more actively (13, 14). The proximity of a ρ^- marker to the deletion end points of the conserved sequence decreases its recombination ability (16). However in some $\rho^+ \times \rho^-$ crosses very high marker transmission cannot readily be explained only by the general recombination (heteroduplex formation between long homologous sequences), so that the existence of specific recognition sites between ρ^+ and ρ^- genomes has been postulated (16).

We have addressed ourselves the following questions. To what extent the general recombination can explain the variability in the marker transmission frequency ? Are there specific sequences or structures which modulate the recombination between ρ^+ and ρ^- genomes ? Can oligomers be considered as recombination intermediates as it is the case for plasmids ?

Results

The characterization of mutants affected in the recombination of the mtDNA should help to answer these questions. The difficulty of isolating such mutants comes from the necessity to screen, after mutagenesis, a potential recessive recombination-deficient mutant with a recombination-proficient wild-type strain. For this reason, we have crossed 30,000 mutagenized colonies from *Saccharomyces cerevisiae* D273-10B with a ρ^- strain which has retained the O_1^R marker and carries the *kar*1 mutation (17) (Fig.1). Using this selection procedure, we have isolated about 12 mutants that might be affected in the recombination of the mtDNA (18). We have characterized three nuclear allelic mutants that have been assigned to the *pif* locus (*pif* for "petite integration frequency").

Fig. 1. Selection procedure of mutants deficient in mtDNA recombination

Mutagenized D273-10B colonies x JC7/103-2

met2 $can1^R$ ρ^+ mit^+ — a leu1 kar1 ρ^- O_1^R

Mating in complete glucose medium

↓

HETEROKARYON (1)

Possible recombination events

↓

2 successive replications on canavanine + methionine minimum medium

↓

Selection of CYTODUCTANTS (2) with D273-10B nucleus

↙ ↘

Recombination positive colony: possible recombination events.	Recombination-deficient colony: No recombination event.
↓	↓
Replication on oligomycin	Replication on oligomycin
GROWTH	If (2) > (1) NO GROWTH OR PAPILLAE

Analysis of the recombination data of <u>PIF</u> ρ^+ x <u>pif</u> ρ^- (heterozygous) and <u>pif</u> ρ^+ x <u>pif</u> ρ^- (homozygous) crosses has led to the following conclusions.

In PIF ρ^+ x pif ρ^- crosses, the transmission frequency of a ρ^- marker to the ρ^+ diploid progeny depends on the arrangement of the repeating-unit (Table 1). In tandemly-arrayed ρ^-, depending on the conserved sequence, the recombination frequencies vary from 10 to 80 %. However in ρ^- with long repeating units ($\geq$ 5 kb), the recombination frequency is always in the range of 50-60 %. The decreased marker transmission linked to the presence of a replication origin is corroborated by our results (Table 1). In contrast, the recombination between ρ^+ and palindromic ρ^- is always very high ($\geq$ 80 %). The frequency does not depend on the location of the markers, the length of the repeating-unit or the presence of specific sequences (AT or GC rich regions) (Table1).

Table 1. Recombination of tandemly-arrayed and palindromic ρ^- in *PIF* ρ^+ x *pif* ρ^- crosses.

ρ^- clone	Retained marker	Repeat (kb)	Structure	Recombinants %
394/E41-1	E514	3.2	Palindrome	93.3
JF11/A8	D1	7.2	Palindrome	92.8
394/R53	E514	11.0	Palindrome	89.1
JF11/A1-3	D1	4.5	Palindrome	84.4
JF11/6-32	0625	0.8	Palindrome	80.8
JF2/2-31	0625	3.1	Tandem	80.2
394/DS14	0118	4.1	Tandem	68.5
JF11/6-32/1	0625	0.8	Tandem	54.1
394/F11	E514	12.0	Tandem	52.2
394/E41-2	E514	3.2	Tandem	47.6
JF11/B17	D1	7.5	Tandem	38.0
JF11/6-32/3	0625	0.7	Tandem	13.0

* Contains the *ori2* origin (15).

Table 2 shows that in some *pif* x *pif* crosses, the recombination frequency is similar to that reached in PIF x pif crosses. However this frequency, in other crosses, can be decreased down to a hundred of times. Analysis of the structure of the ρ^- emphasizes the existence of two recombination systems depending on the orientation of the repeating-unit. Tandem organization prevents recombination at all tested loci of all ρ^- clones except one (Table 2) : The tandemly-arrayed ρ^- 394/E1 located in the 21 S rRNA gene (6) does recombine and we have no explanation for this anomaly.
In contrast all palindromic ρ^- isolated so far recombine in the *pif* as well as in the *PIF* nuclear background (Table 2).

No influence of the AT spacers and GC clusters has been noted. Interestingly those ρ^- clones which are organized in "mixed-repeats" do not recombine, although half of their DNA is in a palindromic configuration. In this respect, it may be significant that agarose gel electrophoresis shows the presence of oligomeric series in the native DNA of palindromic pif ρ^- whereas the DNA of ρ^- genomes organized in "mixed-repeats" is only present in the multimer form. In literature, oligomers are mentioned for tandem and palindromic ρ^-, while no information is available about "mixed-repeats" ρ^-.

Table 2. Recombination in pif ρ^+ x pif ρ^- crosses

ρ^- clone	Retained marker	% Recombinants in pif x pif	Pif x pif	Structure
394/E41-1	E514	83.8	93.3	Palindrome
JF11/6-32	0625	81.0	80.8	Palindrome
JF11/A1-3	D1	75.2	84.4	Palindrome
394/E1	E514	31.8	26.7	Tandem
394/E41-2	E514	3.1	47.6	Tandem
JF11/B17	D1	1.8	38.0	Tandem
JF11/6-32/1	0625	0.5	52.2	Tandem
JF11/6-32/3	0625	0.5	13.0	Tandem
JF2/1-41	0625	0.4	78.6	Mixed

Conclusion

In ρ^+ x ρ^- crosses, transmission of ρ^- markers to the ρ^+ diploid progeny requires heteroduplex formation between homologous DNA sequences. The rather constant recombination frequency (50-60 %) obtained with long repeating-units suggests that the general recombination should play the prominent role. However the large variations in the marker transmission frequency obtained from one ρ^- to another when they have short repeating-units indicate that the recombination between ρ^+ and ρ^- is controled by specific factors. In this work, we did not determine unique sequences involved in recombination. However we have demonstrated that the secondary structure of the ρ^- clones plays a major role in the recombination process(Table3). Indeed, in the _pif_ mutants, the _tandem_ structure prevents recombination, eventhough _palindromic_ structure allows normal recombination to proceed. Moreover although "mixed-repeats" ρ^- are composed of tandem and palindromic sequences, they do not recombine, suggesting that their secondary structure prevents them from recombination. In this respect, it may be significant that the mtDNA of palindromic ρ^- is present in the form of oligomers, whereas the mtDNA organized in "mixed-repeats" is present only as a multimer. The oligomers might be recombination intermediates required for the recombination in _pif_ mutants. It cannot be ruled out also that the ability of some palindromes to produce cruciform structures (19) which are more easily recognized in the recombination process (20) is the cause of the high recombination frequency observed with palindromic ρ^- clones in the _pif_ as well as in the _PIF_ nuclear background.

Table 3. Relationship between the recombination and organization of the ρ^- genome.

Structure	Recombination in _pif_ x _pif_ crosses	Oligomeric series
Palindrome	from 60 to 90 %	Yes
Tandem	$<$5 %	No : multimer
Mixed	$<$1 %	No : multimer

*Suppressiveness is expressed as the percentage of zygotic clones composed only of ρ^- cells.

References

1. Gaillard, C., Strauss, F. and Bernardi, G. (1980) Nature 283, 218-220
2. Zamaroczy, M., Faugeron-Fonty, G. and Bernardi, G. (1983) Gene 21, 193-202
3. Locker, J., Lewin, A. and Rabinowitz, M. (1979) Plasmid 2, 155-181
4. Bos, J.L., Heyting, C., Van der Horst, G. and Borst, P. (1980) Curr. Genet. 1, 233-239
5. Heyting, C., Menke, H.H. (1979) Mol.Gen.Genet. 168, 279-291
6. Sor, F. and Fukuhara, H. (1983) Cell, 32, 391-396
7. Locker, J., Rabinowitz, M. and Getz, G.S. (1974) Proc. Natl.Acad.Sci.USA 71, 1366-1370
8. Lazowska, J. and Slonimski, P. P. (1976) Mol. Gen. Genet. 146, 61-78
9. Potter, H. and Dressler, D. (1977) Proc. Natl. Acad. Sci. USA 74, 4168-4172
10. Kolodner, R. (1980) Proc. Natl. Acad. Sci. USA 77, 4847-4851
11. Perlman, P.S. (1976) Genetics 82, 645-663
12. Bolotin-Fukuhara, M. and Fukuhara, H. (1976) Proc. Natl. Acad. Sci. USA 73, 4608-4612
13. Dujon, B. (1981) In the Molecular Biology of the Yeast Saccharomyces Cerevisiae, eds. Strathern, J. N., Jones, E. W., Broach, J. R. (Cold Spring Harbor Laboratory) pp. 505-635
14. Blanc, H. and Dujon, B. (1980) Proc. Natl. Acad. Sci.USA 77, 3942-3946
15. Zamaroczy, M., Marotta, R., Faugeron-Fonty, G., Goursot, R., Mangin, M., Baldacci, G. and Bernardi, G. (1981) Nature 292, 75-78
16. Michel, F., Grandchamp, G. and Dujon, B. (1979) Biochimie 61, 985-1010

17. Conde, J. and Fink, G.R. (1976) Proc. Natl. Acad. Sci. USA 73, 3651-3655
18. Foury, F. and Kolodynski, J. (1983) Proc. Natl. Acad. Sci. USA In the press
19. Panayotatos, N. and Wells, R.D. (1981) Nature 289, 466-470
20. Mizuuchi, K., Kemper, B., Hays, J. and Weisberg, R. A. (1982) Cell, 29, 357-365

GENETIC CONTROL OF THE BIOGENESIS OF CYANIDE INSENSITIVE RESPIRATION IN *Neurospora crassa*.

Helmut Bertrand, Carole Ann Argan and Norma Ann Szakacs
Department of Biology, University of Regina, Regina, Sask., Canada

Introduction

The obligately aerobic fungus *Neurospora crassa* has a branced respiratory pathway consisting of a cyanide-sensitive electron-transport chain that contains two b-type cytochromes, cytochromes c and c1 and cytochrome aa_3 and an inducible, cyanide-resistant, hydroxamate-sensitive "alternative oxidase" that catalyzes a reaction by which reducing units are transferred from ubiquinone to molecular oxygen (1-7), as illustrated in Fig. 1 by an updated model of the electron-transport system proposed by Lambowitz and Zannoni (7). Whereas cyanide-sensitive, cytochrome mediated respiratory activity is manifested when wild-type *Neurospora* cells are maintained under "normal" growing conditions, alternative-oxidase mediated cyanide (CN) resistant respiration predominates in cells with impaired cytochrome-dependent respiratory systems, as in the

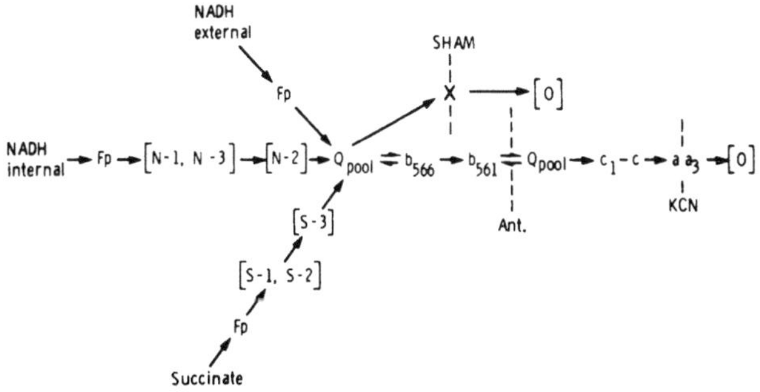

Fig. 1. The electron transport system of *Neurospora*

Mitochondria 1983

case in cytochrome-deficient mutants (1,2,5,6,8), in wild-type cells grown in medium containing inhibitors of cytochrome-mediated respiration (2,9), or inhibitors of mitochondrial protein synthesis (2,10), in copper-deprived wild-type cells (11), and in choline-deprived cells of a choline-requiring mutant (12). Cyanide-insensitive, hydroxamate-sensitive alternative oxidases have been observed in a variety of tissues from numerous higher plants as well as in several eukaryotic microorganisms (13-18).

In comparison with the large amount of information about the biogenesis and polypeptide composition of the cytochromes that has been generated in recent years (19-23), very little is known about the genetics, biogenesis and polypeptide composition of mitochondrial alternative oxidases. At present, the sum of the information on the genetics of the biogenesis of the alternative respiratory pathway consists of preliminary reports on the isolation and partial characterization of a few alternative oxidase deficient mutants in *Neurospora* (9,24-27) and *Ustilago myadis* (28). Although there have been numerous speculations about the characteristics of the alternative oxidase (reviewed in ref. 10), the nature and polypeptide composition of this membrane associated complex remains an enigma.

In order to develop a biological system suitable for studies on the biogenesis and physiological function of different respiratory pathways, we have selected mutants of *Neurospora crassa* which are alternative oxidase deficient (aod) in the sense that they do not induce CN-resistant respiration under any one of a multitude of conditions that cause the induction of this activity in wild-type cells. This respiratory phenotype can not be ascribed to resistance of the mutants to inhibitors commonly used to induce the alternative oxidase. We present evidence that the biogenesis of the alternative oxidase is controlled by two nuclear genes, and that it is unlikely that mitochondrial genetic information is required for the induction of CN-insensitive respiration. Finally, we describe findings that indicate that the induction of the alternative oxidase occurs parallel with the induction of a single polypeptide that appears to be associated tightly with the mitochondrial succinic dehydrogenase complex.

Results and Discussion

Selection of Mutants. Since cyanide-insensitive respiration is effectively induced in *Neurospora* by treatments of growing cells with antimycin A or chloramphenicol (2), it was reasoned that inositol-requiring mutants lacking a functional alternative oxidase system would escape inositolless death when incubated in inositol-deficient medium supplemented with antimycin A. On the basis of this consideration, conidia from a sn cr al-3 inl A culture were treated with nitrosoguanidine and subjected to inositolless-death selection as illustrated in Fig. 2. This procedure is similar to the one described by Edwards et al. and yielded 27 antimycin-sensitive mutants. These isolates all were morphologically indistinguishable from the sn cr al-3 inl A parent strain. Of these mutants, 14 did not induce appreciable amounts of cyanide-insensitive respiratory activity when grown in chloramphenicol (4.0 mg/ml) supplemented medium (respiratory phenotype illustrated in Fig. 3, B), 5 induced very low levels of this activity (Fig. 3, C and D), whereas 8 had very "leaky" alternative oxidase deficiency phenotypes (Fig. 3, E) in spite of being as sensitive to antimycin as all the other isolates.

Genetic Transmission of the Alternative Oxidase Deficiency Phenotype. Each of the 27 presumptive alternative-oxidase deficient isolates was crossed as a conidial parent with 50a wild-type. From 80 to 100 random ascospore progeny were collected from each cross and tested for antimycin sensitivity and for induction of CN-resistant respiration by chloramphen-

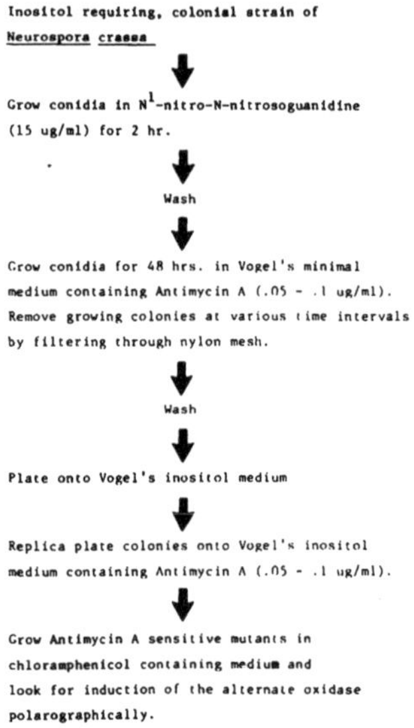

Fig. 2. Procedure for the isolation of alternative oxidase deficient mutants of *Neurospora*.

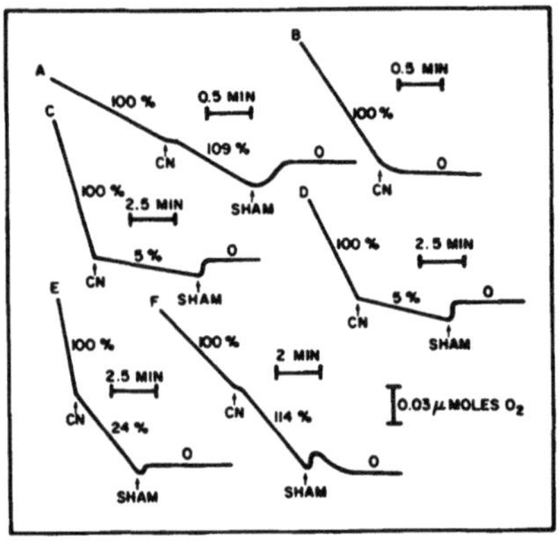

Fig. 3. Oxygen electrode recordings of the respiration of wild-type and antimycin A sensitive strains grown in chloramphenicol medium. (A) wild-type (B) aod-1-1 (C) ANT-1, (D) aod-2-1, (E) aod-2-3, (F) sn cr al-3 inl A.

icol. The ascospore progeny from 21 of the 27 isolates manifested 1:1 phenotypic segregation for antimycin resistance and sensitivity. Upon growth in chloramphenicol supplemented medium, none of the antimycin-sensitive progeny induced appreciable levels of CN-insensitive respiration, whereas all the antimycin-resistant clones induced high levels of CN-resistant respiration.

The antimycin-sensitivity phenotype was not transmitted to any of the ascospore progeny derived from crosses involving any one of the remaining six putative alternative-oxidase deficient mutants as the conidial parent. These six strains all had "leaky" alternative oxidase deficiency phenotypes when grown in chloramphenicol medium, but induced wild-type equivalents of CN-resistant respiration when grown in medium supplemented with oligomycin (1.0 μg/ml). All six isolates, and single-conidia subclones thereof, invariably lost the antimycin-sensitivity phenotype when they were maintained by serial transfers of asexual spores. Attempts to explore the possibility that these strains had sustained mutations in extra-nuclear genetic elements were foiled by the fact that the *sn* and *cr* chromosomal mutations precluded tests for maternal inheritance and the phenotypes in question were too refractive for the performance of heterokaryon tests of the type described by Bertrand and Pittenger (29). These unusual phenotypic characteristics lead to the conclusion that an antimycin-binding protein, rather than the alternative oxidase, had been altered by extrachromosomal mutations in the six unstable isolates.

Genetic Mapping and Complementation Tests. The genetic map locations of the 21 putative *aod* mutants obtained during the course of this study, and three of the four mutants described by Edwards *et al*. (25) were determined through crosses with *alcoy* linkage tester strains as well as crosses with strains that carried markers in linkage groups II (*trp-3* and *thr-3*) and IV (*trp-4* and *pan-1*). Briefly, it was found that 17 of the 21 *aod* mutants and the ANT-1, ANT-4, and ANT-6 mutants kindly provided by Dr. D.L. Edwards represent mutations that have occurred in one or more genes located within ten map units on either side of the centromere in linkage group IV. The remaining four *aod* mutants have genetic defects in a gene that is very closely linked to *thr-3* in linkage group II.

Tests for allelism between different alternative oxidase deficient mutants were conducted through the use of forced heterokaryons involving pairwise combinations of strains requiring nicotinamide and pantothenate, as described elsewhere (30). Briefly, heterokaryons containing any two of the 20 different mutants that had been assigned to linkage group IV did not induce CN-resistant respiration upon growth in medium supplemented with either chloramphenicol, ethidium bromide (5.0 ug/ml) or oligomycin. Similar results were obtained with heterokaryons consisting of any pairwise combination of the four mutants that were assigned to linkage group II. In contrast, all the heterokaroyns composed of one mutant from the set assigned to linkage group IV, the other from the four that were assigned to linkage group II, induced high levels of CN-resistant respiration upon growth in medium containing any one of the above mentioned inhibitors. Therefore, the 24 mutational events that in effect are represented by the mutants used in this study have occurred in only two functionally distinct genetic loci, which we have tentatively named aod-1 (linkage group IV), and aod-2 (linkage group II).

Mitochondrial Cytochrome Complements. Conceivably, a mutation which conveys antimycin sensitivity could simultaneously effect resistance to agents commonly used to induce CN-resistant respiration, such as chloramphenicol, ethidium bromide and oligomycin, thus preventing the induction of the alternative oxidase in strains having either aod-1 or aod-2 mutant alleles. Since chloramphenicol and ethidium bromide both inhibit mitochondrial protein synthesis and thereby cause mitochondrial cytochrome aa_3 and b deficiencies in cells grown for extended periods in medium containing either one of these two compounds (31-33), the possibility that lack of induction of the alternative oxidase was due to a multiple resistance phenotype could be tested readily. The pertinent cytochrome absorbance spectra for a select number of aod-1 and aod-2 mutants that were grown in unsupplemented and chloramphenicol-supplemented media are shown in Fig. 4. Two important observations were made by performing similar tests on the 24 aod mutants: 1) they all have normal cytochrome complements when cells are grown in medium without inhibitors, and 2) the addition of either chloramphenicol or ethidium bromide to the growth medium caused mitochondrial cytochrome aa_3 and b deficiencies in the mutants comparable in severity to the deficiencies observed in treated wild-type cells. Clearly, the mutant alleles of the two aod genes neither affect mitochondrial cytochrome contents, nor do they convey resistance to chloramphenicol and ethidium bromide.

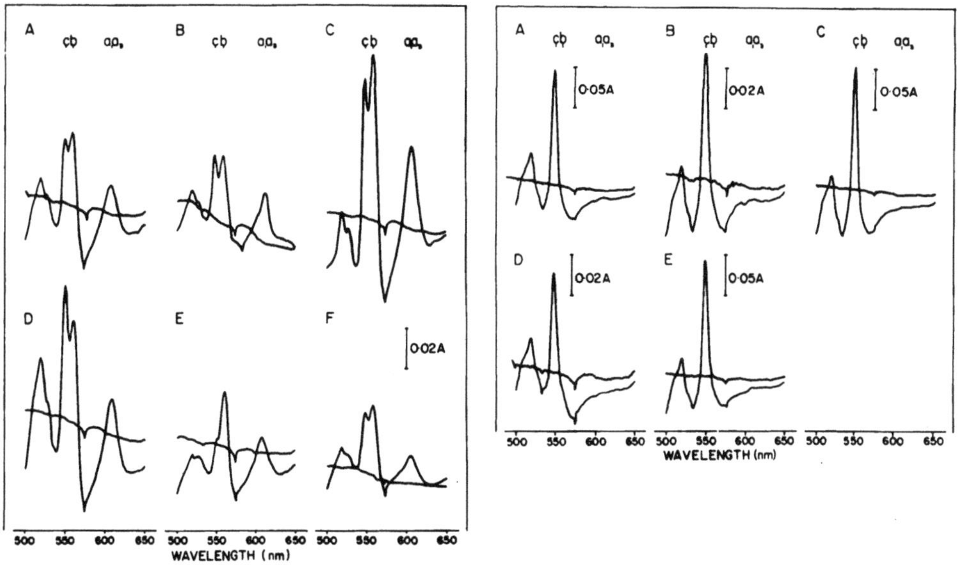

Fig. 4. Cytochrome spectra for wild-type and alternative oxidase deficient mutants. Left panel: untreated cultures. Right panel: chloramphenicol treated cultures. A: wild type, B: aod-1-1, C: ANT-1, D: aod-2-1, E: aod-2-4, F: aod-1-9.

A Putative Alternative Oxidase Polypeptide. Since the CN-resistant respiratory pathway is membrane bound, those polypeptides that are specific components of the alternative oxidase potentially should be present in relatively large amounts in mitochondrial membranes from "induced" cells, whereas membranes from "uninduced" cells might be anticipated to lack the same components. Furthermore, it can be anticipated that inducible alternative oxidase polypeptides are synthesized very rapidly during the induction process, i.e. soon after the addition of oligomycin or antimycin A to a growing culture of wild-type *Neurospora* cells. On the basis of these speculations, we have devised a procedure for pulse-labelling proteins by exposure of "induced" and "uninduced" cells to ^{35}S-sulfate, as shown diagrammatically in Fig. 5.

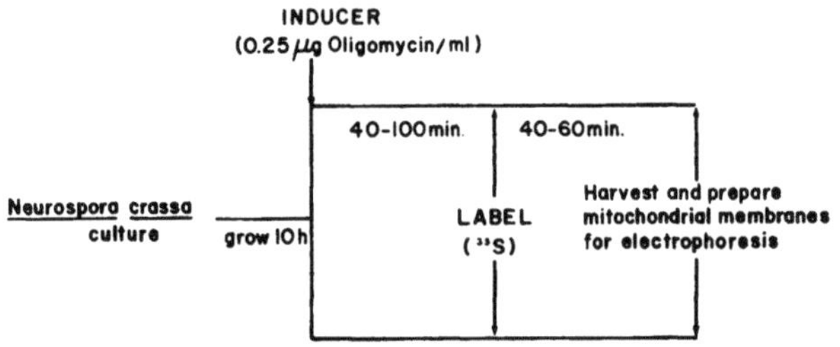

Fig. 5. Procedure used to induce and label putative alternative oxidase proteins.

Mitochondrial membranes were prepared from the labelled cells as described elsewhere (34). The membrane proteins were separated by electrophoresis on SDS-polyacrylamide gradient gels (35) and visualized by autoradiography. As can be seen from Fig. 6 lanes A, B, and C, the treatment of wild-type cells with oligomycin or antimycin resulted in the synthesis of a large amount of one polypeptide that appears to be made either in very low amounts or not at all in untreated cells. This polypeptide also was induced when exponentially growing cells from a temperature-sensitive cytochrome oxidase mutant, cya-7-13 (36,37), were shifted from 25° to 41° (Fig. 6, lanes D and E). In addition to the polypeptide marked by an arrow in Fig. 6, the synthesis of at least one of another three proteins was stimulated by each of the three different treatments that were employed to induce CN-resistant respiration. However, none of these additional proteins was induced by all three treatments. Therefore, only the induction of the major polypeptide indicated by the arrow parallels the induction of alternative oxidase activity.

The induction and labelling procedures that led to the identification of a putative alternative oxidase polypeptide in wild-type *Neurospora* also were applied to all the available alternative oxidase deficient mutants. Representative results derived from experiments were oligomycin was used as the

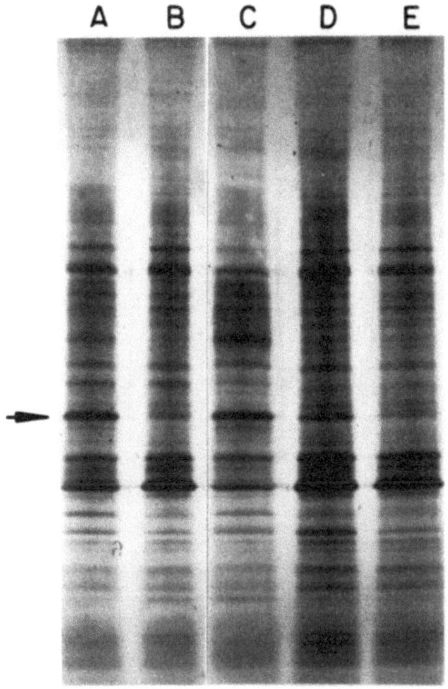

Fig. 6. Mitochondrial proteins from wild-type cells (A) induced with antimycin A, (B) uninduced, (C) induced with oligomycin and cya-7-13 cells (D) grown at 41° and (E) at 25°.

inducing agent are shown in Fig. 7. Briefly, it was observed that none of the four aod-2 mutants induced high amounts of the putative alternative oxidase protein upon treatment with oligomycin or antimycin A, whereas 19 out of 20 aod-1 mutants induced the synthesis of the protein. The exceptional aod-1 mutant, ANT-1, did not produce the protein at least not in a form that was integrated into mitochondrial membranes. In summary, results of the induction experiments are consistent with the interpretation that the aod-1 locus is the structural gene for the putative alternative oxidase polypeptide and that biologically inactive proteins with altered primary structures are produced in aod-1 mutants. The function of the aod-2 gene is far from obvious, but it clearly relates to the induction of the putative alternative oxidase polypeptide. It is possible that aod-2 mutants are deficient in a process that is coupled to cytochrome-mediated electron-transport and results in the synthesis of a diffusible effector required for the induction of alternative oxidase polypeptide synthesis, possibly by controlling the transcription of the aod-1 locus.

To date, attempts to extract an active alternative oxidase from *Neurospora* mitochondria have not been successful. However, using ubiquinol-15 as a substrate for the polarographic

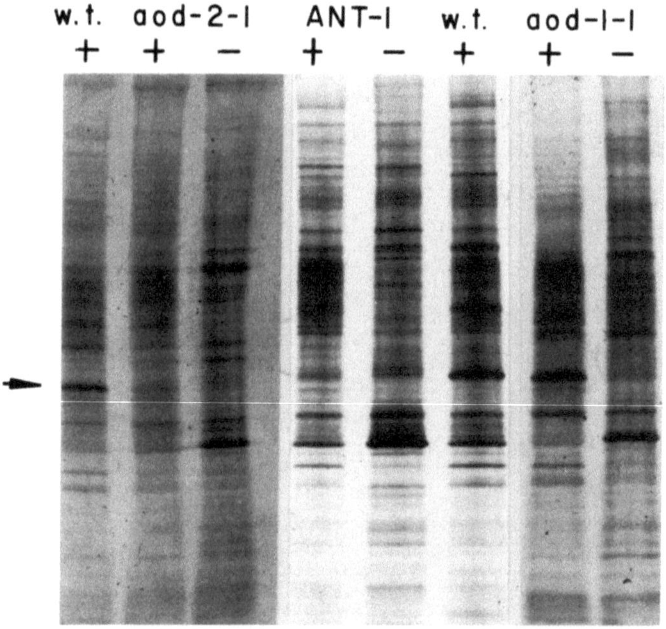

Fig. 7. Mitochondrial polypeptides induced in untreated (-) and oligomycin-treated (+) alternative oxidase deficient mutants.

assay of the enzyme (18), we were able to isolate activity-enriched particulate fractions from mitochondrial membranes that had been partially disrupted by ultrasound treatment in a solution containing 0.001% octyl-β-D-glucopyranoside or 0.01% deoxycholate as emulsifying agents. The use of higher concentrations of detergents invariably led to the complete loss in enzyme activity. The active fractions had 3-5 fold higher alternative oxidase and succinic dehydrogenase activities than intact membranes, and contained at least 40 to 60 different polypeptides. Interestingly, when these fractions were prepared from pulse-labelled mitochondria of oligomycin-induced wild-type cells, the putative alternative oxidase polypeptide was recovered as one of the major components of the complex (Fig. 8). While none of the above experiments actually prove that the polypeptide in question is the alternative oxidase or

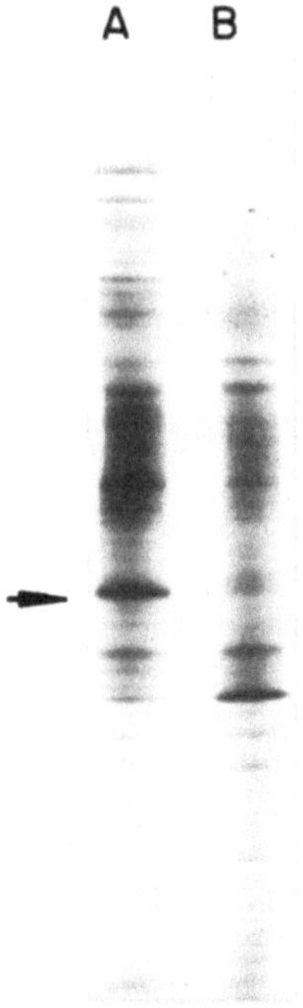

Fig. 8. Putative alternative oxidase polypeptide in mitochondrial particles from (A) oligomycin-treated and (B) untreated wild-type cells.

one of its subunits, they do provide circumstantial evidence indicating that CN-resistant respiration may be a very simple membrane associated protein complex.

Mitochondrial Genes and Alternative Oxidase Activity. As part of an ongoing study on the genetics of the biogenesis of mitochondrial respiratory complexes, we have obtained a series of mitochondrial deletion mutants in *N. crassa* and *N. intermedia*. While each mutant lacks only a portion of the mitochondrial genome, collectively they have deletions that cover the entire mt DNA, except for a 14 kilobasepair segment that encodes both the large and small mitochondrial rRNAs, several tRNAs, and the oxi-2 locus (for a diagram of this region see reference 38). These deletion mutants all have highly active CN-resitant respiratory pathways. Therefore, it is unlikely that mtDNA codes for any cellular component that is part of the active alternative oxidase complex. These observations do not preclude the possibility that mitochondrial genetic information is necessary for the maintenance of the repressed state of cells with respect to the production of the alternative oxidase, as proposed by Rosenberg et al. (39). The experimental verification or rejection of this hypothesis requires unambiguous proof that mitochondrial mutants that are constitutive for CN-resistant respiration have completely normal mitochondrial systems for electron transport and ATP synthesis -- a requirement that can not be met because of the prevailing uncertainties about the nature and function of mitochondrial enzyme complexes.

References

1. Tissieres, A., Mitchell, H.K., Haskins, F.A.: J. Biol. Chem. 200, 819-826 (1953).
2. Lambowitz, A.M., Slayman, C.W.: J. Bacteriol. 108, 1087-1096 (1971).
3. Lambowitz, A.M., Smith, E.W., Slayman, C.W.: J. Biol. Chem. 247: 4850-4858 (1952).
4. Lambowitz, A.M., Smith, E.W., Slayman, C.W.: J. Biol. Chem. 247, 4859-4865 (1972).
5. Jagow, G. von, Weiss, H., Klingenberg, M.: Eur. J. Biochem. 33, 140-157 (1973).
6. Slayman, C.W.: In Functions of Alternative Terminal Oxidases (Degn, H. Lloyd, D., Hill, G.C., eds.), Pergamon Press, pp. 159-167 (1977).
7. Lambowitz, A.M., Zannoni, D.: In Plant mitochondria (Ducet, G., Lance C., eds.), Elsevier/North Holland Biomedical Press, pp. 283-291 (1978).
8. Edwards, D.L., Kwiecinski, F.: J. Bacteriol. 116, 510-618 (1973).
9. Szakacs, N.A., Bertrand, H.: Can. J. Genet. Cytol. 18, 574 (1976).
10. Edwards, D.L., Rosenberg, E.: Eur. J. Biochem. 62, 217-221 (1976).
11. Schwab, A.J.: FEBS Lett. 35, 63-66 (1973).
12. Jurétic, D.: J. Bacteriol. 126, 542-543 (1976).
13. Palmer, J.N.: Ann. Rev. Plant Physiol. 27, 133-157 (1976).
14. Henry, M.P., Nyns, E.J.: Sub-cell. Biochem. 4, 1-65 (1975).
15. Solomos, T.: Ann. Rev. Plant Physiol. 28, 279-297 (1977).
16. Degn, H., Lloyd, D., Hill, G.C., eds.: Functions of Alternative Terminal Oxidases, Pergamon Press, Oxford 1977.
17. Laties, G.G.: Ann. Rev. Plant Physiol. 33, 519-555 (1982).
18. Vanderleyden, J.: Agricultura 29, 121-309 (1981).
19. Schatz, G., Mason, T.L.: Ann. Rev. Biochem. 43, 51-87 (1974).
20. Bücher, Th., Neupert, W., Sebald, W., Werner, S., eds.: Genetics and Biogenesis of Chloroplasts and Mitochondria, North Holland Publishing Company, Amsterdam 1976.

21. Bandlow, W., Schweyen, R.J., Wolf, K., Kaudewitz, F., eds.: Mitochondria 1977. Genetics and Biogenesis of Mitochondria, Walter de Gruyter, Berlin 1977.

22. Kroon, A.M., Saccone, C., eds.: The Organization and Expression of the Mitochondrial Genome, Elsevier/North Holland Biomedical Press, Amsterdam 1980.

23. Slonimski, P., Borst, P., Attardi, G., eds.: Mitochondrial Genes, Cold Spring Harbor Laboratory, New York, 1982.

24. Szakacs, N.A., Bertrand, H.: Genetics 91, s96 (1979).

25. Edwards, D.L., Chalmers, J.H. jr., Guzik, H.J., Warden, J.T.: In Genetics and Biogenesis of Chloroplasts and Mitochondria (Bücher, Th., Neupert, W., Sebald, W. Werner, S., eds.), North Holland Publishing Company, Amsterdam, pp. 865-872 (1978).

26. Bertrand, H., Argan, C.A., Vanderleyden, J., Szakacs, N.A.: Genetics 91, s9 (1979).

27. Edwards, D.L., Unger, B.W.: FEBS Lett. 85, 40-42 (1978).

28. Georgopolous, S.G., Sisler, H.D.: J. Bacteriol. 103, 745-750 (1970).

29. Bertrand, H., Pittenger, T.H.: Genetics 71, 521-533 (1972).

30. Bertrand, H., Pittenger, T.H.: Molec. Gen. Genetics 117, 82-90 (1972).

31. Rifkin, M.R.: Ph.D. Thesis, Rockefeller University, New York, 1969.

32. Wood, D.D., Luck, D.J.L.: J. Cell Biol. 51, 249-264 (1971).

33. Bertrand, H.: Can. J. Genet. Cytol. 14, 720 (1972).

34. Bertrand, H., Werner, S.: Eur. J. Biochem. 79, 559-606 (1977).

35. Chua, N.H., Bennoun, P.: Proc. Natl. Acad. Scie. USA 72, 2175-2179 (1975).

36. Nargang, F.E., Bertrand, H.: Molec. Gen. Genet. 166, 15-23 (1978).

37. Nargang, F.E., Bertrand, H., Werner, S.: Eur. J. Biochem. 102, 297-307 (1979).

38. Bertrand, H., Bridge, P., Collins, R.A., Garriga, G., Lambowitz, A.M.: Cell 29, 517-526 (1982).

39. Rosenberg, E., Mora, C., Edwards, D.L.: Genetics 83, 11-24.

THE ROLE OF HEME AND OXYGEN IN THE REGULATION OF MITOCHONDRIAL CYTOCHROMES

Thomas Mason, Henry N. Little, Camilla VanSickle
Department of Biochemistry, University of Massachusetts
Amherst, Massachusetts 01003

Leonard Guarente
Department of Biology, Massachusetts Institute of Technology
Cambridge, Massachusetts 02139

Introduction

Research in the past decade has provided a clear understanding of the major contributions of nuclear and mitochondrial genes to mitochondrial biogenesis in yeast (1,2). An important challenge for the future will be to unravel the complexities of the regulatory mechanisms that coordinate the expression of these various genetic elements. It will clearly be an advantage to study the regulation of genes coding for mitochondrial proteins in a system where a well-defined physiological variable triggers an on-off response, preferably in both nuclear and mitochondrial genes. In this context, two variables, oxygen tension in the growth medium and the availability of heme, are of particular interest because they both elicit significant changes in the cellular concentration of mitochondrial cytochromes.

Mitochondria 1983

Several laboratories have shown that yeast grown anaerobically and heme biosynthetic mutants grown without a heme supplement lack spectrally detectable mitochondrial cytochromes and are also deficient in many, but not all, of the corresponding apoprotein subunits (3-8). It is not clear, however, where heme and oxygen affect the pathways of biosynthesis and assembly of these proteins. Furthermore, it has been difficult to distinguish a direct effect of oxygen on cytochrome biosynthesis from secondary effects due to the requirement of molecular oxygen for heme biosynthesis in yeast. In order to better understand the roles of heme and of oxygen in the regulation of mitochondrial cytochromes, we have begun to analyze the expression of the iso-1-cytochrome c gene (CYC1) of S. cerevisiae In recent experiments, which were facilitated by the use of a fusion of CYC1 to the E. coli lacZ gene, we have shown that CYC1 is transcriptionally inactive in heme-deficient cells and that heme-dependent induction of CYC1 is mediated by an upstream activation site (UAS_c) centered about 275 bp upstream from the site where transcription initiates (8). These experiments, however, did not answer two important questions: 1) Is O_2 directly involved in the control? and 2) Is there an absolute requirement for metal porphyrin or is the porphyrin nucleus alone sufficient for induction?

Since β-galactosidase synthesis directed by the CYC1-lacZ fusion is a sensitive indicator of CYC1 promoter activity (8,9), it is now possible to systematically study the physiological parameters of CYC1 induction. In this paper, we show that this control does not require the direct involvement of O_2 and is is absolutely dependent on the presence of a metal porphyrin. In addition we present evidence that suggests a possible role for cytochrome c in the biosynthesis and assembly of cytochrome c oxidase.

Materials and Methods

Strains and plasmids. Yeast strains employed in this study were: TM1 (hem1 ura3-52 gal^- ino1); BWG2-9A (αhis4-519 ade ura3-52 gal4); BWG1-7A (a leu2-3 leu2-112 his4-519 ade1-100 ura3-52); TD28 (a ura3-52 ino1 can^R); GM3C-2 (αleu2-3 trp1-1 his4-519 cyc1-1 cyp3-1 (CYC7) gal^-); LG32-B (a leu2-3 leu2-112 his4-519 ade1-100 ura3-52 hem5). The plasmid used in this study was pLGΔ-312 (UAS_C TATAc CYC1-lacZ amp^R $URA3^+$ 2 μori colE1 ori); its construction has been described (8). LG32-B is classified as a ferrochelatase deficient mutant (hem5), on the basis of its ability to grow with heme supplements and its inability to use protoporphyrin IX. (10 and cf. Table I below).

Growth conditions. Strains carrying plasmids were routinely grown at 30°C in a liquid culutre medium containing 0.7% yeast nitrogen base, 0.5% casamino acids, 40 μg/ml adenine, 40 μg/ml inositol and 1% glucose. Media for anaerobic growth or for growth of hem^- mutants was supplemented with either 0.1% Tween 80-20 μg/ml ergosterol or a heme supplement (either 40 μg/ml δ-ALA or 30 μg/ml of either protoporphyrin IX or heme). Strains without plasmids were grown either on the same medium supplemented with uracil or on 1% yeast extract, 2% protease peptone, 1% glucose and other nutritional supplements as required.

Analytical and preparative methods. $NaDodSO_4$/polyacrylamide gel electrophoresis was carried out in 12.5% slab gels (11). The methods used for preparation of yeast extracts, enzymatic assay, of β-galactosidase and cytochrome c oxidase, immune replication and protein determination are as described earlier (8). Succinate-cytochrome c reductase was determined by the method of Tisdale (12). Monoclonal antibodies against subunits II,

III, and V of yeast cytochrome oxidase were obtained by injecting the holoenzyme into BALB/c mice and fusing spleen lymphocytes with X63-Ag8 myeloma cells (13). Mitochondrial translation products were labelled with [^{35}S]-methionine in cells poisoned with cycloheximide (11)

Induction experiments. The induction of CYC1-*lac*Z was routinely measured with cells that were harvested in mid log phase, washed twice with 0.9% NaCl, 0.1% BSA and once with 0.9% NaCl and then resuspended at 15-30 mg wet wt./ml in 0.7% yeast nitrogen base, 0.5% casamino acids, 1% glucose. After incubation for 5 min at 30°, 5x10^{-5}M inducer was added and the incubation continued for 90 min with vigorous mixing. For anaerobic induction experiments, all solutions were purged of oxygen by repeated degassing under vacuum and flushing with oxygen-free N_2 (14). Anaerobic incubations were carried out with 2.5 ml of the cell suspension in 30 ml Thunberg tubes. The inducer was added from the sidearm after flushing the tube free of O_2 by a series of three consecutive evacuations with intermediate flushing with N_2. Stock solutions of inducers were prepared by dissolving porphyrins and metal porphyrins in a minimum amount of 0.1 N NaOH and KOH, respectively. Water and ethanol were added to give a final concentration of 10^{-3}M inducer and 10% ethanol.

Results

Induction of CYC1-lacZ by porphyrins and metal porphyrins. When used as supplements in aerobic culture media, δ-aminolevulinic acid (δ-ALA), protoporphyrin IX, and heme all effectively induce the CYC1-lacZ fused gene in the *hem1* mutant strain (8). In short term induction experiments, however, δ-ALA is inactive and while protoporphyrin and heme are active, they are not as effective as a porphyrin analog, deuteroporphyrin. Fig. 1 shows that deuteroporphyrin elicits a rapid synthesis of β-galactosidase in the heml strain and also stimulates β-galactosidase production during the aerobic adaptation of anaerobically grown HEM1$^+$ cells. Although deuteroporphyrin is converted to deuteroheme by ferrochelatase, the analog lacks the two vinyl side chains necessary for formation of the thioether linkages between the porphyrin ring and the polypeptide chain of C-type cytochromes. Thus no functional cytochrome c is formed upon deuteroporphyrin induction of CYC1 in heml mutants. In vitro, this analog is known to block the formation of cytochrome c in mitochondria isolated from Neurospora crassa (15), but it does not block growth of wild-type yeast on nonfermentable carbon sources and thus does not appear to inhibit cytochrome c formation in vivo.

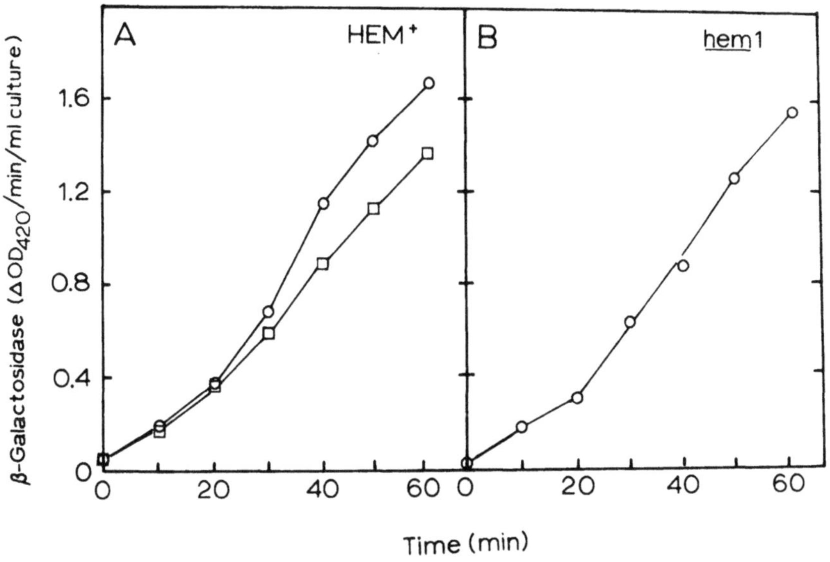

Figure 1. Induction of CYC1-lacZ by deuteroporphyrin. BWG2-9A (HEM1$^+$, pLGΔ-312) and TM1 (heml, pLGΔ-312) were grown anaerobically and aerobically, respectively, in media containing 1% glucose and Tween-ergosterol. The cells were induced aerobically with (O—O) and without (□—□) $5x10^{-5}$ M deuteroporphyrin.

Several porphyrins and metalporphyrins have been tested as inducers of CYC1-lacZ in the heml strain. Deuteroporphyrin consistently produced the strongest induction, even better than deuteroheme. Although several factors including cell permeability properties could easily account for the different levels of induction with these compounds, the results also suggest that metal-free porphyrin might be the active intracellular inducer. In order to test this possibility, we have compared induction by metal-substituted and free porphyrins in a ferrochelatase deficient mutant, LG32-B(pLGΔ-312). The results in Table I convincingly demonstrate a metalloporphyrin requirement for CYC1 induction. The strongest inducer in the heml strain, deuteroporphyrin, is completly inactive in the ferrochelatase deficient strain whereas significant levels of β-galactosidase are obtained in the presence of iron and cobalt substituted porphyrins. It seems likely that the normal intracellular inducer is heme, but the iron atom does not appear to be essential. This conclusion is strongly supported by the ability of Co-porphyrin to induce in the ferrochelatase minus strain where it is unlikely that an iron for cobalt substitution could occur.

Table I. Induction of CYC1-lacZ by Porphyrins and Metal Porphyrins

Inducer	β-galactosidase[a] LG32-B(hem5$^-$)	TM1(heml$^-$)
Protoprophyrin	0.1	4.7
Deuteroporphyrin	0.1	132
Mesoporphyrin	0.1	3.8
Fe-protoporphyrin	6.4	9.4
Fe-deuteroporphyrin	1.8	2.8
Co-protoporphyrin	2.6	5.0

[a] nanomoles ONPG hydrolyzed/min/OD_{600}

Induction of CYC1-lacZ does not require oxygen. An attractive model for the regulation of CYC1 is that a hemoglobin-like regulatory protein binds to UAS_c and activates transcription only in its oxygenated state. A simple test for such a direct role of O_2 is to measure the extent of CYC1

Table II. Induction of CYC1-*lacZ* in the Absence of O_2

Strain	Growth	Induction	β-galactosidase[a]
BWG2-9A	Anaerobic	-	12
HEM1[+]		Aerobic	380
(pLGΔ-312)		Aerobic + deuteroporphyrin	674
		Anaerobic	21
		Anaerobic + deuteroporphyrin	150
TM1	Aerobic	-	7
hem1[-]		Aerobic	5
(pLGΔ-312)		Aerobic + deuteroporphyrin	549
		Anaerobic + deuteroporphyrin	414

[a]nanomoles ONPG hydrolyzed/min/mg protein in whole cell homogenate

induction in cells under strict anaerobiosis but still with an ample intracellular pool of a suitable metal porphyrin inducer. Our strategy for such an experiment was to generate heme-deficient cells, by anaerobic growth of a HEM1[+] strain and aerobic growth of a *hem1* mutant without a heme supplement, and then to challenge these cells with the strong inducer, deuteroporphyrin, in aerobic and anaerobic incubations. An important control for anaerobiosis during the induction is is provided by the HEM[+](pLGΔ-312) strain, which can form endogenous heme, and consequently β-galactosidase, only when oxygen is available. As can be seen from the data in Table II, deuteroporphyrin induction in the hem1 strain proceeded to nearly the same extent under anaerobiosis as under full aeration, 59-fold and 78-fold increases in β-galactosidase, respectively. There was a similar, if less dramatic, anaerobic induction by deuteroporphyrin in the $HEM1^+$ strain; a 12-fold increase under anaerobic conditions as compared to a less than 2-fold increase in unsupplemented anaerobic cells. The results were no different when respiratory competent cells, which did not carry CYC1-lacZ, were added to the incubations in order to deplete any residual O_2. Despite such precautions, it is difficult to be absolutely certain that our conditions of anaerobiosis are sufficient to deoxygenate a system with an extremely high affinity for O_2. We nevertheless interpret these results as a strong indication for the lack of a direct involvement of O_2 in the control of CYC1.

Mutations at CYC1 and CYC7 affect cytochrome c oxidase. We surmise that CYC1 is a member of a family of coordinately regulated genes and that coordination of expression is mediated through the same heme control system we have described for CYC1. On the other hand, it is conceivable that some of the observed effects of heme-deficiency on other cytochromes represent secondary responses to reduced levels of cytochrome *c*. Such a possibility is supported by the observation that cyc1 cyc7 double mutants are pleiotropic, lacking not only iso-1 and iso-2-cytochrome *c* but also cytochrome aa_3 (16,17). We have decided therefore, to compare the effects of heme-deficiency and cyc1 cyc7 mutations on the biosynthesis and assembly of cytochrome *c* oxidase.

In order to be reasonably confident that any phenotype observed in cytochrome *c* mutants could be correlated with a known genotype, the cyc1 cyc7 double mutant was crossed

Table III. Analysis of Mutations at CYC1 and CYC7

Strain	Genotype[a]	Growth on Glycerol-ETOH	Enzyme[b] Succinate-Cyt. *c* Reductase	Cyt. *c* Oxidase
TD28	$CYC1^{+}CYC7^{+}$	+	92	652
GM3C-2	*cyc1 cyc7*	-	26	120
10A	*cyc1 cyc7*	-	20	71
10B	$CYC1^{+}CYC7^{+}$	+	71	860
10C	$CYC1^{+}$*cyc7*	+	61	505
10D	*cyc1* $CYC7^{+}$	+	56	284
20A	*cyc1 cyc7*	-	51	112
20B	$CYC1^{+}CYC7^{+}$	+	76	742
20C	*cyc1* $CYC7^{+}$	+	51	307
20D	$CYC1^{+}$ *cyc7*	+	48	416

[a]The assignment of genotypes is based on cytochrome *c* content of whole cells as estimated by the spectroscopic method of Sherman *et al*. (18). The cytochrome *c* content of spores 10A-10D was confirmed by immunoblot analysis

[b]Succinate-cytochrome *c* reductase and cytochrome *c* oxidase activities are expressed as nanomoles of cytochrome *c* oxidized or reduced/min/mg. Mitochondria were isolated from cells grown to late log phase on yeast extract/bactopeptone/1% glucose.

to a wild-type strain, the resulting diploid was sporulated and two tetrads with tetratype segregation of *cyc1* and *cyc7* were chosen for comparison. An analysis of these segregants, shown in Table III, reveals a clear correlation between the loss of cytochrome *c* oxidase activity and the extent of the lesion in the ability of the cell to produce cytochrome *c*. The double mutant retains approximately 10-20% of the wild-type level of cytochrome oxidase. The single mutant strains have intermediate levels of oxidase and, at least for the two

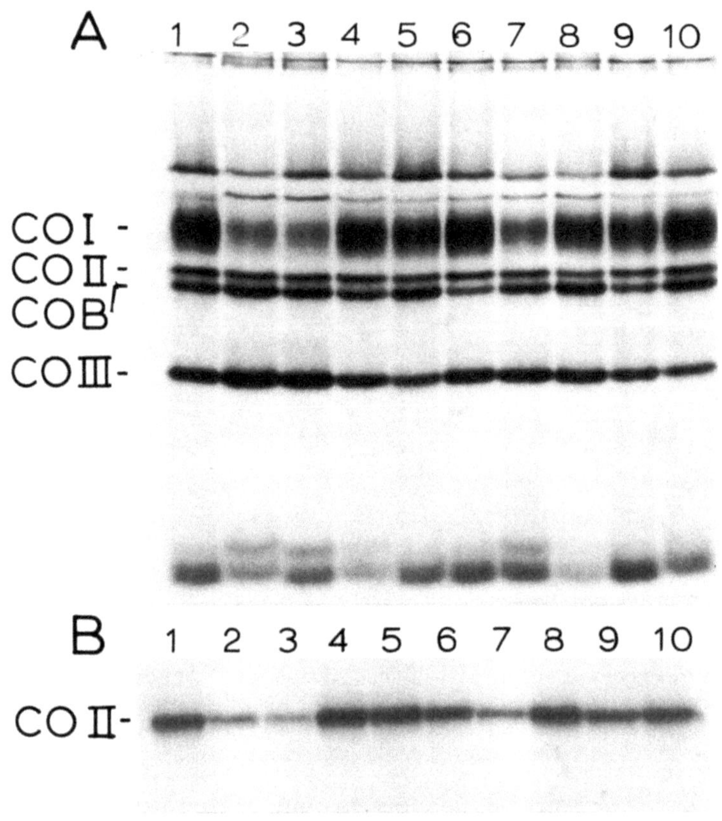

Figure 2. Mitochondrial translation products and cytochrome c oxidase subunit II in cytochrome c mutants. A. The strains listed in Table II were labeled *in vivo* with [^{35}S]-methionine in the presence of cycloheximide. A sample of SDS-solubilized mitochondrial proteins containing 100,000 cpm of TCA-precipitable radioactivity was applied to each track. B. The relative amount of cytochrome c oxidase subunit II in 50 µg of mitochondrial protein was determined by immune replication. The samples are: 1, TD28; 2, GM3C-2; 3-6, spores 10A-D; 7-10, spores 20A-D.

tetrads shown here, CYC1 mutants are lower in oxidase activity than CYC7 mutants. Further correlation between the CYC genotype and cytochrome oxidase can be seen in the profile of mitochondrial translation products and in the concentration of cytochrome oxidase subunits in the

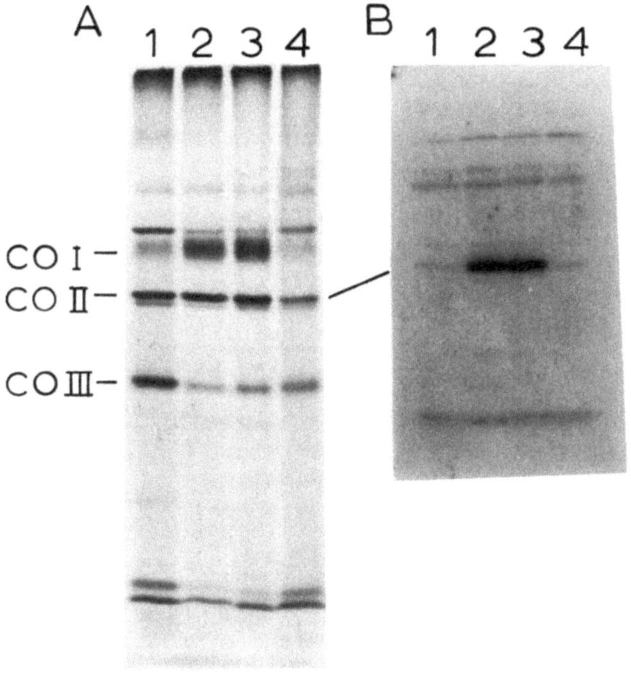

Figure 3. Mitochondrial translation products and cytochrome *c* oxidase subunit II in TM1 (*hem1*, pLGΔ-312). Cells were grown on media containing 1% glucose supplemented with either (Lane 1) Tween-ergosterol, (Lane 2) 40 μg/ml δ-ALA, (Lane 3) $5x10^{-5}$ M protoporphyrin IX, or (Lane 4) $5x10^{-5}$ M deuteroporphyrin. Mitochondrial translation products (A) and cytochrome *c* oxidase II (B) were analyzed as described in the legend for Fig. 2.

cell (Fig. 2). The relative amount of radioactivity associated with subunit I in the *in vivo* labeling experiment is distinctly lower in the double mutant strains than in any of the other segregants. The labeling of cytochrome *c* oxidase subunits II and III and cytochrome *b* is not discernably different in the different strains. The steady state concentrations of cytochrome *c* oxidase subunit II (Fig. 2B) and of subunits III, V, and VI (not shown) as determined by the immune replica procedure are in contrast to the *in vivo* labeling data and more closely reflect the levels of oxidase detected by enzyme activity measurements. Thus, at least for subunits II and III it appears that the stability of the polypeptides, not their synthesis, is affected by variations in cytochrome c.

Similar effects on cytochrome c oxidase have been observed in heme-deficient cells. These include reduced synthesis of subunit I, relatively normal synthesis of subunits II and III, and greatly reduced cellular concentrations of all cytochrome oxidase subunits (4). The results shown in Fig. 3. confirm the published observations and show in addition that deuteroporphyrin does not alleviate the effects of heme deficiency on cytochrome c oxidase.

Discussion

There has been a long standing interest in the possible roles of heme and oxygen in controlling the expression of genes for mitochondrial cytochromes (see ref. 19 for review). Studies on this subject are complicated by the difficulty in treating heme and oxygen as separate variables, since heme deficiency is a necessary consequence of anaerobic growth in yeast. For example, we have previously shown that the transcription of CYC1 is blocked by either anaerobic growth or by heme deficiency imposed by mutation, but we failed to determine whether CYC1 induction required only heme or heme and oxygen (8). Here we obtained evidence in support of a primary role for heme in the regulation of CYC1 with no direct involvement of O_2. Oxygen regulation of CYC1 thus appears to be mediated indirectly through the requirement for O_2 as a substrate in heme biosynthesis.

Since the apoproteins corresponding to several heme proteins fail to accumulate in heme-deficient cells, it is tempting to assume that there is a coordinately regulated family of genes corresponding to these proteins. So far, however, only the synthesis of iso-1-cytochrome *c* and catalase (20) are known to be regulated by heme at the level of transcription. We are currently trying to identify other genes for mitochondrial proteins that might form a regulatory family with CYC1. At the same time we will continue to explore the possibility that cytochrome *c* modulates the synthesis and/or assembly of cytochrome *c* oxidase.

Acknowledgment

We thank Dr. Donna L. Montgomery, University of Texas Health Science Center, San Antonio, Texas, for providing the GM3C-2 strain of yeast and Ms. Cathy Hayward for help in preparing the manuscript. This work was supported by National Institutes of Health grant to Leonard Guarente and National Science Foundation grant to Thomas Mason.

References

1. Tzagoloff, A., Macino, G., Sebald, W., Ann. Rev. Biochem. 48, 419-441 (1979).
2. Dujon, B. in The Molecular Biology of the Yeast Saccharomyces (Strathern, J.N., Jones, E.W., Broach, J.R., eds) pp 505-635, Cold Spring Harbor Laboratory, New York, 1981.
3. Schatz, G., Mason, T.L., Ann. Rev. Biochem. 43, 51-87 (1974).
4. Saltzgaber-Muller, J., Schatz, G., J. Biol. Chem. 253, 305-310 (1978).
5. Djavadi-Ohaniance, L., Rudin, Y., Schatz, G., J. Biol. Chem. 253, 4402-4407 (1978).
6. Clejan, L., Beattie, D. S., Gollub, E.G., Liu, K.-P., Sprinson, D.B., J. Biol. Chem. 255, 1312-1316 (1980).
7. Matner, R.R., Sherman, F.J., Biol. Chem. 257, 9811-9821 (1982).
8. Guarente, L., Mason, T., Cell 32, 1279-1286 (1983).
9. Guarente, L., Ptashne, M., Proc. Natl. Acad. Sci. USA 78, 2199-2203 (1981).

10. Gollub, E.G., Liu, R.-P., Dayan, J., Adlersberg, M., Sprinson, D.B., J. Biol. Chem. 252, 2846-2854 (1977).

11. Douglas, M. G., Butow, R.A., Proc. Natl. Acad. Sci. USA 73, 1083-1086 (1976).

12. Tisdale, H.D., Meth. Enzymol. 10, 213-215 (1967).

13. Kimura, K., Mason, T.L., Khorana, H.G., J. Biol. Chem. 257, 2859-2867 (1982).

14. Schatz, G., Kovac, L. Meth. Enzymol. 31, 627-632 (1974).

15. Hennig, B., Neupert, W., Eur. J. Biochem. 121, 203-212 (1981).

16. Downie, J.A., Stewart, J.W., Brockman, N., Schweingruber, A.M., Sherman, F., J. Mol. Biol. 113, 369-384 (1977).

17. Faye, G., Leung, D.W., Tatchell, K., Hall, B.D., Smith, M., Proc. Natl. Acad. Sci. USA 78, 2258-2262 (1981)

18. Sherman, F., Lawrence, C.W. in Handbook of Genetics (King, R., ed) pp. 359-393, Plenum Press, New York, 1974.

19. Matoon, J.R., Lanashire, W.E., Sanders, H.K., Carvajal, E., Malamud, D., Brag, G.R.C., Panek, A.D., in Biochemical and Clinical Aspects of Oxygen (Caughey, W.J., ed) pp. 421-435, Academic Press, New York, 1979.

20. Richter, K., Ammerer, G., Hartler, E., Ruis, H., J. Biol. Chem. 255, 8019-8022 (1980).

ABOUT THE REGULATION OF PROTOHEME SYNTHESIS IN THE YEAST SACCHAROMYCES CEREVISIAE

Rosine Labbe-Bois, Danièle Urban-Grimal, Christiane Volland,
Jean-Michel Camadro and Pierre Dehoux *

Laboratoire de Biochimie des Porphyrines, Université Paris VII
75251 Paris Cedex 05, France

Introduction

The biosynthetic pathway leading to protoheme formation plays an important role in cellular metabolism and biogenesis of mitochondria. In yeast, the final product protoheme is utilized as the prosthetic group of a variety of hemoproteins (b type cytochromes, cytochrome P 450, catalases, cytochrome c peroxidase, etc) ; it (or protoporphyrin) is the precursor of hemes c and a found in c and a type cytochromes. The pathway also provides the precursor (uroporphyrinogen) for the synthesis of siroheme, prosthetic group of sulfite reductase. On the other hand protoheme has been shown to control (directly or indirectly ?) the synthesis, proteolytic maturation and/or assembly of the protein moieties of many hemoproteins either mitochondrial or cytosolic (see for example 1-5). Therefore it appears important to know how protoheme synthesis is regulated and what is the metabolic disposition of newly synthesized protoheme. In the present article we have summarized the salient features of the protoheme pathway and its regulation in yeast, documented by physiological, enzymatic and genetic approaches, with special emphasis on the first and last steps which are located in the mitochondria.

* Present address : Centre de Génétique Moléculaire, C.N.R.S.,
Gif sur Yvette, France

Mitochondria 1983

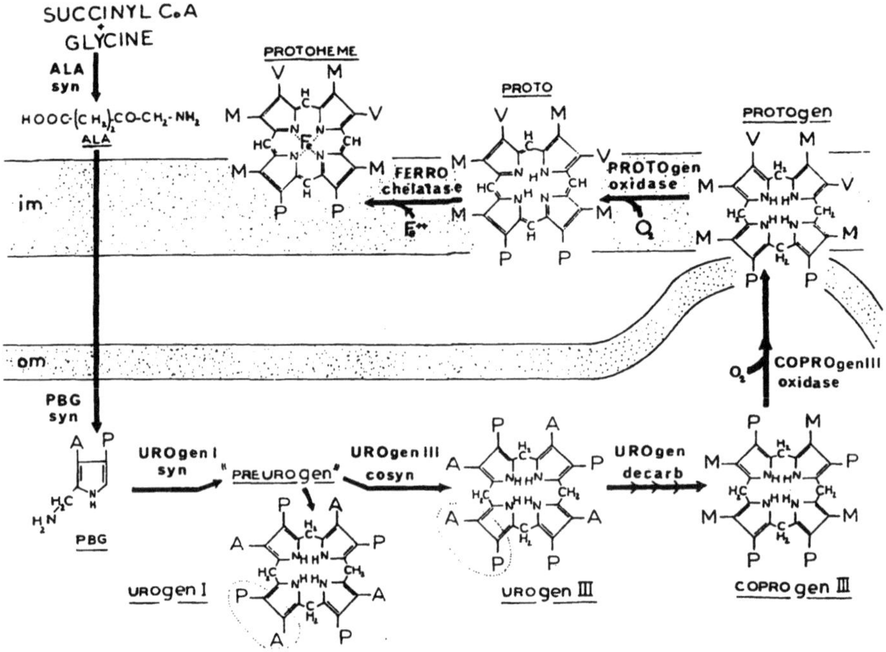

Figure 1. Pathway of protoheme synthesis in yeast.
ALA : δ aminolevulinic acid ; PBG : porphobilinogen ; uro-, copro-, protogen : uro-, copro-, protoporphyrinogen ; proto : protoporphyrin ; syn : synthase ; cosyn : cosynthase ; decarb : decarboxylase ; A, P, M, v : acetyl, propionyl, methyl, vinyl.

Results and Discussion

1) The metabolic pathway for protoheme synthesis in yeast (as in mammalian cells) involves integration of 8 mitochondrial and cytoplasmic enzyme activities (Fig. 1). The first commited reaction unique to porphyrin biosynthesis takes place in the mitochondrial matrix where the enzyme ALA synthase catalyses the condensation of succinyl-CoA supplied by the Krebs cycle and glycine. How does ALA go out of the mitochondria is not yet decided, however there is much evidence for a simple diffusion transport. The subsequent 5 reactions that convert ALA into protoporphyrinogen occur in the cytosol, and the final steps catalysed by protogen oxidase and ferrochelatase are associated with the mitochondrial membranes, probably the inner one. (In rat liver

mitochondria, coprogen oxidase activity has been described in the intermembrane space (6) and ferrochelatase suggested to be located on the inner face of the inner mitochondrial membrane (7)). That compartmentation of the 3 last steps implies that a close association must exist between the soluble cytosolic coprogen oxidase and the membranous enzymes, especially if one remembers that both protogen and protoporphyrin are lipophilic, protogen is highly autoxidizable and protoporphyrin a potent photosensibilizer. Fig. 1 shows also that besides the substrates for ALA synthesis, two other substrates are required for protoheme formation : molecular oxygen and ferrous iron. That will be discussed later.

2) To gain an insight into the regulatory mechanisms of the protoheme pathway in yeast, we undertook a systematic study of the parameters of this pathway in cells grown under a variety of physiological conditions (Table 1). The specific activities of 7 enzymes are measured in acellular extracts prepared from cells harvested at the end of the exponential phase of growth (line b). We also measured during the growth the amounts of heme precursors accumulated in the cells (ALA, PBG, porphyrins) or excreted in the medium (porphyrins) and of the end products hemes b + c + a (based on cytochromes spectra and on pyridine hemochromes) (line c). From these data, the rates of production of these metabolites in vivo could be estimated, assuming there is no catabolism of heme in growing cells (line d). (Methods described in (8-11) were used for those measurements). The results reported in Table 1 represent the range of values obtained in many experiments with different wild type strains. The main conclusions are the following : i) PBG synthase (also named ALA dehydratase) appears to be the rate-limiting step of the pathway, due to the low level of its activity and to its low affinity for its substrate ; this, taken together with the higher level of ALA synthase activity, explains the accumulation of a large pool of "free" intracellular ALA ; ii) there is no general correlation between the level of ALA synthase activity and glucose, the state of derepression of the mitochondria or the amount of total heme formed (galactose or raffinose derepressed cells have a higher ALA synthase activity as compared to ethanol grown cells) ; it must be

Table 1

Km µM	gly 4 000 succ CoA ⩽2	500	6-40[a]	0.4-1[a]	0.05- 0.1	1	Zn^{++} 1.5 Fe^{++} 1.6 proto5.9
	→ 8 ALA	→ 4 PBG	→ URO III gen	→→→→ COPRO III gen	→→ PROTO gen	→ PROTO	→ PROTOHEME
O_2 - Eth	0.25-0.4[b]	0.25-0.5	0.12-0.15	0.3-0.4	0.2-2	3-6	7-10
	(90-150)[c]	(<10)		(15-25)			(150-250)
	0.21-0.35[d]	0.14-0.23	0.14-0.22	0.14-0.22	0.13-0.21	0.13-0.21	0.12-0.20
O_2 - Glu	0.5-0.80	0.12-0.25	0.06-0.10	0.2	0.2-2	2-4	7-10
	(60-120)	(<10)		(5-10)			(60-120)
	0.23-0.44	0.13-0.23	0.12-0.22	0.12-0.22	0.11-0.21	0.1-0.2	0.1-0.2
N_2 - Glu	0.5-0.75	0.10-0.15	0.06-0.10		10-15	2-4	5-8
	(100-200)	(<10)		(20-30)		(5-10)	(25-50)
	0.27-0.48	0.1-0.14	0.09-0.14	0.09-0.14	0.05-0.1	0.05-0.1	0.04-0.08

a. Values taken from the litterature for erythrocyte or liver enzymes.
The enzymatic activities -b- measured in cell-free extracts and -d- estimated in vivo, are expressed in nmol equivalent porphyrin/h/mg protein.
c. Heme precursors and products measured during the growth are in nmol equivalent porphyrin/g DW.
See text for details.

stressed that these "physiological" variations of ALA synthase activity are strain-dependent ; iii) ALA synthase and all cytosolic enzymes function in vivo at nearly maximum velocity (except for coprogen oxydase in some strains) ; iv) that is not the case for the two last membranous enzymes which work in vivo at only a few percents of the maximum velocity measured in vitro ; v) when grown under anaerobiosis, the cells excrete coprogen III and contain some Zn protoporphyrin and protoheme detected as cytochromes b_5, P_{450} and a pigment X422 absorbing at 420-422 nm in the CO-difference spectrum. This indicates that the activities of the last enzymes are limiting in vivo, although the coprogen oxidase activity, when measured aerobically in cell-free extract, is much higher in these cells. The synthesis of protoporphyrin in anaerobic cells is puzzling since the enzymes coprogen and protogen oxidases are known to require O_2. Either there is some entry of oxygen during the culture, in spite of the cares taken to avoid it, or the enzymes utilize in vivo other hydrogen acceptors but with a very low efficiency. Concerning the coprogen oxidase, we could measure in acellular extracts only a residual activity (0.5 to 1 %) in absence of O_2 and in presence or not of the system described by Poulson (12) as an alternate acceptor : NAD^+ or $NADP^+$, + ATP and methionine. With regards to the protogen oxidase, we found that fumarate, used as an alternate acceptor by the E. coli enzyme (13) could not replace O_2 in the yeast extracts. Anyhow that will be technically difficult to show whether these residual "anaerobic" activities observed both in vivo and in vitro are due to contaminating traces of oxygen, or to some others acceptors which remain to be discovered.

The pigment absorbing at 503 nm and present in whole cells, P503, is not included in Table 1, although it has been proposed to be protoporphomethene, a semi-oxidized intermediate between protogen and protoporphyrin (14). This is because the following results (unpublished) obtained during these 3 last years led us to reconsider the nature of P503 : i) in growing cells, the adddition of methionine and aminopterine in the medium prevented the presence of P503 while the addition of glycine increased it ; ii) in resting cells the addition of glucose caused a rapid increase of P503 in 5 to 10 minutes up to about 300 nmole/g D.W. (using the extinction coefficient given in (15)), whether the cells contained it or not initially ; here also, methionine prevented it and glycine increased it ; iii) it has never been possible to extract P503 as protoporphyrin ; iv) to explain the rapid glucose-induced increase of P503 in resting cells, it was necessary to postulate the existence of a pool of porphyrinogens ; however all the attemps to extract them have been unsuccessful ; v) all the hem mutants isolated in the laboratory and which are blocked in different enzymic steps of the heme pathway contained P503 which could be further increased by glucose as in wild type strains. All these data are inconsistent with the proposal that P503 is protoporphomethene. Since they show some relationship between the presence of P503 and the metabolism of tetrahydrofolate, one could propose now that this absorption at 503 nm is due to the quinonoid intermediates formed by many enzymes with pyridoxal phosphate, tetrahydrofolate and substrates as has been well documented in the case of the enzyme serine hydroxymethyltransferase.

3) Genetic studies. Mutants totally heme deficient have been isolated which are each completely blocked in one enzymic step of the pathway(8):

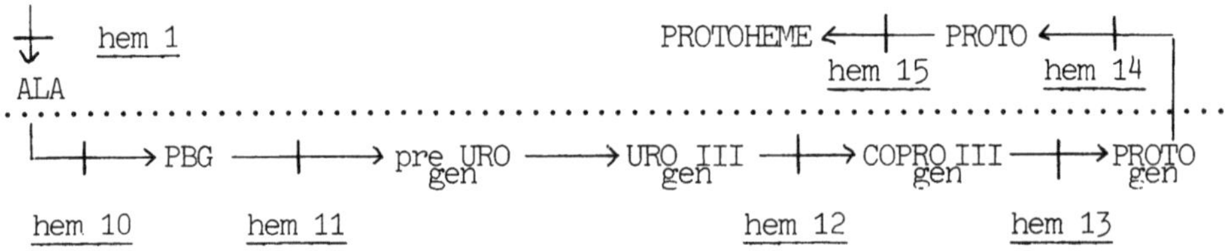

For each mutant, the phenotype is under the control of a single recessive nuclear gene designated hem 1, hem 10... hem 15. These loci are unlinked except for hem 12 and hem 13 which are separated by an interval of 6.7 cM. We have some evidence that hem 1, hem 12, hem 13 are structural genes and it is highly probable that the other loci are structural rather than regulatory. Up to now we have not found mutants with decreased activity for all enzymes which would have indicated coordinated regulation of the whole pathway, neither mutants greatly overproducing porphyrins. Also there is no evidence for more that one enzyme catalyzing each step of the pathway.

4) Concerning ALA synthase, biochemical analysis of these heme less mutants revealed that the absence of formation of protoheme does not lead to an increased level of its activity, suggesting no end-product feedback repression. However a 3 fold increase in that activity has been observed recently in partially blocked mutants at the hem 12 or hem 15 loci which synthesized about 1/3 of total heme as compared with the parental strain (11). The event underlying that increase is unclear at the present time.

ALA synthase has been purified from yeast mitochondria and found to be a dimer with a subunit molecular weight of 53 000 daltons (16). In order to get some information about its synthesis, mitochondrial import and functionning, we isolated new mutants affected in ALA synthesis and started to clone HEM 1 gene by functional complementation. 10 mutants were obtained, devoid of ALA synthase activity. The genetic analysis of these mutants and of 2 others, ole 3 and GL1 previously isolated in other laboratories (17, 18 respectively), revealed a phenomenon of functional interallelic complementation between some of them at the hem 1 locus : the products of the 2 different hem 1 alleles can form an hybrid enzyme with restoration of part of the activity, the defects in each subunit correcting each other. It is worth to mention that the 4 hem 1 mutations which did not complement any other one, led to the absence of

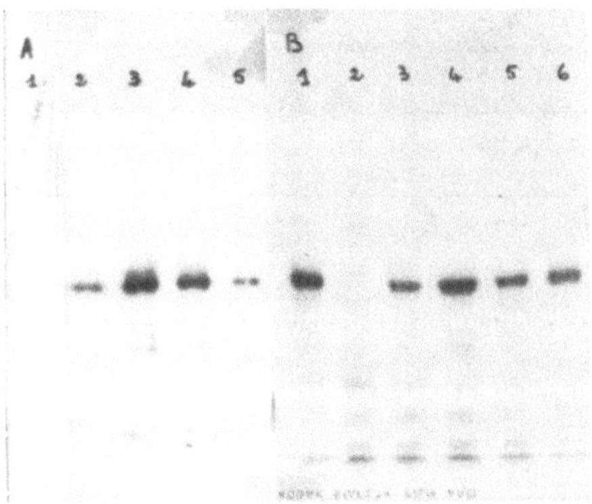

A-1. G204 (hem 1-5)
2. FL 200 (HEM 1)
3. Transformant T1
4. Transformant T11
5. purified ALA synthase

B-1. G101 (hem 1-8)
2. G102 (hem 1-9)
3. G104 (hem 1-11)
4. ole 3 (hem 1-12)
5. GL 1 (hem 1-3)
6. FL 200 (HEM 1)

Figure 2. Analysis of ALA synthase in wild type, hem 1 mutants and hem 1^+ transformed strains. The proteins present in protoplasts (A-2-3-4) or extracted in TCA from whole cells (A-1, B-1 to 6) were resolved by SDS-PAGE and then tested for reaction against the anti-serum by an immune replica technique (ref. in 3).

polypeptide immunodetectable with the ALA synthase anti-serum, as shown in Fig. 2 for 2 of them G 204 and G 102. The mutant strain A16 (*hem 1-5*, *ura 3*, *leu 2*) has been transformed by a pool of yeast genomic cosmids constructed in M. Guerineau laboratory. Two transformants able to grow on glycerol, have been analysed : the ALA synthase activity was increased by a factor of 2 and 5 respectively in T11 and T1, as compared to the wild type parental strain FL 200 and was found in the mitochondria (specific activity in FL 200, T1 and T11 : 4.9, 23 and 11 nmole ALA/h/mg protein in acellular extract). This correlates well with the amount of immunodetectable protein as shown in Fig. 2. The insert of 30 Kb carried by the cosmid in T11 transformant has been subcloned and restriction mapping of DNA fragments carrying HEM 1 is under investigation. It is hoped that the use of a transformant with a very high ALA synthase activity, together with pulse-labelling and in vitro translation studies will help to decide whether or not that protein is synthesized as a higher molecular weight precursor in yeast, as it has been clearly demonstrated in the case of chick embryo liver cells (19 and ref. therein).

Table 2

ALA synthase, α ketoglutarate and pyruvate dehydrogenases activities in FL 200 parental and G 201 mutant strains.

Growth conditions		Strain	Cyto. chromes	ALA Synthase succ. CoA	ALA Synthase α ketoglu	α ketoglu tarate DH	Pyruvate DH
Glu 2 %	exp.	FL 200	+	5.9	5.6	120	144
		G 201	-	4	⩽ 0.1	⩽ 10	⩽ 10
	late stat.	FL 200	+ +	3.8	3.9	500	288
		G 201	+ +	4.2	4.5	205	480
Glu 0.2 %	exp.	FL 200	+	6.6	7.2	605	2 208
		G 201	traces	4.8	0.5	⩽ 10	⩽ 10

The enzymatic activities measured in cell-free extracts are expressed in nmol product/h/mg prot.

One mutation at a locus different and unlinked to hem 1 has been obtained which interferes with the activity of ALA synthase. We isolated a mutant G201 carrying a single recessive nuclear mutation hem 16 which grew very slowly on glycerol (to be published). When grown on glucose, it had no cytochromes during the exponential and early stationary phase of growth but recovered them in late stationary (not because of reversion). The addition of ALA in the medium could restore the cytochromes to normal. Surprisingly, a normal level of ALA synthase activity was measured and furthermore found located in the mitochondria. The simplest explanation of that phenotype is a defect in the supply of the substrates for ALA synthase. And indeed it was found that α ketoglutarate dehydrogenase activity was undetectable in exponential cells, whether measured directly or coupled with ALA synthase activity. As shown in Table 2, pyruvate dehydrogenase activity was also deficient in the mutant. These results suggest that some regulatory mechanism common to both dehydrogenases is affected in the mutant, leading to a lack of succinyl CoA for ALA synthesis.

5) Other nuclear gene products are expected to interfere with protoheme and cytochrome synthesis when considering the last steps of the pathway as proposed in Fig. 3. One of the basic prerequisite for normal protoheme synthesis is a normal supply of both protoporphyrin and Fe^{++} to ferrochelatase and a normal release and transport of the end-product. The exact localisation of protogen oxidase and ferrochelatase within the membrane is unknown but one can anticipate a rather close association between the both. Recently we found that the affinity of ferrochelatase for protoporphyrin was 100 times higher when the protoporphyrin was generated within the membrane as compared to exogenous protoporphyrin (unpublished). On the other hand the availability of metal ion plays a crucial role in controlling the activity of ferrochelatase. The enzyme can incorporate Zn^{++} or Fe^{++} with an almost similar efficiency and both metals act as reciprocal competitors : Fe^{++} is a potent competitive inhibitor of Zn^{++} chelation (ki = 0.05 μM) while Zn^{++} has a smaller effect on Fe^{++} chelation (ki = 1.5 μM) ; Fe^{+++} which is not substrate is a very weak competitive inhibitor for both chelatase activities (ki = 115 - 140 μM) (10). These competitions are likely to work in vivo.

That might explain why yeast cells accumulate Zn-protoporphyrin (in the mitochondrial membranes) when incubated as resting cells in the absence of glucose (ref. in 10), while they make protoheme in the presence of glucose (21) : the absence of an energy source would lead to a depletion of the mitochondrial Fe^{++}, rendering Zn^{++} chelation operative if enough Zn^{++} is available. A major question is how does Fe^{++} reach the site of ferrochelatase ? Actually nothing is known about the metabolism of iron in the yeast Saccharomyces cerevisiae. Considering that "free" Fe^{+++} is very insoluble at neutral pH, that Fe^{++} is very rapidly autoxidized, and by analogy with what has been described for bacteria and mammalian cells, there should exist in yeast some chelating agent(s), binding protein(s) and reductase(s) to account for iron assimilation by the cells, transport across the cytosol, stockage, reduction in the cytosol or in mitochondrial membranes and eventually its transport across these membranes. We have measured an "iron-reductase" activity in mitochondrial membranes (specific activity = 3 nmol Fe^{++}/h/mg prot ; Km Fe^{+++}= 28 μM) which depends upon the functionning of the NADH or succinate dehydrogenase of the respiratory chain. Such an activity had been previously described in liver mitochondria and bacteria (7, ref. in 10). We hope that the detailed study of new yeast mutants (in collaboration with

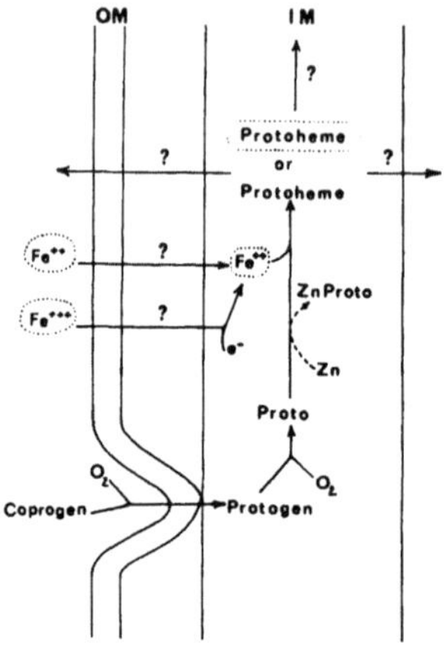

Figure 3. Simplest organization of the last enzyme steps of protoheme synthesis in yeast mitochondria, proposed as a working model. ( ) means a chelation.

J. Rytka who isolated them) will help to clarify the problem of the accessibility of iron to ferrochelatase : they are allelic or not to hem 15, have a normal ferrochelatase activity in vitro, althought in vivo they make less heme and accumulate protoporphyrin.
The last important question shown in Fig. 3 concerns the fate of protoheme once made in the membranes, since protoheme has been involved in processus located in different cellular compartments (1-5). The presence of cytosolic heme-binding proteins has been reported which stimulate the transport of protoheme from mitochondria to microsomes in mammalian cells (ref. in 20 and 21). The most conclusive evidence for a small cytosolic pool of newly synthesized heme has been obtained from pulse-labeling studies in cultured hepatocytes : they showed that the earliest heme detectable was localized almost entirely to the cytosolic fraction of the cell, what implies immediate transfer of newly synthesized protoheme from mitochondria to the cytosol (20). Such dynamic studies are lacking in yeast. However we have some indication that a pool of soluble "free" protoheme might exist in yeast (21) : i) anaerobic, glucose-repressed and ρ^- cells contain a pigment named X422 absorbing at 420-422 nm in the reduced-CO difference spectrum ; ii) X422 increases in resting cells incubated in conditions preventing protein synthesis and Zn-protoporphyrin formation ; iii) X422 is found for about 85 % in the acellular soluble fraction ; iv) during the in vitro assay of ferrochelatase with mitochondrial membranes there is an increase of X422 concomittant with the synthesis of protoheme detected by its pyridine hemochrome. These data tell in favour of X422 representing unassigned "free" protoheme, i.e. protoheme associated with unspecified ligands, peptides or proteins. The exact intracellular compartmentalization of that protoheme remains to be done. Also it would be important to know if that pool of "free" protoheme is accessible for metabolic functions or whether it only represents firmly bound extra-protoheme made in certain conditions.

Acknowledgments. P. Labbe is gratefully acknowledged, who initiated most of these researches. We are indebted to Nicole Brouillet, Henriette Chambon, Françoise Felix and Monique Casteras-Simon for their participation in some part of these studies. The work has been supported over the years by grants from CNRS, University Paris VII, DGRST (n° 77.7.0257), MIR (n° 82.L.1119) and Fondation pour la Recherche Médicale.

References

1. Guarente, L., Mason, T. : Cell 32, 1279-1286 (1983).
2. Hörtner, H., Ammerer, G., Hartter, E., Hamilton, B., Rytka, J., Bilinski, T., Ruis, H. : Eur. J. Biochem. 128, 179-184 (1982).
3. Ohashi, A., Gibson, J., Gregor, I., Schatz, G. : J. Biol. Chem. 257, 13042-13047 (1982).
4. Saltzgaber-Müller, J., Schatz, G. : J. Biol. Chem. 253, 305-310 (1978).
5. Lin, C.P., Gollub, E.G., Beattie, D.S. : Eur. J. Biochem. 128, 309-313 (1982).
6. Grandchamp, B., Phung, N., Nordmann, Y. : Biochem. J. 176, 97-101 (1978).
7. Barnes, R., Connelly, J.L., Jones, O.T.G. : Biochem. J. 128, 1043-1055 (1972).
8. Urban-Grimal, D., Labbe-Bois, R. : Mol. Gen. Genet. 183, 85-92 (1981).
9. Camadro, J.M., Urban-Grimal, D., Labbe, P. : Biochem. Biophys. Res. Commun. 106, 724-730 (1982).
10. Camadro, J.M., Labbe, P. : Biochim. Biophys. Acta 707, 280-288 (1982).
11. Rytka, J., Bilinski, T., Labbe-Bois, R. : Submitted to Biochem. J. (1983).
12. Poulson, R., Polglase, W.J. : J. Biol. Chem. 249, 6367-6371 (1974).
13. Jacobs, N.J., Jacobs, J.M. : Biochim. Biophys. Acta 449, 1-9 (1976).
14. Labbe, P., Volland, C., Chaix, P. : Biochim. Biophys. Acta 143, 70-78 (1967).
15. Poulson, R., Polglase, W.J. : Biochim. Biophys. Acta 329, 256-263 (1973).
16. Volland, C., Felix, F. : Submitted to Eur . J. Biochem. (1983).
17. Resnick, M.A., Mortimer, R.K. : J. Bacteriol. 92, 597-600 (1966).
18. Gollub, E.G., Lin, K., Dayan, J., Adlersberg, M., Sprinson, D.B. : J. Biol. Chem. 252, 2846-2854 (1977).
19. Srivastava, G., Borthwick, I.A., Brooker, J.D., May, B.K., Elliott, W.H. : Biochem. Biophys. Res. Commun. 110, 23-31 (1983).
20. Grandchamp, B., Bissell, D.M., Licko, V., Schmid, R. : J. Biol. Chem. 256, 11677-11683 (1981).
21. Dehoux, P. : Thèse de 3ème cycle, Université Paris VII (1978).

MONOCLONAL ANTI-H^+-ATPase ANTIBODIES IN THE STUDY OF THE ASSEMBLY OF THE YEAST MITOCHONDRIAL H^+-ATPase.

Sangkot Marzuki, Richard G. Hadikusumo, Wan Mee Choo, Linton Watkins, H. Bruce Lukins and Anthony W. Linnane.

Department of Biochemistry, Monash University, Clayton, Victoria, Australia 3168

1. Introduction.

Recent studies on the biosynthesis of the mitochondrial H^+-ATPase have been focused mainly on the characterization of the molecular structure of nuclear and mitochondrial genes coding for subunits of the enzyme complex, and on the elucidation of processes involved in the import of subunits which are synthesized in the extra-mitochondrial cytoplasm into the organelle. Thus, the mictochondrial genes coding for subunits 6 and 9 of the membrane F_o sector of the enzyme complex have now been completely sequenced, and a partial sequence of the nuclear gene for the subunit of the F_1 sector has been reported [1,2,3,4]. Furthermore, we now have detailed knowledge on how the three largest subunits of the F_1 sector are synthesized on the cytoplasmic ribosomes as larger precursors, the transport of these precursors across the mitochondrial membranes and their processing into mature subunits [5,6].

In contrast, very little is known about the assembly events that follow the import of the cytoplasmically synthesized subunits. Information regarding these events can be deduced from the extent to which the assembly process can proceed in mutants in which one of the H^+-ATPase subunits is not

Mitochondria 1983

synthesized or synthesized in a grossly altered form. Several of such mutants have previously been reported [7,8], but since the defective ATPase complex in these mutant strains are unstable, it has not been possible in the past to isolate the mutant enzyme complex in an intact form, to allow the nature of their assembly defect to be determined.

In this communication we describe the property of several monoclonal antibodies against the β subunit of the H^+-ATPase which have recently been produced in our laboratory, as part of our general strategy to elucidate the mechanism of the assembly of the H^+-ATPase in yeast. Two of these antibodies are very effective in immunoprecipitating intact ATPase complex from triton extracts of yeast mitochondria, and have been used to define the nature of the H^+-ATPase assembly defect in mutants which lack various mitochondrially synthesized subunits of the enzyme complex. From the results of this investigation an assembly pathway is proposed involving sequential addition of the mitochondrially synthesized subunits to the F_1 sector, leading to the formation of a functional H^+-ATPase complex.

2. Monoclonal Antibodies Against Subunits of Yeast Mitochondrial H^+-ATPase and the Definition of the Subunit Structure of the Enzyme Complex.

H^+-ATPase immunoprecipitates obtained with antisera against purified F_1 sector of the enzyme complex frequently do not contain one or more of the mitochondrially synthesized subunits of the F_o sector [9,10,11]. The reason for the apparent inability of polyclonal anti-F_1-subunits antibodies to reproducibly bring down some of the F_o subunits is not clear. One possible explanation is that the binding of an antibody molecule to certain antigenic determinants on the

H^+-ATPase complex might induce conformational changes in the subunit(s) concerned, resulting in the dissociation of the F_o subunits from the enzyme complex.

We have recently isolated a series of stable lines of myeloma-spleen cell hybrids, producing antibodies against various subunits of yeast mitochondrial H^+-ATPase. The procedures used for the isolation and the characterization of the hybridoma are described elsewhere [12]. A total of fourteen hybridoma cell lines were found to produce anti-H^+-ATPase antibodies. Characterization of these antibodies by immunoblotting against components of the purified yeast H^+-ATPase revealed that one of the antibodies reacts with the α subunit of the enzyme complex, nine with the β subunit and four with a 25,000 dalton subunit which has previously been reported [10], but has not been characterized.

The monoclonal anti-β-subunit antibodies are probably directed against different antigenic determinants on the β subunit because they exhibit a wide range of effects on the H^+-ATPase activity. Some of these antibodies inhibit the enzyme activity but some have no effect or even stimulate the activity.

Two of the monoclonal anti-β-antibodies which do not inhibit the enzyme activity (clone RH 48.6 and RH 51.4) are of a particular interest because they were found to be very effective in immunoprecipitating the intact H^+-ATPase complex. Electrophoretic analysis of the anti-β-subunit immunoprecipitates on SDS-polyacrylamide gels reveal the presence of fourteen protein bands (Fig 1a), but only ten of these are probably genuine subunits of the enzyme complex; the other four protein bands could be shown to be antifacts which arise as the result of aggregation or proteolytic breakdown of some of the other polypeptides. Thus, the

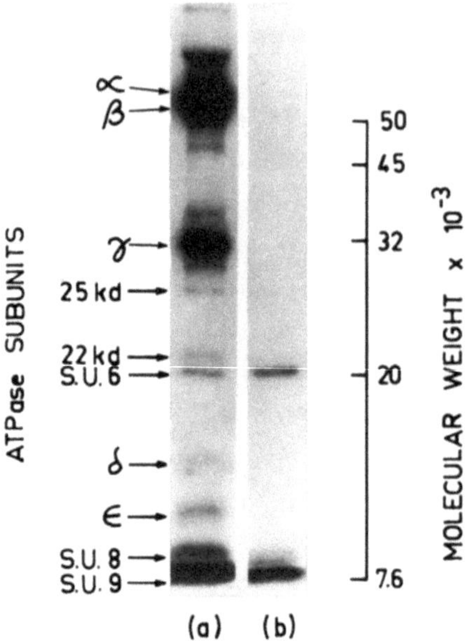

Fig.1.Monoclonal Anti-β antibody Immunoprecipitate of Yeast H^+-ATPase Complex.

Yeast cells (wild-type strain J69-1b) were labelled with [^{35}S]sulphate at 28•C in the absence (4 h incubation) or presence (1 h incubation) of cylcoheximide [13]. Mitochondria were isolated and H^+-ATPase was immunoprecipitated from the triton extract [13] of the mitochondria by using monoclonal anti-β antibody RH 48.6 coupled to Sepharose 4B beads [12]. Protein subunits immunoprecipitated by the monoclonal antibody were separated by SDS-polyacrylamide gel electrophoresis [13] and visualized by fluorography [14].

(a) Autoradiogram of immunoprecipitate from cells labelled in the absence of cylcoheximide.

(b) Autoradiogram of immunoprecipitate from cells labelled in the presence of cycloheximide.

highest molecular weight band observed on the gel (apparent Mr 61,000) can be attributed to the subunit of the F_1 sector (Mr 56,000), since this protein band was also observed when purified α subunit was electrophoresed on a similar polyacrylamide gel (Orian, J.M. and S. Marzuki, unpublished observation). The protein bands with the mobilities of a 47,000, a 37,000 and a 29,000 dalton proteins are probably the breakdown products of the α and β subunit described previously by Ryrie and Gallagher [15]; the intensity of these bands varied from one immunoprecipitate to another and were greatly reduced by the addition of protease inhibitors to the solutions used throughout the mitochondrial isolation and immunoprecipitation steps.

Five of the ten presumptive ATPase subunits can be identified as the subunits of the F_1 sector of the enzyme complex by comparing their mobilities on an SDS-polyacrylamide gel to those of purified F_1-ATPase (α subunit-Mr 56,000, β subunit-Mr 54,000, γ subunit-Mr 31,000, δ subunit-Mr 14,000 and ε subunit-Mr 12,000. In addition three of the other subunits can be shown to be mitochondrially synthesized, by analysing on the SDS-polyacrylamide gel a similar immunoprecipitate but obtained from cells which have been labelled with [^{35}S]sulphate in th presence of cycloheximide (Fig 1b). These subunits include the two subunits of the F_o sector which are relatively well characterized (subunit 6, apparent Mr 20,000 and subunit 9, Mr 7,600) [9,16], as well as a mitochondrial translation product with a mobility of a 10,000 dalton protein which in the past has been the subject of conflicting reports. As discussed in the subsequent section, we now believe that this polypeptide is a genuine subunit of the yeast H^+-ATPase complex.

The identity of the other two polypeptides (apparent Mr 22,000 and 25,000) has not been established. However, these polypeptides reproducibly co-purify with the anti-β subunit immunoprecipitate, which strongly indicates that they are genuine subunits of the yeast H^+-ATPase. Monoclonal antibodies against these proteins would be very useful for their identification and characterization.

The best immunoprecipitate of assembled yeast H^+-ATPase which has been reported to date is that of Todd et al [10] obtained with the use of a polyclonal anti-β antiserum. The immunoprecipitate contains all of the cytoplasmically synthesized subunits observed in this communication but lacks the mitochondrially synthesized subunit 8. In addition the variation in the amount of subunit 9 associated with various immunoprecipitates obtained with the polyclonal anti-β antibodies, indicates that some of this subunit might be dissociated during the extraction and immunoprecipitation of the complex [10].

3. Yeast Mitochondrial H^+-ATPase contains a Third Mitochondrially Synthesized Subunit.

The exact number of the mitochondrially synthesized subunits associated with yeast H^+-ATPase has in the past been the subject of uncertainty. It was initially suggested that the enzyme complex contains four mitochondrially synthesized subunits: subunit 5, 6, 8 and 9, which in our gel system have apparent molecular weights of 35,000, 20,000, 10,000, and 8,000 respectively. Subunit 5 has been shown in our laboratory to be a contaminant, consisting of newly sythesized subunit II of the cytochrome oxidase complex and the apoprotein of the cytochrome b, which copurify with the H^+-ATPase [9,16]. Subunit 6 and 9 are now well established to be functional subunits of the enzyme complex. Until recently, however, the status of subunit 8 has remained obscure.

We now have two lines of evidence which strongly indicate that the 10,000 dalton mitochondrial translation product is a genuine subunit of the enzyme complex. Firstly, in contrast to the previous reports [9,10,11], our data on the subunit composition of the H^+-ATPase, obtained with the use of the monoclonal anti-β antibodies RH 48.6 and RH 51.4, suggest that the 10,000 dalton polypeptide is assembled into the H^+-ATPase since the 10,000 dalton polypeptide is always associated with the anti-β subunit immunoprecipitate. One possible explanation for the apparent inability of some polyclonal anti-F_1 antibodies used in previous studies [9,10] to immunoprecipitate the 10,000 dalton subunit, is that these antibodies might have caused the dissociation of the 10,000 dalton polypeptide from the enzyme complex. The binding of monoclonal antibodies such as RH 48.6 and RH 51.4, on the other hand, do not affect the stability of the enzyme complex.

Secondly, our laboratory has recently isolated several mit⁻ mutants of yeast which lack the 10,000 dalton mitochondrial translation product [8]. The mutations in these strain have been mapped into a previously unidentified reading frame on the mtDNA denoted aapl which is located between the oli2 and the oxi3 regions. Mitochondria isolated from these mutants contain a significant level of ATPase activity, but the activity shows little or no oligomycin sensitivity (data not shown), indicating that the F_1 sector in these mutants is associated with a defective F_o sector.

The nucleotide sequence of the aapl gene, the amino acid sequence of the predicted gene product and the nature of the mutations in several aapl mit⁻ have been reported by us [17,18]. Furthermore we have reported also that H^+-ATPase isolated from the aapl mutants lacks the mitochondrially synthesized subunit 8 [17]. We have now comfirmed this observation by using a published chloroform-methanol extraction procedure which is designed for the isolation of proteolipids from the mitochondrial membrane [19]. As shown in Fig 2, the chloroform-methanol extract from mitochondria of a wild-type strain contains two mitochondrial translation

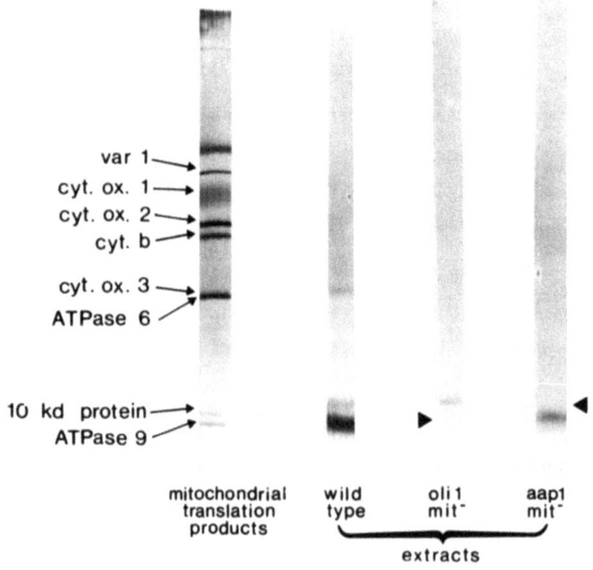

Fig.2. Proteolipid Extract from an aap1 mit⁻ Mutant Does Not Contain the 10,000 Dalton Mitochondrial Translation Product.

Mitochondrial translation products in a wild-type strain J 69-1b, an oli1 mit⁻ strain 5726 and an aap1 mit⁻ strain M26-10 were labelled with [^{35}S]sulphate as previously described [13]. Proteolipids were extracted with a chloroform-methanol mixture from the mitochondria essentially as described by Verlour et. al. [19]. The proteolipid extracts were then analysed by electrophoresis on a 12.5% polyacrylamide gel in the presence of SDS [13], and mitochondrial translation products visualized by fluorography [14]. Total mitochondrial translation products of the wild-type strain are also shown.

products with apparent molecular weights of 10,000 and 8,000. The 8,000 dalton mitochondrial translation product can be identified to be subunit 9 of the H^+-ATPase because this product is missing in an extract obtained from an oli1 mit⁻(Fig. 2) which cannot sythesize this subunit (Fig. 3). The 10,000 dalton mitochondrial translation product is missing in the extract obtained from an aap1 mit⁻ (Fig. 2).

The amino acid composition of the 10,000 dalton proteolipid, purified from the chloroform-methanol extract of a wild-type strain mitochondria by the combination of chromatography on LH60 column and reverse phase HPLC, has previously been determined [20]. The comparison between this aminoacid composition and that predicted by the nucleotide sequence of the aap1 gene shows that the 10,000 dalton subunit 8 [17] is the product of the aap1 gene.

It is of interest to note that the predicted gene product of the aap1 gene is only 48 amino acid residues in length. The calculated molecular weight of this predicted gene product is only 5,800 which is much smaller than that of the apparent molecular weight of subunit 8 on SDS-polyacrylamide gel. However, it is well documented now that many hydrophobic membrane proteins behave abnormally in SDS-polyacrylamide gels, presumably due to the unusual interaction between these hydrophobic proteins and SDS [21,22]. SDS-protein complexes, which are formed primarily by hydrophobic interaction, have been shown to assume the hydrodynamic properties of rodlike particles which are uniquely constant in diameter, but vary in length in proportion to the molecular weight of the proteins [23]. The rodlike particles begin to approximate a sphere at a molecular weight of around 15,000 when the length approaches the diameter in magnitude. Erratic behaviour can be expected as this limit is approached [23], especially if there is any abnormal interactions between the detergent and the proteins. In addition, it is also possible that the 5,800 dalton gene product exists as a dimer which is not dissociated by SDS, similar to the oligomeric form of subunit 9.

4. Defect in the Assembly of the Mitochondrially Synthesized Subunits of H^+-ATPase in mit$^-$ Mutants of Yeast.

The synthesis of H^+-ATPase subunits in the extramitochondrial cytoplasm can proceed in the absence of mitochondrial protein synthetic activity. Results of recent investigations in our laboratory indicate that under the above conditions, the cytoplasmically synthesized subunits are not only imported into the mitochondria, but are also assembled into a membrane bound complex which consist of a functional F_1 sector, as well as the 22,000 and the 25,000 dalton subunits described in a previous section [24]. Thus it appears that the mitochondrially synthesized subunits are not essential for the assembly of subunits imported from the cytoplasm.

We have investigated the assembly of the mitochondrially synthesized subunits to the above complex, by analysing the assembly defects in various mit⁻ mutants which lack one or more of the mitochondrially synthesized H^+-ATPase subunits as the result of the mutations. These mutants include an oli1 mit⁻ mutant which lack ATPase subunit 9, an oli2 mit⁻ mutant which does not synthesize subunit 6 [7] and an aap1 mutant which does not contain the 10,000 dalton subunit 8 [8] (Fig 3). For this purpose, the mitochondrial translation products in the mutant strains were labelled with $[^{35}S]$sulphate in the presence of cyclocheximide. H^+-ATPase complex was then immunoprecipitated from triton extracts of the mitochondria by using the monoclonal anti-β-subunit antibody RH 48.6. As shown in Fig. 4, the three ^{35}S-labelled mitochodrially synthesized ATPase subunits are associated with the immunoprecipitate of the wild-type ATPase complex, which indicate that our labelling conditions allow the assembly of the newly synthesized subunits 6, 8 and 9 into the H^+-ATPase.

The absence of a normal subunit 6, as in the oli2 mit⁻ strain does not appear to affect the assembly of subunits 8 and 9, because these subunits are still associated with the mutant

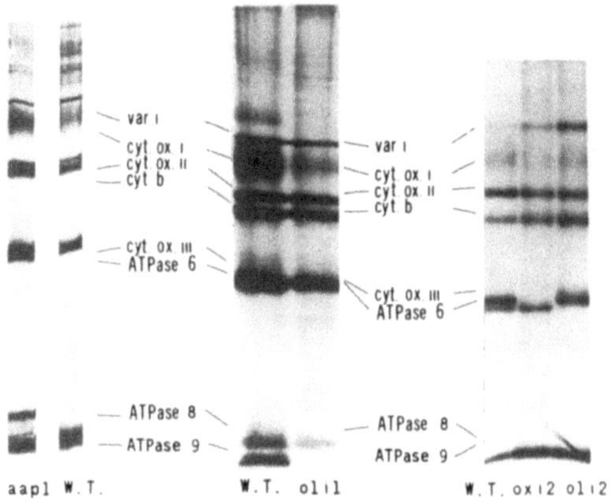

Fig.3. Mitochondrial Translation Products in an aapl, an oli1 and an oli2 mit⁻ Mutants.
Mitochondrial translation products of strains J69-1b [rho⁺], M26-10 [aapl mit⁻],5726 [oli1 mit⁻], M10-7 [oli2 mit⁻] and 2008 [oxi2] were labelled in the presence of cylohexamide with [^{35}S]sulphate [13]. Mitochondria were isolated and the mitochondrial translation products were separated by electrophoresis in the presence of SDS in a 12.5% polyacrylamide (strains M26-10 and 5726) or a 11% polyacrylamide (strains 2008 and M10-7) gel [13]. The electrophoretic conditions used for strains M26-10 and 5726 are optimal for the separation of ATPase subunits 8 and 9, whereas those used for strain M10-7 and 2008 were utilized to obtain the separation of ATPase subunit 6 from cytochrome oxidase subunit III. Mitochondrial translation products were visualized by fluorography [14].

immunoprecipitate (Fig. 4). The presence of subunit 8, on the other hand, is essential for the assembly of subunit 6 but not for the assembly of subunit 9. Thus, although the aapl mutant, which lacks subunit 8, still synthesizes subunit 6 (albeit in reduced amount, data not shown), this subunit could not be detected when the ATPase complex immunoprecipitated from the mutant strain was analysed on an SDS-polyacrylamide gel (Fig. 4). This immunoprecipitate contains the ATPase subunit 9. The absence of subunit 9 appears to prevent the assembly of subunits 6 and 8. Neither of these subunits could be detected in the ATPase immunoprecipitate obtained from the oli1 mit⁻ strain which

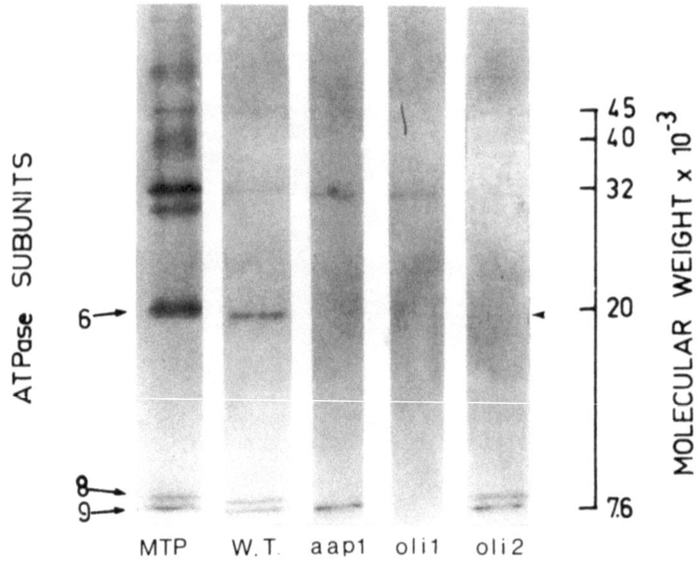

Fig.4. Mitochondrially Synthesized Subunits Associated with Monoclonal Anti-β antibody Immunoprecipitates of H^+-ATPase Complex in an aap1, an oli1 and an oli2 mit^- Mutants.

Mitochondrial translation products in strain J69-1b (rho^+), M26-10 (aap1 mit-), 5726 (oli1 mit^-) and M10-7 (oli2 mit^-) were labelled with [^{35}S]sulphate as previously described [13]. Mitochondria were then isolated and H^+-ATPase was immunoprecipitated from triton extracts of the mitochondria by using monoclonal anti-β antibody RH48.6 (coupled to sepharose 4B beads) as described elesewhere [12]. Mitochondrial translation products associated with the immunoprecipitates were analysed by SDS-polyacrylamide gel electrophoresis [13] followed by fluorography [14]. MTP = total mitochondrial translation products in strain J69-1b (rho^+) [13].

cannot synthesize subunit 9 (Fig. 4).

A tentative scheme can be drawn from the above results with regard to the sequence of assembly of the mitochondrially synthesized H^+-ATPase subunits to the enzyme complex, as outlined in Fig 5. According to this scheme, subunit 9 plays a key role in the assembly of the F_o sector, in that subunit 9 is the first mitochondrial translation product to be

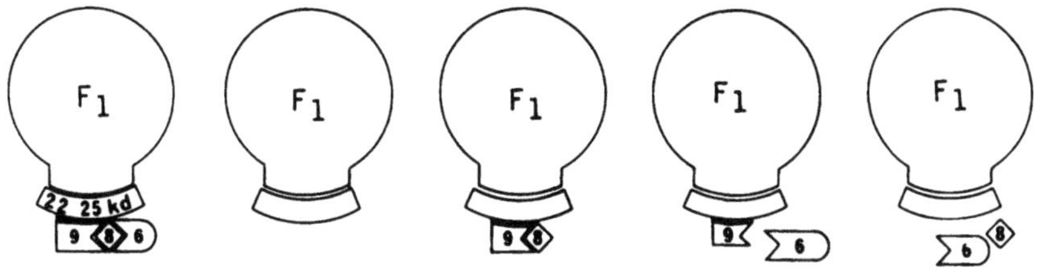

	rho$^+$	rho^0	oli2 mit$^-$	aapl	oli1 mit$^-$
Not synthesized	-	Su 6,8,9	Su 6	Su 8	Su 9, 6
Not assembled	-	6,8,9	6	8,6	9,6,8

Fig.5. H^+-ATPase Assembly Defects in Mitochondrial Mutants of Yeast.
The assembly defects in the oli1, aapl and oli2 mit$^-$ mutants were deduced from results presented in Fig.4. In the rho^0 strain the cytoplasmically synthesized subunits are assembled into a membrane bound complex consisting of the F_1 subunit and the 22,000 and 25,000 dalton subunits [24].

assembled to the F_1 sector and that without subunit 9 the assembly of subunit 6 and 8 is not possible. The assembly of subunit 9 appears to be followed by subunit 8 because in the absence of this subunit, the ATPase complex still contains subunit 9 but not subunit 6. Subunit 6 is the last mitochondrially synthesized subunit to be assembled, since the absence of this subunit does not affect the assembly of subunit 8 or 9.

5. Concluding Remarks

The assembly pathway for the mitochondrial H^+-ATPase outlined in Fig. 5 is very different from that suggested for E. coli [25], in that in E. coli the assembly of the F_1 sector is tightly coupled to that of the F_o sector. Thus, in E. coli the assembly pathway is initiated by the formation of a complex which consists of an F_o subunit a (equivalent to subunit 6 in yeast) and a β subunit. Other F_1 and F_o subunits are then assembled in turn to this complex, and in contrast to the yeast mitochondrial H^+-ATPase, the assembly of the E. coli F_1 sector can only be completed after the F_o sector has been completely assembled.

Although the assembly defects reported in this communication have been confirmed in several additional oli1, oli2 and aap1 mit^- mutants, the number of mutants which have been analysed is still limited. More mutants in each group of mit^- strains would obviously have to be analysed to substantiate and refine the tentative assembly pathway outlined in Fig. 5. These should include not only mutants which cannot synthesize various mitochondrial translation products but also those which can synthesize these products but in an altered form. More aap1 mit^- mutants have to be analysed, for example, to establish whether the observed reduction in the amount of subunit 6 synthesized in the subunit 8-less mutant reported in this communication is a general phenomenon. Preliminary results in our laboratory indicate that the synthesis of subunit 6 might also be affected by oli1 mit^- mutations. Together with results of similar analysis on mutants which lack various cytoplasmically synthesized H^+-ATPase subunits, and that of pulse chase experiments, the information on H^+-ATPase assembly defect in the mit^- strains should provide us with a detailed picture of the mitochondrial H^+-ATPase assembly pathway.

REFERENCES

1. Macino, G., Tzagoloff, A.: Cell 20, 507-519 (1980).
2. Hensgens, L.A.W. Grivell, L.A., Borst, P., Bas, J.L.,: Proc,. Natl. Acad. Sci. U.S.A. 76, 1663-1667 (1979).
3. Macino. G., Tzagoloff, A.: Proc. Natl. Acad. Sci. U.S.A. 76, 131-135.(1979).
4. Douglas, M.G., Saltzgaber, J., Kunapuli, S., Boutry,

M.: In "Manipulation and Expression of Genes in Eukaryotes". (P. Nagley, A.W. Linnane, W.J. Peacock, and J.A. Pateman eds). pp. 151-158 Academic Press, Australia 1983.

5. Lewin, A.S., Gregor, I., Mason. T.L., Nelson, N., Schatz, G.: Proc. Natl. Acad. Sci. U.S.A. 77, 3998-4002 (1980).
6. Gasser, S.H., Daum, G., Schatz, G.: J. Biol. Chem. 257, 13034-13041 (1982).
7. Roberts, H., Choo, W.M., Murphy, M., Marzuki, S., Lukins, H.B., Linnane, A.W.: FEBS. Lett. 108, 501-504 (1979).
8. Macreadie, I.G., Choo, W.M., Novitski, C.E., Marzuki, S., Nagley, P., Linnane, A.W., Lukins, H.B.: Biochem. Int. 5, 129-136 (1982).
9. Orian, J.M., Murphy, M., Marzuki, S.: Biochim. Biophys. Acta. 652, 234-239 (1981).
10. Todd, R.D., Griesenback, T.A., Douglas, M.G.: J. Biol. Chem. 255, 5461-5467 (1980).
11. Tzagoloff, A., Meagher, P.: J. Biol. Chem. 247, 594-603 (1972).
12. Hadikusumo, R.G., Hertzog, P., Marzuki, S. Submitted for publication.
13. Murphy, M., Choo, K.B., Macreadie, I., Marzuki, S., Lukins, H.B., Nagley, P., Linnane, A.W.: Arch. Biochem. Biophys. 203, 260-270 (1980).
14. Chamberlain, J.P.,: Anal. Biochem. 98, 132-135 (1979).
15. Ryrie, I.J., Gallagher, A.: Biochem. Biophys. Acta. 463, 1-27 (1979).
16. Orian, J.M., Marzuki, S.: J. Bacteriol., 146, 813-815 (1981).
17. Novitski, E.C., Macreadie, I.G., Maxwell, R.J., Lukins, H.B., Linnane, A.W., Nagley, P.: In "Manipulation and Expression of Genes in Eukaryotes". (P. Nagley, A.W. Linnane, W.J. Peacock, J.A. Pateman eds). pp 257-267 Academic Press, Australia (1983).
18. Macreadie, I.G., Novitski, C.E., Maxwell, R.J., John, U., Ooi, B.G., McMullen, G.C., Lukins, H.B., Linnane, A.W., Nagley, P.: Nucleic Acids. 11, 4435-4451 (1983).
19. Velours, J., Guerin, M., Guerin, B.: Arch. Biochem, Biophys. 201, 615-628 (1980).
20. Velours J., Esparza, M., Guerin, B.: Biochem. Biophys. Res. Commun. 109, 1192-1199 (1982).
21. De Jong, W.W., Zweers, A., Cohen, L.H.: Biochem. Biophys. Res. Commun. 82, 532-539 (1978).
22. Tung, J.S., Knight, C.A.: Anal. Biochem. 48, 153-163 (1972).
23. Reynolds, J.A., Tanford, C.: J. Biol. Chem. 245, 5161-5165 (1970).
24. Orian, J.M., Marzuki, S., Linnane. A.W., In preparation.
25. Gibson, F.: Proc. R. Soc. Lond. B 215, 1-18 (1982).

TRANSPORT OF CYTOCHROME C INTO MITOCHONDRIA:
INVOLVEMENT OF SPECIFIC RECEPTORS AND OF CYTOCHROME C HEME LYASE.

Bernd Hennig, Helmut Köhler and Walter Neupert
Institut für Biochemie, Universität Göttingen

The individual steps in the transport of mitochondrial proteins to their final submitochondrial destinations may differ for the various proteins (c.f. ref.1). Cytochrome c is assembled in the intermembrane space of mitochondria. It is coded by a nuclear gene, translated as an extramitochondrial precursor, apocytochrome c, on free cytoplasmic ribosomes and post-translationally transported into mitochondria. Its transport pathway is distinct from pathways used by other extramitochondrial precursors (2) and has several remarkable features: 1) The precursor becomes covalently linked to protohemin; 2) It is not proteolytically processed; 3) Its transport into mitochondria does not depend on the energized state of the inner membrane.

With aid of specific antibodies that discriminate between apo- and holocytochrome c we are investigating the details of transport and assembly of cytochrome c in cell-free, reconstituted systems prepared from Neurospora hyphae (2-7).

Previously we reported that in the cell-free system apocytochrome c is rapidly transported into mitochondria and no intermediate accumulation of apocytochrome c in mitochondria is observed (5). However, with the heme analogue deuterohemin reversible inhibition of transport of apocytochrome c has been achieved. Two steps have thus been resolved: 1) Specific binding of apocytochrome c to the mitochondrial surface. 2) Translocation of apocytochrome c across the outer membrane and attachment of heme. We have determined some characteristics of these steps. The results strongly support our suggestion that specific apocytochrome c receptors and a special converting enzyme are involved in the assembly of cytochrome c.

Mitochondria 1983

Apocytochrome c is recognized by specific receptors at the mitochondrial surface.

Apocytochrome c obtained by translation in a cell-free system from Neurospora hyphae firmly binds to mitochondria in the presence of deuterohemin. It is not removed from mitochondria by washing with isotonic buffer. By a series of experiments we have demonstrated that this binding is specific, binding sites are located at the surface of mitochondria, and binding is related to transport of apocytochrome c into mitochondria and conversion to holocytochrome c (5, 6).

First, binding is specific: Bound labelled apocytochrome c is displaced from Neurospora mitochondria by increasing amounts of unlabelled Neurospora apocytochrome c, but not by holocytochrome c. Apocytochromes c from different eucaryotic species displace bound Neurospora apocytochrome c with different efficiency (ref.6 and Tab.I). This efficiency is not correlated to the net charge of the various apocytochromes. Rather it seems to be related to an adressing signal (8) which might recognize the receptor. This assumption is supported by the finding that apocytochrome c from the procaryotic species Paracoccus denitrificans does not displace bound Neurospora apocytochrome c from Neurospora mitochondria despite the close homology of Paracoccus cytochrome c with eucaryotic cytochromes c. It is remarkable that a hydrophobic sequence which may form the addressing signal and which is invariably present in the C-terminal half of all eucaryotic cytochromes c, is lacking in Paracoccus cytochrome c .

Second, apocytochrome c is bound at the surface of mitochondria: When mitochondria with bound apocytochrome c were gently treated with trypsin, the apocytochrome c associated with mitochondria was degraded whereas endogenous holocytochrome c contained in the intermembrane space remained unaffected (5).

Third, binding is related to transfer of apocytochrome c into mitochondria: A mixture of deuterohemin-treated mitochondria containing bound apocytochrome c and of mitochondria not pretreated with deuterohemin was prepared and incubated under conditions that were appropriate for uptake and concersion of apocytochrome c. It was observed that apocytochrome c was not detached from the deuterohemin-treated mitochondria and was not transferred to the non-inhibited mitochondria where it would have been converted to holocytochrome _c_. However, when protohemin had been added to the mixture, conversion of apocytochrome _c_ was taking place. Control experiments showed that the non-inhibited mitochondria retained their activity for transport and conversion of apocytochrome _c_ in the mixture: when freely dissolved, labelled apocytochrome _c_ was added to the mixture, the added apocytochrome c was converted in contrast to the bound apocytochrome _c_ (6).

Tab. I. Specificity of apocytochrome c binding to mitochondria.

Unlabelled protein used for displacement of bound labelled apocytochrome _c_ — Species	Net charge at pH 7	Relative affinity (% displacement of labelled apocytochrome _c_)
Neurospora apocytochrome _c_	+4	100
Horse apocytochrome _c_	+9	70
Rat apocytochrome _c_	+9	70
Yeast iso-1-apocytochrome _c_	+8	70
Parsnip apocytochrome _c_	+7	50
Paracoccus denitrificans apocytochrome _c_	-4	0
N-terminal fragment (Pos.-5/+59) of _Neurospora_ apocytochrome _c_	+4	0
Neurospora holocytochrome _c_	+7	0
Polylysine (1 nmol/ml) (average chain length 60 residues)	+60	0

Neurospora ^{3}H-apocytochrome _c_ was synthesized by cell-free translation and bound to deuterohemin-treated _Neurospora_ mitochondria. Samples were mixed with 10 - 10^4 pmoles unlabelled apocytochrome _c_/ml and incubated 15 min at 25°C for equilibration between free and bound apocytochrome _c_. Mitochondria were reisolated by centrifugation, lysed with Triton X-100 and the ^{3}H-apocytochrome _c_ was adsorbed onto Protein A-agarose using specific antibodies. The immunoprecipitate was subjected to electrophoresis by SDS-PAGE and radioactivity determined in gel slices. The displacement measured at 10 nmol unlabelled protein/ml is shown (c.f. also ref. 6).

In order to determine precisely the binding parameters it is necessary to have sufficiently large amounts of apocytochrome *c* with known specific radioactivity. Reductive methylation of apocytochrome *c* by ^{14}C-formaldehyde and cyanoborohydride results in modification of all but a few lysine residues by mono- and dimethylation. This was found to be a convenient method as methylation does not distort binding (6, 7).

Mitochondrial binding sites were titrated employing this apocytochrome *c* until complete saturation was obtained. Free and bound apocytochrome *c* was determined after equilibration. The results were transformed into a Scatchard plot in order to calculate binding parameters. A biphasic curvature is observed demonstrating the presence of two kinds of binding sites (ref.6 and Fig. 1): some 90 pmol high-affinity binding sites per mg mitochondrial protein with an association constant of $K_a = 2.2 \times 10^7\ M^{-1}$ and a larger number of low-affinity binding sites with $K_a = 4 \times 10^6\ M^{-1}$. It is the high-affinity binding sites that represent the specific binding sites involved in transport of apocytochrome *c* into mitochondria. The low-affinity binding sites presumably represent unspecific binding.

The assay for the presence of the high-affinity binding sites is a suitable experimental approach for defining the binding and for isolation of the corresponding receptor. The interaction between apocytochrome *c* and its mitochondrial receptor appears to be essentially hydrophobic. The presence of very low amounts of non-ionic detergents (0.01 % Triton X-100), that did not solubilize the mitochondrial membranes, significantly reduced binding of apocytochrome *c* to mitochondria, but binding was recovered after removal of the detergent from the mitochondrial suspension. In contrast, neither the polycationic peptide polylysine (1 nmol/ml) nor low concentrations of the divalent cations Mg^{2+} and Ca^{2+} (1-5 mmol/l) did interfere with apocytochrome *c* binding. Gentle treatment of mitochondria with trypsin completely abolished the high-affinity binding (Fig.1). This indicates that the receptor is a protein.

Extraction of the receptor from mitochondria has been achieved using the detergent octylglucoside (0.8-1.7 %). When the extracted material was inserted into liposomes prepared from asolectin, the specific high-affinity binding sites were recovered. High-affinity binding of apocytochrome *c* is not an intrinsic property of the liposomes themselves. Differential extraction of mitochondria and purification of the apocytochrome *c* receptor by column chromatography is now under study and so far has succeeded up to a 45-fold enrichement.

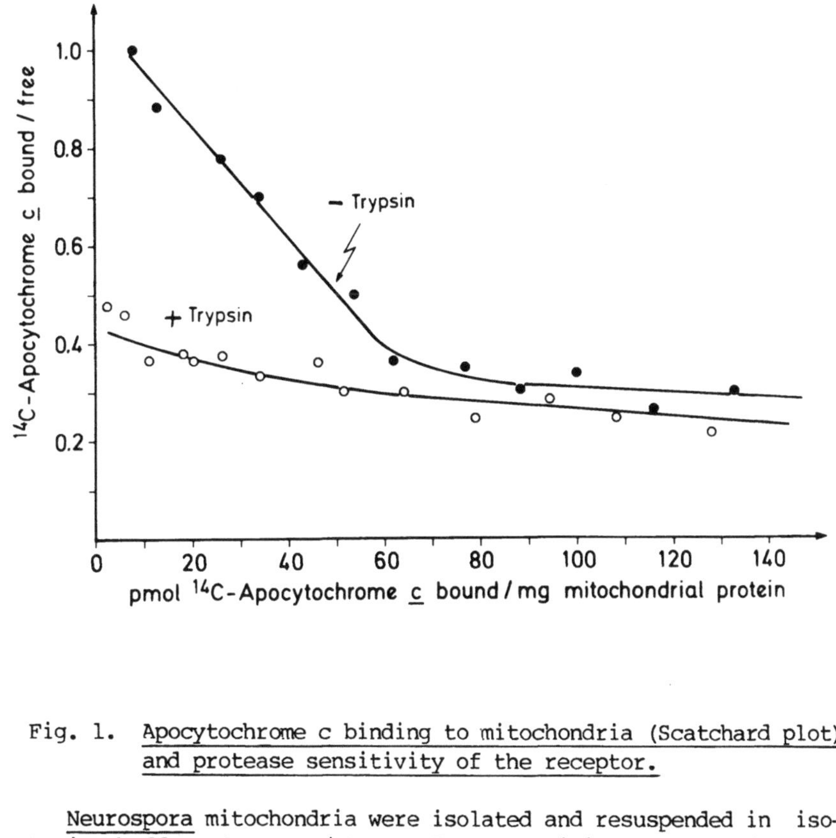

Fig. 1. Apocytochrome c binding to mitochondria (Scatchard plot) and protease sensitivity of the receptor.

Neurospora mitochondria were isolated and resuspended in isotonic buffer (sucrose/MOPS pH 7.5) containing 30 nmol deuterohemin/ml. Samples were titrated with increasing amounts of Neurospora ^{14}C-methylated apocytochrome c (0.9 Ci/mmol protein). After 15 min at 25°C mitochondria were spun down. Free and bound apocytochrome c was determined by scintillation counting of radioactivity in the supernatant and the mitochondria, respectively. Trypsin sensitivity of binding sites was investigated by treatment of mitochondria with 40 ug trypsin/ml x mg mitochondrial protein for 15 min at 25°C. After addition of soybean trypsin inhibitor mitochondria were titrated with apocytochrome c as above.

Heme attachment to apocytochrome c is mediated by an enzyme.

The conversion of apo- to holocytochrome c, i.e. the covalent attachment of protohemin to apocytochrome c, does not proceed spontaneously. This reaction rather requires certain conditions which strongly suggest that it is catalyzed by an enzyme. This enzyme must be either contained as a soluble protein in the intermembrane space or located at the inner side of the outer membrane (5, 9).

First, the insertion of protohemin in cytochrome c is theoretically possible in two alternative ways, but actually takes place stereospecifically (Fig. 2): the vinyl group at carbon atom C 3 of protohemin is only attached to the cysteine at position 17 of cytochrome c and the vinyl group at carbon atom C 8 to the cysteine at position 14. Only this particular stereospecific insertion leads to the native, functional conformation of holocytochrome c (10).

Second, certain heme analogues, which completely lack vinyl groups and therefore cannot be linked to apocytochrome c, compete with the physiological substrate, protohemin, and inhibit conversion. In addition, there is a remarkable structural specificity determining whether a heme derivative competes with protohemin or does not interfere with conversion (Fig. 2): Deuterohemin and mesohemin efficiently compete with protohemin. In contrast, hematohemin, which has two hydroxyethyl sidechains, does not compete at all with protohemin and does not interfere with conversion. A similar but distinct competition of porphyrin substrates is known with the ferrochelatase (11). It seems as if the hematohemin would not fit into the substrate binding site of the putative converting enzyme.

Fig. 2. Stereospecificity and substrate specificity for covalent attachment of the heme group to apocytochrome c.

^{3}H-apocytochrome c was synthesized in a cell-free system prepared from *Neurospora* hyphae. Mitochondria were preincubated with the various heme derivatives (10 nmol/ml) for 5 min at 25°C and then mixed with the postribosomal supernatant containing the ^{3}H-apocytochrome c. ^{3}H-holocytochrome c formed after 30 min at 25°C was immunoprecipitated and radioactivity determined as described in Tab. I.

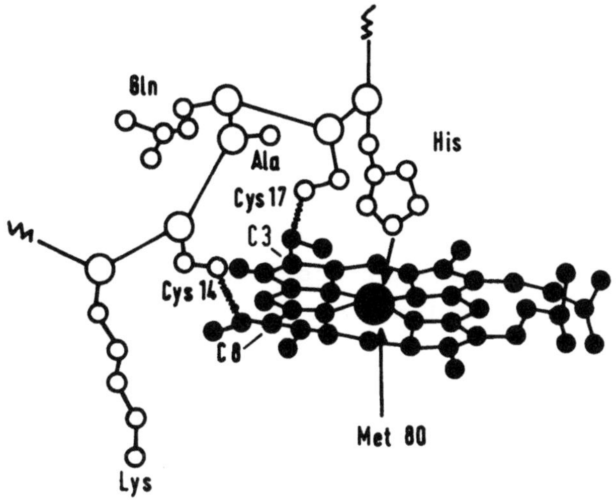

Heme derivative added	Substitution at C 3	at C 8	Holocytochrome c formed (%)
None (= control)			100
Protohemin	$-CH{=}CH_2$	$-CH{=}CH_2$	100
Deuterohemin	-H	-H	20
Mesohemin	$-CH_2-CH_3$	$-CH_2-CH_3$	19
Hematohemin	$-CH(OH)-CH_3$	$-CH(OH)-CH_3$	95
Pemptohemin	-H	$-CH{=}CH_2$	20
Isopemptohemin	$-CH{=}CH_2$	-H	20
Deuterohemin + Protohemin			86
Deuterohemin + Protoporphyrin IX			18

Third, the simultaneous presence of two vinyl groups at the heme substrate is required for linkage to apocytochrome c . The two heme derivatives, pemptohemin and isopemptohemin, have only a single vinyl group at the carbon atoms C 8 and C 3, respectively. Theoretically, covalent attachment of pemptohemin to apocytochrome c via a thioether linkage between the single vinyl group at C 8 and the cysteine 14 would appear possible, as well as of isopemptohemin via the vinyl group at C 3 and the cysteine 17. In fact, neither pemptohemin nor isopemptohemin are used as substrates for the attachment to apocytochrome c but rather inhibit conversion like deuterohemin (Fig. 2).

Furthermore, the presence of two vinyl groups in the porphyrin system is a necessary requirement, but in addition also the iron must be present in the porphyrin molecule: protoporphyrin IX is not used as a substrate (Fig. 2) although it is otherwise identical to protohemin and artificially can be introduced in cytochrome c (12).

Tab. II. Requirement of a cytosolic factor for conversion.

Neurospora mitochondria were incubated with deuterohemin (10 nmol/ml) and mixed with a postribosomal supernatant containing ^{3}H-apocytochrome c. After incubation for 15 min at 25°C mitochondria were isolated by centrifugation and resuspended in the different media as indicated. Formation of ^{3}H-holocytochrome c was determined as described in Fig. 3. The amount is given as counts/ min x ml assay x mg mitochondrial protein:

	Treatment of mitochondria	Holocytochrome c formed
1.	Resuspension in unlabelled postribosomal cytosol	
	+ protohemin (10 nmol/ml)	3,230
2.	Resuspension in isotonic buffer (sucrose/MOPS, pH 7.5)	
	+ protohemin (10 nmol/ml)	0
3.	Resuspension in isotonic buffer (sucrose/MOPS, pH 7.5)	
	+ protohemin (10 nmol/ml)	
	+ cytosolic extract	2,045

Finally, a cytosolic cofactor is required to promote conversion. When deuterohemin-treated mitochondria associated with bound apocytochrome c were resuspended in unlabelled cytosol with addition of protohemin, conversion was resumed. However, when they were resuspended in defined isotonic buffer and excess protohemin was added, no conversion took place. Conversion was then resumed only after addition of a deproteinated extract prepared from the extramitochondrial cytosol (Tab. II). A similar finding was previously reported by others (9). We assume that the cytosolic extract contains a cofactor which is required by the putative enzyme to activate either the vinyl groups of the protohemin or the cysteine SH-groups of apocytochrome c in order to initiate the subsequent covalent coupling.

The putative enzym has to be named cytochrome c heme lyase according to the recommendations given by the IUPAC/IUB Committee on Enzyme Nomenclature. Experiments to locate the enzyme in submitochondrial fractions or even to isolate it, have failed as yet. The converting activity has been lost whenever the intact structure of mitochondria was disrupted by sonication, osmotic swelling - shrinking or lysis with non-ionic detergents.

Concluding remarks: Transfer of apocytochrome c across the outer membrane of mitochondria.

Protohemin is covalently attached to apocytochrome c within mitochondria. The reaction does obviously not take place at the mitochondrial surface. Otherwise holocytochrome c would be formed at the cytoplasmic side of the outer membrane and holocytochrome c but not apocytochrome c would be translocated across the outer membrane. This is clearly not the case. The findings unequivocally demonstrate that apocytochrome c and not holocytochrome c competes for binding and transport (2, 5, 6).

On the other hand, covalent attachment of protohemin to apocytochrome c and translocation of the protein across the outer membrane must be intimately coupled events. When conversion of apo- to holocytochrome c is prevented by deuterohemin, apocytochrome c does not accumulate in the intermembrane space but remains firmly bound to the receptor sites (5). Transport into mitochondria only proceeds when conversion is resuming.

Considering these two aspects we have arrived at the following hypothesis for the assembly of cytochrome c: Binding of apocytochrome c to its receptor may arrange the precursor at the outer membrane in such away that it becomes competent for transfer across the membrane. That part of apocytochrome c, which carries the heme binding region, might then protrude through the outer membrane, perhaps via a pore close to the receptor or via a channel formed by the receptor itself. Thus the cysteine SH-groups of the heme-binding region might become exposed at the inner side of the outer membrane where they would be accessible to the converting enzyme. Attachment of protohemin by cytochrome c heme lyase would force the polypeptide to refold around the heme group. This refolding is a spontaneous process (13, 14). It may pull the polypeptide chain of cytochrome c completely across the outer membrane. It is known that refolding of a polypeptide around the heme group proceeds with a half-time of considerable length (14, 15). Refolding of the polypeptide chain after heme attachment and concomitant completion of its transfer across the outer membrane may therefore be the rate limiting step in the assembly process.

ACKNOWLEDGEMENTS

This work was supported by the Deutsche Forschungsgemeinschaft (grant He 875/4). We are grateful to Leopoldine Sander for skillful technical assistance.

REFERENCES

1. Hennig, B. and Neupert, W. (1983): Assembly of mitochondrial proteins. In: Horizons in Biochemistry and Biophysics, vol.7 , (eds. E. Quagliariello, F. Palmieri and T.P.Singer), J. Wiley / Chichester, New York (in press)

2. Zimmermann, R., Hennig, B., and Neupert, W. (1981): Different transport pathways of individual precursor proteins in mitochondria. Eur. J. Biochem. 116, 455-460

3. Korb, H. and Neupert, W. (1978): Biogenesis of cytochrome c in Neurospora crassa. Synthesis of apocytochrome c, transfer to mitochondria,

and conversion to holocytochrome c. Eur. J. Biochem. 91, 609-620

4. Zimmermann, R., Paluch, U., and Neupert, W. (1979): Cell-free synthesis of cytochrome c. FEBS Letters 108, 141-146

5. Hennig, B. and Neupert, W. (1981): Assembly of cytochrome c. Apocytochrome c is bound to specific sites on mitochondria before its conversion to holocytochrome c. Eur. J. Biochem. 121, 203-212

6. Hennig, B., Köhler, H., and Neupert, W. (1983) : Receptor sites involved in posttranslational transport of apocytochrome c into mitochondria: Specificity, affinity and number of sites. Proc. Natl. Acad. Sci. USA 80, 4963-4967

7. Hennig, B. and Neupert, W. (1983) : Biogenesis of cytochrome c in Neurospora. Methods in Enzymology 97 (in press)

8. Matsuura, S., Arpin, M., Hannum, C., Margoliash, E., Sabatini, D.D., and Morimoto, T. (1981): In vitro synthesis and posttranslational uptake of cytochrome c into isolated mitochondria: role of a specific addressing signal in the apocytochrome c. Proc. Natl. Acad. Sci. USA 78, 4368-4372

9. Basile, G., DiBello, C. and Taniuchi, H. (1980): Formation of an iso-1-cytochrome c-like species containing a covalently bonded heme group from the apoprotein by a yeast cell-free system in the presence of hemin. J. Biol. Chem. 255, 7181-7191

10. Takano, T. and Dickerson, R.E. (1981): Conformation change of cytochrome c. II. Ferricytochrome c refinement at 1.8 A and comparison with the ferrocytochrome c structure. J. Mol. Biol. 153, 95-115

11. Porra, R.J. and Jones, O.T. (1963): Studies on ferrochelatase. 2. An investigation of the role of ferrochelatase in the biosynthesis of various haem prosthetic groups. Biochem. J. 87, 186-192

12. Brems, D.N., Liu, Y.C. and Stellwagen, E. (1982): The conformational transition of horse heart porphyrin c. J. Biol. Chem. 257, 3864-3868

13. Stellwagen, E. (1968): The reversible unfolding of horse heart ferricytochrome c. Biochem. 7, 2893-2898

14. Leutzinger, Y. and Beychok, S. (1981): Kinetics and mechanism of heme induced refolding of human a-globin. Proc. Natl. Acad. Sci. USA 78, 780-784

15. Ridge, J.A., Baldwin, R.L., and Labhardt, A.M. (1981): Kinetics of the fast and slow refolding reactions of iron(III) cytochrome c. Biochem. 20, 1622-1630

MITOCHONDRIA AND THE IMPORT PROBLEM. GENES FOR IMPORTED SUBUNITS OF THE UBIQUINOL-CYTOCHROME *c* REDUCTASE

L.A.Grivell, A.P.G.M.Van Loon, M.De Haan, A.C.Maarse and A.Dekker

Section for Molecular Biology, Laboratory of Biochemistry, University of Amsterdam, Kruislaan 318, 1098 SM Amsterdam, The Netherlands

Introduction

Most mitochondrial proteins are encoded by nuclear genes and have to be imported by the organelle after their synthesis in the cytoplasm. The path(s) followed by several imported proteins are now reasonably well defined (see ref. 1 for review) and this leaves two major areas of ignorance. First, how does the cell tag proteins for import into the mitochondrion rather than into any other organelle? Second, how is the expression of genes coding for such proteins coordinated and how does the nucleus control mitochondrial biogenesis?

The ubiquinol-cytochrome *c* reductase (complex III) of the mitochondrial inner membrane is an attractive model system for the study of both questions. In yeast, this complex consists of 8 to 10 subunits of which all but one are encoded by nuclear genes (Table I). Like other respiratory enzymes, levels of the complex respond to the need for mitochondrial function and steady-state levels of the constituent subunits change in parallel, suggesting that they are subject to coordinate regulation [7,8].

We have isolated genes coding for the 11, 14, 17, 40 kDa and FeS subunits of the complex [8; Van Loon, A.P.G.M.,

Abbreviations: kb, kilo base pair(s); kDa, kilo Dalton(s); mRNA, messenger RNA; SDS, sodium dodecylsulphate.

Mitochondria 1983

TABLE I

BIOSYNTHESIS OF COMPLEX III IN YEAST

Subunit	Site of synthesis	Precursor extension (kDa)*	Ref.
44 kDa	Cytoplasm	0.5	2
40 kDa	Cytoplasm	0.5	2
Cyt. b	Mitochondrion	None	3
Cyt. c_1	Cytoplasm	6	4
FeS	Cytoplasm	2	5
17 kDa	Cytoplasm	9	2
14 kDa	Cytoplasm	Undetectable	2
11 kDa	Cytoplasm	Undetectable	2
<11 kDa**	Cytoplasm	Not determined	6

*Estimated from electrophoretic mobility.

**Two incompletely characterized polypeptides with M_r 5 and 4 kDa.

manuscript in preparation]. Our hope is that their characterization will give clues to the ways in which their expression is regulated and will lead to the identification of the topogenic elements responsible for directing their products to the mitochondrion. A preliminary report of our results is presented.

Features of the cloned genes

Restriction maps of the genes under study are presented in Fig. 1. The cloned inserts range in length from approximately 3 to 6 kb. With the exception of a small part of p209 (17 kDa), they are co-linear with unique segments of chromosomal DNA. Physical mapping of chromosomal re-integrants for 11, 14, 17 and 40 kDa genes shows there to be only a single copy of each per haploid genome; for the FeS gene, mapping is still in progress. In common with other coordinately-expressed genes in yeast, the genes are not clustered and are even located on different chromosomes. Genetic mapping of chromosomal re-integrants, carried out according to the procedure devised by Falco et al. [9] puts the 11 and 17 kDa and FeS

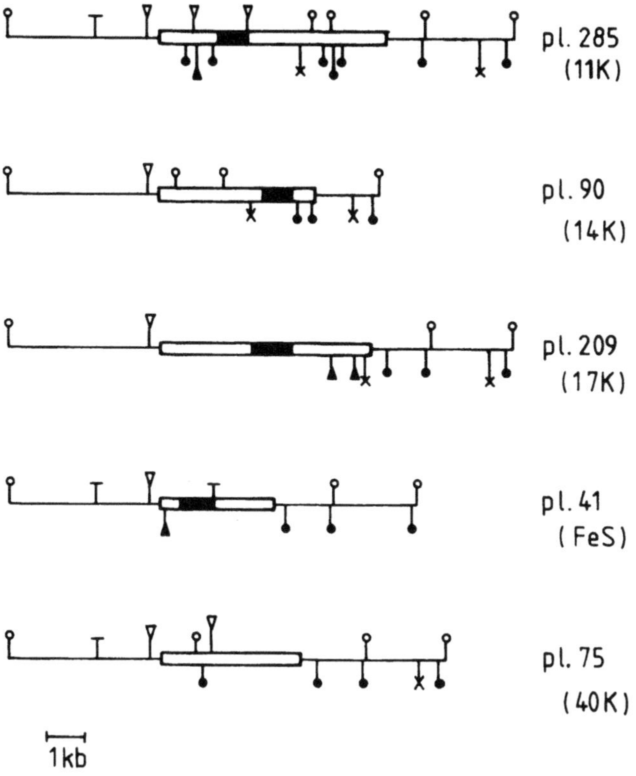

Fig. 1. Restriction maps of recombinant plasmids encoding subunits of complex III. The vector used is pFL1 (see ref. 7 for details) and its sequences are indicated by a thin line. Plasmids 285 and 90 lack (part of) the URA3 region. However, in plasmid 285 the gene is still functional. The black areas represent the sequences coding for a subunit of the complex. Restriction sites are indicated by the following symbols: , EcoRI; , PvuII; , SalI; , HindIII; , KpnI; , XbaI.

genes on chromosomes XII, VII and VIII, respectively, while the 14 and 40 kDa genes are so far ambiguously located (Van Loon, A.P.G.M., manuscript in preparation).

Positions of sequences coding for 11, 14, 17 kDa and FeS subunits are also shown in Fig. 1. They were determined by a combination of several approaches, including comparative mapping of overlapping clones, transcript mapping and *in vitro* translation of messenger RNA (mRNA) selected by hybridization to isolated restriction fragments. In most cases cloned inserts contain transcribed regions additional to the gene of interest, but their identity has not been investigated.

Gene regulation: Discoordinate synthesis of subunits in response to increased gene dosage

Our previous work [7,8] has shown that steady-state levels of the subunits of complex III are markedly reduced during growth of cells on glucose. Levels of translatable mRNAs show parallel changes, suggesting that synthesis of the subunits is subject to a common regulatory mechanism. By analogy with other glucose-repressible genes in yeast [10], this could be at the level of transcription. To investigate the mechanism involved, cloned sequences for 11, 17, 40 kDa and FeS subunits have been re-introduced into wild-type cells by transformation with free-replicating plasmids and the effects of increased gene dosage on complex synthesis and assembly analysed. Three questions were of especial interest:

1. Does increased dosage lead to over-production of the protein involved? If so, what is the effect of this on the synthesis of other subunits and does the cell attempt to compensate for over-production?

2. What is the fate of over-produced protein?

3. What are the effects - if any - of over-production on mitochondrial function?

The results obtained can be briefly summarized as follows:

1. For the 17 kDa gene no increase in subunit level is observed; results obtained are discussed below. For the 11, 40 kDa and FeS genes, increase in dosage leads to a 5-20 fold increase in steady-state level of the subunit concerned (Fig. 2), which is somewhat lower than expected from measurements of plasmid copy number ($\gg$30). For the FeS subunit, increase in protein level corresponds roughly with the level of mRNA (data not shown). For the 11 kDa subunit, however, quantitative studies [8] suggest that a lower translational efficiency of over-produced mRNA and an increased turnover of excess protein are responsible for lowered steady-state levels. Both effects

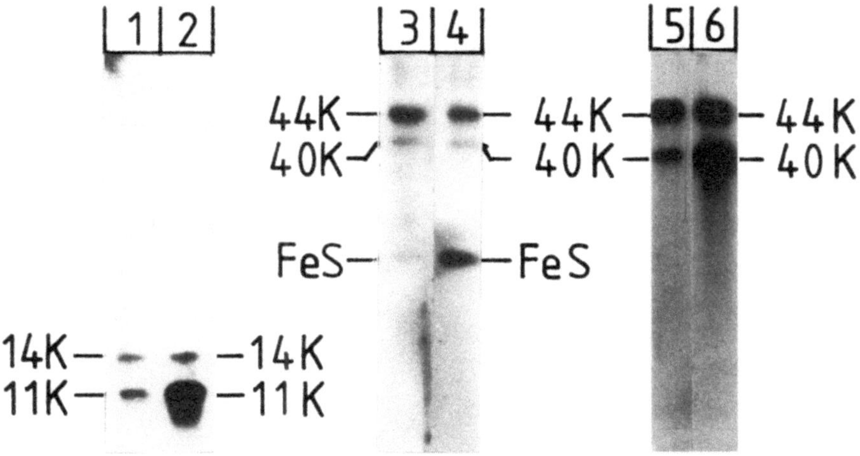

Fig. 2. Increased steady-state concentrations of subunits in yeast cells in response to increased gene dosage. Wild-type cells of the Saccharomyces cerevisiae strain DL1 [8] were transformed with recombinant plasmids carrying genes for subunits of complex III (see Fig. 1) and grown on 0.67% yeast nitrogen base containing 2% galactose plus 20 µg per ml of histidine and leucine. Total cellular proteins were then resolved on 10-17% (lanes 1 and 2) or 10% (lanes 3-6) SDS-polyacrylamide gels. Proteins were blotted to nitrocellulose filters and subunits of complex III detected as described [4], using antibodies against the 11 and 14 kDa subunits (lanes 1 and 2), against the FeS, 40 and 44 kDa proteins (lanes 3 and 4), or against the 40 and 44 kDa proteins (lanes 5 and 6). The plasmids used for transformation were p285, encoding the 11 kDa protein (lane 2), p41 which contains the gene for the FeS subunit (lane 4), or p75 which codes for the 40 kDa protein (lane 6). Lanes 1, 3 and 5 contain lysates of cells transformed with the vector pFL1.

may form part of a compensatory mechanism (cf. refs 11,12). Comparable data for the 40 kDa subunit are as yet lacking.

2. Over-produced protein is associated with mitochondria. This is most clearly seen for 11 and 40 kDa subunits, for which the bulk of the excess protein co-purifies with the mitochondrial fraction (Fig. 3). Proteinase K protection experiments, carried out with the 40 kDa subunit (data not shown), rule out aspecific interaction and suggest that these proteins have in fact been imported. Results for the FeS subunit are less clear-cut due to a tendency of this protein to leak from damaged mitochondria. However, the fact that transformants contain only mature subunits, while processing can be obtained by action of the chelator-sensitive protease located in the mitochondrial matrix (Van Loon, A.P.G.M.,

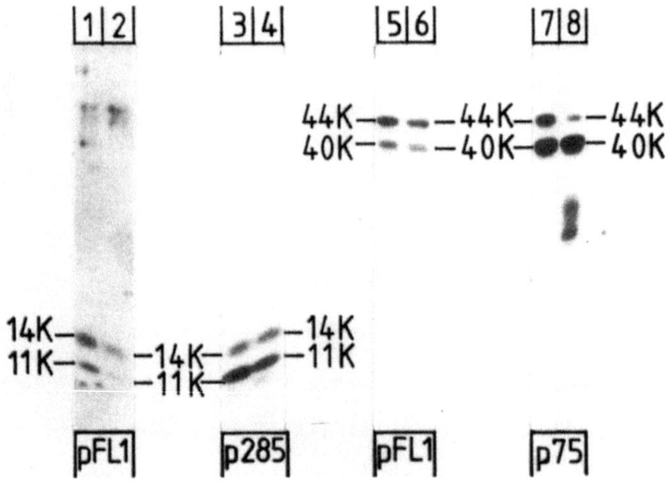

Fig. 3. Over-produced subunits are associated with the mitochondrial fraction. Yeast cells transformed with the vector pFL1 (lanes 1, 2, 5 and 6), p285 (lanes 3 and 4), or p75 (lanes 7 and 8) were grown as described in the legend to Fig. 2. The mitochondrial pellet, prepared by differential centrifugation, was washed repeatedly as described [8]. Samples of total cell lysates (lanes 1, 3, 5 and 7) or of the mitochondrial fractions (lanes 2, 4, 6 and 8) were resolved on a 10-17% (lanes 1-4) or a 10% (lanes 5-8) SDS-polyacrylamide gel. Subunits were visualized on protein blots as described in the legend to Fig. 2 using antisera against the 11 and 14 kDa proteins (lanes 1-4) or against the 40 and 44 kDa subunits (lanes 5-8). Levels of the 14 kDa protein (lanes 1-4) or of the 44 kDa subunit (lanes 5-8) are used as internal standards in each fraction.

manuscript in preparation), suggests that import has taken place.

3. In no case, as shown by Figs 2 and 4, does over-production affect levels of other subunits in the complex, implying that effects on their synthesis, transport or assembly are absent. Further, as shown by Table II, transformants exhibit normal growth rates on non-fermentable media so that over-production cannot grossly interfere with mitochondrial function.

As stated above, transformants carrying additional copies of the 17 kDa gene do not contain increased levels of this subunit. Fig. 5 shows that this is not due to lack of transcription, although the increase in mRNA concentration is

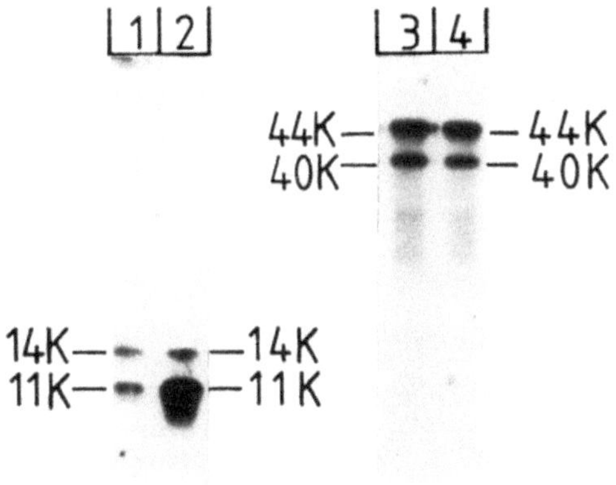

Fig. 4. An increase in concentration of the 11 kDa subunit does not influence steady-state levels of other subunits. Transformed cells were grown as described in the legend to Fig. 2. Equal amounts of total cellular protein of wild-type cells that contain the vector pFL1 (lanes 1 and 3) or p285 (lanes 2 and 4) were sized in different lanes of a 10-17% (lanes 1 and 2) or a 10% (lanes 3 and 4) SDS-polyacrylamide gel. Protein blots were incubated with antisera against the 11 and 14 kDa proteins (lanes 1 and 2) or against the 40 and 44 kDa subunits (lanes 3 and 4) and subunits were visualized with ^{125}I-labelled protein A.

significantly lower than seen for the 11 kDa subunit. Instead, main limitation of over-production seems to lie at a later stage. As shown by Fig. 6, pulse-chase analysis of transformants reveals a higher-than-normal synthesis of the precursor to the 17 kDa subunit, but this is not matched by an increase in the level of mature subunit. A rapid and specific degradation

TABLE II

GROWTH RATES OF TRANSFORMANTS ON NON-FERMENTABLE MEDIA

Transformed cells were grown on selective medium with glycerol, ethanol and lactate as carbon sources and then transferred to yeast extract with the same carbon sources. Generation times were measured within the first 7-8 h after transfer to avoid possible effects of different levels of the _URA3_ gene product on growth rates.

Transformant	Generation time (min)
pFL1	297 ± 12 (n = 6)
11 kDa	308 ± 4 (n = 4)
FeS	300 ± 23 (n = 4)
40 kDa	296 ± 18 (n = 4)

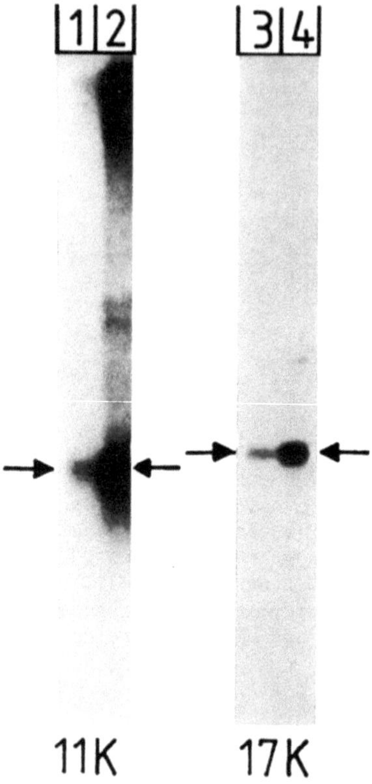

Fig. 5. Increased steady-state concentrations of the mRNA for the 17 kDa subunit in cells containing extra copies of its gene. Wild-type cells containing the vector pFL1 (lanes 1 and 3), p285 (lane 2), or p209 (lane 4) were grown as described (see Fig. 2). Total RNA was prepared, sized on an agarose gel and blotted to DBM-paper. Equal amounts of total cellular RNA were used in each slot. mRNAs encoding the 11 kDa protein (lanes 1,2) or 17 kDa protein (lanes 3 and 4) were detected using gene-specific probes. Hybridization to high molecular weight material in lane 2 is due to the presence of plasmid DNA.

of precursor would account for this, but degradation of excess, mature subunit cannot be excluded. Such degradation could be related to the highly unusual character of this protein (see below), to an inability to import excess subunit into mitochondria, or to lack of integration into complex.

These findings lead to the conclusion that synthesis of the various subunits of the complex is not tightly coupled. They tend to exclude models of control involving common positive or negative regulators, as found for genes involved in sugar or amino acid metabolism [13,14]. How comparable levels of synthesis are achieved and how these levels are modulated in response to growth consitions are thus open questions for the future.

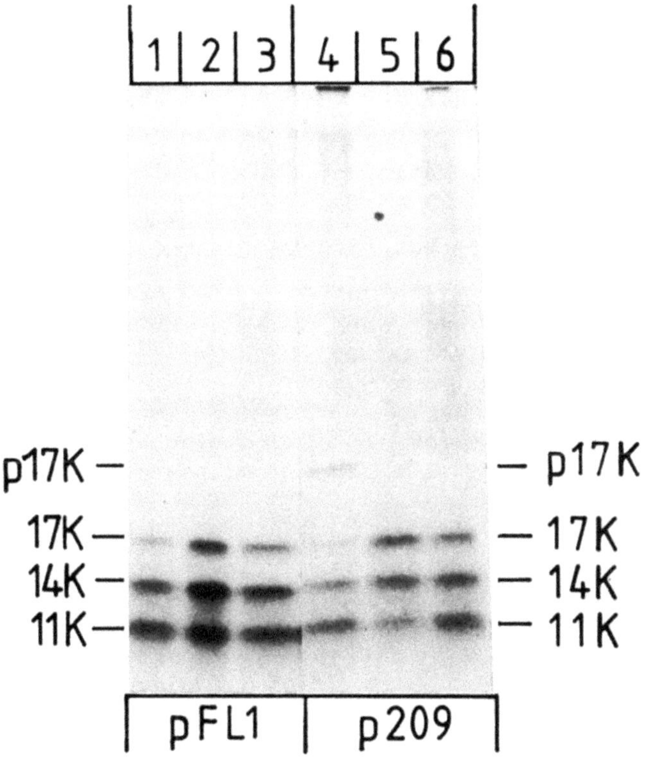

Fig. 6. Increased synthesis of only the precursor of the 17 kDa subunit (p17K) and not of the mature subunit. Transformed wild-type yeast cells were grown as described (see Fig. 2). [^{35}S]Methionine was added to samples of the logarithmically growing cultures and proteins were pulse-labelled (lanes 1 and 4) at 28°C for 30 min (1/7th to 1/10th of the generation time). After addition of unlabelled methionine and cas-amino acids to 10 mM and 0.5% (w/v), respectively, incubation was continued for 90 min (lanes 2 and 5), or 5 h (lanes 3 and 6). Samples were withdrawn and of each 4.5×10^6 cpm of hot trichloroacetic acid-insoluble material was used for quantitative immunoprecipitation with an antiserum against the 17, 14 and 11 kDa subunits. The precipitates were resolved on a 10-17% SDS-polyacrylamide gel. A fluorograph of the dried gel is shown.

Protein topogenesis

Most proteins imported by mitochondria are synthesized with extensions that are proteolytically cleaved during or after import. The extensions vary more widely in lengths than the signal peptides of secretory proteins. Their importance for import is demonstrated by the fact that proteins processed *in vitro* are neither bound nor taken up by isolated mitochondria [15]. As yet, however, there is no direct evidence that topogenic signals are confined to the pre-sequence, nor is there an obvious correlation between the size of the extension and the final location of the protein within the mitochondrion.

As a first step towards the identification of topogenic signals, sequences of the genes coding for the 14 and 17 kDa subunits of complex III have been determined. We have previously shown that the 14 kDa subunit is synthesized both *in vivo* and *in vitro* without detectable extensions [2]. The signals responsible for protein routing are thus most likely located internally. In contrast, the 17 kDa subunit is synthesized as a precursor with an apparent molecular weight of 26 kDa so that the extension is predicted to be unusually long in relation to the mature protein [2]. Whether this is in fact the case is discussed more fully below.

The 14 kDa subunit

The amino acid sequence of this subunit, derived from the nucleotide sequence of its gene [16], is shown in Fig. 7. It is predicted to be a predominantly hydrophilic protein of 126 amino acids with a calculated molecular weight of 14,561. Charged residues are distributed fairly evenly throughout its length, but a short apolar segment located between positions 19-38 could be capable of spontaneous insertion into a lipid environment.

We have examined the sequence for regions of similarity with the cleavable extensions of two other imported mitochondrial proteins: ATPase subunit 9 and cytochrome *c* peroxidase, located in the inner membrane and the intermembrane space, respectively. In both proteins, the extensions which may be responsible for protein targetting are amino-terminal, relatively long (66 and 68 amino acids, resp.) and positively-charged residues predominate. For ATPase subunit 9, Viebrock

M P Q S F T S I A R I G D Y I L K S P V
1 10 20

L S K L C V P V A N Q F I N L A G Y K K
21 30 40

L G L K F D D L I A E E N P I M Q T A L
41 50 60

R R L P E D E S Y A R A Y R I I R A H Q
61 70 80

T E L T H H L L P R N Q W I K A Q E D V
81 90 100

P Y L L P Y I L E A E A A A K E K D E L
101 110 120

N I E V S K
121

Fig. 7. Amino acid sequence of the 14 kDa subunit of yeast complex III. The sequence shown is that predicted from the DNA sequence of the gene [16].

et al. [17] have suggested that a clustering of positive charge might function in driving the protein into and across the inner membrane,
in the direction of the potential gradient (positive outside for mitochondria). For cytochrome *c* peroxidase short stretches of positive residues flank a hydrophobic region that is thought to span the inner membrane and a function of these as stop-transfer sequences have been proposed [18]. No part of the 14 kDa protein bears any obvious similarity to these pre-sequences. The subunit carries only a low net positive charge and those charges present are neither clustered, nor are they positioned in any regular fashion relative to the predicted secondary structure [16]. The mechanism by which all three proteins are addressed to and traverse the mitochondrial outer membrane is thus unclear. Furthermore, import by electrophoresis of protein down an electrochemical gradient seems unlikely, but cannot be entirely discarded until the location of the 14 kDa protein within complex III is more accurately known.

The 17 kDa subunit

The amino acid sequence of the precursor polypeptide predicted from the DNA sequence of the gene for this subunit is given in Fig. 8. It has a length of 145 amino acids, corresponding to a calculated molecular weight of 17,394 and is clearly an extremely unusual protein. 59 residues are either glutamic or aspartic acids and the most prominent feature is a stretch of 25 acidic residues between positions 49 and 73. This stretch and the conserved sequences downstream of it immediately identify the 17 kDa protein as a homologue of the so-called 'hinge' protein, a 9 kDa subunit of complex III in beef, which binds strongly to cytochrome $\underline{c}_1$ and is thought to mediate binding of cytochrome $\underline{c}$ to the complex [19]. Lack of amino acid sequence information on the mature subunit precludes an exact definition of its limits within the sequence shown. Homology with the beef-heart protein begins at

M D M L E L V G E Y W E Q L K I T V V P
1 10 20

V V A A A E D D D N E Q H E E K A A E G
21 30 40

E E K E E E N G D E D E D E D E D E D D
41 50 60

D D D D D E D E E E E E E V T D Q L E D
61 70 80

L R E H F K N T E E G K A L V H H Y E E
81 90 100

C A E R V K I Q Q Q Q P G Y A D L E H K
101 110 120

E D C V E E F F H L Q H Y L D T A T A P
121 130 140

R L F D K L K
141

Fig. 8. Amino acid sequence of the precursor of the 17 kDa 'hinge' protein of yeast complex III. The sequence shown is that predicted from the gene (Van Loon, A.P.G.M. et al., manuscript in preparation).

position 48 and continues as far as the C-terminus. However, removal of a pre-sequence at any point after position 3 would result in a protein without methionine and this is not in accordance with its labelling behaviour. This raises some intriguing questions concerning the identity of the 26 kDa precursor. While the strongly acidic nature of the protein may underlie an anomalous electrophoretic behaviour due to low binding of sodium dodecylsulphate (*cf*. ref. 20), this cannot explain the large difference in apparent size between the precursor and mature subunit. One possibility is that the newly-synthesized protein is firmly associated with other molecules that compensate for its negative charge and such an association could be of importance for transport into the mitochondrion (see below).

The probable association of the 17 kDa subunit with cytochromes *c* and c_1 (*cf*. ref. 4) suggests that it is located on the outer side of the inner membrane. However, pulse-labelling experiments carried out with wild-type cells or mit^- mutants that tend to accumulate precursors to imported mitochondrial proteins have given no indication of import via the two-step pathways described for cytochrome c_1 [21] or intermembrane proteins like cytochrome *c* peroxidase [22]. No part of the sequence resembles the pre-sequence of this latter protein and long apolar segments are lacking. As for the 14 kDa protein, the mechanism by which this subunit is addressed to and traverses the outer mitochondrial membrane is unclear. In this case, however, the preponderance of negative charge creates additional difficulties in the movement of the protein through a phospholipid bilayer and this re-emphasizes the question raised above whether transport of the protein is obligatorily coupled to the movement of an oppositely-charged molecule.

Prospects: The import problem

Synthesis of a functional mitochondrion requires the

coordinate expression of many, perhaps hundreds of nuclear genes and the ordered movement of the proteins they encode to the correct locations within the organelle. We have begun to address the questions still outstanding on both points by isolation and characterization of genes coding for the imported subunits of a single enzyme complex - the ubiquinol-cytochrome c reductase. This complex has the attractive features that it is well defined, that the topography of the major subunits has been established [23] and that in yeast its synthesis and assembly is regulated in response to the need for mitochondrial function. In three of the four cases studied, yeast cells containing extra copies of a gene for one of the subunits of the complex synthesize increased levels of that subunit and the over-produced protein is associated with the mitochondria. In the remaining case, over-production of mature subunit is prevented by some form of compensation at the level of import, processing or specific degradation. In all cases the system promises to be of use in the detailed analysis of the expression of individual genes and of the import process. As shown above, the identity of topogenic elements in imported proteins is unlikely to emerge from simple sequence comparisons: such signals may comprise sequences internal to the mature protein in addition to part or all of the cleavable extensions and they may be specifically created by folding of the precursor protein (cf. ref. 24). This identification must, therefore, be based on assay for function and attempts to do this, using a combination of gene fusion and site-directed mutagenesis, are now under way.

Acknowledgements

We thank Mr R.J.De Groot, Mr E.Van Eijk, Mr G.T.J.Van der Horst, Dr. J.Kreike, Mr R.T.M.J.Vaessen and Mr J.E.Polman for their considerable help in the various phases of the work reported here and Prof. P.Borst and Dr. H.F.Tabak for valuable suggestions and criticism. This work was supported in part by a grant to LAG from The Netherlands Foundation for Chemical Research (SON) with financial aid from The Netherlands Organization for the Advancement of Pure Research (ZWO).

References

1. Schatz, G., Butow, R.A.: Cell 32, 316-318 (1983).

2. Van Loon, A.P.G.M., Kreike, J., De Ronde, A., Van der Horst, G.T.J., Gasser, S.M., Grivell, L.A.: Europ.J.Biochem. (1983) in press.

3. Nobrega, F.G., Tzagoloff, A.: J.Biol.Chem. 255, 9828-9837 (1980).

4. Ross, E., Schatz, G.: J.Biol.Chem. 251, 1991-1996 (1976).

5. Côté, C., Solioz, M., Schatz, G.: J.Biol.Chem. 254, 1437-1439 (1979).

6. Marres, C.A.M.: Subunit Structure and Function of QH_2:Cytochrome c Oxidoreductase, Ph.D.Thesis, Rodopi, Amsterdam (1983).

7. Van Loon, A.P.G.M., De Groot, R.J., Van Eijk, E., Van der Horst, G.T.J., Grivell, L.A.: Gene 20, 323-327 (1982).

8. Van Loon, A.P.G.M., Van Eijk, E., Grivell, L.A.: EMBO J. (1983) in press.

9. Falco, S.C., Li, Y., Broach, J.R., Botstein, D.: Cell 29, 573-584 (1982).

10. Zitomer, R.S., Montgomery, D.L., Nichols, D.L., Hall, B.D.: Proc.Natl.Acad.Sci.U.S. 76, 3627-3631 (1979).

11. Pearson, N.J., Fried, H.M., Warner, J.R.: Cell 29, 347-355 (1982).

12. Osley, M.A., Hereford, L.M.: Cell 24, 377-384 (1981).

13. Johnston, S.A., Hopper, J.E.: Proc.Natl.Acad.Sci.U.S. 79, 6971-6975 (1982).

14. Penn, M.D., Galgoci, B., Greer, H.: Proc.Natl.Acad.Sci.U.S. 80, 2704-2708 (1983).

15. Gasser, S.M., Daum, G., Schatz, G.: J.Biol.Chem. 257, 13034-13041 (1982).

16. De Haan, M., Van Loon, A.P.G.M., Kreike, J., Vaessen, R.T.M.J., Grivell, L.A.: Europ.J.Biochem. (1983) submitted.

17. Viebrock, A., Perz, A., Sebald, W.: EMBO J. 1, 565-571 (1982).

18. Kaput, J., Goltz, S., Blobel, G.: J.Biol.Chem. 257, 15054-15058 (1982).

19. King, J.E.: Advan.Enzymol. 54, 267-366 (1983).

20. Wakabayashi, S., Takeda, H., Matsubara, H., Kim, C.H., King, T.E.: J.Biochem. (Tokyo) 91, 2077-2085 (1982).

21. Ohashi, A., Gibson, J., Gregor, I., Schatz, G.: J.Biol.Chem. 257, 13042-13047 (1982).

22. Reid, G.A., Yonetani, T., Schatz, G.: J.Biol.Chem. 257, 13068-13074 (1982).

23. Karlsson, B., Hovmöller, S., Weiss, H., Leonard, K.: J.Mol.Biol. 165, 287-302 (1983).

24. Matsura, S., Arpin, M., Hannum, C., Margoliash, E., Sabatini, D.D., Morimoto, T.: Proc.Natl.Acad.Sci.U.S. 78, 4368-4372 (1981).

A MODEL SYSTEM FOR THE STUDY OF TRANSPORT OF PROTEINS INTO MITOCHONDRIA

Ersilia Marra, Salvatore Passarella, Margherita Greco, Cecilia Saccone, Ernesto Quagliariello

Istituto di Chimica Biologica e Centro di Studio sui Mitocondri e Metabolismo Energetico CNR, Università di Bari
70126 Bari, Italia

Shawn Doonan

Department of Biochemistry, University College
Cork, Ireland

Introduction

With the exception of a small number of components of the inner membrane, mitochondrial proteins are synthesized in the cell cytosol as completed translation products and subsequently translocated into their proper sites within the mitochondrion.
In the case of matrix proteins this process, at least formally, requires transport of proteins across two membranes. The object of the present paper is to describe a model system that we have developed to study this process and the results that have been obtained by its use. It will also be necessary, however, to examine the extent to which the model system properly reflects the events occurring *in vivo* and to justify its use for gaining insights into the process *in vivo*.

The Model Systems

We have used in our studies the cytosolic and mitochondrial isoenzymes of aspartate aminotransferase (AAT) and malate dehydrogenase (MDH). The initial choice of AAT was based on the fact that the primary structures of both isoenzymes from pig heart are known (1) and hence the possibility arose of

Mitochondria 1983

interpreting the topogenesis of the two isoenzymes in terms of their primary structures.

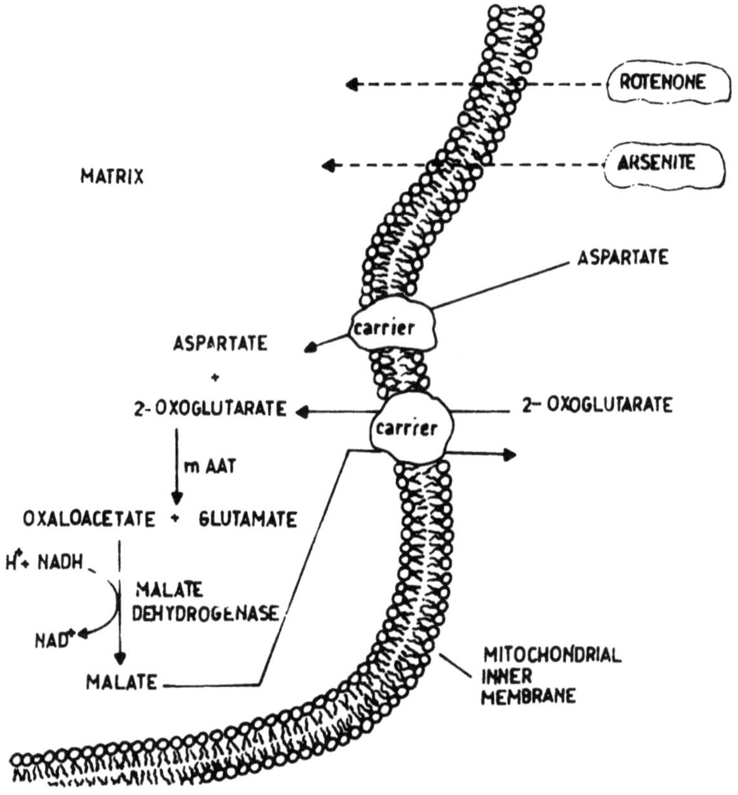

Fig. 1 Measurement of intramitochondrial AAT activity.
Mitochondria are loaded with rotenone and sodium arsenite to prevent oxidation of intraorganellar NAD(P)H by terminal electron transport and of 2-oxoglutarate by oxoglutarate dehydrogenase respectively. Aspartate is then added to the mitochondrial suspension and allowed to accumulate inside after which 2-oxoglutarate is added. Rapid import of this substance provided the substrate pair for AAT and enzyme activity generates oxaloacetate as one of the products. This substance is then reduced by endogenous MDH with consequent oxidation of NAD(P)H. The measured parameter is then the rate of decrease of fluorescence of NAD(P)H which provides a direct measure of AAT activity, the rate limiting step in the process.

Some time ago we developed a system which allowed measurement of uptake of native purified AAT into isolated mitochondria with total exclusion of the cytosolic form (2). The method essentially involves determination of the intramitochondrial activity of the enzyme in the presence and absence of purified enzyme added externally and is described in the legend to Fig. 1.

The type of result obtained is shown in Fig. 2 (Trace A). In trace B is shown the effect of addition of purified mitochondrial AAT to the suspension 1 min before the addition of 2-oxoglutarate. The rate of change of fluorescence was increased by 50% due to uptake of the enzyme into the organelles. In contrast (Trace C) addition of purified cytosolic aspartate aminotransferase (cAAT) to the suspension did not change the observed rate. This is important in two respects. Firstly it confirms that the increased rate of change of fluorescence observed in trace B was not due to any effect of enzyme remaining outside the organelles since this would have occurred equally with external cAAT.

Secondly it established the specificity of the model system in that import of mAAT occurs but import of cAAT does not; this immediately, of course, reflects the situation *in vivo*.

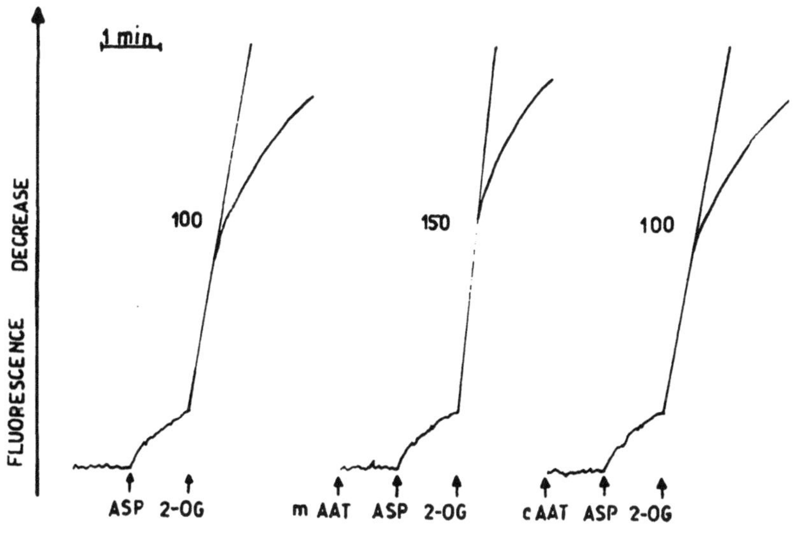

Fig. 2 Selective uptake of mAAT into mitochondria

Subsequently we developed a similar assay system for import of malate dehydrogenase into mitochondria (3).
This assay is in principle simpler since MDH has a single substrate, oxaloacetate. In practice, it was found that, because of the high activity of the endogenous malate dehydrogenase and the relatively slow rate of import of oxaloacetate into mitochondria, uptake of substrate was rate-limiting with native organelles. This problem was overcome by pretreating the mitochondria with sulphite which partially inhibits the endogenous enzyme. Under these conditions, when oxaloacetate is added to the mitochondrial suspension, the rate of change of fluorescence of NAD(P)H provides a true measure of residual enzyme activity; this activity is increased by preincubation of the organelles with mMDH but not by preincubation with cMDH. Hence in this case, as well as with AAT, isolated mitochondria are able to internalize the mitochondrial isoenzyme but not the cytosolic form.

Using these model systems some characteristics of the uptake process were determined and are summarized in Table 1.

TABLE 1

CHARACTERISTICS OF THE UPTAKE OF

ASPARTATE AMINOTRANSFERASE	and MALATE DEHYDROGENASE
1. Hyperbolic dependence of externally added enzyme	Hyperbolic dependence of externally added enzyme
2. High speed of import	High speed of import
3. Dependence on temperature	not determined
4. Protease resistance	Protease resistance
5. Energy dependence	Energy dependence
6. Inhibition by mercurial reagents	Inhibition by mercurial reagents
7. Inhibition by 2-mercapto-ethanol	Inhibition by 2-mercapto-ethanol
8. Inhibition by MDH	Inhibition by AAT
9. Uptake into mitoplasts	Uptake into mitoplasts

From the mechanistic point of view, characteristics 5-8 are of particular interest and will be returned to later. The rate of the process (item 2), which might be considered strange, will also be commented on below.
Closer examination of the protease resistence of the imported AAT (item 4) yielded some interesting results when studied using a different method to measure import. In this method the AAT was radiolabelled by reduction of the apoenzyme-cofactor linkage with sodium borotritiide and uptake of the enzyme assessed by measurement of radioactivity incorporated into the mitochondria after incubation with labelled isoenzyme (4). Using this method it was shown that mAAT interacts with mitochondria in two distinct ways. At high ionic strength only import of enzyme into the organelles was observed. At low ionic strength, however, both import and binding of the enzyme to the outside of the mitochondria occurs. These are clearly different phenomena as shown, for example, by the fact that the compounds tetraphenylboron and bromothymol blue both inhibit external binding but not uptake into the mitocondria (5).
Interestingly, using this system it was shown that whereas imported enzyme was rendered protease-resistant, externally bound enzyme was much more sensitive to proteases than was the native mAAT in solution (6). The implication of this result is that binding of native AAT to the outside of mitochondria induces a conformational change such as to render the enzyme much more sensitive to proteases; this may suggest that a binding event associated with a conformational change is also involved in the uptake process.

Precursors of mitochondrial proteins

Before proceeding to describe some more recent results obtained with the model systems it is important to discuss one aspect in which the model system and the processes <u>in vivo</u> clearly differ, and hence to question the validity of the model system as a true reflection of the processes occurring <u>in vivo</u>.

It is now known that most mitochondrial proteins are synthesized as precursors with molecular weights larger than those of the mature forms (for refs see 7-9). Aspartate aminotransferase is no exception to this as shown by several groups (10-12). We have obtained similar results (13, 14).

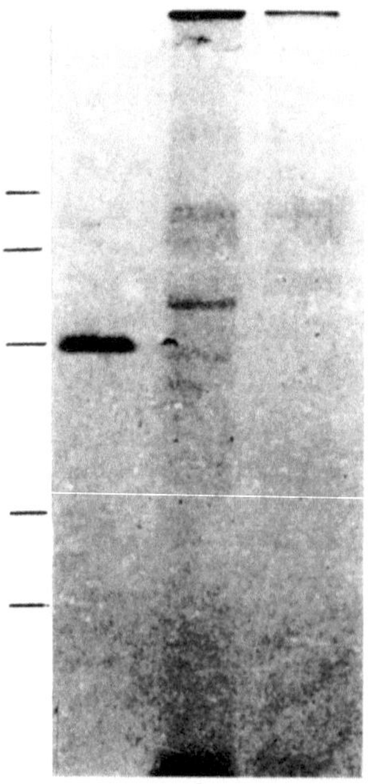

Fig. 3 Fluorography after SDS-PAGE of the in vitro translation products

PolyA$^+$RNA from free or membrane-bound polysomes was translated in vitro and the products were analyzed by immunoprecipitation and SDS-PAGE.

Lane a, mature ^{3}HmAAT immunoprecipitated with specific antibody; lane b, polyA$^+$RNA from free polysomes added and immunoprecipitated with specific antibody; lane c polyA$^+$RNA from membrane-bound polysomes added and immunoprecipitated with specific antibody.

The molecular weight markers are: phosphorilase B (95.2K), BSA (66.2K), ovalbumin (45K), carbonic anydrase (31K), soybean trypsin inhibitor (21.5K).

Free and membrane-bound polysomes were isolated from rat liver by differential centrifugation and were used as sources of polyA$^+$RNA. The polyA$^+$RNA was translated in a cell-free system derived from rabbit reticulocytes and the products examined by SDS-polyacrylamide gel electrophoresis after precipitation with a specific antiserum to mAAT. Fig. 3 shows the results obtained.

Track a contains native radiolabelled mAAT. Track b shows the pattern obtained from the immunoprecipitate after translation of polyA$^+$RNA obtained from free polysomes. A band with molecular weight greater than that of the native enzyme is clearly visible; in a control experiment (not reported) it was found that this band of higher molecular weight was not precipitated by anti-mAAT antiserum in the presence of excess native enzyme. In track c it is shown that the precursor form of mAAT was not synthesized by polyA$^+$ RNA obtained from membrane-bound ribosomes. In another experiment not reported in detail, polyA$^+$ RNA from rat liver polysomes was subjected to polyacrylamide gel electrophoresis and the RNA subsequently eluted from gel slices and tested for its ability to support synthesis of the precursor of mAAT. It was found that the precursor is synthesized from messenger RNA of about 14-15S (1200-1300 bases).

Comparison of the Model System with the Process Occurring *in vivo*

The results quoted above confirm the observation reported by other authors that mAAT is synthesized by free cytosolic polysomes as a precursor of molecular weight greater than that of the native form. The same is true for mMDH (15) and indeed for the vast majority of other mitochondrial proteins examined to date (see 7). It is than legitimate to enquire as to whether uptake of native proteins into mitochondria *in vitro* as studied by us has any relevance to the process occurring *in vivo*. We would maintain that the two processes are extremely similar and moreover that comparison of them throws some light on the role of the extra piece of sequence found in precursor proteins. The following comparison can be made.

a. Specificity. The subcellular localization of proteins is obviously specific. For example, in the case of AAT the two isoenzymes show non-overlapping distributions in the cytosol and mitochondria (16). This specific distribution is reflected by the model systems described here in that uptake by native cytosolic enzyme into the mitochondria does not occur (see Fig. 2).

b. Speed of import. Uptake of native proteins into mitochondria occurs very rapidly, the process being complete in less than 1 min. This was originally seen as a difficulty for the following reason. Many studies have been reported where the uptake into mitochondria of precursor proteins synthesized in cell-free systems was observed (17-20). In all cases, extended incubation periods were used, typically of 30-60 min, and this has led to the view that uptake of precursors into mitochondria in vivo is also a slow process. Recent work has shown, however, that this is not so. For example Jaussi et al. (21) have studied the biosynthesis of mAAT in chicken embryo fibroblasts by pulse labelling of the cells and find that in a subsequent chase the precursor was converted into mature isoenzyme with a half-life of 30-60 sec. Similar rates of uptake of mitochondrial proteins in vivo have been found in other cases (e.g. 22-24). Hence the picture is emerging of an extremely rapid uptake process in vivo the rate being similar to that found in our model system. The reasons for the very slow rate of uptake and processing of precursor proteins in in vitro systems are obscure, but may be related to lack of integrity of the mitochondria used in those studies.

c. Energy dependence. We have shown that uptake of mAAT (25) and of mMDH (26) into mitochondria in vitro is energy-dependent and moreover that the source of energy required is the transmembrane pH gradient. Energy dependence of uptake of proteins into mitochondria in vivo has been demonstrated by experiments in which cells treated with uncouplers of oxidative phosphorilation have been shown to accumulate precursors in the cytosol (21, 27). These experiments did not, however allow a distinction to be drawn between requirement for ATP or for a transmembrane electrochemical gradient in the uptake process. This question has been recently addressed by studying the effects of ionophores on the uptake of precursor proteins into isolated mitochondria (20, 28). In both cases, a requirement for transmembrane electrochemical gradient was established, but whereas Gasser et al. (20) did not claim to be able to distinguish between the two components of the gradient, Schleyer et al. (28) claimed that the ionic component ($\Delta\psi$) of the gradient is required and not ΔpH, in contrast to our results. We would maintain that this discrepancy arises because of the difficulties of controlling experimental conditions in the systems used by Schleyer et al.. Be that as it may, the fact remains that the model system reflects the situation in vivo in that protein transport into mitochondria in both situations requires a transmembrane electrochemical gradient.

If the uptake of native proteins into mitochondria is an accurate model for the process in vivo what role can be ascribed to the extra piece of sequence, presumibly N-terminal which exists for the vast majority of mitochondrial precursors? It is important to recall that the process of import of proteins into mitochondria is fundamentally different from the process of export of proteins from cells in that the latter is co-translational whereas the former is post-translational. Hence the primary role of N-terminal extensions of excreted proteins in passage of these proteins through the endoplasmic reticular membrane may not be shared by the N-terminal extensions of mitochondrial proteins. Indeed our results suggest that N-terminal extension of mitochondrial proteins do not function in this respect. Other roles are not hard to find. In some cases the extension may function in solubilizing otherwise very hydrophobic proteins (see, for example the work of Viebrock et al. on the subunit IX of the ATPase from N. crassa (29)). Alternatively, precursors may function to prevent premature assembly of newly-synthesized proteins into multimeric native structures. Thirdly, the pre-piece may be involved in maintaining a non-native configuration particularly adapted for transmembrane movement or membrane insertion, the contact between the protein and the mitochondrial membrane system being provided by regions other than the N-terminal extension. In this case, our results would require that the native mAAT and mMDH could attain the correct conformation for import on interacting with the mitochondrial membrane. As described above, we have direct experimental evidence that a major change in conformation does occur when mAAT binds to the outer face of mitochondria leading to enhanced protease sensitivity. Similarly, Webster et al. (30) have shown that conformational changes occur when mMDH binds to liposomes. In summary, therefore, we would suggest that the signals for translocation of proteins across mitochondrial membranes are contained within the part of polypeptide chain retained in the mature protein. Interaction of these signal sequences with mitochondria may require a non-native conformation but this conformation can be induced, at least with mAAT and mMDH, by interaction of mature proteins with mitochondria.

Receptors for Import of Proteins into Mitochondria

Since our model reflects accurately the features of protein import into mitochondria in vivo it provides a convenient system for probing the mechanism of the process at the molecular level. Our recent work has been directed to works demonstrating the existence of receptors for protein import.

Early evidence for such receptors came from the work of Zimmerman and Neupert (31). They showed that when mitochondria were added to a cell-free system containing the precursor form of the ADP/ATP translocator protein, then the precursor bound rapidly to the mitochondria in a protease-sensitive fashion, but only more slowly become integrated into the inner membrane. Similar results were obtained with cytochrome c (32). It was also shown that the receptors for the precursors of the ADP/ATP translocator and of cytochrome c were distinct (33). This work, however, provided little information about the nature of the receptors.

We have used two different approaches to this problem. The first arose out of the observation that blockage with mersalyl of a single protein thiol group in mAAT and of one or a small number of such groups in mMDH blocks import of the proteins into mitochondria (34, 26); in the former case the residue involved is probably cysteine-166 (1). As well as providing strong support for the view that internal amino acid residues play critical roles in the import process, these results suggested that import involves interaction of protein thiols with specific binding sites in membrane receptors. Confirmation of this hypothesis was provided by experiments in which it was shown that 2-mercaptoethanol inhibited uptake of both isoenzymes when added to the model systems. Inhibition was not due to interaction with the isoenzymes themselves as shown by experiments in which the isoenzymes were preincubated with 2-mercaptoethanol before addition to the mitochondrial suspension; no inhibition occurred. On the other hand, mitochondria preincubated with 2-mercaptoethanol failed to internalize the isoenzymes (35).

The kinetics of the process were examined and interestingly the precise nature of the inhibition differed for the two isoenzymes. In the case of AAT inhibition of uptake by 2-mercaptoethanol was competitive whereas with mMDH inhibition was of a mixed type. These results indicate that whereas 2-mercaptoethanol interacts only with the unoccupied receptor for AAT, in the case of MDH the inhibitor can react with both the unoccupied receptor and with the receptor-enzyme complex. It was also found that inhibition of uptake of mMDH was less sensitive to 2-mercaptoethanol than was uptake of mAAT.

Both the varying sensitivity to inhibition by 2-mercaptoethanol and the difference in kinetic form of the inhibition for the two isoenzymes suggests that the receptors for the two isoenzymes are distinct, although this evidence is not conclusive. More definitive evidence on this point was obtained by studying the mutual inhibition of the two isoenzymes for uptake into mitochondria. In these experiments, the rate of uptake process was measured for various concentrations of the test isoenzyme in the presence or absence of the other isoenzyme as an inhibitor. In both cases, the double-reciprocal plots of the results obtained were parallel, indicative of purely uncompetitive inhibition. This type of inhibition is difficult to explain since it implies that neither isoenzyme binds to the receptor for the other, but that binding of the inhibitory isoenzyme to the complex between test isoenzyme and its receptor occurs with blockage of the uptake process. This mechanism seems inherently unlikely. We prefer

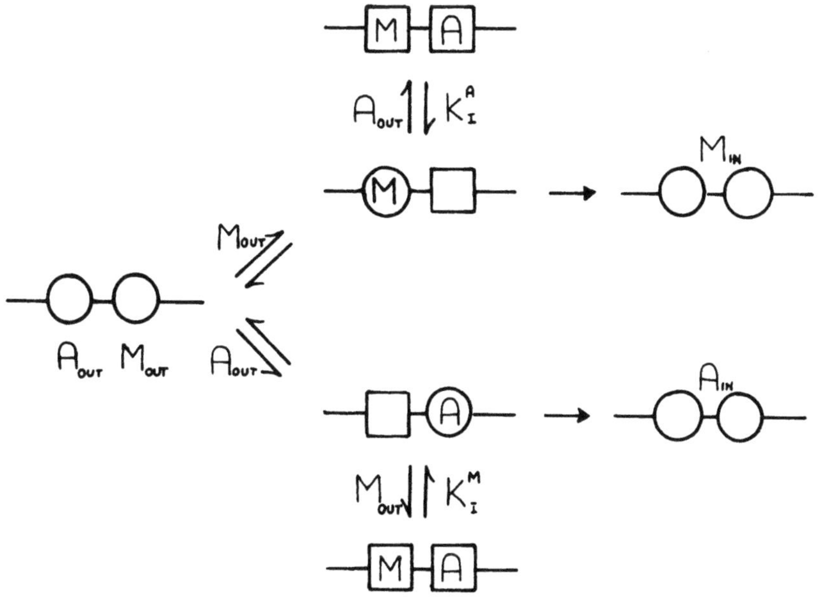

Fig. 4 Hypothetical mechanism for mutual inhibition of uptake of isoenzymes into mitochondria

Circles represent receptor for the two isoenzymes in the mitochondrial membrane system; aspartate aminotransferase (A) does not interact with the receptor for malate dehydrogenase (M) or *vice versa*. Squares represent high-affinity modified form of receptors, modification being induced by binding of an isoenzyme to the adjacent site. This modified receptors are not competent in protein import into mitochondria. Import is represented by movement of an isoenzyme from the cytosolic side (out) to the matrix side (in) of the membrane.

an interpretation in which the two isoenzymes have separate but interacting receptor sites as shown diagramatically in Fig. 4. When only one site is occupied then uptake occurs whereas when the adjacent site is also occupied by the other isoenzyme uptake of neither isoenzyme can occur. To account for the observed kinetics it is necessary to further postulate that occupation of one binding site by its isoenzyme changes the adjacent binding site in such a way as to increase the affinity for the second isoenzyme (represented in Fig. 4 by a change to square shape).

Acknowledgements
This work was partially supported by a grant from the Progetto Finalizzato "Ingegneria Genetica e Basi Molecolari delle Malattie Ereditarie" C.N.R. Italy.

References

1. Barra, D., Bossa, F., Doonan, S., Fahmy, H.M.A., Hughes, G.J., Martini, F., Petruzelli, R., Wittman-Liebold, B.: Eur. J. Biochem. 108, 405-414 (1980).

2. Marra, E., Doonan, S., Saccone, C., Quagliariello, E.: Biochem. J. 164, 685-691 (1977).

3. Passarella, S., Marra, E., Doonan, S., Quagliariello, E.: Biochem. J. 192, 649-658 (1980).

4. Marra, E., Doonan, S., Saccone, C., Quagliariello, E.: Eur. J. Biochem. 83, 427-435 (1978).

5. Marra, E., Passarella, S., Doonan, S., Saccone, C., Quagliariello, E.: Function and Molecular Aspects of Biomembrane Transport (Quagliariello et al., eds.) 459-462 (1979) Elsevier North-Holland.

6. Marra, E., Passarella, S., Doonan, S., Quagliariello, E., Saccone, C.: FEBS Lett. 122, 33-36 (1980).

7. Ades, I.Z.: Mol. Cell. Biochem. 43, 113-127 (1982).

8. Neupert, W., Schatz, G.: Trends Biochem. Sci. 6, 1-4 (1981).

9. Schatz, G., Butow, R.A.: Cell 32, 316-318 (1983).

10. Sonderegger, P., Jaussi, R., Christen, P.: Biochem. Biophys. Res. Commun. 94, 1256-1260 (1980).

11. Sannia, G., Abrescia, P., Colombo, M., Giardina, P., Marino, G.: Biochem. Biophys. Res. Commun. 105, 444-449 (1982).

12. Sakakibara, R., Huynh, A.K., Nishida, Y., Watanabe, T., Wada, H.: Biochem. Biophys. Res. Commun. 95, 1781-1788 (1980).

13. Marra, E., Greco, M., Quagliariello, E., Saccone, C.: Boll. Soc. It. Biol. Sper. (in press).

14. Greco, M., Marra, E., Guanti, G., Saccone, C., Quagliariello, E.: Boll. Soc. It. Biol. Sper. (in press).

15. Mihara, K., Omura, T., Harano, T., Brenner, S., Fleischer, S.,Rajagopalan, K.V., Blobel, G.: J. Biol. Chem. 257, 3355-3358 (1982).

16. Baumber, M.E., Doonan, S.: Int. J. Biochem. 7, 119-124 (1976).

17. Harmey, M.A., Neher, E.M., Zimmerman, R., Neupert, W.: BioSystems 12, 283-287 (1980).

18. Sakakibara, R., Kamisaki, Y., Wada, H.: Biochem. Biophys. Res. Commun. 102, 235-242 (1981).

19. Maccecchini, M.L., Rudin, Y., Blobel, G., Schatz, G.: Proc. Natl. Acad. Sci. 76, 343-347 (1979).

20. Gasser, S.M., Daum, G., Schatz, G.: J. Biol. Chem. 257, 13034-13041 (1982).

21. Jaussi, R., Sonderegger, P., Fluckiger, J., Christen, P.: J. Biol. Chem. 257, 13334-1340 (1982).

22. Mori, M., Morita, T., Ikeda, F., Amaya, Y., Tatibana, M., Cohen, P.P.: Proc. Natl. Acad. Sci. 78, 6056-6060 (1981).

23. Reid, G.A., Schatz, G.: J. Biol. Chem. 257, 13062-13067 (1982).

24. Ades, I.Z., Butow, R.A.: J. Biol. Chem. 255, 9925-9935 (1980).

25. Passarella, S., Marra, E., Doonan, S., Languino, L.R., Saccone, C., Quagliariello, E.: Biochem. J. 202, 353-362 (1982).

26. Passarella, S., Marra, E., Doonan, S., Quagliariello, E.: Biochem. J. 210, 207-214 (1983).

27. Reid, G.A., Yonetani, T., Schatz, G.: J. Biol. Chem. 257, 13068-13074 (1982).

28. Schleyer, M., Schimdt, B., Neupert, W.: Eur. J. Biochem. 125, 109-116 (1982).

29. Viebrock, A., Perz, A., Sebald, W.: EMBO Journal 1, 565-571 (1982).

30. Webster, K.A., Patel, H.V., Freeman, K.B., Papahadjopoulos, D.: Biochem. J. 178, 147-158 (1979).

31. Zimmerman, R., Neupert, W.: Eur. J. Biochem. 109, 217-229 (1980).

32. Hennig, G.B., Neupert, W.: Eur. J. Biochem. 121, 203-217 (1981).

33. Zimmerman, R., Hennig, B., Neupert, W.: Eur. J. Biochem. 116, 455-460 (1981).

34. Marra, E., Passarella, S., Doonan, S., Saccone, C., Quagliariello, E.: Arch. Biochem. Biophys. 195, 269-279 (1979).

35. Passarella, S., Marra, E., Casamassima, E., Perlino, E., Doonan, S., Saccone, C., Quagliariello, E.: Structure and Function of Membrane Proteins (Quagliariello E. and Palmieri F. eds.) Elsevier, North-Holland (in press).

BIOSYNTHESIS AND TOPOGENESIS OF CYTOSOLIC AND MITOCHONDRIAL ASPARTATE AMINOTRANSFERASE

Philipp Christen, Rolf Jaussi, Peter Sonderegger, Heinz Gehring, Ursula Graf-Hausner, Renata Behra, Jörg Flückiger, and Radek C. Skoda

Biochemisches Institut der Universität Zürich
CH-8028 Zürich, Switzerland

Summary

The cytosolic and mitochondrial isoenzyme of aspartate aminotransferase are homologous proteins and are both coded for by nuclear DNA. Their spatial structures are virtually superimposable. The duplication of the ancestral gene is estimated to have occurred 1000 million years ago, i.e. at the time of the emergence of eucaryotic cells. Comparison of amino acid sequences allows the identification of amino acid residues conserved specifically in the mitochondrial isoenzyme during phylogenetic evolution. These residues appear to cluster in certain surface regions.

Mitochondrial aspartate aminotransferase is synthesized as a higher molecular weight precursor (pre-mAspAT) on free polysomes in the cytosol. Pre-mAspAT which has been synthesized by *in vitro* translation of polysomes is taken up by exogenous mitochondria and processed to mature mAspAT. Apparently, the translocation of pre-mAspAT occurs after completion of translation. The mRNA of pre-mAspAT has a sedimentation coefficient of ∿16 S and contains relatively large noncoding regions. Pre-mAspAT was also detected in cultured chicken embryo fibroblasts. The rate of *in vivo* import corresponds to a half-life of the precursor of 0.5 min (in chicken embryo fibroblasts at 37°C). The proteolytic processing step in the mitochondrial matrix is accompanied by a change in conformation. On treatment of the cells with carbonyl cyanide m-chlorophenylhydrazone (CCCP), a protonophore dissipating the electrochemical gradient over the mitochondrial inner membrane, the import of pre-mAspAT is blocked. The precursor accumulates in the cytosol and is subject to rapid degradation ($t_{1/2}$ ∿ 5 min). This degradation takes place only in intact cells, in cell homogenates pre-mAspAT is stable.

Mitochondria 1983

The cytostatic agent and cytoskeleton inhibitor vinblastine inhibits the maturation of the precursor. In the presence of 50 μM vinblastine, the steady-state pool of pre-mAspAT is increased 3-fold and its half-life is found to be prolonged correspondingly. None of 7 other cytoskeleton inhibitors tested showed this effect. Apparently, it does not relate to the inhibiting action of vinblastine on the polymerization of tubulin.

The conclusions drawn from the experimental evidence available to date are summarized in the Scheme below.

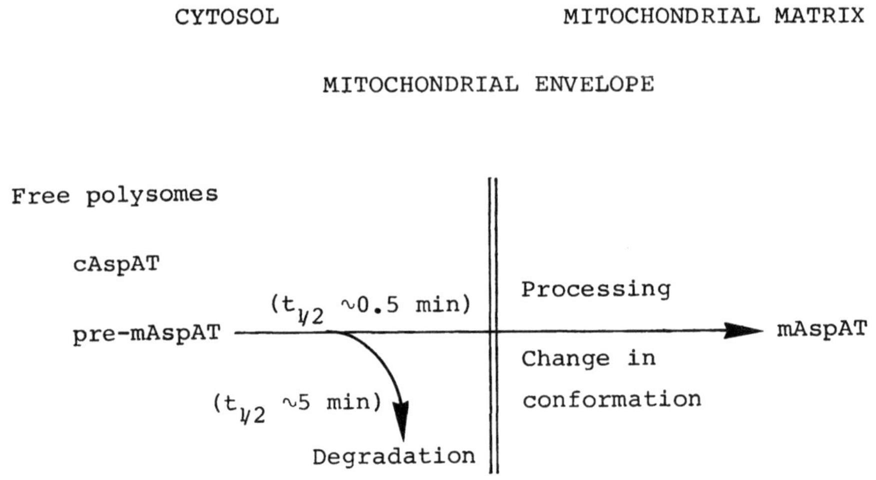

Two isoenzymes of aspartate aminotransferase (AspAT) occur in animal cells, one being located exclusively in the cytosol, the other in the mitochondria. Not only the amino acid sequences (1-6), but also the three-dimensional structures (7-10) of the two isoenzymes have been determined. AspAT is a dimeric protein, consisting of two identical subunits with ∿ 400 amino acid residues (M_r ∿ 45,000).

Mitochondrial AspAT is an attractive object for studying the import of proteins into mitochondria, not only because of the extensive structural information available but also because of the occurrence of a closely related protein, cytosolic aspartate

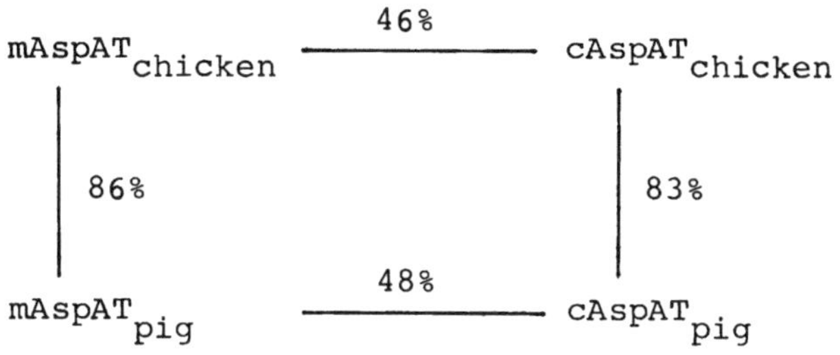

Fig.1. *Intra*- and *inter*-species amino acid sequence relationships of mitochondrial and cytosolic AspATs from chicken and pig. The figures indicate the degree of sequence identity. The sequence information is taken from Refs. 1-6.

aminotransferase (cAspAT), which is not taken up by mitochondria. Thus, a comparison of the two isoenzymes might contribute to the identification of structural features responsible for their intracellular heterotopism.

In two species, chicken and pig, the amino acid sequences of both the mitochondrial and cytosolic isoenzyme have been determined. The sequence identity between mAspAT and cAspAT is close to 50% (Fig. 1), clearly the two isoenzymes are homologous proteins. Correspondingly, their chain folding patterns have been found to be virtually superimposable with the exception of a few deletions in the mitochondrial isoenzyme. The degree of *inter*-species similarity of the mAspATs and of the cAspATs by far exceeds the *intra*-species similarity of the mitochondrial and cytosolic isoenzyme. The difference indicates that the gene duplication that gave rise to the development of the two isoenzymes occurred long before the phylogenetic divergence of birds and mammals. The gene duplication is estimated to have taken place about 1000 million years ago, *i.e.* at the time of the emergence of eucaryotic cells (6).

To our knowledge, a comparison of the amino acid sequences of a pair of two homologous but heterotopic proteins from two species is to date only possible in the case of AspAT. Such a comparison allows the identification of those residues that are invariant specifically in the mitochondrial isoenzyme. The selection criterion for identifying such residues was identity in the two mAspATs coincident with nonidentity between the mAspATs and cAspATs and between the two cAspATs themselves ($mAspAT_{chicken} = mAspAT_{pig} \neq cAspAT_{chicken} \neq cAspAT_{pig}$). The criterion of nonidentity excludes residues that are essential for structural or functional features common to both isoenzymes. In the total sequence there are 36 residues (9% of total) that comply with the above selection criterion; henceforth we denote them "m-residues" because they are conserved specifically in mAspAT. For comparison, only 10 residues (2.4% of total) are identical specifically in the two cAspATs ($cAspAT_{chicken} = cAspAT_{pig} \neq mAspAT_{chicken} \neq mAspAT_{pig}$). Thus there are almost 4 times more "m-residues" than "c-residues". This difference might have arisen not by chance but by necessity and might indicate that the mitochondrial isoenzyme was subject to additional evolutionary constraints.

The "m-residues" are not distributed at random in the spatial structure of mAspAT; they cluster in certain surface regions in the "lower" half of the subunit (6). The selective uptake into the mitochondria being the most distinctive feature of mAspAT in comparison to cAspAT, we may speculate that the "m-residues" might be involved in the translocation of mAspAT through the mitochondrial membrane. Perhaps they constitute recognition sites responsible for the specific binding to the mitochondria. Such features might relate to the reported interaction of mature enzyme with mitochondria (11).

Like all other proteins of the mitochondrial matrix tested so far (for reviews, see Refs. 12-14), the primary translation product of mAspAT in a cell-free protein synthesizing system is a higher molecular weight precursor. Pre-mAspAT with Δmolecular weight ∿ 2,500 has been detected after translation of chicken liver mRNA in a rabbit reticulocyte lysate (15). The primary translation product of cAspAT has the same apparent molecular weight as the mature enzyme. Synthesis of a higher molecular weight precursor of mAspAT was also found in rat (16) and pig (17).

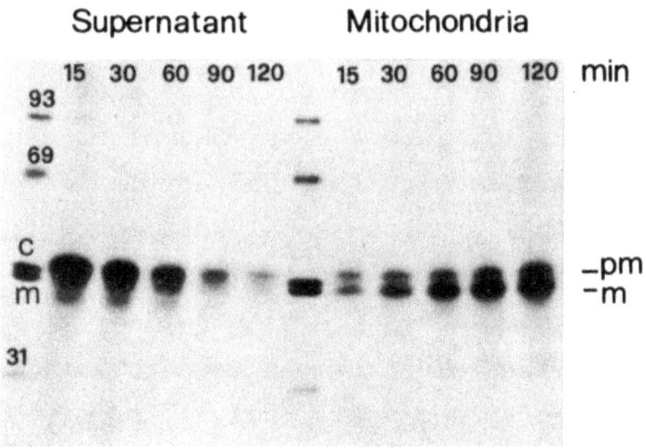

Fig. 2. In vitro import of pre-mAspAT into mitochondria. Free polysomes from chicken liver were translated in a rabbit reticulocyte lysate in the presence of 1 mCi ml^{-1} of [^{35}S]methionine (10^3 Ci $mmol^{-1}$) for 30 min (18). Freshly prepared chicken heart mitochondria (19) were added to the translation products and incubated at 30°C (per sample 50 µl of translation mixture plus 250 µg mitochondrial protein). At the times indicated, samples were removed and chilled on ice. After a short spin (15 min, 15,000 x g, 4°C) the supernatant and the mitochondrial fraction were processed separately. The immunoprecipitates with anti-denatured mAspAT antiserum were analyzed by sodium dodecylsulfate polyacrylamide gel electrophoresis. The fluorogram of the gel is shown. The methods used are described in detail elsewhere (18). The positions of the bands of mAspAT (m), cAspAT (c), and pre-mAspAT (pm) are indicated.

On incubation of the *in vitro* translation mixture with isolated chicken heart mitochondria, the precursor was converted in a time-dependent process to mature enzyme that cosedimented with mitochondria (Fig. 2). The estimated half-life of the precursor in the *in vitro* system is ∿ 30 min. The *in vitro* import process is much less efficient than that *in vivo* (see below). Thus, the question was raised whether the *in vivo* import occurs cotranslationally or posttranslationally, *i.e.* whether the synthesis of pre-mAspAT takes place on membrane-bound or on free polysomes.

Membrane-bound and membrane-free polysomes from chicken liver were separated by subcellular fractionation. The products of read-out translation of the two polysome fractions were immunoprecipitated by antisera against mAspAT or cAspAT and analyzed by sodium dodecylsulfate polyacrylamide gel electrophoresis. Both cAspAT and pre-mAspAT were found to be synthesized exclusively by free polysomes (18).

A preparation of free polysomes was used to purify and characterize the mRNA of pre-mAspAT. $(PolyA)^{+}$ mRNA was extracted and fractionated by ultracentrifugation in a sucrose gradient. The centrifugal fractions were analyzed by translation in a rabbit reticulocyte lysate. Pre-mAspAT was found to be synthesized by the fraction containing mRNA of ∿16 S (Fig. 3). This S value is higher than that expected for the coding region of a protein of the size of the precursor. The presence of relatively large noncoding regions in the message for pre-mAspAT is also indicated by the fact that the majority of the proteins coded for by this mRNA fraction proved to have higher molecular weights than pre-mAspAT. The mRNA fraction in which the specific messenger was enriched about 20fold served to construct a cDNA library with the goal of isolating a pre-mAspAT clone for studies of gene structure and expression

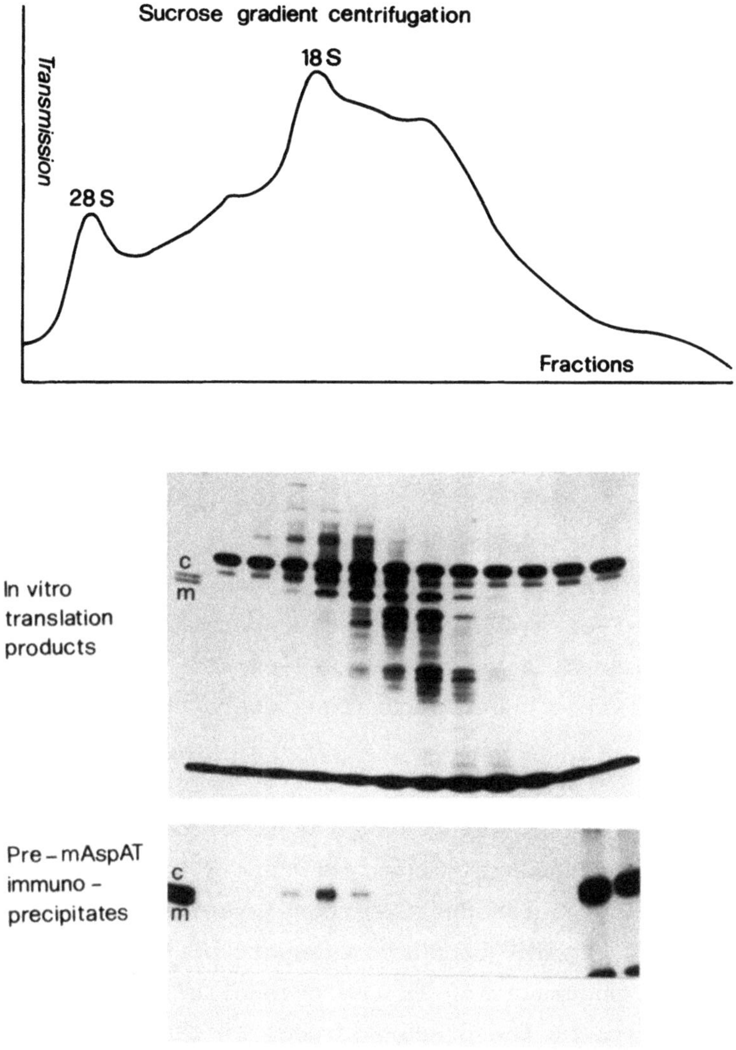

Fig. 3. Purification of mRNA of mAspAT. (PolyA)$^+$mRNA (150 µg) from free polysomes of chicken liver was centrifuged on an isokinetic sucrose gradient (20). Fractions of the elution profile on top of the figure were collected and the RNA translated in reticulocyte lysate. The fluorograms after sodium dodecylsulfate polyacrylamide gel electrophoresis of the total translation products and the immunoprecipitates obtained with anti-denatured mAspAT antiserum are aligned below the elution profile. The methods used are described elsewhere (18).

as well as for enzyme mechanistic studies by site-directed mutagenesis.

In accordance with synthesis of the precursor on free polysomes, *i.e.* with its posttranslational uptake into the mitochondria, a small pool of pre-mAspAT could be detected in the supernatant of homogenized chicken embryo fibroblasts (21). The half-life of pre-mAspAT in the cells was estimated to be ∿ 0.5 min, *i.e.* one to two orders of magnitude shorter than that observed on *in vitro* import (see Fig. 2). At present, the reasons for the discrepancy are unknown.

A consequence of posttranslational uptake is that the precursor might encounter the mitochondrion as an at least partially folded protein. In order to compare the conformation of pre-mAspAT with that of mAspAT, their susceptibility toward proteases was compared. Both trypsin and papain completely digested the precursor under conditions where the mature enzyme was stable.

The import of proteins depends on the electrochemical gradient over the mitochondrial inner membrane (22, 23). Addition of the uncoupling protonophore carbonylcyanide m-chlorophenylhydrazone (CCCP, 20 μM) blocks the import of pre-mAspAT and results in its accumulation in the cytosol of the cells (21). However, the size of the precursor pool in CCCP-treated cells is limited to about ten times the size of that in untreated cells. The limitation in pool size is due to a rapid degradation of the precursor. The degradation can be demonstrated if the pulse under CCCP is followed by a chase in the continued presence of the uncoupler. The precursor band disappears with a half-life of 5 min (Fig. 4). At present, it is unknown whether this degradation occurs with the same rate in untreated cells. If the degradation ($t_{1/2}$ ∿ 5 min) occurred,

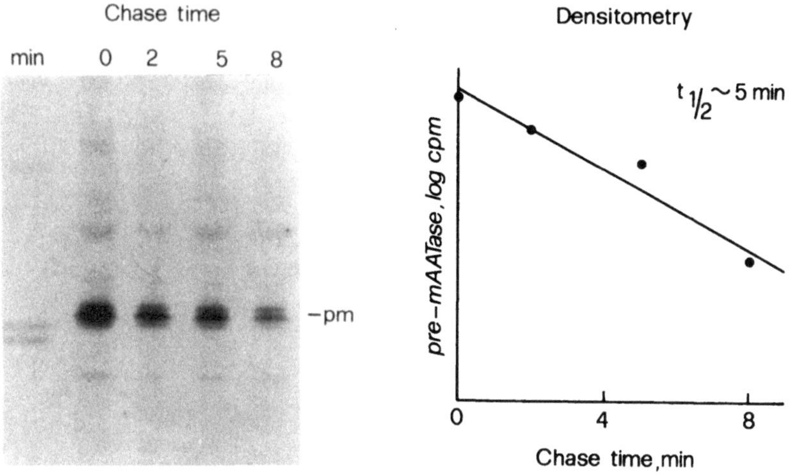

Fig. 4. Degradation of pre-mAspAT in CCCP-treated cells. Chicken embryo fibroblasts were prelabeled with [^{35}S]methionine in the presence of 20 μM CCCP (see Ref. 21). The cells were then chased by replacing the pulse medium with fresh medium containing 10 mM unlabeled methionine and 20 μM CCCP. At the indicated times, samples were withdrawn and analyzed for their content of pre-mAspAT and mAspAT by specific immunoprecipitation (see Ref. 18). The estimated half-life of pre-mAspAT is ∿5 min.

only ∿ 90% of newly synthesized precursor would be taken up into mitochondria ($t_{1/2}$ ∿ 0.5 min) and 10% would be lost by degradation. We are currently engaged in characterizing the responsible proteolytic system and in assessing the physiological rôle of the degradative process. To date it has not been possible to conduct experiments on the degradation of pre-mAspAT in a cell-free system. Pre-mAspAT produced and kept in a rabbit reticulocyte lysate as well as pre-mAspAT in a homogenate of chicken embryo fibroblasts was not degraded. Degradation perhaps depends on intact cellular structures.

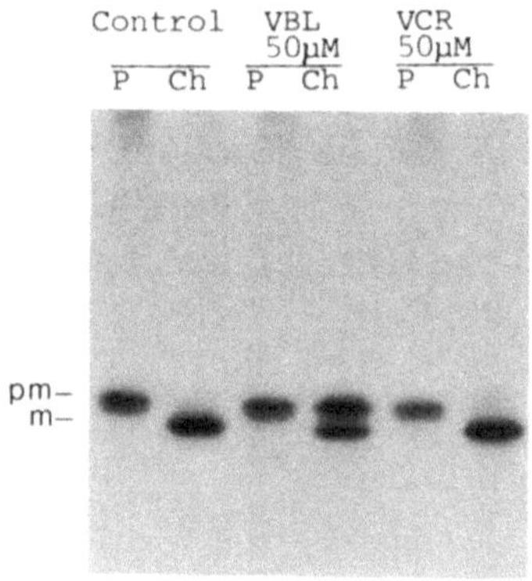

Fig. 5. Effect of vinblastine (VBL) and vincristine (VCR) on the rate of maturation of pre-mAspAT. Two Petri dishes at a time were preincubated during 1 h with vinblastine or vincristine at the indicated final concentrations. The inhibitors were also present in all subsequent stages of the experiment. The cells were then pulsed for 15 min with [^{35}S]methionine in the presence of 20 μM CCCP. The fluorograms of the immunoprecipitates after sodium docecylsulfate polyacrylamide gel electrophoresis are shown. Lanes P: After a 15 min pulse one Petri dish for each cytoskeleton inhibitor and one control without inhibitor were harvested without chase. Lanes Ch: The other Petri dish and one control were subjected to a 3 min chase in CCCP-free medium.

Recent experiments have indicated that the cytostatic agent vinblastine which is known to act as a cytoskeleton inhibitor retards the maturation of pre-mAspAT (31). There is ample experimental evidence that mitochondria are associated with the cytoskeleton (24). Both morphological and biochemical data suggest that mitochondria interact with microtubuli (25, 26), intermediate filaments (27-29) or γ-actin (30). We have tested several such inhibitors in chicken embryo fibroblasts for possible effects on the import of the precursor of mAspAT into mitochondria. Of all inhibitors tested only vinblastine inhibited the maturation of pre-mAspAT in pulse-chase experiments (Fig. 5). Chicken embryo fibroblasts were pulsed with radioactive methionine in the presence of the uncoupler CCCP. No conversion of pre-mAspAT to mAspAT occurred under these conditions. If the cells were chased under dilution of CCCP, pre-mAspAT was rapidly processed to mature enzyme with a half-life of ∿ 0.5 min. However, in the cells treated with vinblastine the half-life was prolonged to 3 min. The structurally closely related vincristine as well as 6 other cytoskeleton inhibitors, i.e. colchicine (50 μM), nocodazole (50 μM), podophyllotoxin (50 μM), taxol (45 μM), cytochalasin D (20 μM) and phalloidin (25 μM) showed no difference to the controls. In pulse experiments without CCCP, vinblastine increased the pool of precursor almost 3fold (Table 1).

Table 1. Effect of vinblastine on the steady-state pool of pre-mAspAT. Three Petri dishes were incubated with vinblastine at the indicated final concentrations for 1 h. The cells were then pulsed with [^{35}S]methionine for 7 min. The fluorograms of the immunoprecipitates after sodium dodecylsulfate polyacrylamide gel electrophoresis were estimated by densitometry.

Vinblastine (μM)	pre-mAspAT	mAspAT
	(Relative densitometer readings)	
0	1.0	13.5
50	2.8	12.5

The inhibitory effect of vinblastine is probably not mediated by the cytoskeleton. All other microtubuli disrupting agents tested proved ineffective, and the concentration of vinblastine needed (50 μM) was higher than that necessary for inhibition of polymerization of microtubuli in vitro or for mitotic arrest.

We have measured a slight decrease in oxygen consumption (to 90% of the initial value) in vinblastine-treated fibroblasts. This minor effect most likely cannot explain the observed inhibition of the import. However, it might indicate that vinblastine interacts with the mitochondrial inner membrane. Vinblastine is more hydrophobic than vincristine (Fig. 6) which affected neither the maturation of pre-mAspAT nor the oxygen consumption. Conceivably, vinblastine might be inserted into the mitochondrial membranes.

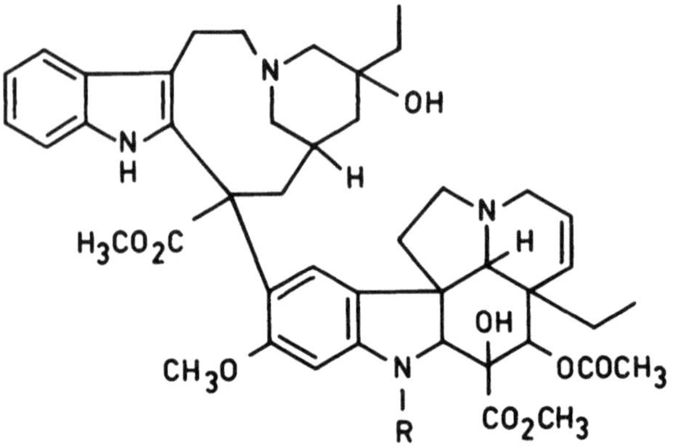

Fig. 6. Structure of vinblastine (R = CH_3) and vincristine (R = CHO).

References

1. Ovchinnikov, Yu.A., Egorov, C.A., Aldanova, N.A., Feigina, M.Yu., Lipkin, V.M., Abdulaev, N.G., Grishin, E.V., Kiselev, A.P., Modyanov, N.N., Braunstein, A.E., Polyanovsky, O.L., Nosikov, V.V.: FEBS Lett. 29, 31-34 (1973)

2. Doonan, S., Doonan, H.J., Hanford, R., Vernon, C.A., Walker, J.M., Lucia, P., da Airoldi, S., Bossa, F., Barra, D., Carloni, M., Fasella, P., Riva, F.: Biochem. J. 149, 497-506 (1975)

3. Shlyapnikov, S.V., Myasnikov, A.N., Severin, E.S., Myagkova, M.A., Torchinsky, Yu.M., Braunstein, A.E.: FEBS Lett. 106, 385-388 (1979)

4. Barra, D., Bossa, F., Doonan, S., Fahmy, H.M.A., Hughes, G.J., Martini, F., Petruzzelli, R., Wittmann-Liebold, B.: Eur. J. Biochem. 108, 405-414 (1980)

5. Kagamiyama, H., Sakakibara, R., Tanase, S., Morino, Y., Wada, H.: J. Biol. Chem. 255, 6153-6159 (1980)

6. Graf-Hausner, U., Wilson, K.J., Christen, P.: J. Biol. Chem., in press

7. Borisov, V.V., Borisova, S.N., Kachalova, G.S., Sosfenov, N.I., Vainshtein, B.K., Torchinsky, Yu.M., Braunstein, A.E.: J. Mol. Biol. 125, 275-292 (1978)

8. Arnone, A., Briley, P.D., Rogers, P.H., Hyde, C.C., Metzler, C.M., Metzler, D.E.: in Molecular Structure and Biological Activity (Griffen, J.F., Duax, W.L., eds.), pp. 57-74, Elsevier North-Holland Inc., New York (1982)

9. Harutyunyan, E.G., Malashkevich, V.N., Tersyan, S.S., Kochkina, V.M., Torchinsky, Yu.M., Braunstein, A.E.: FEBS Lett. 138, 113-116 (1982)

10. Ford, G.C., Eichele, G., Jansonius, J.N.: Proc. Natl. Acad. Sci. USA 77, 2559-2563 (1980)

11. Marra, E., Doonan, S., Saccone, C., Quagliariello, E.: Eur. J. Biochem. 83, 427-435 (1978)

12. Sabatini, D.D., Kreibich, G., Morimoto, T., Adesnik, M.: J. Cell. Biol. 92, 1-22 (1982)

13. Ades I.Z.: Mol. Cell. Biochem. 43, 113-127 (1982)

14. Schatz, G., Butow, R.A.: Cell 32, 316-318 (1983)

15. Sonderegger, P., Jaussi, R., Christen, P.: Biochem. Biophys. Res. Commun. 94, 1256-1260 (1980)

16. Sakakibara, R., Huynh, Q.K., Nishida, Y., Watanabe, T., Wada, H.: Biochem. Biophys. Res. Commun. 95, 1781-1788 (1980)

17. Sannia, G., Abrescia, P., Colombo, M., Giardina, P., Marino, G.: Biochem. Biophys. Res. Commun. 105, 444-449 (1982)

18. Sonderegger, P., Jaussi, R., Christen, P., Gehring, H.: J. Biol. Chem. 257, 3339-3345 (1982)

19. Behra, R., Christen, P., Spycher, M.A., Sonderegger, P.: Comp. Biochem. Physiol. 71B, 187-191 (1982)

20. McCarty, K.S. jr., Vollmer, R.T., McCarty K.S.: Anal. Biochem. 61, 165-183 (1974)

21. Jaussi, R., Sonderegger, P., Flückiger, J., Christen, P.: J. Biol. Chem. 257, 13334-13340 (1982)

22. Schleyer, M., Schmidt, B., Neupert, W.: Eur. J. Biochem. 125, 109-116 (1982)

23. Gasser, S.M., Ohashi, A., Daum, G., Böhni, P.C., Gibson,J., Reid, G.A., Yonetani, T., Schatz, G.: Proc. Natl. Acad. Sci. USA 79, 267-271 (1982)

24. Wolosewick, J.J., Porter, K.R.: J. Cell Biol. 82, 114-139 (1979)

25. Ball, E.H., Singer, S.J.: Proc. Natl. Acad. Sci. USA 79, 123-126 (1982)

26. Bernier-Valentin, F., Rousset, B.: J. Biol. Chem. 257, 7092-7099 (1982)

27. Smith, D.S., Järlfors, U., Cayer, M.L.: J. Cell Sci. 27, 255-272 (1977)

28. Toh, B.D., Lolait, S.J., Mathy, J.P., Baum, R.: Cell-Tissue Res. 211, 163-169 (1980)

29. Mose-Larsen, P., Bravo, R., Fey,S.J., Small, J.V., Celis, J.E.: Cell 31, 681-692 (1982)

30. Pardo, J.V., Pittenger, M.F., Craig, S.W.: Cell 32, 1093-1103 (1983)

31. Skoda, R.C., Jaussi, R., Christen, P.: Biochem. Biophys. Res. Commun., in press

IMPORT OF ATP2-LacZ GENE FUSION PROTEINS INTO MITOCHONDRIA

Bruce L. Geller, Marjorie L. Britten, Clyde M. Biggs and Michael G. Douglas

Department of Biochemistry, UT Health Science Center, San Antonio, Texas 78284

Scott D. Emr

Division of Biology, California Institute of Technology, Pasadena, CA 91125

Introduction

The combination of *in vitro* and *in vivo* studies have described in some detail the mitochondrial import of the F_1-ATPase β-subunit to its final destination on the interior surface of the mitochondrial inner membrane (for reviews see 1,2). These studies have demonstrated that the transient amino terminal presegment of the β-subunit precursor is required for mitochondrial uptake. In order to extend the analysis of the specific targeting signals which direct the β-subunit to its correct mitochondrial compartment, we have constructed gene fusions between portions of the *ATP2* gene (3) encoding different lengths of the amino-terminal end of the β-subunit protein and a portion of *lacZ* encoding a large carboxy-terminal active fragment of β-galactosidase (4). This family of hybrid genes efficiently expresses β-galactosidase activity in yeast which is targeted to mitochondria where it is bound most likely to the inner membrane. These studies suggest that a minimum of two sequences on the hybrid protein product are necessary for the delivery of the β-subunit protein into the mitochondrial interior. The first sequence is a transient β-subunit presequence which targets the protein to the mitochondrial inner membrane (5). The second sequence which may participate in the complete or partial translocation across the inner membrane occurs ~243 to ~277 residues from the amino terminal end of the β-subunit precursor. *ATP2-lacZ* gene fusion products containing greater than ~277 aa of the β-subunit protein prevent the growth of otherwise wild-type yeast cells on a nonfermentable carbon source.

Mitochondria 1983

These initial observations open up the mitochondrial import pathway to the genetic characterization of the signals and the cellular machinery responsible for mitochondrial delivery and uptake.

Results

1) Construction of ATP2-lacZ gene fusions.

An E. coli-yeast shuttle vector pSEY101 carrying a truncated lacZ gene was used for the ATP2-lacZ gene fusion constructions. This plasmid contains selectable markers and origins for replication in both E. coli and yeast (Figure 1). It also carries a lacZ gene which lacks all of its 5'

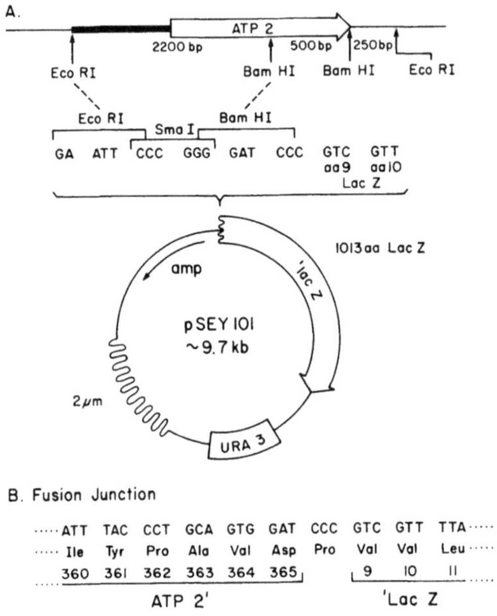

Figure 1. Gene Fusions Between ATP2 and lacZ in the Yeast-E. coli Shuttle Vehicle pSEY101. The 2.2 kb EcoRI-BamHI fragment containing approximately 1.1 kb of DNA to the 5' side of the ATP2 structural gene and 1.1 kb coding the first ~365 amino acids of the F_1-ATPase β-subunit was ligated in SEY101 (A). This construction yields pβ-Z1 containing a fusion with an uninterrupted reading frame across the BamHI site (B). Bal 31 nuclease digestion of a larger DNA fragment from the BamHI site of the ATP2 structural gene followed by digestion with EcoRI generated fragments containing shorter amino-terminal coding regions of ATP2. In frame fusions were selected following ligation into EcoRI, SmaI digested pSEY101.

regulatory sequences as well as the coding sequences for the first 8 aa of the lacZ gene product, β-galactosidase. In their place, unique restriction sites for EcoRI, SmaI and BamHI are present. Expression of β-galactosidase from pSEY101 in yeast or E. coli requires the presence of the correct transcriptional and translational signals.

The ATP2 gene coding the F_1-ATPase subunit 2 is contained within a 3.3 kb EcoRI fragment (3) with approximately 100 bp of DNA to its 5' side and 700 bp of DNA to its 3' side. DNA sequence analysis of the ATP2 gene indicated that an in frame ATP2-lacZ fusion could be generated with pSEY101 at the BamHI site corresponding to amino acid position ~365 of the β-subunit. The resulting construction, pβ-Z1 contained a 2.2 kb EcoRI-BamHI ATP2 DNA fragment carrying ~1100 bp of DNA upstream of the β subunit coding sequences as well as ~1100 bp of coding DNA.

Transformation of the ura3 strain SEY2102 with the pβ-Z1 plasmid yields Ura^+ transformants that express β-galactosidase. The rapid loss of the Ura^+ phenotype simultaneously with β-galactosidase activity which occurs in the absence of Ura^+ selection indicates that pβ-Z1 is maintained as a plasmid in these yeast cells. Furthermore, the levels of β-galactosidase activity in the pβ-Z1 transformant are repressed 4-5 fold upon shift from a galactose to glucose carbon source indicating that the level of expression is under the control of the ATP2 promoter.

In order to construct ATP2-lacZ gene fusions coding shorter segments of the amino terminal end of the β-subunit, Bal 31 nuclease resection experiments were performed. In these constructions ATP2 DNA fragments of variable lengths containing a common 5' EcoRI end with ~1100 bp of DNA upstream from the ATP2 initiator codon and different 3' ends located within the ATP2 coding DNA were ligated into EcoRI-SmaI restricted pSEY101. In frame fusions were determined in E. coli on indicator plates. Retention of the BamHl site of pSEY101 in these ATP2-lacZ fusions allows the convenient determination of the size of the ATP2 DNA present by codigestion with EcoRI and BamHI (Figure 2). The constructions selected for further analysis are designated pβ-Z1 containing ~365 aa of ATP2 DNA through pβ-Z8 encoding ~225 aa of the β-subunit.

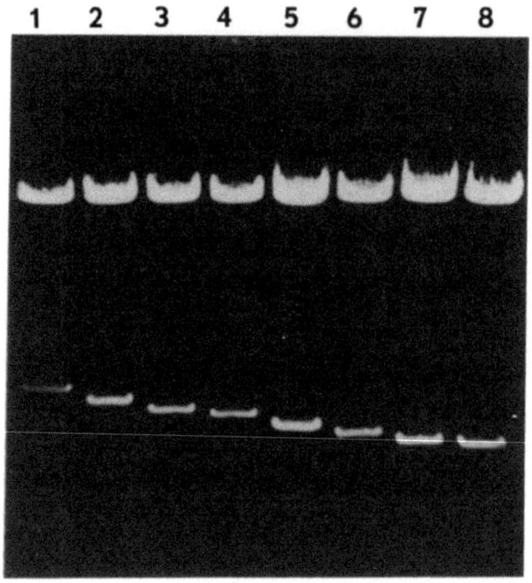

Figure 2. Restriction Analysis of ATP2-lacZ Gene Fusions. Analysis of the ATP2 insert length in selected pβ-Z constructions utilized sequential digestion of the plasmid with BamHI and EcoRI. The length of the respective EcoRI-BamHI fragments and expected ATP2 gene product for pβ-Z1 (lane 1) to pβ-Z8 (lane 8) are 1. 2200 bp (~365 aa), 2. 2100 bp (~327 aa), 3. 2020 bp (~300 aa), 4. 1985 bp (~288 aa), 5. 1950 bp (~277 aa), 6. 1850 bp (~243 aa), 7. 1810 bp (~230 aa) 8. 1795 bp (~225 aa).

TABLE I

β-Subunit-β-galactosidase Hybrid Proteins are Targeted to the Mitochondria

	Whole Cell (units/mg)	Cell Fractions (total units)	
Fusion Plasmid		9K Pellet	9K Supernatant
pβ-Z1	152	17,004 (89.6%)	1,970
pLG669-Z	1006	1,827 (1.8%)	100,013

An identical wet weight of cells (SEY2102 harboring pβ-Z1 or pLG669-Z) previously grown on 2 x SD-ura medium were spheroplasted (6). After a debris spin at 3000 x g for 3 min, the supernatant was resolved into a 9000 x g (10 min) supernatant and pellet fraction then assayed for β-galactosidase (7). (1 unit is the amount of enzyme which hydrolyzes 1 n mol of o-nitrophenol-β-D galactoside per minute at 28°C.)

2) Cytoplasmic targeting of ATP2-lacZ hybrid gene products to mitochondria. All of the ATP2-lacZ gene fusions described above express β-galactosidase activity to about the same extent (120 to 200 n mol of o-nitrophenol-β-D-galactoside per minute per mg of whole cell protein). This enzyme activity provides a convenient indicator for the localization of the β-subunit-β-galactosidase hybrid proteins in subcellular fractions. The soluble β-galactosidase activity expressed from a plasmid carrying a CYC1-lacZ gene fusion pLG669-Z served as a control (5). In each case whole cell homogenates were prepared from equal amounts of yeast strain SEY2102 carrying different pβ-Z or pLG669-Z plasmids and resolved into mitochondria and supernatant fractions. The data illustrated in Table I for pβ-Z1 is essentially the same for pβ-Z2 through pβ-Z8. The mitochondrial pellet contains greater than 85% of the β-galactosidase activity expressed from the ATP2-lacZ gene fusion and less than 5% of the CYC1-lacZ encoded enzyme activity. Thus, the protein sequences responsible for targeting the hybrid protein to the mitochondria appear to reside at least within the first ~225 aa of the β-subunit. In addition, sequences upstream of the structural gene for CYC1 appear not to direct targeting to mitochondria.

The combination of submitochondrial fractionation and protease accessibility studies strongly indicate that the hybrid proteins of the larger size classes (pβ-Z1 to pβ-Z5) reside in the mitochondrial interior in a non-soluble form. These conclusions are based on the following observations: a. When mitochondrial inner and outer membranes are resolved on sucrose gradients β-galactosidase activity expressed from pβ-Z1 - pβ-Z5 fractionates almost entirely with the inner membrane marker enzyme, cytochrome oxidase. b. The β-subunit-β-galactosidase activity cannot be released from submitochondrial preparations by treatment with high salt. c. Proteinase K digestion of freshly isolated mitochondria containing the hybrid proteins indicates that a large part of the outer membrane marker enzyme or residual CYC1-lacZ gene fusion product is accessible to the external probe, whereas, a small percentage of the ATP2-lacZ hybrid gene protein or cytochrome oxidase is digested (Table II). β-subunit-β-galactosidase gene products of the shorter fusions (pβ-Z6 to pβ-Z8) appear to be more exposed to the protease under these conditions.

TABLE II

Accessibility of Mitochondrial β-galactosidase
Activities to Proteinase K

CONDITIONS		ACTIVITY REMAINING (% ± S.D.)			
Proteinase K (0.1 mg/ml)	Triton X-100 (0.5% w/v)	pβ-Z1	pβ-Z5	pβ-Z6	pLG669-Z
		β-galactosidase			
+	-	88.6 ± 10.6	73.7 ± 28.3	38.7 ± 19.0	12.3 ± 5.7
+	+	1.7 ± 1.2	0.9 ± 0.8	1.5 ± 1.5	2.0 ± 2.6
		Cytochrome Oxidase			
+	-	81.0 ± 16.5	78.3 ± 2.8	83.3 ± 9.3	81.3 ± 23.5
+	+	15.0 ± 7.5	11.0 ± 3.9	8.7 ± 1.9	7.7 ± 7.2
		Antimycin Insensitive NADH cyt c Reductase			
+	-	10.4 ± 3.17	11.1 ± 7.7	10.5 ± 8.2	10.9 ± 7.2
+	+	-	-	-	-

Strain SEY2102 harboring the indicated plasmids were grown in 2 x SD-ura medium. Mitochondria were prepared (6) and immediately resuspended in isotonic buffer at 5 mg/ml in triplicate. Proteinase K and Triton X-100 were added to the samples as indicated followed by incubation at 30°C for 30 min. Each sample was made 0.5mM in PMSF followed by analysis of enzymatic activities. In each case the value of the control (absence of detergent or protease) was set at 100%. The values shown are from 3 separate determinations.

3) Conditional phenotype associated with cells harboring some pβ-Z plasmids.

In addition to directing the synthesis of hybrid proteins, the ATP2-lacZ gene fusions on plasmids pβ-Z1 through pβ-Z5 confer a no growth phenotype to otherwise wild-type yeast cells on a nonfermentable carbon source. These growth inhibitory ATP2-lacZ gene fusions code for β-subunit-β-galactosidase hybrid proteins containing from ~365 aa (pβ-Z1) to ~277 aa (pβ-Z5) of the ATP2 gene product. In contrast gene fusions containing shorter lengths of the β-subunit protein, pβ-Z6 (~240 aa) to pβ-Z8 (~225 aa) grow normally on a nonfermentable carbon source.

The respiration negative phenotype (Gly^-) appears to result from the import of a β-subunit-β-galactosidase hybrid gene product containing $> {\sim}277$ aa of the β-subunit protein. Expression of β-galactosidase from another hybrid gene like CYC1-lacZ or ATP2-lacZ coding $< {\sim}243$ aa of F_1-β does not lead to this phenotype. In addition, the presence of the ATP2 gene alone on a multicopy plasmid in yeast also does not confer this phenotype. This Gly^- phenotype does not appear to result simply from the presence of a nonfermentable carbon source in the growth medium since cells containing the pβ-Z plasmids grow fine in a combined medium containing both glucose and glycerol. Finally, the Gly^+ transformants containing plasmids pβ-Z1 through pβ-Z5 are unstable in the absence of Ura^+ selection. Ura^- segregants, all of which have simultaneously lost detectable β-galactosidase activity are restored for growth on glycerol.

As illustrated in Figure 3 the β-subunit-β-galactosidase activity is localized in the mitochondria for all of the pβ-Z hybrid genes examined. Thus, the signal(s) required for mitochondrial delivery of the hybrid gene product are still present within the first ~225 aa of the β-subunit (pβ-Z8 gene product). For the ATP2-lacZ gene products coded in pβ-Z1 to pβ-Z5, the β-subunit-β-galactosidase activity remains firmly associated with the submitochondrial membrane preparations and cannot be released by mild detergent treatment which solubilizes essentially all of the mitochondrial oligomycin-sensitive ATPase complex. Thus, unlike the wild-type β-subunit protein which is associated with a readily solubilized complex of the inner membrane, the β-subunit-β-galactosidase hybrid protein of pβ-Z1 to

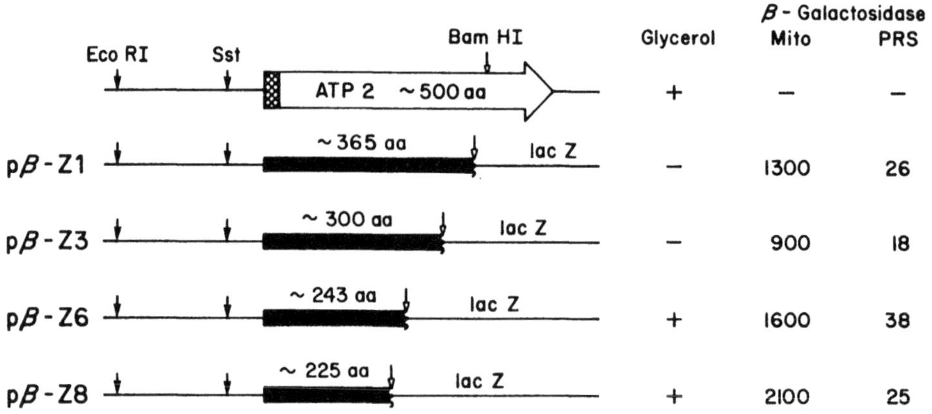

Figure 3. Cells Containing Shorter β-Z Gene Fusion Plasmids Localize the Hybrid Protein in Mitochondria Yet Grow on Glycerol. This figure illustrates the results of growth analysis of yeast strain SEY2102 harboring the indicated plasmids on 2 x SG-ura medium (4) with 3% glycerol as the carbon source. Mitochondria and post ribosomal supernatant fractions (PRS) prepared as described (6) from cells grown on 2 x SD-ura medium were assayed for β-galactosidase activity as published (7). Activities are expressed as n mol of o-nitrophenol-β-D galactoside/min · mg.

pβ-Z5 remains tightly associated with the organelle possibly sequestered in the inner membrane. On the other hand, the hybrid protein products of pβ-Z6, 7 and 8 appear to be less firmly associated with the mitochondrial membrane. Although the β-subunit-β-galactosidase activities of these fusions fractionate predominantly with the cytochrome oxidase of the inner membrane they are more easily solubilized by detergent than ATP2-lacZ gene products containing > ~277 aa of the β-subunit protein.

In the studies described above, β-galactosidase activities were used to mark the location of the hybrid gene product. In order to confirm the presence of the hybrid protein, cells harboring different pβ-Z or control plasmids were first radioactively labeled with $^{35}S-SO_4$ then immunoprecipitated with antiserum directed against either β-galactosidase or against the F_1-ATPase β-subunit (Figure 4). Both antisera specifically recognize proteins with apparent molecular weights of 158,000 (pβ-Z1 gene product) and 144,000 (pβ-Z5 gene product). These proteins were found to be predominantly localized to the mitochondrial enriched fraction of these cells.

Cells containing the control plasmid pLG669-Z show a single immunoprecipitable product of 120,000 MW which can be detected with the β-galactosidase antiserum but not with the anti β-subunit serum. Thus, ATP2-lacZ gene fusions direct the synthesis of hybrid proteins consisting of both F_1-ATPase β-subunit and β-galactosidase polypeptide sequences. Analysis of immunoprecipitates in this manner indicates that the level of the β-subunit-β-galactosidase protein in the cell is comparable to that of the F_1-β subunit protein. Furthermore, the amount of immunodetectable β-subunit in the cell is not significantly changed in the presence of the different hybrid gene products.

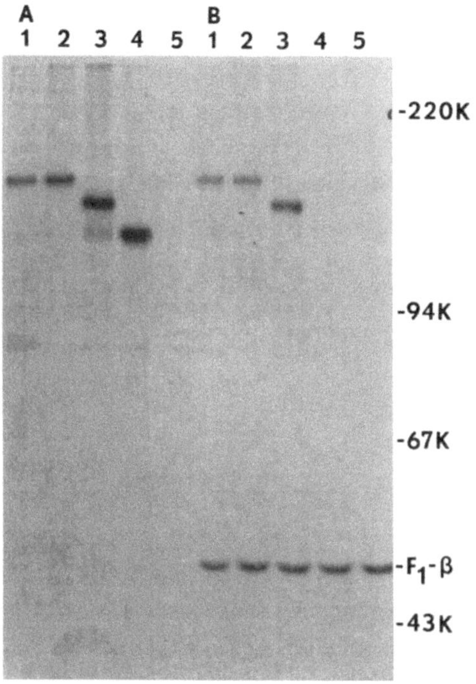

Figure 4. Identification of ATP2-lacZ Hybrid Proteins in Yeast. Yeast strains SEY2102 harboring the indicated plasmids were grown on 2 x SD - ura medium and labeled with $^{35}S-SO_4$ (8). The labeled cell homogenate was divided in half and immunoprecipitated (9) with either anti β-galactosidase serum, A. or anti F_1-ATPase β-subunit serum, B. the labeled cell homogenates were prepared from cells containing the following plasmids: 1 and 2 pβ-Z1, 3. pβ-Z5, 4. pLG669-Z and 5. pSEY101.

In order to determine if the hybrid protein interferes with either the import or function of other mitochondrial proteins we assayed the level of mitochondrial matrix, inner membrane, outer membrane and intermembrane space marker enzyme activities in cells harboring either pβ-Z1 or pLG669-Z plasmids. These results (Table III) indicate that the ATP2-lacZ gene product does not completely block the import of other components into the mitochondria. The reduced levels of the different inner membrane respiratory complexes is probably not sufficient to account for the Gly^- phenotype observed. These results further indicate that the ATP synthesizing capacity of the energy transducing enzyme as measured by $^{32}P_i$-ATP exchange activity is reduced ten fold. This marked decrease in the level of the organelle ATP synthesizing capability may be responsible for the failure of yeast strains carrying the pβ-Z1 plasmid and other inhibitory plasmids, pβ-Z2 to pβ-Z5, to grow on a nonfermentable carbon source. The precise mechanism by which β-subunit-β-galactosidase hybrid proteins containing greater than ~277 aa of the β-subunit protein lead to this effect is not yet known, however, the data support the localization of these proteins in a manner which may prevent the proper function of the energy transducing enzyme.

Discussion

It has been proposed that the presequence of the β-subunit may function to bind the precursor protein specifically to the mitochondrial import machinery thus initiating the translocation process. The present data concerning the ATP2 gene product is consistent with such a model. Carboxy-terminal amino acids are apparently not required for directing mitochondrial protein import. The shortest ATP2-lacZ gene fusion is missing the coding sequences for ~277 carboxy-terminal amino acid residues of the β-subunit protein. In addition, replacement of these amino acids with a large heterologous polypeptide like β-galactosidase appears not to block, at least, initiation of import of the hybrid polypeptide. Analysis of the shorter fusion size classes indicates that additional protein sequences within the β-subunit preprotein may participate in the translocation event. The gene products of all of the fusions are targeted to the

TABLE III

Marker Enzyme Activities in Mitochondria

	pβ-Z1	pLG669-Z	%
Matrix			
Malate dehydrogenase	2.11 ± 1.19	3.11 ± 1.85	67.8
Inner Membrane			
Succinate cyt c reductase	2.23 ± 1.93	9.94 ± 3.63	22.4
Cytochrome c oxidase	40.7 ± 22.5	48.3 ± 23.0	84.3
Oligomycin sensitive ATPase	0.076 ± 0.04	0.087 ± 0.04	87
$^{32}P_i$-ATP exchange	1.15 ± 0.21	12.28 ± 1.30	9.3
Inter Membrane Space			
Cytochrome c peroxidase	2.35 ± 0.4	1.89 ± 0.3	124
Adenylate kinase	77.2 ± 9.7	90.7 ± 4.8	85.1
Outer Membrane			
Antimycin insensitive NADH cyt c reductase	21.0 ± 14.0	16.5 ± 8.4	127
Kynurinine hydroxylase	1.84 ± 0.80	0.74 ± 0.25	248

Mitochondria were prepared from SEY2102 containing the indicated plasmids. The strains were grown to 2 x 10^7 cells/ml in 2% glucose maintaining uracil selection. Average β-galactosidase activities of the mitochondrial preparations were 1499 and 170 units per mg for pβ-Z1 and pLG669-Z respectively. Enzymatic activities were determined as described in (4). Specific activities are expressed in n mol/min·mg for malate dehydrogenase, succinate cytochrome c reductase, cytochrome oxidase, cytochrome c peroxidase, adenylate kinase, antimycin insensitive NADH cyt c reductase and kynurine hydroxylase; μmole/min·mg for ATPase and uncoupler sensitive $^{32}P_i$-ATP exchange. The percentage of each enzyme activity present in cells harboring plasmid pβ-Z1 vs. cells harboring pLG669-Z also is indicated.

mitochondrial inner membrane, however, the hybrid gene products discriminated by gene fusions pβ-Z6 and pβ-Z5 (~243 aa to ~277 aa respectively of the β-subunit protein) exhibit different properties with regard to solubility, protease accessibility and growth on a nonfermentable carbon source. These properties are consistent with a model in which protein sequences within the β-subunit protein participate in the translocation of the precursor subunit across the inner membrane. This "translocation signal" is either missing or not available due to conformational restraints of the hybrid molecule in gene products of the shorter gene fusion classes.

The Gly^- phenotype due to the presence of the longer ATP2-*lacZ* gene fusion proteins could provide important genetic handles for extending studies on the import of proteins into mitochondria. We have found that it is possible to isolate Gly^+ mutants in yeast cells carrying different pβ-Z plasmids. Most of the Gly^+ mutants no longer express any detectable β-galactosidase activity, further suggesting that it is the β-subunit-β-galactosidase hybrid protein that leads to the Gly^- phenotype. We have selected 3 Gly^+ mutants that still express active β-galactosidase. Further, analysis of mutants of this type should uncover molecular features of protein targeting and cellular components which participate in the mitochondrial import pathway.

Acknowledgements

We would like to acknowledge support from the NIH, GM-26713, and the Robert W. Welch Foundation, AQ-814, to M.G.D. and The Miller Institute for Basic Research in Science of the University of California, Berkeley, to S.D.E. A special thanks to Janet Kendall for preparing this manuscript.

References

1. Schatz, G., Butow, R.: Cell 32, 316-318 (1983).
2. Neupert, W., Schatz, G.: Trends in Biochem. Sci. 6, 1-4 (1981).

3. Saltzgaber, J., Kunapuli, S., Douglas, M.: J. Biol. Chem. 258, in press (1983).

4. Douglas, M., Geller, B., Emr, S.: Proc. Natl. Acad. Sci. 80, in press (1983).

5. Maccecchini, M., Rudin, Y., Blobel, G., Schatz, G.: Proc. Natl. Acad. Sci. 76, 343-347 (1979).

6. Daum, G., Bohni, P., Schatz, G.: J. Biol. Chem. 257, 13028-13033 (1982).

7. Miller, J.: Experiments in Molecular Genetics, Cold Spring Harbor Laboratory, Cold Spring Harbor, New York 1972.

8. Douglas, M., Finkelstein, D., Butow, R.: Methods in Enzymol. 56G, 58-65 (1979).

9. Reid, G., Schatz, G.: J. Biol. Chem. 257, 13056-13061 (1982).

ASSEMBLY OF THE MITOCHONDRIAL ATPASE IN SACCHAROMYCES CEREVISIAE

A.S. Lewin, L.J. Wells, D.K. Norman
Department of Chemistry
Indiana University
Bloomington, Indiana 47405, USA

Introduction

In essence, mitochondria are assembled from the outside-in. Proteins, lipids and small molecules are ferried through existing membranes in order to generate new organelles during cell growth. The import of proteins is particularly complex. Many mitochondrial proteins are made as larger precursors in the cytoplasm on either free or on membrane-bound ribosomes (1,2,3). The transport of mitochondrial proteins does not depend on concurrent protein synthesis, and therefore differs from the vectorial translation of proteins through the endoplasmic reticulum (4).

The proton ATPase (F_1F_0) complex provides a good model for studying the assembly of a mitochondrial enzyme. Most of the enzyme's 9 different subunits are made in the cytoplasm under the control of nuclear genes, but two subunits of the membrane-embedded domain of the enzyme (F_0) are made in the mitochondrion in yeast (5). The largest subunits of the F_1 component are made as larger precursors which are cleaved to their "mature" (active) sizes during, or immediately following, their transport from the cytoplasm (6).

We have been studying the importation and assembly of F_1 ATPase subunits in isolated yeast mitochondria. Our work has employed three types of respiratory deficient mutation of yeast (pet$^-$, mit$^-$, and ρ^-) in order to explore the energy requirements for protein transport into mitochondria. We conclude that ATPase subunits are assembled into F_1 ATPase in isolated

Mitochondria 1983

mitochondria and that either an active respiratory chain or an active proton pump are required to energize protein uptake by mitochondria.

Results

Assembly of F_1 ATPase in isolated mitochondria

The uptake of proteins into mitochondria has been studied *in vitro* by labeling precursors of mitochondrial proteins in a cell-free translation system and incubating them with mitochondria (6, 7). Proteins which are imported by isolated mitochondria sediment with mitochondria and become resistant to added proteases. If membranes are then dissolved under conditions in which the ATPase complex should not dissociate (150 mM KCl, 1 mM EDTA, 0.2 mM ATP, 10 mM Tris HCl, pH 7.4, 1% Triton X-100), subunit specific antisera can be employed to determine if newly imported (labeled) subunits have been assembled with each other to form new enzyme. Figure 1, lane B shows the radioactive proteins immunoprecipitated with antiserum specific for only the largest ATPase subunit following such an import reaction. The three largest subunits are co-precipitated together with some contaminant bands. The labeled subunits are associated with the inner membrane fraction of mitochondria but tend to dissociate in the cold (8). These results suggest that newly imported ATPase subunits do join together, but it is uncertain whether the membrane bound or only the F_1 ATPase is formed.

The objection might be raised that the imported ATPase subunits are not associating with each other, but are exchanging with subunits of pre-existing ATPase complexes. In order to rule-out this possibility, we have determined that ATPase subunits made *in vitro* do not associate with sub-mitochondrial particles, even when they are processed to their mature forms by the matrix protease which normally fulfills this function (8,9,10). In addition, the assembly of ATPase subunits in isolated mitochondria can be competed by excess unlabeled precursors only if the competing "cold" proteins are added before the labeled F_1 precursors (Fig. 2). In this experiment, precursors

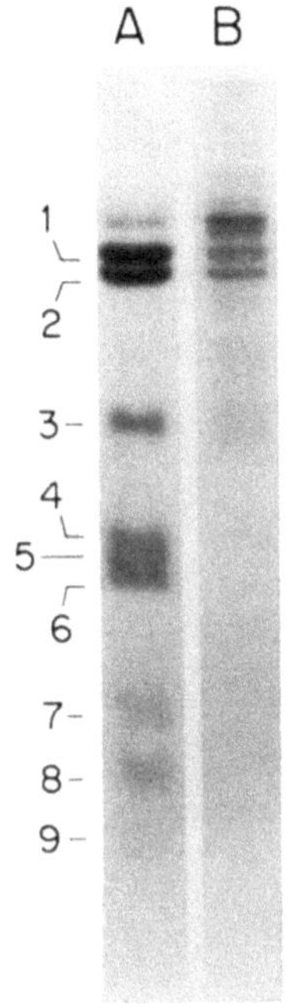

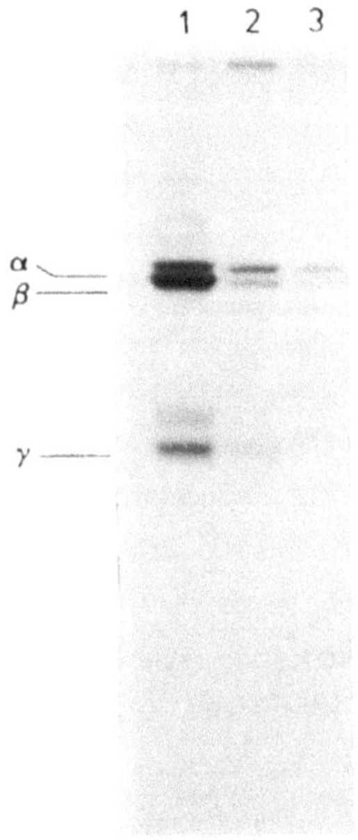

Figure 1: Import and assembly of F_1 ATPase *in vitro*. Mitochondria were incubated with ^{35}S-labeled precursors of mitochondrial proteins as described in ref. 7 and were monitored for ATPase assembly by immunoprecipitation with subunit specific antiserum as described in (8). (A) Oligomycin sensitive ATPase complex isolated from yeast mitochondria, separated on an SDS polyacrylamide gel and stained with Coomassie Blue. (B) Radiolabeled subunits immunoprecipitated following import using anti-α serum and separated on the same gel as (A). (autoradiogram)

Figure 2: Competition of ATPase assembly in isolated mitochondria. Lane 1, ^{35}S-labeled F_1 -ATPase standard made from continuously labeled yeast cells (6). Lane 2, immunoprecipitation with α subunit specific serum following import of labeled proteins first followed by import excess unlabeled proteins. Lane 3, unlabeled proteins first followed by labeled proteins. Autoradiogram of an acrylamide gel.

were made in a reticulocyte lysate programmed with yeast mRNA either in the presence or absence of ^{35}S-methionine. In one case (lane 2), mitochondria were first incubated with labeled precursors under import conditions. The mitochondria were then recovered and incubated with a 10-fold excess of unlabeled lysate. In the second case (lane 3) unlabeled precursors were added first, followed by the labeled proteins. Import of the α subunit occurs in both incubations, but the coprecipitation of the β and γ subunits is competed when cold precursors come first. This result suggests that exchange of subunits does not occur, but that newly imported precursors associate with each other.

We can also inhibit import of a single ATPase subunit in order to demonstrate that we are examining *de novo* assembly. Uptake of the β subunit of the ATPase complex is prevented by preincubation of the labeled reticulocyte lysate with antibodies specific for the β subunit (Fig. 3A). Under these circumstances, neither the labeled β nor the labeled γ subunit associate with the α subunit immunoprecipitated from mitochondria following import (Fig. 3B). Non-immune serum does not affect import or assembly of ATPase.

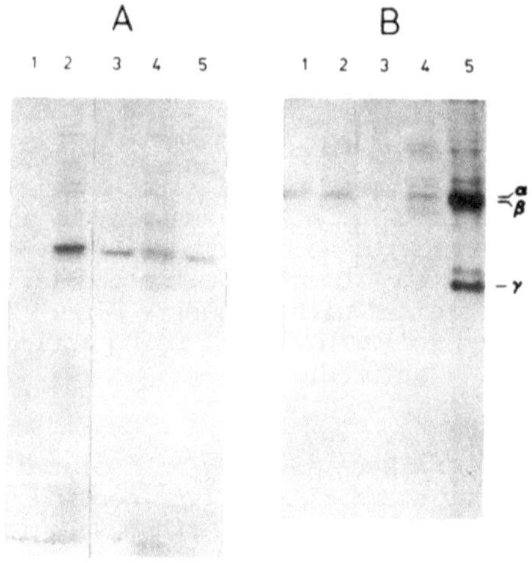

Figure 3: Antibody inhibition of β subunit and of ATPase assembly. Samples were treated with antibody to the β subunit (A, lanes 1 and 2; B, lanes 3 and 4) or non-immune serum (A, lanes 3 and 4; B, lanes 1 and 2) prior to import reactions. Lanes 1 and 3 are mitochondria fractions in A and B; lanes 2 and 4 are supernatant fractions. In A, mitochondria supernatants were lysed with SDS and immunoprecipitated with anti-β serum. In B, mitochondria were dissociated with Triton X100 and treated with a anti-α serum. A, lane 5 is "mature" β standard. B, lane 5 is F_1 ATPase standard.

There is a delay between the import of ATPase subunits into mitochondria and their association into F_1 ATPase (Fig. 4). We have followed the time-course of β subunit import into isolated mitochondria and compared this with the relative amount of β subunit which co-immunoprecipitates with the α subunit. At 10 min. of incubation, a substantial fraction of the β subunit precursor has been processed to its mature form and imported, whereas little of the imported subunit is associated with the α and γ subunits. By 30 min. the level of import is beginning to level-off, but the extent of ATPase assembly is still increasing linearly. These findings suggest that F_1 ATPase (at least) is assembled in isolated mitochondria from a transient pool of newly imported subunits. This conclusion is supported by the finding that newly-made F_0 is assembled with F_1 in isolated mitochondria (Fig. 5).

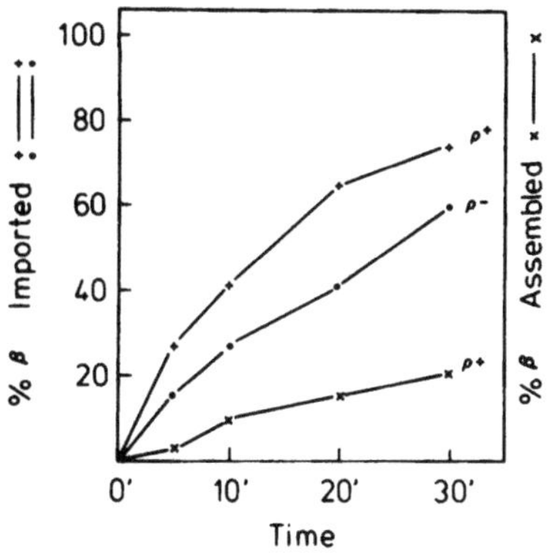

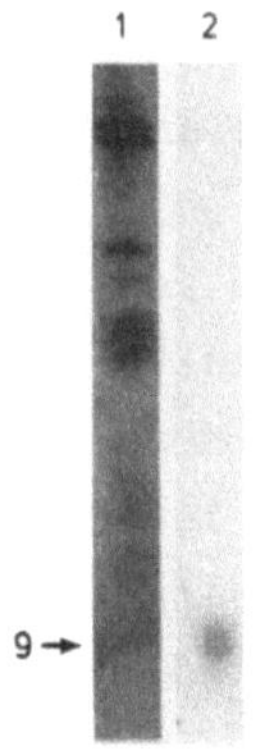

Figure 4: Time course of β subunit import and assembly. Percentage of import is based on the fraction of recovered β subunit associated with the mitochondria. Percent assembled is based on the amount of β subunit recovered as a fraction of the amount of α subunit immunoprecipitated. Import into ρ^- mitochondria was at pH 5 (see below).

Figure 5: F_0 is assembled in isolated mitochondria. Protein synthesis was conducted in isolated mitochondria (18) and then the ATPase complex was isolated (8). The proteolipid subunit was then extracted from the purified ATPase by chloroform-methanol extraction (5) and then run a 15% SDS polyacrylamide gel. Lane 1, isolated ATPase complex (Coomassie blue stain); lane 2, radiolabeled subunit 9 (autoradiogram) extracted from the purified ATPase.

The import of mitochondrial proteins in respiratory deficient mutants

The energy for the transport of proteins across the mitochondrial membranes is thought to be generated by the organelle itself in the form of an energized inner mitochondrial membrane (7,11,12). Experiments to establish this fact have relied upon the use of metabolic inhibitors and uncouplers of oxidative phosphorylation. We have re-examined the energetics of protein import in mitochondria using nuclear and mitochondrial mutations of well characterized biochemical consequences. Specifically, we have used *mit*$^-$ mutations which block the respiratory chain at cytochrome *b* or at cytochrome oxidase, and we have studied protein import in a nuclear petite mutant which contains no functional F_1 ATPase. In addition, we have examined the uptake of mitochondrial proteins by a cytoplasmic *petite* (ρ^-) strain of yeast. This paper presents data only for the β subunit of the ATPase complex, but similar results were found for citrate synthase and other ATPase subunits.

Mitochondria from three mit$^-$ strains which we examined import the ATPase β subunit normally. After a 45 min incubation at 30°C, most of the precursor is converted to mature β and sediments with mitochondria (data not shown). As in mitochondria isolated from respiring yeast, import of proteins into mit$^-$ mitochondria is completely inhibited by the uncoupler carbonylcyanide-m-chlorophenylhydrazone (CCCP). The same result is found when mitochondria isolated from a nuclear petite mutant are used for import (Fig. 6). The particular strain we have employed is E2-126, one of a series of mutants with altered mitochondrial ATPase which were generously donated to us by Dr. Alexander Tzagoloff of Columbia University. This strain has no ATPase activity (13) and lacks the β subunit of the enzyme (data not shown). Nevertheless, mitochondria from strain E2-126 import the β subunit normally and convert it into the protease resistant "mature" form (Fig. 6). Preliminary evidence suggests that the newly imported ATPase subunits assemble with each other to form F_1 in isolated mitochondria in this mutant. If these experiments are corroborated, then the possibility of imported subunits exchanging with pre-assembled F_1 subunits is completely excluded--since there is no active ATPase in this strain.

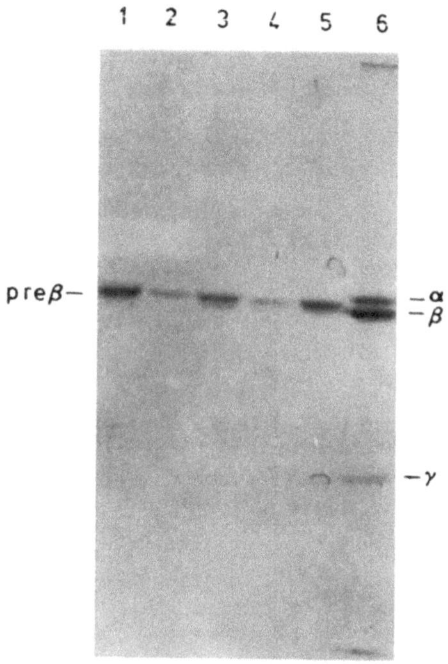

Figure 6: Import and protection in an ATPase defective *pet* mutant (E2-126). Lane 1 ^{35}S-labeled F_1 standard; lane 2, mitochondrial fraction following import, immunoprecipitated with antibody directed against the β subunit; lane 3, β immunoprecipitate from the supernatant following import. (Mature β in the supernatant is probably the result of processing protease leaking from the mitochondria.) Lanes 4 and 5, as in lanes 2 and 3, but 50 μg/ml of trypsin was added for 20' at 0°C following the import reaction.

Figure 7: Isolated mitochondria from ρ^- cells do not import and process precursor to the β subunit. Lane 1, β subunit precursor standard labeled *in vitro* with ^{35}S-methionine (6); lane 2, mitochondrial pellet following import reaction with ρ^- mitochondria, immunoprecipitated with anti-β serum; lane 3, supernatant following import; lanes 4 and 5, as in lanes 2 and 3, but import was in the presence of 10 μM CCCP; lane 6, F_1 ATPase standard.

In contrast to the mitochondria from nuclear petites or from mit^- strains, mitochondria from cytoplasmic petite yeast do not import proteins under our normal import conditions (Fig. 7). The petite mitochondria which we isolate contain near normal levels of mitochondrial marker enzymes such as citrate synthase and cytochrome b_5 (lactate dehydrogenase), so that their inability to import protein is not attributable to structural damage during organelle isolation. Importantly, ρ^- mitochondria retain the o-phenanthroline sensitive protease which converts pre-β into the mature protein (Fig. 8).

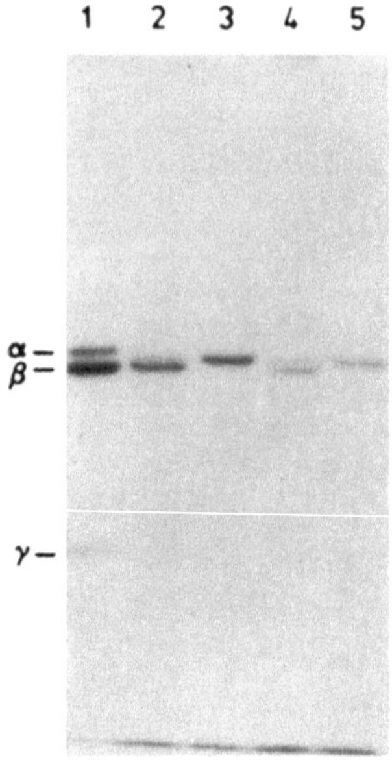

Figure 8: Hypotonic extracts from both ρ^+ and ρ^- mitochondria contain an o-phenanthroline sensitive processing protease. Precursors to mitochondrial proteins were made *in vitro* and exposed to hypotonic extracts (9) of ρ^+ or ρ^- mitochondria in the presence (lanes 3 and 5) or absence (lanes 2 and 4) of 1mM o-phenanthroline. Lane 1, F_1 standard; lanes 2 and 3, extract from *grande* cells; lanes 4 and 5, extract from *petite* cells.

Mitochondrial protein uptake by ρ^- mitochondria: pH jump experiments

A likely explanation for the inability of *petite* mitochondria to import proteins is that they cannot provide the driving force for this process. The mitochondrial inner membrane is energized either by the chemiosmotic gradient generated during electron transport or by proton pumping by the mitochondrial ATPase complex. In mitochondria from petite yeast, neither mechanism is available; there is no "motor" to drive protein uptake.

We have attempted to restore a motor to ρ^- mitochondria by applying an artificial proton gradient across the membrane. Mitochondria were isolated as normal in the presence of 20 mM HEPES-KOH, pH 7.4, but were shifted to pH 5.0 during the import reaction. Under these circumstances, precursor to the β subunit does associate with the mitochondrial fraction even after washing at pH 7.4 but is not processed into the mature subunit. Interestingly, the precursor which associates with mitochondria at pH 5 is resistant to added trypsin at pH 7.4, indicating that it is internalized by mitochondria (Fig. 9). When mitochondria are osmotically shocked and mitoplasts are isolated (14), precursor to β is seen to associate with this fraction (Fig. 9, lane 1). However, the precursor is sensitive to added trypsin in mitoplasts. When mitochondria were fractionated according to method of Daum *et al.* (14) most of the "imported" precursor was associated with the mitochondrial matrix fraction (data not shown).

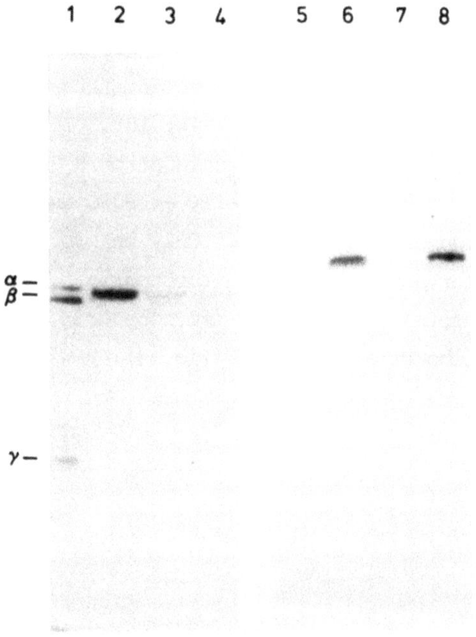

Figure 9: Import of β into ρ^- mitochondria at pH 5. Lane 1, F_1 standard. Lane 2, mitoplast fraction isolated after mitochondria from ρ^- cells were prepared at pH 7.4 and incubated with precursors at pH 5.0 in citrate buffer; immunoprecipitated with β subunit specific antiserum. Lane 3, intermembrane space fraction from import reaction; lane 4, mitochondrial supernatant from the same import reaction. Lane 7, supernatant from a similar ρ^- import reaction; lane 8, mitochondria from this reaction; lane 5 and 6 as in 7 and 8, except 50 μg per ml of trypsin was added to the import reaction after import for 20' at 0°C.

We interpret these experiments to suggest that in this pH-jump experiment, precursors to the β subunit of ATPase are somehow imported into mitochondria, but that they cannot be processed or assembled with the inner membrane. The lack of proteolytic processing is not unexpected under these conditions. At pH 5, the o-phenanthroline sensitive processing protease is inactivated (ref. 9 and unpublished observations). It is also not surprising that the precursor remains in the matrix fraction, since there is no F_0 component in ρ^- mitochondria with which it could associate.

We were worried by the possibility that the labeled precursor might be non-specifically and irreversibly adhereing to the mitochondrial membranes at pH 5. However, we found that the association of β subunit precursor with the mitochondria is not rapid, as might be expected if the association were simply a matter of electrostatic attraction (Fig. 10). In this experiment, ρ^- mitochondria were mixed with precursors at pH 5 and were incubated at 30°C for 45 min. or were mixed briefly at 30°C and then chilled on ice before sedimentation of mitochondria. Only on prolonged incubation at 30°C did the β precursor associate with mitochondria. Indeed, we find that the time course of protein uptake by ρ^- mitochondria at pH 5 roughly parallels that of ρ^+ mitochondria at pH 7.4 (Fig. 4).

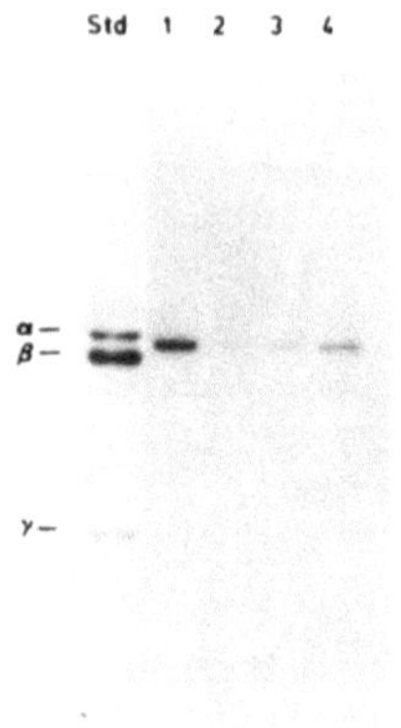

Figure 10: Import of pre-β in ρ^- mitochondria requires prolonged incubation. Std., F_1 ATPase standard. Lane 1, mitochondrial pellet from import reaction at pH 5.0. Incubation was 45 min. at 30°C. Lane 2, supernatant from this reaction. Lane 3, mitochondrial fraction from a similar import incubation, except mitochondria were mixed briefly and then chilled rather than incubated at 30°C. Lane 4, supernatant from lane 3. Lanes 1-4 are all immunoprecipitates with β specific antiserum.

It seems possible, then, to supply the driving force for protein uptake to ρ^- mitochondria, by artificially generating a proton gradient across the mitochondrial inner membrane. However, we cannot rule out that the mildly acid pH is more important than the pH differential across the mitochondrial membranes in our experiment. The gradual acidification of mitochondria might alter the membranes in manner which permits protein import *in vitro* but does not necessarily reflect the process occurring in living cells.

Time course of import *in vivo*

If isolated mitochondria from petite cells are unable to import and process ATPase subunits, how are they able to do so *in vivo* (12)? The answer is unknown, but petite cells are less efficient than wild type cells in converting precursors of mitochondrial proteins into their mature forms. Fig. 11 compares the time course for β subunit processing by ρ^+, mit^-, and ρ^- cells in a kinetic labeling experiment. Although the rate of protein synthesis (based on acid precipitable ^{35}S) was the approximately equal in all strains, precursor form is much longer-lived in petite yeast, even in the absence of uncoupler. In the presence of CCCP, high steady-state levels of β subunit precursor accumulates in ρ^- cells (16), though some import and processing occurs.

The uptake of mitochondrial proteins does not seem to require a great deal of metabolic energy. Therefore, it occurs slowly in ρ^- cells and at normal rates in pet^- and mit^- cells (Fig. 11 and unpublished results). In ρ^- yeast, the requisite membrane potential can probably be generated by an electrogenic carrier system operating in the direction of its substrates' concentration gradient. A likely candidate is the adenine nucleotide transporter. Since most ATP is generated outside the mitochondrion in ρ^- cells, the adenine nucleotide transporter might transport ATP from the cytoplasm into the matrix space in exchange for matrix ADP (the opposite of its normal exchange polarity) without consuming metabolic energy. Because of the differential charges of ATP and ADP, such reversed translocation would have the effect of moving positive charges out of the organelle. That the adenine nucleotide transporter may be involved in protein uptake in petite mitochondria is suggested by the fact that bonkrekic acid inhibits the import of mitochondrial proteins in petite cells (12).

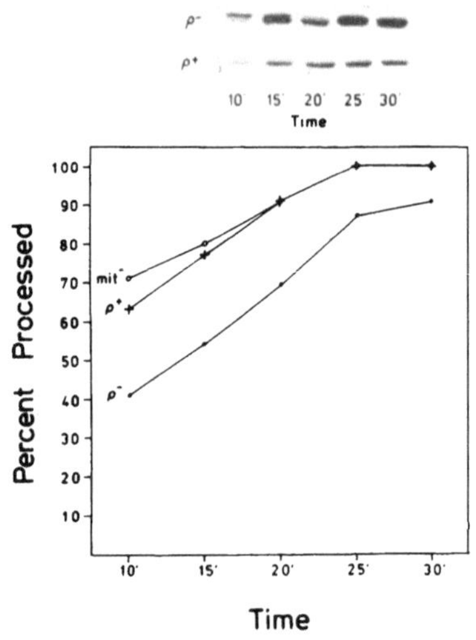

Figure 11: Time course of conversion of pre-β to mature subunit *in vivo*. Cells were pulse labeled for 10 min. at 12°C as described in ref. 12. Casamino acids were added to a final concentration of 1% which stopped the incorporation of ^{35}S-methionine into protein. Samples were drawn at the times indicated, lysed in the presence of o-phenanthroline and immunoprecipitated with antiserum to the β subunit. The extent of conversion to mature subunit was determined by scanning autoradiograms of the proteins separated on SDS polyacrylamide gels using a Joyce-Loebl microdensitometer.

Discussion

The assembly of F_1 ATPase in isolated mitochondria suggests that the system we have used to study the transport of proteins into mitochondria reflects processes occurring within the living cell. This system permits us to vary experimental parameters in order to determine what conditions favor assembly of the complex *in vitro*. Because ATP may be tightly bound to the enzyme at loci other than the catalytic site (15), it is of interest whether ATP is required for ATPase assembly. The time differential between protein uptake by mitochondria and the assembly of the ATPase may permit us to resolve this question by reducing matrix ATP (by addition of CCCP) at a time when most of the precursors of ATPase subunits have been imported but little of the enzyme has been assembled.

Of course we intend to study the interaction of the nuclear encoded, imported subunits with those made inside the mitochondrion. Because there may be a transient pool of free subunits inside the mitochondrion (Fig. 4 and ref. 16) it is possible that regulation of ATPase synthesis occurs at the level of assembly of the enzyme. Similarly, it has been suggested that the availability of oxygen stimulates the assembly of cytochrome *c* oxidase (17).

Our results employing respiratory deficient mutants to explore the energetics of mitochondrial protein import confirm the conclusions of other investigators (7, 11) that an energized mitochondrial inner membrane is required for protein transport. We find that either an intact respiratory chain or a functional ATPase complex is sufficient to provide electrochemical gradient required.

That ρ^- mitochondria take-up proteins at low pH seems to echo the findings of workers investigating other forms of protein-membrane interaction. For example, bacterial toxins which bind to and depolarize plasma membranes only do so *in vitro* at pH 5 or below (19). Similarly, transferrin receptor requires pH below 5.5 in order to bind apotransferrin during its recycling (20). In addition, membrane fusion proteins, such as the hemagglutinin of influenza or Semliki Forest viruses require exposure to mildly acid conditions ($<$ pH 5.5) for activity (21,22). In our system as well, protein import does not occur at pH $\geq$ 5.5.

It has been suggested that membrane fusion sites between the inner and outer mitochondrial membranes are employed for protein transport into mitochondria (2). Since the outer mitochondrial membrane is permeable to protons, there is no reason to expect that the intermembrane space of mitochondria is particularly acidic even during rapid respiration. It is not unreasonable to suggest, however, that local membrane inhomogeneties could permit localized acidification and thus activate a receptor or stimulate membrane fusion events between the two mitochondrial envelopes. Such transient fusion sites probably occur normally during the expansion and division of respiring mitochondria. Working out the details of this, or whatever mechanism for protein import corresponds to reality, remains a challenging object for research.

Acknowledgements

This work was supported by grants from the U.S. Public Health Service (No. GM29387-02) and The Research Corporation. A.S.L. is a Junior Faculty Research Fellow of the American Cancer Society (JRFA 55). The authors wish to thank Drs. G. Schatz, H. Mahler and A. Tzagoloff for providing yeast strains used in these experiments.

References

1. Neupert, W., Schatz, G.: Trends. Biochem. Sci. 6, 1-4 (1981).
2. Schatz, G., Butow, R.A.: Cell 32, 316-318 (1983).
3. Suissa, M., Schatz, G.: J. Biol. Chem. 257, 13048-13055 (1982).
4. Blobel, G., Dobberstein, B.: J. Cell Biol. 67, 835-851 and 852-862 (1975).
5. Tzagoloff, A., Meagher, P.: J. Biol. Chem. 246, 7328-7336 (1971).
6. Maccechini, M.L., Rudin, Y., Blobel, G., Schatz, G.: Proc. Natl. Acad. Sci. USA 76, 343-347 (1979).
7. Gasser, S.M., Daum, G., Schatz, G.: J. Biol. Chem. 257, 13034-13041 (1982).
8. Lewin, A.S., Norman, D.: J. Biol. Chem. 258, 6750-6755 (1983).
9. Böhni, P., Daum, G., Schatz, G.: J. Biol. Chem. 258, 4937-4943 (1983).
10. McAda, P., Douglas, M.: J. Biol. Chem. 257, 3177-3182 (1982).
11. Schleyer, M., Schmidt, B., Neupert, W.: Eur. J. Biochem. 125, 109-116 (1982).
12. Nelson, N., Schatz, G.: Proc. Natl. Acad. Sci. USA 76, 4365-4369 (1979).
13. Tzagoloff, A., Akai, A., Needleman, R.B.: J. Biol. Chem. 250, 8228-8235 (1975).

14. Daum, G., Bohni, P.C., Schatz, G.: J. Biol. Chem. 257, 13028-13033 (1982).

15. Cross, R.L.: Ann. Rev. Biochem. 50, 681-714 (1981).

16. Todd, R.D., Buck, M.A., Douglas, M.G.: J. Biol. Chem. 256, 9037-9043 (1981).

17. Ried, G.A., Schatz, G.: J. Biol. Chem. 257, 13056-13061 (1982).

18. Woodrow, G., Schatz, G.: J. Biol. Chem. 254, 6088-6093 (1979).

19. Cramer, W.A., Dankert, S., Uratani, Y.: Biochim. Biophys. Acta 731, 173-193 (1983).

20. Dautry-Varsat, A., Crechanover, A., Lodish, H.: Proc. Natl. Acad. Sci. USA 80, 2258-2262 (1983).

21. White, J., Helenius, A., Gething, M.J.: Nature 300, 658-659 (1982).

22. White, J., Helenius, A.: Proc. Natl. Acad. Sci. USA 77, 3273-3277 (1980).

AUTHOR INDEX

Agostinelli, M., 285
Agsteribbe, E., 343
Ahne, F., 313
Argan, C.A., 495
Arnberg, A., 343
Attardi, G., 107

Baldacci, G., 57, 165
Behra, R., 593
Benne, R., 285
Bergantino, E., 227
Bernardi, G., 57
Bertrand, H., 495
Biggs, M., 607
Biswas, T., 69
Bolotin-Fukuhara, M., 455
Borst., P., 285
Britten, M., 607
Brown, T.A., 303
Burger, G., 331
Butow, R.A., 95

Camadro, J.-M. 523
Cantatore, P., 127
Carignani, G., 277
Choo, W.M., 535
Christen, Ph., 593
Christianson, T., 69
Colson, A., 481

Davies, R.W., 179, 303
Dawson, A.J., 269
De Haan, M., 563
Dehoux, P., 523
De Jong, J.C., 343
Dekken, A., 563
De Vries, B.F., 285
De Vries, E., 79
De Vries, H., 343
De Zamaroczy, M.,
Distler, S., 313
Douglas, M., 607
Dujardin, G. 233
Dujon, B., 1, 389

Edwards, J., 69
Emr, S., 607
Esser, K., 251

Fauconnier, A. 481
Faugeron-Fonty, G.,

Faye, G., 433
Feldmann, H., 141
Flagella, Z., 127
Flückinger, J., 593
Foury, F., 487
Fox, T.D., 441
Francisci, S., 165
Frezza, D., 227
Frontali, L., 165
Fukuhara, H., 455

Gadaleta, M.N., 127
Gaines, G., 107
Gargouri, A., 259, 405
Gehring, G., 593
Geller, R., 607
Görtz, G. 141
Graf-Hausner, U. 593
Greco, M. 579
Grivell, L.A., 25, 593
Groot Koerkamp, M.J.A. 79
Grosskopf, R. 141
Groudinski, O., 227
Guarente, L., 509

Hack, E., 269
Hadikusumo, R.G., 535
Hanson, D.K., 211
Hawthorne, D.C., 449
Helmer-Citterich, M., 357
Hennig, B., 551
Holl, J., 191

Isaac, P.G., 269
Jacquier, A., 389
Jaussi, R., 593
Jones, V.P., 269

Kaudewitz, F., 191, 313, 411
Klaver, B., 285
Klobučniková V., 65
Kobayashi, M., 371
Köhler, H., 551
Koike, K., 371
Kolodynski, J., 487
Kovác, L., 65
Kreike, J., 411
Kroon, A.M.,
Krupp, J. 151
Kuchino, Y., 371

Labbe-Bois, R., 523
Labouesse, M., 233
Lang, B.F., 313, 411
Lazowska, J., 259, 405
Leaver, C.J. 269
Levens, D., 69
Lewin, A.S., 621
Linnane, A., 535
Little, H.M., 509
Lukins, H.B., 535

Maarse, A.C., 563
Macino, G., 357
MacKellar, W., 211
Mahler, H., 211
Mannhaupt, G., 449
Marra, E., 579
Martin, N.C., 151
Marzuki, S., 535
Mason, T.L., 509
Meunier, B., 481
Michaelis, G., 449
Miller, D., 151
Montoya, J., 107
Morelli, G., 357
Mueller, D., 69

Naggert, J., 425
Najarian, D.R., 151
Nelson, M.A., 357
Netter, P., 233
Neupert, W., 551
Norman, D.K., 621

O'Brien, Th, W. 469
Palleschi, C., 165
Peijnenburg, A., 343
Perlman, P.S. 211
Pesole, G., 127
Pillar, T., 411
Pratje, E., 449

Quagliariello, E., 579

Rabinowitz, M., 69
Ray, J.A., 303
Reif, M.K., 441
Rödel, G., 191

Saccone, C., 127, 579
Sass, P.M., 211
Scazzocchio, C., 179, 303
Schiavon, E., 227
Schieber, G., 469
Schmelzer, C., 191
Schmidt, C., 191
Schroeder, R., 203
Schweizer, E., 425, 449
Schweyen, R.J., 191
Sekiguchi, T., 371
Simon, M., 433
Skoda, R., 593
Slonimski, P.P., 203, 227, 233, 259, 405
Sonderegger, P., 593
Sor, F., 455
Stahl, U., 251
Szakacs, N.A., 495

Tabak, H.F., 79
Taira, M., 371
Thurlow, D.,
Trinkl, H. 313
Tudzynski, P., 251

Underbrink-Lyon, K., 151
Urban-Grimal, D., 523

Van den Burg, J., 285
Van der Bliek, A.M., 79
Van der Horst, G.T.J., 79
Van der Horst, G., 79
Van Loon, A.P.G.M., 563
Van Sickle, C., 509
Volland, C., 523

Waring, R.B., 179, 303
Watkins, L., 535
Weiss-Brummer, B., 191
Wells, L.J., 621
Werner, S., 331
Wettstein, J., 69
Wolf, K., 313

Yaffe, M., 47
Yaginuma, K., 371

Zarley, C.D., 211
Zassenhaus, H.P., 95
Zennaro, E., 165
Zwarthoff, E.C., 79

SUBJECT INDEX

Acanthamoeba 25-29

ALA synthase
-mutants 529-581
-protein 529

Alternative oxidase
-genetic control 497-501,505
-isolation 499-501
-mutants 497-501
-putative polypeptide 501-505
-topology 501-505

Amino acid sequence
-yeast
14 kDA subunit,complex III 573
17 kDA subunit,complex III 574
-Trypanosoma,rf1,rf2,rf3 296

Antimycin
-resistance of diu^r 6,481-485
-mutants 6,481-485

Aspartate aminotransferase
-cytosolic isoenzyme
amino acid sequence 594-596
evolution of the gene 595-596
topogenesis
-mitochondrial isoenzyme
amino acid sequence 594-598
degradation of precursor 601
evolution of gene 595-596
measurement of intramitochondrial activity 580-583
mRNA 597-599
precursor 583-585,593-606
receptors for import 587-590
topogenesis 593-606
-transport into mitoch. 579-606
energy dependence 586-587
model for 579-592
specificity of 585
-vinoblastine, effect on precursor maturation 602-606

Aspergillus amstelodami 10

Aspergillus nidulans
-DNA duplications 310-311
-genome organization 8-11, 303-311
-rRNA genes 304,309-310
-tRNA genes 304,309-310
-protein coding genes 306-309
-unassigned reading frames 308-310
-predicted amino acid homology to gene products of other organisms 303-310
-introns 179-189,307
conservation of domains 180-189
open reading frames 180-181
RNA secondary strucure 181-189
-mitochondrial mutations 303-306
drug resistance 10, 188, 303-305
cold sensitivity 10,305

ATPase, A. nidulans
-aap protein 309
-subunit 6 306
-subunit 9 (proteolipid) 307-309
-nuclear and mitochondrial mutations 307

ATPase, N. crassa
-subunit 6 (oli2) gene 357-369
-urfx 359
-mal gene (mitoch. proteolipid-like) 350-354

ATPase, S. cerevisiae
-antibodies against subunits 536-547,621-635
-assembly of subunits 543-548,621-635
-ATP2-lacZ fusions 607-619
-definition of subunits 536-543
-import of subunits 47-55 607-619,621-635
-subunit 6 (oli2 gene)
mRNA and precursor 436-439
mit-mutants 543-548
-subunit 8
abnormal behaviour in gel electrophoresis 543
mitoch. gene product 540-549

ATPase, S. cerevisiae
mit⁻ mutants 541-549
-subunit 9 (oli1 gene)
initiation for transcription 82-83
mit⁻ mutants 541-549
ATPase, S. pombe
-genes for 313-322
ATPase, Plants 14,19,286-288
Bovine mt ribosomes 469-479
Brettanomyces 25-29
Chlamydomonas 14,25-29
Chloramphenicol
-resistance mutation in A. nidulans 305
-resistance mutation in rat 11,383-387
-resistance mutation in S. cerevisiae 7,389-403
Chritidia 25-29
Codon usage
-N. crassa 337
-plants 15,271-274
-S.cerevisiae 31-32
-S.pombe 318-321
-Trypanosoma 294-297
-various organisms 1-32
Complementation
-alternative oxidase mutants 499-500
-cob intron 4 mutants 196-198,211-217,264-266
-in vitro complementation 203-209
-oxi3 intron 1 mutants 230
-pet complementation by transformation 418
Complex III, S. cerevisiae
-amino acid sequence of
precursor of 14 kDa protein 572-573
precursor of 17 kDa protein 574-575
-cloned genes 564-566
-import of subunits into mitochondria 563-577
-protein topogenesis 561-562
-regulation of synthesis 566-571
effect of gene dosage 566-571
effect of glucose 566
Cyanide insensitive respitation: see alternative oxidase
Cytochrome b, A. nidulans
-cob gene 305-307
Cytochrome b, Neurospora crassa
-cob gene 357-369
-intron RNA 359-362
-maturase 362
Cytochrome b, Plants
-nucleotide sequence 272-274
-transcripts 273-274
Cytochrome b, S. cerevisiae
cob-box gene
-effect on bc1 complex 217-226
-in vitro complementation 203-209
-maturases 211-217,244-248, 265-267
-mutations,DNA sequence of 194-199,408-410
-mutants, phenotypes of 211-226
-NAM 2 suppressor 233-250
-novel proteins in mutants 212-217
-RNA splicing 191-201 211-226
-DNA splicing 259-268
-precursors of mRNA 433-439
Cytochrome b, S. pombe
-gene for 314-320
Cytochrome b, Tyrpanosomes
-amino acid sequence 296
-codon usage 294-296
-conservation of sequence 295-298
-rf3 294-296
-transcript 291
Cytochrome c, N. crassa
-assembly 551-561
-cytochrome c heme lyase 556-559
-import into mitoch. 551-561
-receptor for 552-555

Cytochrome c, S. cerevisiae
-CYC1,CYC7 mutations 517-520
-CYC1-lacZ fused gene 509-522,611-618
import of protein 611-618
induction by porphyrins 513-515,520
regulation by heme and oxygen 515-517,520
Cytochrome c oxidase, A. nidulans
-subunit I,II,III genes 306-308
Cytochrome c oxidase, N. crassa
-subunit I gene 331-342, 343-350
sequence 333,345
prepiece 332-336
transcripts and processing 333-341,346-350
-subunit II gene
transcript mapping350-355
Cytochrome c oxidase Rat
-transcript concentrations 132-138
Cytochrome c oxidase, S. cerevisiae
-effect of CYC1,CYC7 mutations 516-520
-subunit I (oxi3)gene
initiation of transcriptions 82
mutations 227-232
NAM2 suppressor 233-250
precursors of mRNA 240-241,411-417,433-439
-subunit II (oxi1) gene
pet mutations, effect on expression 449-454
post-translational processing 449-454
presequence 450
-subunit III (oxi2) gene
pet mutation, effect on expression 441-448
transcript variants 442-445
Cytochrome c oxidase, S. pombe
-genes for 314-320
Cytochrome c oxidase Plants, Z. mays
-subunit I gene 272-274
-SU I polypeptide 274
-subunit II gene 271-274
-SU II intron 271
-SU II polypeptide 274
Diuron
-resistance in S. cerevisiae 481-485
cross-resistance to antimycin 482-484
mitochondrial mutations 482-483
nuclear mutations 483-484
DNA splicing
S. cerevisiae, cob gene 259-268
Evolution
-Aspartate amino transferases 595-596
-Genetic code 30-32,40
-Introns 30,318-321
-Var 1 gene 3, 10 96-99
Genetic code, see codon usage
Genomes (size,structure,map, genes)
-Acanthamoebia 25-29
-Aspergillus 8-10,25-29 303-311
-Brettanomyces 25-29
-Cabbage 25-29
-Chlamydomonas 14,25-29
-Chrithidia 25-29
-Cucurbits 14,25-29
-Human 11-13,25-29
-Insects 25-29
-Klyveromyces 8,25-29
-Maize 25-29,269-283
-Malaria 25-29
-Neurospora 8-10,25-29
-Paramecium 13,25-29
-Plants 13-16
-Podospora 8-10,25-29,253
-Rat 11-13,129-131
-Saccharomyces 4-7,25-29 35,57
-Schizosaccaromyces 7-8 25-29,314-318

Genomes
-Tetrahymena 13,25-29
-Torulopsis 8,25-29
-Trypanosoma 25-29,290

Glucose repression
-effect on gene 165-170
-effect on prototheme biosythesis 523-534

Heme
-attachment to cytochrome c 556-560
-derivatives 556-560
-effect on gene expression 513-517
-lyase 556-560
-mutants 513-517,529-531
-protoheme synthesis 523-534
ALA synthase 525-530
ferrochelatoase 531-532
regulation 523-532
-transport to microsomes 533

Hela cell, see Human mtDNA

Human mtDNA 11-13,25-29, 107-125
-dual mode of transcription 107-112
-effect of intercalatin drugs and of low temperature on transcription 114-119
-genetic map 12
-transcription 114-119
-processing of RNA precursors 107-125
-polycistronic transcripts 107-125
-regulation of RNA sythesis 107-125
-termination of transcription 115-116
-transcription in isolated organelles 115-116
-transcription in vitro 142-145

Import of proteins in mitochondria
-Aspartate aminotransferase 579-606
-ATPase subunits47-55,607-635
-ATP2-LacZ fusion protein 607-619
-Cytochrome b2 47-55

Import
-Cytochrome c 48,551-561
-Cytochrome c lyase, function in 556-561
-energy dependence 586-589 626-633
-effect of mit, pet-, rho 626-633
-inner membrane 47-55,551-558,607-635
-inter membrane space 48-52
-malate dehydrogenase 579-592
-outer membrane 48-50,559-560
-porteases 49
-proteins of complex III 563-577
-precursor proteins 47-55 563-577,583-585,593-619
-receptors for protein 50,523-561,587-591
-translocation signal 618

Insects 25-29

Intron
-circular RNA 38-39,168-169, 365
-class I 30,38-40,179-188, 195-201,317-322,359-365
-class II 30,38-40,192-195, 254,318-320
-cob gene 179-188,191-201, 203-209,245-248, 318-329,359-362
-COI gene (oxi3) 168-169, 192-195,239-248, 318-320,331-350
-COII gene (oxi1) 271
-encoded maturase, 39,211-226,231,233-250, 266-267, 362
-internal guide 39,187-188,359-362
-large rRNA 179-189
-mini-intron 361-362
-mutations 191-250,405-410
-oli2 gene 357-363
-secondary structure 38-39, 179-201,318-320, 340,359-362
-sequence homology 38-39 340,359, 179-201,318-320,
-urf1 339-341

Klyveromyces lactis 8,25-29
-initiation of transcription 79-82,90-91

Kinetoplast DNA, see Trypanosoma

Malate dehydrogenase
-Receptors for import into mitochondria 587-591
-transport into mitochondria 579-592

Maturase, see mRNA maturase

Mitochondrial genes
See: -ATPase (oli1,oli2,aap, mal)
-Cytochrome b (cob-box)
-Cytochrome c oxidase
-COI (oxi3), COII (oxi1), COIII (oxi2)
-mRNA maturase
-rRNA
-tRNA
-urf
-var1

mtDNA sequences, N. crassa
-region 5'to COI gene 333,345

mtDNA sequences, Podospora
-mt plasmid insertion sites 255

mtDNA sequences, Rat tumor cells
-segment $tRNA^{Tyr}$ to $tRNA^{Trp}$ 374

mtDNA sequences, S. cerevisiae
-cob intron 1 194,408
-cob mutations 194,408
-comparison of transcription initiation sites 82
-in vitro mutatet rRNAs 85-86
-rRNA21s, mutations in 392
-$tRNA^{met}$gene 156
-$tRNA^{pro}$gene 156
-tRNA sythesis locus 156

mtDNA sequences, Z. mays
-apocytochrome b gene 273

mRNA maturase, A. nidulans 180-185

mRNA maturase, N. crassa 362-363

mRNA maturase, S. cerevisiae
-cob 211-226,233-250,266-267,316-318,411-423
-NAM 233-250
-oxi3 227-250

Mutants, A. nidulans 8-11,305
-chloramphenicol resistance 305
-syn, cs 305

Mutants, N. crassa
-alternative oxidase (aod) 497-503

Mutants, Rat 11
-chloramphenicol resistance 383-387
-rRNA 383-387
-tRNA 372-379

Mutants, S. cerevisiae 4-7
-aap 541-547
-C^R and suppressors 339-403
-cob 191-226,233-250,405-410
-CYC 517-521
-hem 511-515,528-532
-mss 433-439
-NAM 2 233-250
-nuclear ts, defective mtRNA 74-78
-oli1 543-547
-oli2 543-547
-oxi3 mutants in introns 227-232
-pet, defective in RNA processing 411-423,425-431
-pet, defective in protein processing 449-454
-pet, defective in CO III expression 441-448
-pif 487-494
-ribosomal RNA mutations
in RNA 87,389-403,455-467
nuclear suppressors 463-466
-syn, in tRNA 152
in rRNA 455-467

Mutants, Plants
-Cytoplasmic male sterility 278-279

Mucidin, A. nidulans, resistance mutation 305

Neurospora crassa
-Alternative oxidase 495-507
genetic control of 497-500 505

Neurospora crassa
-Alternativ oxidase
mutants (aod) 497-500,
isolation of 503-506
putative polypeptide 501-506
-topology 503-505
-codon usage in mtDNA 337
-COB gene 357-369
intron RNA, secondary structure 360-362
-COI gene 331-342,345-350
codon usage 337
nucleotide sequence 333,345
prepiece 332-334
transcript and processing 333-341,347-350
-COII gene 350
-import of proteins to mitochondria 551-561
-mal gene (mtATPase proteolipid like)
expression in various N.c. strains 350-351
transcript mapping 350-354
-mitoch. genome 331
-oli2 gene 357-369
intron RNA, secondary structure 360-362
mini intron 360-362
-Pst Palindromes 37-38, 338-339,351-352,365-367
-urf1 331-342

N. intermedia 351
N. sitophila 351
N. tetrasperma 351

Nuclear genes
-aod 497-503
-aspartate amino transferases 595-597
-ATPase subunits 607-619
-Complex III subunits 563-577
-CYC 517-521
-hem 511-515,528-532
-mitochondrial ribosomal proteins 463-465
-mss 433-439
-NAM2 233-250
-pet 411-454
-pif 487-494

Oli1,Oli2, see ATPase

Origin of replication, see Replication

Oxygen
-effect on heme and cytochrome synthesis 509-522,524-528,533
-respiratory induction 165-170

Plants
-apocytochrome b nucleotide sequence, Z.mays 273
-codon usage in mitochon. genes 15,271-274
-cytochrome c oxidase subunits 278-282
-cytoplasmic male sterility 15,277-278
-F_1-ATPase 14
-genes in Z.mays 13-16,271-274
-genome organization 13-16, 269-274
-mitoch.mutations 15,278-282
-mitoch.translation products 14,269-283
-rRNAs 271

Podospora anserina
-senescence 15,251-257
intron1 of COI gene 254-256
mitochondrial plasmid 253-257
nuclear control 252-254

Promoter 35,69,82,101,142-145 113-115,155-157,171-177,350

Punctuation in RNA processing
-Pst Palindromes, N. crassa 8,37-38,338-339,353-354, 365-367
-tRNAs, S. cerevisiae 37-38 51-52
-tRNAs, S. pombe 36,323-326

Rat
-mutations in mDNA of tumor cells 371-387
chloramphenicol resistance 383-387

Rat
-Mutations in mtDNA of tumor cells
rRNA 383-387
tRNA 373-379

-transcription of mtDNA in rat liver 127-140,141-149
-concentration of RNA species 127-140
-cytochrome b gene,cloned in pBr322 141-145
-heterologous in vitro transcription 141-149
-mRNA 127-140
-polymerase III 144-146
-regulation 127-140
-rRNA 127-140
-variation in development 136-138

Recombination 1-16,96-98, 310-311,397-400,487-494

Replication, mammalian mitochondria
-replication origin 379-382

Replication, S. cerevisiae
-initiation of transcription in ori (rep) sequences 57-63,71-72,84
-map of ori sequences 57
-RNA primers 62-63

Regulation
-genes for complex III proteins 566-571
-heme biosynthesis 523-534
-mitochondrial cytochromes 509-522
-RNA sythesis 90-91,107-125 127-140,325

Ribosomes
-Bovine mitoch. ribosomes
protein pattern 471-474
cytoplasmic synthesis 469-479

Ribosomes, S. cerevisiae
-ratio of subunits in mutants 458

RNA processing, splicing 28,34-38
-COI (oxi3) 338,345-350,411-439
-COII (oxi1) 350-354
-COIII (oxi2) 441-448
-cob/box 363-367,411-425
-in vitro complementation 203-209
-mal 350-354
-maturase, see RNA maturase
-mutants 191-250,259-268 405-410
-oli1 69-76,95-106
-oli2, 325,363-367
-pet⁻mutants 411-448
-Pst palindromes 37-38, 338-339,353-354,365-367
-punctuation by tRNA 36-37,99-100,323-326
-rRNA 107-125,324
-tRNA 36,69-77,90-91, 151-164,171-177
-tRNA synthesis locus 35,156
-var1 35-38,99-103

RNase P
-E. coli 162
-S. cerevisiae, RNase P like enzyme 36,158-162

RNA secondary structure
-intron RNA 38-40,179-201 318-320,340,361
-rRNA 386,394
-9sRNA (tRNA synthesis locus) 162
-tRNA 335,375

rRNA, A. nidulans
-genes 309-310

rRNA, Human
-polycistronic precursors 107-125
-processing site 121-123
-regulations of synthesis 113-115
-termination ot syntheses 115-117

rRNA, Rat
-new chloramphincol resistance locus 383
-secondary structure, 16s 386
-transcript concentrations 135-138

rRNA
-Trypanosomes 25-29,285-302
lack of intron 292
map position 290-291

rRNA, S. cerevisiae
-mutation in rRNA 85-86
394,461-463
-location of mutations in
21S rRNA 392-394,460
-nuclear suppressors 463-466
-phenotype of mutants
389-403,457-459
-transcription initiation
69-78
rRNA, S. pombe, genes for
320-321
Saccharomyces cerevisiae
-action of ionophores on
mtDNA 65-68
-ATPase
assembly of subunits
543-547,621-635
import of subunits
47-55,607-635
mit⁻mutants 541-547
subunit composition
535-549
-C^R mutations 389-403
-CYC1,CYC7 mutants 517-520
-CYC1-LacZ fused gene
509-522,611-612
import of protein 611-617
induction by porhyrins
513-515
regulation by heme and
oxygen 515-517,520
-cloned nuclear genes 418-422
433,563-577,607-609
-cob-box mutations
complementation in vitro
of slicing deficiency
203-209
defect in splicing 191-201
211-217
DNA sequence 194,408-410
effect on bc 1 biogenesis
217-225
effect on RNA folding
191-201
-diu^rmutations 481-485
-DNA sequence of bI1 194,408-
410

S. cerevisiae
-DNA splicing 259-268
-initiation of transcription
69-93,99-103,151-164,171-177
-hem mutants 528-533
-mtRNA synthesis mutants
75-77
-nuclear suppressors
of bJ1 mutation 195
of box7 mutation (NAM2)
233-250
of rRNA mutations 463-465
-omega locus, nature of
397-401
-oxi3 mutants defective in
splicing 227-232
complementation 230
novel polypeptides 230
RNA splicing 228
-pet mutants, defective in
processing
of cob RNA 411-423
of oxi3 RNA 411-439
-pet mutants defective in
COX II 449-454
-pet-mutants defective in
COX III expression 441-448
-pif mutants 487-494
-protoheme synthesis 523-534
-recombination 487-494
-tRNA processing 151-178
-ubiquinol cytochrome c
reductase genes
amino acid sequence of
precursors (14kd,17kd)
572-575
cloned genes 564-566
import into mitochondria
575-576
regulation of synthesis
topogenesis 571-572
-var 1 gene
evolutionary origin 96-99
103
transcription 99-104
RNA processing 99-104

Schizosaccharomyces pmbe
-codon usage 321-322
-evolution of genome 326-327
-gene organization 313-322
-introns 318-321
-map of genome 315
-transcription 322-327

Senescence, P. anserina 10, 251-257

Splicing, see RNA processing

Suppressor mutations
S. cerevisiae
-mitochondrial 198,392-394
-nuclear 195,233-250, 463-466

Tetrahymena 13,25-29,183

Torulopsis 8,25-29

Transcription
-Hela mtDNA 34-35
diffential synthesis of mRNA and rRNA 107-125
dual mode of transcription 107-115
isolated organelles 116-118
effect of drugs 118-120
effect of low temperature 118-120
regulation 107-125
termination attenuation 108,115-116

Transcription, N. crassa
-initiation site 350

Transcription, Rat mtDNA
-concentration of RNA species 127-140
-cytochrome b gene 141-144
-heterologous in vitro systems 141-149
-regulation 128-138
-tRNA genes144-148,378-379
-variation in development 136-138

Transcription, S. cerevisiae
-co-transcription 35-36 69-78,87,99,153-155,171-177,437-438
-effect of glucose repression 165-170
-guanylyl transferase assay 69-73,91,155
-homologous in vitro system 63-66,80-86
-initiation sites 35-36, 69-77,80-86,91-93,101,155-157
-on mutated templates 65,74-75

Transcription, S. pombe, 322-327

Transcription, Trypanosomes
-direction of 292
-map of 290-291

Tetrahymena
-intron in nuclear rRNA 38,180-185,196-197

tRNA, A.nidulans
-genes 309-310

tRNA, Human
-steady state amounts 115

tRNA, N. crassa
-mtDNA sequence tRNA$^{arg}_{min}$ 335

tRNA, Rat
-mtDNA-sequence of Tyr-Cys-Asn-Ala-Trp 372-375
-mutation in tumor cells 372-375
-transcription 144-148 378-379
-secondary structure 375

tRNA, S. cerevisiae
-DNA sequence tRNA^{met}-tRNA^{pro} 156
-polycistronic transcripts 35-36,69-78,90-91,171-177
-processing 151-164,171-177
-synthesis locus, DNA sequence 156
-transcription initiation 69-78,83-84,155-157,171-177

tRNA, S. pombe
-function in RNA-processing 322-327
-genes for 321-322

Trypanosomes
-genes for apocytochrome b 295-299
-genome organisation 285-302
-intensitivity to antimitoch. drugs 286-288
-transcripts 288-294
direction of synthesis 292
map 290
poly A + 291
-urfs
amino acid sequence 296
amino acid homology 295-298
codon specificity 294,297

Unidentified reading frames (urf)
urfs, A. nidulans 1-13,28,308
urfs, Drosophila, 308
urfs, Human 308
urfs, N.crassa
-codon usage 337
-transcripts 333-341
-urf1, urfN, urfU 346-350
-urfx 359
urfs, Rat
-transcripts 132-135
urfs, S. pombe 316-317
urfs, Trypanosomes
-amino acid sequence 296
-codon usage 294-297
-identification of proteins 294-298
-transcripts 288-294

Ubiquinol-Cytochrome c Reductase, see Complex III,

Var 1, S. cerevisiae
-evolutionary origin of the gene 96-99,103
-RNA processing 99-103
-transcription 99-103

Zea mays, see plants

Bei Fragen zur Produktsicherheit wenden Sie sich bitte an:
If you have any questions regarding product safety,
please contact:

Walter de Gruyter GmbH
Genthiner Straße 13
10785 Berlin
productsafety@degruyterbrill.com